THERMODYNAMICS OF THE ATMOSPHERE: BASIC PHYSICS AND APPLICATIONS TO EARTH'S SCIENCE

THERMODYNAMICS OF THE ATMOSPHERE: BASIC PHYSICS AND APPLICATIONS TO EARTH'S SCIENCE

Edited by:

Stefano Spezia

ARCLER
PRESS

www.arclerpress.com

Thermodynamics of the Atmosphere: Basic Physics and Applications to Earth's Science

Stefano Spezia

Arcler Press
224 Shoreacres Road
Burlington, ON L7L 2H2
Canada
www.arclerpress.com
Email: orders@arclereducation.com

ISBN: 978-1-77469-873-0 (Hardcover)

DECLARATION

ABOUT THE EDITOR

Stefano Spezia was born in Erice (Italy) in 1981. He obtained a master's degree in Electronic Engineering (Telecommunications) at the University of Palermo in 2006, and in 2012, at the same university, he got a PhD degree in Applied Physics. From 2007 to 2014, he carried out research in the Physics of Complex Ecological Systems, Semiconductor Spintronics, Nonlinear Optics and Quantum Optics, publishing several works in international journals and books. Since 2014, high school teacher of Mathematics and Physics. He is also an amateur mathematician interested in integer sequences.

TABLE OF CONTENTS

Section 5 Thermal Radiation

LIST OF CONTRIBUTORS

Jacques Arnaud
Mas Liron, F30440 Saint Martial, France

Laurent Chusseau
IES, UMR n°5214 au CNRS, Université Montpellier II, F34095 Montpellier, France

Fabrice Philippe[3]
LIRMM, UMR n°5506 au CNRS, 161 rue Ada, F34392 Montpellier, France

Yunus A. Çengel
Department of Mechanical Engineering, University of Nevada, Reno, NV 89557, USA

Mehmet Kanoglu
Department of Mechanical Engineering, Alanya Alaaddin Keykubat University, 07425 Antalya, Turkey

Ying Liu
State Key Laboratory of Severe Weather, Chinese Academy of Meteorological Sciences, Beijing 100081, China

Chongjian Liu
State Key Laboratory of Severe Weather, Chinese Academy of Meteorological Sciences, Beijing 100081, China

Donghai Wang
State Key Laboratory of Severe Weather, Chinese Academy of Meteorological Sciences, Beijing 100081, China

Gabriele Vissio
CEN, Meteorological Institute, University of Hamburg, Hamburg, Germany
Institute of Geosciences and Earth Resources (IGG) - National Research Council (CNR), Turin, Italy

Valerio Lucarini
Department of Mathematics and Statistics, University of Reading, Reading, UK
Centre for the Mathematics of Planet Earth, University of Reading, Reading, UK

Stephen I. Thomson
College of Engineering, Mathematics and Physical Sciences, University of Exeter, UK

Geoffrey K. Vallis
College of Engineering, Mathematics and Physical Sciences, University of Exeter, UK

Andrew L. Stewart
Department of Atmospheric and Oceanic Sciences, University of California, Los Angeles, CA, USA.

James C. McWilliams
Department of Atmospheric and Oceanic Sciences, University of California, Los Angeles, CA, USA.

O. Onishchenko
Institute of Physics of the Earth, 123995, 10 B. Gruzinskaya str., Moscow, Russia
Space Research Institute, 117997, 84/32 Profsoyuznaya str., Moscow, Russia

O. Pokhotelov
Institute of Physics of the Earth, 123995, 10 B. Gruzinskaya str., Moscow, Russia

V. Fedun
Dept. of Automatic Control and Systems Engineering, The University of Sheffield, Mappin Street, Sheffield, S1 3JD, UK

Theresa A. McHugh
U.S. Geological Survey, Southwest Biological Science Center, Moab, UT, USA.
Center for Ecosystem Science and Society, Northern Arizona University, Flagstaff, AZ, USA.

Ember M. Morrissey
Center for Ecosystem Science and Society, Northern Arizona University, Flagstaff, AZ, USA.

Sasha C. Reed
U.S. Geological Survey, Southwest Biological Science Center, Moab, UT, USA.

Bruce A. Hungate
Center for Ecosystem Science and Society, Northern Arizona University, Flagstaff, AZ, USA.
Department of Biological Sciences, Northern Arizona University, Flagstaff, AZ, USA.

Egbert Schwartz
Center for Ecosystem Science and Society, Northern Arizona University, Flagstaff, AZ, USA.
Department of Biological Sciences, Northern Arizona University, Flagstaff, AZ, USA.

Mikiko Fujita
Japan Agency for Marine-Earth Science and Technology, 3173-25 Showa-machi, Kanazawa-ku, Yokohama, 236- 0001, Japan.

Tomonori Sato
Faculty of Environmental Earth Science, Hokkaido University, Kita-10, Nishi-5, Sapporo, 060–0810, Japan

Agnieszka Wypych
Jagiellonian University, 7 Gronostajowa St., 30-387 Krakow, Poland

Bogdan Bochenek
Institute of Meteorology and Water Management—National Research Institute, 14 Piotra Borowego St., 30-215 Krakow, Poland

Hien Xuan Bui
Department of Atmospheric Sciences, National Central University, Taoyuan City, Taiwan

JiaYuh Yu
Department of Atmospheric Sciences, National Central University, Taoyuan City, Taiwan

Chia Chou
Research Center for Environmental Changes, Academia Sinica, Taipei, Taiwan
Department of Atmospheric Sciences, National Taiwan University, Taipei, Taiwan

Tiina Nygård
Finnish Meteorological Institute, Helsinki, Finland

Michael Tjernström
Department of Meteorology, Bolin Centre for Climate Research, Stockholm University, Stockholm, Sweden

Tuomas Naakka
Finnish Meteorological Institute, Helsinki, Finland

Hubert Luce
Meditterranean Insitute of Oceanography, CNRS/INSU, UMR7294, IRD, Université de Toulon, La Garde, France.

Lakshmi Kantha
Research Institute for Sustainable Humanosphere, Kyoto University, Kyoto, Japan.

Hiroyuki Hashiguchi
Department of Aerospace Engineering Sciences, University of Colorado Boulder, Boulder, CO, USA.

Dale Lawrence
Research Institute for Sustainable Humanosphere, Kyoto University, Kyoto, Japan.

Tyler Mixa
Research Institute for Sustainable Humanosphere, Kyoto University, Kyoto, Japan.

Masanori Yabuki
Department of Aerospace Engineering Sciences, University of Colorado Boulder, Boulder, CO, USA.

Toshitaka Tsuda
Department of Aerospace Engineering Sciences, University of Colorado Boulder, Boulder, CO, USA.

Andrea I. Flossmann
Université Clermont Auvergne, CNRS, LaMP, F-63000 Clermont-Ferrand, France

Wolfram Wobrock
Université Clermont Auvergne, CNRS, LaMP, F-63000 Clermont-Ferrand, France

Hossein Dadashazar
Department of Chemical and Environmental Engineering, University of Arizona, Tucson, AZ, USA

David Painemal
NASA Langley Research Center, Hampton, VA, USA
Science Systems and Applications, Inc., Hampton, VA, USA

Majid Alipanah
Department of Systems and Industrial Engineering, University of Arizona, Tucson, AZ, USA

Michael Brunke
Department of Hydrology and Atmospheric Sciences, University of Arizona, Tucson, AZ, USA

Seethala Chellappan
Rosenstiel School of Marine and Atmospheric Science, University of Miami, Miami, FL, USA

Andrea F. Corral
Department of Chemical and Environmental Engineering, University of Arizona, Tucson, AZ, USA

Ewan Crosbie
NASA Langley Research Center, Hampton, VA, USA
Science Systems and Applications, Inc., Hampton, VA, USA

Simon Kirschler
Institute of Atmospheric Physics, German Aerospace Center, Oberpfaffenhofen, Germany

Hongyu Liu
National Institute of Aerospace, Hampton, VA, USA

Richard H. Moore
NASA Langley Research Center, Hampton, VA, USA

Claire Robinson
NASA Langley Research Center, Hampton, VA, USA
Science Systems and Applications, Inc., Hampton, VA, USA

Amy Jo Scarino
NASA Langley Research Center, Hampton, VA, USA
Science Systems and Applications, Inc., Hampton, VA, USA

Michael Shook
NASA Langley Research Center, Hampton, VA, USA

Kenneth Sinclair
NASA Goddard Institute for Space Studies, New York, NY, USA
Universities Space Research Association, Columbia, MD, USA

K. Lee Thornhill
NASA Langley Research Center, Hampton, VA, USA

Christiane Voigt
Rosenstiel School of Marine and Atmospheric Science, University of Miami, Miami, FL, USA

Hailong Wang
Atmospheric Sciences and Global Change Division, Pacific Northwest National Laboratory, Richland, WA, USA

Edward Winstead
NASA Langley Research Center, Hampton, VA, USA
Science Systems and Applications, Inc., Hampton, VA, USA

Xubin Zeng
Department of Hydrology and Atmospheric Sciences, University of Arizona, Tucson, AZ, USA

Luke Ziemba
NASA Langley Research Center, Hampton, VA, USA

Paquita Zuidema
Rosenstiel School of Marine and Atmospheric Science, University of Miami, Miami, FL, USA

Armin Sorooshian
Department of Chemical and Environmental Engineering, University of Arizona, Tucson, AZ, USA
Department of Hydrology and Atmospheric Sciences, University of Arizona, Tucson, AZ, USA

Yang Yang
Key Laboratory of Meteorology and Ecological Environment of Hebei Province, Shijiazhuang, China
Weather Modification Office of Hebei Province, Shijiazhuang, China

Chuanfeng Zhao
Key Laboratory of Meteorology and Ecological Environment of Hebei Province, Shijiazhuang, China
College of Global Change and Earth System Science, Beijing Normal University, Beijing, China

Jiao Fu
Weather Modification Office of Hebei Province, Shijiazhuang, China

Yi Cui
Weather Modification Office of Hebei Province, Shijiazhuang, China

Xiaobo Dong
Weather Modification Office of Hebei Province, Shijiazhuang, China

Rong Mai
Weather Modification Office of Hebei Province, Shijiazhuang, China

Feng Xu
Weather Modification Office of Hebei Province, Shijiazhuang, China

Jun Li
Shaanxi Key Laboratory of Earth Surface System and Environmental Carrying Capacity, Northwest University, Xi'an, China

Wanyu Liu
Shaanxi Key Laboratory of Earth Surface System and Environmental Carrying Capacity, Northwest University, Xi'an, China

Dimitri Castarède
Department of Chemistry and Molecular Biology, Atmospheric Science, University of Gothenburg, Gothenburg, Sweden

Wenjun Gu
State Key Laboratory of Organic Geochemistry, Guangzhou Institute of Geochemistry, Chinese Academy of Sciences, Guangzhou, China

Linjie Li
Department of Chemistry and Molecular Biology, Atmospheric Science, University of Gothenburg, Gothenburg, Sweden

Takuji Ohigashi
UVSOR Synchrotron, Institute for Molecular Science, Okazaki, Japan

Guiqian Zhang
Shaanxi Key Laboratory of Earth Surface System and Environmental Carrying Capacity, Northwest University, Xi'an, China

Mingjin Tang
State Key Laboratory of Organic Geochemistry, Guangzhou Institute of Geochemistry, Chinese Academy of Sciences, Guangzhou, China

Erik S. Thomson
Department of Chemistry and Molecular Biology, Atmospheric Science, University of Gothenburg, Gothenburg, Sweden

Mattias Hallquist
Department of Chemistry and Molecular Biology, Atmospheric Science, University of Gothenburg, Gothenburg, Sweden

Sen Wang
Shaanxi Key Laboratory of Earth Surface System and Environmental Carrying Capacity, Northwest University, Xi'an, China

Xiangrui Kong
Department of Chemistry and Molecular Biology, Atmospheric Science, University of Gothenburg, Gothenburg, Sweden

Leslie V. Woodcock
Department of Physics, University of Algarve, 8005-139 Faro, Portugal

Hermann Harde
Laser Engineering and Materials Science, Helmut-Schmidt-University Hamburg, Holstenhofweg 85, 22043 Hamburg, Germany

Henrik Svensmark
National Space Institute, Technical University of Denmark, Elektrovej 327, 2800 Kongens Lyngby, Denmark

Jacob Svensmark
Niels Bohr Institute, University of Copenhagen, Niels Bohr Building, Jagtvej 128, 2. floor, 2200 Copenhagen N., Denmark

Martin Bødker Enghoff
National Space Institute, Technical University of Denmark, Elektrovej 327, 2800 Kongens Lyngby, Denmark

Nir J. Shaviv
Racah Institute of Physics, Hebrew University of Jerusalem, Giv'at Ram, Jerusalem, 91904 Israel

LIST OF ABBREVIATIONS

AOD	Aerosol Optical Depth
APE	Atom Percent Excess
CALIOP	Cloud-Aerosol Lidar with Orthogonal Polarization
CALIPSO	Cloud-Aerosol Lidar and Infrared Pathfinder Satellite Observations
CAM3	Community Atmospheric Model
CAM5	Community Atmospheric Model Version
CAO	Cold-air Outbreak
CCN	Cloud Condensation Nuclei
CERES	Clouds and the Earth's Radiant Energy System
CESM	Community Earth System Model
CIN	Convective Inhibition
CIP	Cloud Imaging Probe
CPI	Cloud Particle Imager
ECMWF	European Centre for Medium-Range Weather Forecasts
FCDP	Fast Cloud Droplet Probe
GBRT	Gradient-Boosted Regression Tree
GCM	General Circulation Model
GFD	Geophysical Fluid Dynamics
IGWs	Internal Gravity Waves
INP	Ice Nucleation Particles
IWC	Ice Water Content
LWC	Liquid Water Content
MEPP	maximum entropy production principle
MISR	Multiangle Imaging Spectro-Radiometer
MODIS	MODerate resolution Imaging Spectroradiometer
NCAR	National Center for Atmospheric Research
NEF	Negative Entropy Flow
NPI	Negentropy Principle of Information

OSCAR	Ocean Surface Current Analysis Real-time
PBL	Planetary Boundary Layer
PBLH	Planetary Boundary Layer Height
PWV	Recipitable Water Vapour
RCMs	Regional Climate Models
SAT	Surface Air Temperature
SLP	Sea Level Pressure
SOMs	Self-Organizing Maps
SSF	Single Scanning Footprint
TAMMS	Turbulent Air-Motion Measurement System
TCAP	Two-Column Aerosol Project
TCW	Total Column Cloud Water
TKE	Turbulent Kinetic Energy
TRMM	Tropical Rainfall Measuring Mission
UAV	Unmanned Aerial Vehicle
VOCALS	Variability of the American Monsoon System Ocean-Cloud-Atmosphere-Land Study
WNAO	Western North Atlantic Ocean
ZTD	Zenith Tropospheric Delay

PREFACE

The atmosphere behaves like a giant thermal engine in which disorganized energy is transformed into organized kinetic energy by the uneven absorption of solar radiation. The last one drives the circulation of the atmosphere by temperature differences between the polar and equatorial regions. For such reasons, understanding the fundamental properties of the climate system and its behaviour full benefits from the basic principles of physics, i.e. the thermodynamics laws.

Section 1 of this book introduces the underlying physics of the ideal gases by presenting the main thermodynamics results. It addresses the study of atmospheric behaviour in terms of different entropy measures, including Boltzmann, Gibbs, and Shannon entropies. Moreover, it discusses the interplay between atmosphere thermodynamic and ocean dynamical processes on the trend of marine heatwaves and the aerosol effects on warm convective cloud fields.

Section 2 treats the effects of gravity on the climate and terrestrial atmospheric circulation through geo-fluid-dynamical models. Moreover, employing a closed system of nonlinear equations, it examines the convective cells of internal gravity waves in the Earth's atmosphere layers.

Section 3 focuses on the water content in the atmosphere. In particular, it examines the air as a possible source of water in arid and semiarid regions, the trends of precipitable water vapour and precipitation intensity as a function of surface air temperature. Moreover, it broadly examines the vertical structure of the atmosphere in different layers and zones of the terrestrial planet.

Section 4 discusses the cloud drops and mixtures in the atmosphere. In more detail, it considers the temporal and spatial scales of cloud processing of aerosol particles in marine stratocumulus and the cloud drop number concentrations over the western North Atlantic Ocean.

Moreover, it treats the response of the cloud microphysical properties to the cloud seeding and ice nucleation properties of mixtures originating from East Asian dust storms.

Finally, the last Section 5 studies thermal radiation by revisiting the basic principles of thermodynamic equilibrium in heat transfer within the atmosphere, considering radiation and heat transfer with a molecular basis approach, and the influence of atmospheric ionization by cosmic rays on aerosol and cloud.

SECTION 1
BASIC PHYSICS: IDEAL GASES AND THERMODYNAMICS LAWS

CHAPTER 1

On Classical Ideal Gases

Jacques Arnaud[1], Laurent Chusseau[2] and Fabrice Philippe[3]

[1]Mas Liron, F30440 Saint Martial, France

[2]IES, UMR n°5214 au CNRS, Université Montpellier II, F34095 Montpellier, France

[3]LIRMM, UMR n°5506 au CNRS, 161 rue Ada, F34392 Montpellier, France

ABSTRACT

We show that the thermodynamics of ideal gases may be derived solely from the Democritean concept of corpuscles moving in vacuum plus a principle of simplicity, namely that these laws are independent of the laws of motion, aside from the law of energy conservation. Only a single corpuscle in contact with a heat bath submitted to a z and t-invariant force is considered. Most of the end results are known but the method appears to be novel.

Citation: Arnaud, Jacques, Laurent Chusseau, and Fabrice Philippe. 2013. "On Classical Ideal Gases" Entropy 15, no. 3: 960-971. https://doi.org/10.3390/e15030960.

The mathematics being elementary, the present paper should facilitate the understanding of the ideal gas law and of classical thermodynamics even though not-usually-taught concepts are being introduced.

Keywords: ideal gas law, relativistic gases submitted to gravity, corpuscular concepts, classical gas theory, Democritus physics, canonical single-corpuscle thermodynamics

INTRODUCTION

This paper is devoted to an alternative derivation of the ideal gas law, which agrees with the accepted postulates of statistical mechanics, and may present pedagogical and scientific interest. We postulate the law of thermal equilibrium. By *ideal gas*, we mean a collection of non-interacting corpuscles in some enclosure. These corpuscles may be submitted to an external force, e.g., of electrical or gravitational origin, and the non-relativistic approximation does not have to be made. We will call *perfect gas* an ideal gas with no external force acting upon the corpuscles and with the non-relativistic approximation being made. There is full agreement between our general results and the first and second laws of thermodynamics as they are spelled out in textbooks such as Callen [**1**]. There is also full agreement with the usual *perfect-gas* laws that can be found in the Bernoulli work and elementary text-books, in the limit considered.

The question has been often raised by historians of science as to why the industrial revolution that occurred in the 17th century in Europe did not occur much earlier, for example at the time of Hero of Alexandria (100 BC), who was a disciple of Democritus and invented the first known steam engine called the aeolipile. Some of the possible reasons invoked are lack of interest on the part of the rulers, the paucity of the wood needed to heat water, and the fact that materials, that can withstand high vapor pressures, had not been discovered yet. We will not consider here these sociological or technological problems, but instead concentrate on the conceptual issues. From a pedagogical viewpoint many teachers point out the usefulness of presenting to students the major discoveries in science as they actually occurred, and perhaps also *as they could have occurred*, from a logical standpoint. On the scientific side, we observe that most text-books reproduce the Bernoulli argument based on Galilei–Newton mechanics. But the ideal gas and the barometric laws are not so restricted. In the present paper, we give a generalized and—at least in our opinion—simplified treatment that accounts for the effect of a constant force acting on the corpuscle and special-relativity effects. Some of the results in those cases are known (but not very well-known). They can be generalized to other situations including various dispersion relations (or Hamiltonians), and z-dependent forces (see the last section of our longer paper in [**2**,**3**]).

Let us first quote Feynman [**4**] "If, in some cataclysm, all of scientific knowledge were to be destroyed, and only one sentence passed to the next generations of creatures, what statement would contain the most information in the fewest words? I believe it is the hypothesis that all things are made of atoms. In that one sentence there is an enormous amount of information about the world, if just a little imagination and thinking are applied." In line with this quotation, it seems worthwhile exploring how much physics may be derived from the corpuscular concept. Only some qualitative observations are needed. One is the fact that it takes some time, denoted here $\tau(z_m)$, for a corpuscle thrown upward on earth to reach an altitude z_m and come back to the ground level. We have of course (0)=0. If the corpuscle bounces elastically on the ground, (z_m) represents the oscillation period. The time during which the corpuscle is located above some altitude $z \leq z_m$ during a period is: $(z_m - z)$. Under our assumptions the -function does not depend on the initial altitude or initial time. A second observation is that two bodies left in contact for a sufficient period of time tend to reach the same temperature, as one can judge by our senses.

The purpose of the present paper is thus to show that the thermodynamics of ideal gases and in particular the ideal gas law may be obtained on the sole basis of the Democritean model according to which nature consists of corpuscles moving in a vacuum, plus a principle of simplicity: namely that these fundamental laws are *independent* of the law of corpuscle motion (non-relativistic, special relativistic, or otherwise). According to *general relativity*, thermal energy has weight. But this so-called Tolman effect that entails a temperature variation on the order of 10^{-12} on earth is entirely negligible. The temperature enters solely for dimensional reasons. We later show from the general expressions of the gas internal energy and of the force (or pressure) that the heat delivered by the gas is θdS, an expression of the entropy S being given. This result enables us to prove that the formally-introduced temperature *is* a thermodynamic temperature. Indeed, we recover for ideal gases the general Carnot result asserting that the efficiency of reversible thermal engines is $1-\theta_l/\theta_h$, where θ_l denotes the cold bath temperature and θh the hot bath temperature. Since Kelvin time this expression *defines* the thermodynamics temperature to within an arbitrary proportionality factor, which is fixed by specifying that $\theta \approx 370 \times 10^{-23}$ joules at the water triple point. An alternative derivation of the Carnot result, also solely based on the concept of potential energy, is in [**2**,**5**].

In the title of this paper the word "classical" means "non-quantum" ($\hbar \rightarrow 0$), but the non-relativistic approximation is not made, except in examples. "Ideal Gases" refer to non-interacting identical corpuscles. Accordingly, the corpuscles are independent, and we may consider a single corpuscle. As a result of the slight thermal motion of the container wall, there is an exchange of energy between the corpuscle and the heat bath (for a static wall, even a rough one, the corpuscle energy would be a constant of motion). Thus we are working in the so-called "canonical ensemble". Concretely, the corpuscle may be a cold neutron submitted to the earth gravity and reflected by a diamond at some non-zero temperature [**6**]. The paper [7] "Microcanonical single-particle distributions for an ideal gas in a gravitational field" is valuable because it establishes (within the non-relativistic approximation) the gas law in the micro-canonical ensemble with the effect of gravity taken into account as in the present paper. These authors find that the micro-canonical result approaches the canonical result for few corpuscles only. However that paper is complicated both from mathematical and physical viewpoints. The reader may wish to compare the micro-canonical many-corpuscle paper just cited with the present canonical single-corpuscle paper in terms of pedagogical value. Of course the final results are the same.

The reader may feel that our statement that the above invariance principle *implies* the ideal gas laws, without anything else, is surprising. Yet, we hope that we can convince him/her that this is indeed the case. On the other hand, a quick reading of standard books on thermodynamics may lead other readers to believe that this is a well-known fact. For example, Callen [**1**] states correctly that: "The essence of the ideal-gas law is that molecules of the gas do not interact. This simple fact *implies* that $PV \propto NT$". However, in order to reach this conclusion, the author needs postulate the Newtonian law of motion, quantum theory, and the Boltzmann factor. The claim that corpuscle independence entails the ideal gas law (without anything else) therefore does not appear to have been justified before. For a broad discussion of the laws of thermodynamics, see for example [**8**].

Let us emphasize that our goal is to derive the ideal gas laws from first principles, only conservation of potential energy and thermal equilibrium being assumed (our model cannot describe directly the so-called "adiabatic atmospheres" that are in mechanical but not thermal equilibrium). We do not use the concept of kinetic energy, nor do we postulate any particular law of corpuscle motion. Accordingly, a given potential $(z)=wz$ for a weight w does not imply any specific law of motion $z(t)$. We call "weight" the downward force exerted on the corpuscle, even though this force does not have to be of gravitational origin. Because the presence of a corpuscle affects negligibly the potential, one may consider the latter to be an *external* potential. To our knowledge, the concepts presented in this paper have not been considered before. However, the final expressions relating to the relativistic ideal gas law in the presence of gravity are known, see for example [**9**].

Admittedly the present paper departs much from usually-taught concepts in thermodynamics [**10**]. We indeed ignore the concepts of kinetic energy, of micro-states, and the laws of Physics that would provide the motion $z=z(t)$ (to within arbitrary t and z translations) of corpuscles of given energy. The temperature (to within a constant factor) follows from dimensional considerations and not as usual from the derivative of the average energy (S,h) with respect to the entropy S at constant volume (here denoted h). Quite to the contrary, an explicit expression of the (classical ideal gas) entropy function $S(\theta,h)$ is derived from the average force $\langle F\rangle$ (or pressure) and average energy U. This change of perspective suggests that one should first emphasize, in teaching the subject to students, the Democritean concept of corpuscles moving in vacuum and the concept of potential energy (e.g., from cords and pulleys experiments). Next, consider the up and down motion of a corpuscle thrown upward from $z=0$ in the earth gravitational field (supposed to be static and independent of the vertical z coordinate), the particular function (t) being left undefined. As a result of weak thermal contact with the bath (that does not need to be specified further), the maximum altitude z_m varies slowly in the course of time. This variation is described statistically by some probability law (z_m). It should be explained next that a weight measured by a high-inertia balance does not depend on possible up and down motion of the corpuscle. It is intuitive that the up and down motion of a corpuscle cannot modify the average force that it exerts on the ground plate. The formal proof is as follows: The force is the product of twice the momentum p at the collision time and the number of collisions per unit time, that is: $2p/\tau^2$. The Hamilton equations of motion of a corpuscle are $dz/dt=\partial(z,p)/\partial p$ and $dp/dt=-\partial H(z,p)/\partial z$ for a Hamiltonian $H(z,p)$. If a uniform time-invariant force $-w$ is applied to the corpuscle, we have $(z,p)=H(p)+wz$, where wz represents the potential energy and $H(p)$ is an *arbitrary* function of p. The main contribution of our paper is indeed to show that the barometric and ideal gas laws *do not* *depend* on that (p) function. Application of the Hamilton equations just recalled: $p=p_0-wt=-wt$, with a time translation when w is not strictly equal to zero. If $w=0$, $p=p_0$ is of course a constant of motion, but gets reversed in sign upon a collision. Consider now a full oscillation period beginning at z_m and ending up at z_m, the collision with the ground ($z=0$) occurring half-way. Accordingly we have $p=w\tau/2$ and $2p/\tau=w2$ as asserted. In the non-relativistic approximation, $H(p)=p^2/2m$, where m denotes the corpuscle mass. However, since the barometric and ideal gas laws do not depend on (p), they do not depend on m either. The quantum theory is mathematically feasible for non-relativistic or ultra-relativistic corpuscles, but it becomes very involved for general (p) functions. The semi-classical (Bohr–Sommerfeld) approximation of the quantum theory coincides at the end with our treatment, but it may not be as straightforward.

Accordingly, the impact is equal to w times the motion period. This observation enables us to express the force F in terms of round-trip times when a rigid plate is present at $z=h$, see **Figure 1**. The condition that the average force be *independent* of the (z_m) function introduced above provides the ideal gas law in full generality.

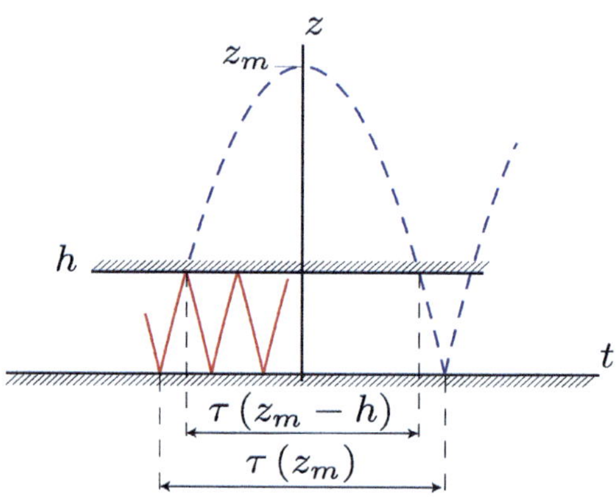

Figure 1. Space-time $(z,)$ trajectory for a corpuscle of weight w bouncing off the ground ($z = 0$). The maximum altitude reached by the corpuscle is $zm=E/w$, where E denotes the energy. The motion is periodic with period (z_m), where $\tau(Z)$ denotes the corpuscle round-trip time at a distance Z from the top of the trajectory. When the altitude is restricted to h by a plate (dashed horizontal line) the motion remains periodic with a period evidently equal to: $(z_m)-\tau(z_m-h)$. Note that this expression holds even if the motion is not symmetric in time.

To conclude, the approach advocated in this paper enables us to recover classical results in thermodynamics in a considerably generalized form (but only for independent classical corpuscles and laws of physics that are independent of time and altitude translations). Admittedly, our presentation goes against the order in which basic concepts are usually taught, but the greater simplicity and generality will perhaps be appreciated.

AVERAGE FORCE EXERTED BY A CORPUSCLE ON A PISTON

We consider a unit-area cylinder with vertical z-axis in uniform gravity resting on the ground ($z=0$) at some temperature. A tight piston is free to move in the vertical direction. The cylinder height is denoted by h and contains a single corpuscle of weight w, or submitted to a force $-w$ constant in space and time. The motion may be relativistic or not. In our one-dimensional model, the pressure P corresponds to the average force $\langle F \rangle$, the volume V to the height h, and $N=1$. Our result provides the ideal-gas law in a generalized form, taking into account gravity. In that case, the pressure varies as a function of altitude. More precisely, the force exerted by the corpuscle on the lower end of the cylinder exceeds the force exerted on the upper end (or piston) by the corpuscle weight. But in the absence of gravity, the forces exerted on both ends would be the same. We are introducing (static and uniform) gravity mainly because this helps clarify the concept of corpuscle energy: the corpuscle energy E is the maximum altitude z_m that the corpuscle would reach in the absence of the piston times w. The corpuscle bounces elastically off the ground and off the piston (if it reaches it), that is, without any loss or gain of energy. (Clarification is perhaps needed here. Some quantities in thermodynamics have a clear experimental significance and are unambiguous. Let us first recall that we are considering a single corpuscle. However, an ideal gas, as we define it, is a collection of non-interacting corpuscles. Accordingly, at a given temperature and at a given value of h, for N identical corpuscles the force and the internal energy, and thus the free energy, are simply multiplied by N).

The force (or pressure) and internal energy have clear meanings. This is the case in particular for the average force exerted by the gas on a piston. The word "average" here enters in two ways, as a mechanical average and as a thermal average. This point is clarified below. Consider a purely mechanical problem: the so-called "bouncer". A corpuscle of energy E is submitted to a constant force. The mechanical average referred to above is the product of $2p$, that is, twice the corpuscle momentum p when it collides with the z=0 plane, times the number of collisions per unit time, which is equal to $1/\tau$, where denotes the corpuscle round-trip time (or motion period), which is $F=2p/\tau$. Physically, it is supposed that the z=0 plane (e.g., the earth) has so much inertia that it does not respond to individual collisions. The second averaging refers to thermal averaging. We postulate a distribution (E) of energy, and the second averaging is taken with respect to this distribution. The point of our paper is that this (E) distribution does not have to be postulated (or derived from considerations a la Boltzmann); it follows from our simplicity principle.

Without piston ($h\to\infty$) we obtain the so-called "barometric law", that is, the decay of the average corpuscle density as a function of altitude z. The density decay is found from our simplicity principle to be: $(z)\propto exp(-wz/\theta)$. This is of course a well-known result, but here the Boltzmann factor does not have to be postulated. Similar considerations apply to the case where there is a piston at altitude h. Both the piston and the ground z=0 are assumed to have great inertia, so that they do not react significantly to individual collisions, and again the mechanical average force for some corpuscle energy E is $2p/\tau^2$. In the presence of a force w acting on the corpuscle, p is of course larger at z=0 than at $z=h$. If there is no force acting on the corpuscle, i.e., w=0, p and thus F are the same on the piston and on the ground. The corpuscle energy in the non-relativistic approximation is $E=p^2/2m$, and thus the mechanical-average force acting on the piston or on the ground are $F=2p/\tau$. Since $\tau=2h/v$-;$=mv$, we obtain $F=2E/h$. Performing now a thermal averaging, we finally obtain $Fh=\langle pv\rangle=\langle 2E\rangle=2U=\theta$ if we define θ as the average corpuscle energy (which reduces here to the kinetic energy, $U=\langle mv^2/2\rangle$. The above discussion applicable to perfect gases, initially due to Bernoulli, is given in elementary textbooks. The purpose of the present note is to clarify that the well-known results applicable to perfect gases are special cases of our more general results, applicable to ideal gases. Consider first the case where h is infinite, that is, in the absence of a piston. The time period is denoted as before by (Z) with $Z=z_m$. The average force exerted on the ground, equal to the corpuscle weight is the impulse i divided by the period. Thus $i=w/(z_m)$. In other words, the impulse transmitted to a piston when the corpuscle impacts on it is equal to the motion period. If the piston is located at $z=h$, the impulse it receives is $ih=w(z_m-h)$ if $z_m\geq h$, and zero otherwise. On the other hand, the motion period is (z_m) if $z_m\leq h$ and $\tau(z_m)-\tau(z_m-h)$ if $z_m>h$.

Accordingly, the force F experienced by the piston when the corpuscle energy is wz_m is the ratio of the impulse and period:

$$\begin{cases} F(z_m) = 0 & z_m \leq h \\ F(z_m) = \frac{i_h}{\tau(z_m)-\tau(z_m-h)} = \frac{w\tau(z_m-h)}{\tau(z_m)-\tau(z_m-h)} & z_m > h \end{cases} \tag{1}$$

Because the cylinder lower end is in contact with a bath, there is a slight thermal motion and the corpuscle energy z_m slowly varies in the course of time. Accordingly, the average force $\langle F \rangle$ experienced by the piston is, if (z_m) denote the distribution:

$$\langle F \rangle = \frac{\int_0^\infty dz_m\, \omega(z_m) F(z_m)}{\int_0^\infty dz_m\, \omega(z_m)} = w \frac{\int_h^\infty dz_m\, \omega(z_m) \frac{\tau(z_m-h)}{\tau(z_m)-\tau(z_m-h)}}{\int_0^h dz_m\, \omega(z_m) + \int_h^\infty dz_m\, \omega(z_m)} \tag{2}$$

Initially, we obtained the form of the force $\langle F \rangle$ from the Bohr–Sommerfeld approximation of the quantum theory. But this amounts to saying that the corpuscle action (area in the position-momentum *z-p* phase-space, where here $p=wt$ is discrete, and evenly spaced in units of the Planck constant. However, because the "bouncer" considered in our paper is not a harmonic oscillator, the discrete corpuscle energies are not evenly spaced. This is why, when going to the continuous limit, converting the sum into an integral, we must introduce the factor (z_m) in the form given. However, after going through all these semi-classical considerations, we discovered that it was sufficient to postulate the simplicity principle given earlier. The latter concept (unlike the quantum theory) could have been understood at the time of the ancient Greece. Going the opposite way, one may say that the simplicity principle suggests the semi-classical quantum theory. This viewpoint is not discussed further here.

According to our simplicity principle, the average force $\langle F \rangle$ must be *independent* of the corpuscle equation of motion, and thus of the $\tau(.)$ function. This condition obtains from Equation (**2**) if one selects the following distribution:

$$\begin{cases} \omega(z_m) = \exp(-wz_m/\theta)\tau(z_m) & z_m \leq h, \\ \omega(z_m) = \exp(-wz_m/\theta)\big(\tau(z_m) - \tau(z_m - h)\big) & z_m > h, \end{cases} \tag{3}$$

where θ has the dimension of an energy. The average force becomes, using Equations (2) and (3):

$$\begin{aligned} \langle F \rangle &= \frac{w \int_h^\infty dz_m \exp(-wz_m/\theta)\tau(z_m - h)}{\int_0^h dz_m \exp(-wz_m/\theta)\tau(z_m) + \int_h^\infty dz_m \exp(-wz_m/\theta)\big(\tau(z_m) - \tau(z_m - h)\big)} \\ &= \frac{w}{\exp(w\,h/\theta) - 1} \end{aligned} \tag{4}$$

In the above integrals going from h to ∞ we have replaced exp($-wzm/\theta$) by exp($-wh/\theta$)exp($-w(zm-h)/\theta$) and introduced the variable $z'_m \equiv z_m - h$, so that all the integrals go from zero to infinity and cancel out. Note that even though integral signs have been introduced, no integration has been performed. We simply consider an integral as a sum of terms and employ the rule of addition associativity. We have employed also the fact that (x)=exp(x) is the only function such that $f(a+b)=f(a)f(b)$.

For a collection of N independent corpuscles having weights wi,=1,...N respectively, the force is a sum of N terms of the form given in Equation (**4**). In the case of zero gravity (w=0 or more precisely: $wh \ll \theta$, the above expression gives $\langle F \rangle h=\theta$. Thus we have obtained the ideal gas law $\langle F \rangle h=N\theta$.

Average force for a three-dimensional space:

We suppose that the cylinder radius is very large compared with h, and we do not consider the force exerted by the corpuscle on the cylinder wall. Motion of the corpuscle along directions perpendicular to z (say, x and y) *does* affect the round-trip time function (Z).

However, since the average force does not depend on this function, the ideal gas law is unaffected. This is so for any physical system involving a single corpuscle provided the physical laws are invariant under a z-translation (besides being static).

The internal energy, to be discussed in the following section, though, is incremented. One can prove that in the non-relativistic approximation and in the absence of gravity the internal energy is multiplied by 3. It would be incremented further by corpuscle rotation or vibration, not considered here. Using conventional methods, Landsberg [9] and Louis-Martinez [**11**] obtain exactly the same result as given above (except for the factor 3 in the expression of the internal energy, relating to the number of space dimensions considered).

INTERNAL ENERGY

The gas internal energy U is the average value of $E \equiv wz_m$, the gravitational energy being accounted for. Note that only corpuscle motion along the z-axis is being considered. In the present section, a non-unity (but constant) weight w is considered to clarify the dimensions of the quantities introduced. The expression of U is, using the energy distribution given in Equation (**3**):

$$
\begin{aligned}
U &= \frac{\int_0^{wh} dE\, E \exp(-E/\theta)\tau(E/w) + \int_{wh}^{\infty} dE\, E \exp(-E/\theta)\big(\tau(E/w) - \tau(E/w-h)\big)}{\int_0^{wh} dE\, \exp(-E/\theta)\tau(E/w) + \int_{wh}^{\infty} dE\, \exp(-E/\theta)\big(\tau(E/w) - \tau(E/w-h)\big)} \\
&= \frac{\int_0^{\infty} dE\, E \exp(-E/\theta)\tau(E/w) - \int_{wh}^{\infty} dE\, (E - wh + wh) \exp(-E/\theta)\tau(E/w-h)}{\int_0^{\infty} dE\, \exp(-E/\theta)\tau(E/w) - \int_{wh}^{\infty} dE\, \exp(-E/\theta)\tau(E/w-h)} \\
&= \frac{(1-\exp(-wh/\theta))\int_0^{\infty} dE\, E \exp(-E/\theta)\tau(E/w) - wh\int_{wh}^{\infty} dE\, \exp(-E/\theta)\tau(E/w-h)}{(1-\exp(-wh/\theta))\int_0^{\infty} dE\, \exp(-E/\theta)\tau(E/w)} \\
&= \frac{(1-\exp(-wh/\theta))\int_0^{\infty} dE\, E \exp(-E/\theta)\tau(E/w) - wh\exp(-wh/\theta)\int_0^{\infty} dE\, \exp(-E/\theta)\tau(E/w)}{(1-\exp(-wh/\theta))\int_0^{\infty} dE\, \exp(-E/\theta)\tau(E/w)} \\
&= \frac{\int_0^{\infty} dE\, E \exp(-E/\theta)\tau(E/w)}{\int_0^{\infty} dE\, \exp(-E/\theta)\tau(E/w)} - \frac{wh}{\exp(wh/\theta) - 1} \qquad (5)
\end{aligned}
$$

The first term in the final above expression, minus θ, corresponds to the kinetic energy K.

The second term, plus θ, corresponds to the potential energy P. In the non-relativistic limit, the first term gives the well-known expression $K=\theta/2\equiv k\mathrm{B}T/2$. Without gravity, $P=0$. In the general case the splitting of U into $K+P$ looks somewhat artificial.

The internal energy divided by the temperature: $U(\theta,h)/\theta$, thus is the sum of a term function of θ but not of h, and a term that tends to $-1-1$ when $wh\ll\theta$. To evaluate the first term, we need to know the round-trip time (Z) to within an arbitrary proportionality factor, and an integration must be performed in that case.

In order to write the above expression in terms of dimensionless quantities, we introduce the corpuscle rest-mass energy μ. Defining: $x=E/\theta$, $\theta'=\theta/\mu$, $h'=wh/\mu$, we have:

$$
\frac{U}{\theta} = \frac{\int_0^{\infty} dx\, x \exp(-x) f(\theta' x)}{\int_0^{\infty} dx\, \exp(-x) f(\theta' x)} - \frac{h'/\theta'}{\exp(h'/\theta') - 1} \qquad (6)
$$

where the function $f(y)$ depends on the round-trip time.

In the special case of non-relativistic motion $y\ll 1$, we have $f(y) \propto \sqrt{y}$. The first term in Equation (**6**) is then equal to 3/2 (note that $2\int_0^{\infty} dx\, x \exp(-x)\sqrt{x} = 3\int_0^{\infty} dx \exp(-x)\sqrt{x}$), so that in the absence of gravity the internal energy U is equal to $3\theta/2-\theta=\theta/2$, a well-known result. More generally, we have $f(y) \propto \sqrt{y^2+2y}$. A plot of U from low ($\theta\ll\mu$) to high temperatures can be seen in [**2**]. A three-dimensional expression of U in terms of Bessel functions can be found, for example, in [**12**], p. 234.

THE ENERGY Θ IS A THERMODYNAMIC TEMPERATURE

We prove in this section that θ, introduced in previous sections on dimensional grounds only, *is* a thermodynamic temperature. We do this by showing that the efficiency of a reversible thermal cycle employing ideal gases is $1-_{\theta l}/\theta h_1$, where θ_l is the cold-bath temperature and θ_h the hot bath temperature—this is the accepted Kelvin definition of absolute temperatures.

The expressions given earlier for the average force $\langle F\rangle$ in Equation (**4**) and the internal energy U in Equation (**5**) may be written, setting $\beta\equiv 1/\theta$, as:

$$\langle F\rangle = \frac{\partial \ln(Z)}{\beta\, \partial h} \qquad U = -\frac{\partial \ln(Z)}{\partial \beta}$$
$$Z(\beta,h) = \left(1-\exp(-\beta\, w\, h)\right)\int_0^\infty dE\, \exp(-\beta\, E)\tau(E/w) \tag{7}$$

Z is essentially the quantity called in statistical mechanics the partition function. It becomes dimensionless if it is divided by the reduced Planck constant $\hbar$, which however plays here no physical role.

If we introduce the Helmholtz free-energy (the letter A is from the German "Arbeit" or work): $(\theta,h)\equiv-\theta\ln(Z(\theta,h))$ the expressions in Equation (7) are conveniently written:

$$\langle F\rangle = -\frac{\partial A}{\partial h} \qquad U = A - \theta\frac{\partial A}{\partial \theta} \tag{8}$$

We have thus obtained from our simplicity principle both the force function $F(\theta,h)$ and the internal energy function $U(\theta,h)$. We noticed that they follow from the partial derivations of a free-energy function that we denote (as some, but not all, authors do): (θ,h). This function has exactly the same physical and mathematical content as the usual $U(S,h)$ internal energy function, where S denotes the entropy, since they are related by a Legendre transform. Here is the definition. One can forget for a moment the h-variable. Given some point A, of the known $A(\theta)$ function, we define a straight line in the $U(S)$ plane as: $U=A(\theta)+\theta S$, and obtain in that way (using as a parameter) a collection of straight lines whose envelop is the $U(S)$ function looked for. Callen [**1**], for example, shows how the stability condition initially defined for the (S,h) function may be transformed into two conditions on the free-energy function $A(\theta,h)$. Discussing this point would, however, lead us too far astray. The enthalpy follows from the same (S,h) function through a Legendre transform with respect to h (rather than S). For our problem, this latter transform does not seem to bring anything new.

From Equation (**8**) we obtain:

$$-\delta Q \equiv dU + \langle F\rangle\, dh = dA - \frac{\partial A}{\partial \theta}d\theta - \frac{\partial A}{\partial h}dh - \theta\, d(\frac{\partial A}{\partial \theta}) = \theta\, dS \qquad S \equiv -\frac{\partial A}{\partial \theta} \quad (9)$$

where δQ represents the heat released by the gas, from the law of conservation of energy. For any function (θ,h) such as U, A, S: $df \equiv \frac{\partial f}{\partial \theta}d\theta + \frac{\partial f}{\partial h}dh$. Note that we employ only two independent variables, namely θ and h, so that partial derivatives are unambiguous. If the gas is in contact with a thermal bath (θ=constant), δQ is the heat gained by the bath. The quantity S defined above is called "entropy". In particular, if heat cannot go through the gas container wall (adiabatic transformation) we have $\delta Q=0$, that is, according to the above result, $dS=0$. Thus reversible adiabatic transformations are isentropic. We have obtained an expression for the free-energy (θ,h) for the special case of a single corpuscle submitted to a constant force in the canonical ensemble. This free-energy function has the same physical content as the often-used energy function $U=(S,h)$, or the entropy function $S=S(U,h)$, see [**1**]. The expressions following Equation (**8**) in this section therefore coincide with the conventional ones, applicable to any working substance, and our entropy S is the same as the one employed, e.g., in [**1**]. For many corpuscles, the postulate of equi-probability of the microstates may be viewed as implicit.

The Carnot cycle: A Carnot cycle consists of two isothermal transformations at temperatures θ_l and θ_h, and two intermediate reversible adiabatic transformations ($dS=0$). After a complete cycle, the entropy recovers its original value and therefore $dS_l+dS_h=0$. According to Equation (**9**): $-\delta Q_l=\theta_l dS_l$, $-\delta Q_h=\theta_h dS_h$ and therefore $\delta Q_l/\theta_l+\delta Q_h/\theta_h=0$. Energy conservation gives the work δW performed over a cycle from: $\delta W+\delta Q_l+\delta Q_h=0$. The cycle efficiency is defined as the ratio of δW and the heating $-\delta Q_h$ supplied by the hot bath. We have therefore $\eta \equiv \frac{\delta W}{-\delta Q_h} = \frac{\delta Q_h+\delta Q_l}{\delta Q_h} = 1-\frac{\theta_l}{\theta_h}$, from which we conclude that θ is the "thermodynamic temperature". Since Kelvin time, thermodynamics temperatures are strictly defined from Carnot (or other reversible) cycles efficiency. In practice, temperatures may be measured by other means and employed in other circumstances.

We have implicitly assumed in the above discussion that the working medium (presently an ideal gas) has reached the bath temperature before being contacted with it. Otherwise, there would be at that time a jump in entropy, and the cycle would no longer be reversible. Given initial θ,h values, the temperature change $d\theta$ for an increment dh in the isentropic regime ($dS=0$) follows from the relation: $d\theta = -\left(\frac{\partial S/\partial h}{\partial S/\partial \theta}\right)dh$, where $S(\theta,h)$ may be expressed in terms of $Z(\theta,h)$, Equation (7), from the above expressions. The details will be omitted. It suffices to know that θ may be varied by varying h, in a calculable manner, in an isentropic transformation.

Practical units: The energy $\theta=\langle F\rangle h$ has been defined so far only to within a multiplicative factor from dimensional considerations. This factor is fixed by agreeing that $\theta=273.16$ exactly when the cylinder is in thermal equilibrium with water at its triple point. Here $k_B=1.38066...10^{-23}$ joules is considered as an energy unit (akin to the calorie = 4.182... joules). This manner of defining θ is equivalent to the usual one, though expressed differently. The dimensionless quantity $T\equiv\theta/k_B$ is the usual unit of thermodynamic temperature, expressed in Kelvin.

Next, measurements have shown that the number of atoms in 0.012 kg of carbon 12 is: $N_A\approx6.0221367 10^{23}$. For this quantity of matter (called a mole), the ideal gas law therefore reads: $\langle F\rangle h=NA\theta$, or: $PV=RT$, with the ideal gas constant: $R\equiv N_A k_B\approx8.31451$joules per Kelvin per mole.

CONCLUSIONS

Let us briefly recall the concepts introduced in the present paper. One can imagine that after having introduced the corpuscular concept, Democritus observed the elastic bounces of a unit weight on a balance and defined the weight "impulse" from the motion period. Not knowing the nature of the motion, he may have thought of introducing a distribution such that the average force $\langle F\rangle$ *does not depend* on the law of motion. This, as we have seen, may be done. This distribution involves for dimensional reasons a quantity θ having the dimension of energy. Considering a thermal engine operating between two baths at temperatures $\theta_{l,h}$ one finds on the basis of the principles just stated that the maximum efficiency is $1-\theta_l/\theta h1$. This allows us to call θ the thermodynamic temperature.

The present paper provides a first-principle proof of the ideal gas law, including a possible effect of uniform gravity with no knowledge of the round-trip time function $\tau(Z)$ being required. Our thesis is that this law may be obtained on the sole basis of the Democritus model of corpuscles and vacuum. It is indeed unnecessary to specify the law $(t;E)$ of corpuscle motion, where E denotes the energy. Explicit expressions of the internal energy are obtained provided the round-trip time is known. One can further show that the ideal-gas internal energy depends only on temperature in the absence of gravity. The theory presented is strictly classical. As such, it does not depend on the numerical value of the universal constant $\hbar$. However, the introduction of this quantity is needed to make the results dimensionless.

REFERENCES

1. Callen, H.C. *Thermodynamics and an Introduction to Thermo-Statistics*, 2nd ed.; Wiley and Sons: New York, NY, USA, 1985.
2. Arnaud, J.; Chusseau, L.; Philippe, F. Democritus and the motive power of fire, 2011. Available online: http://arxiv.org/abs/1104.0836 (accessed on 12 December 2012).
3. Arnaud, J.; Chusseau, L.; Philippe, F. On the ideal gas law, 2011. Available online: http://arxiv.org/abs/1105.2643 (accessed on 12 December 2012).
4. Feyman, R.P.; Leighton, R.B.; Sands, M. *The Feynman Lectures in Physics*, 6th ed.; Addison-Wesley: Reading, PA, USA, 1977.
5. Arnaud, J.; Chusseau, L.; Philippe, F. A simple model for Carnot heat engines. *Am. J. Phys.* 2010, *78*, 106–110.
6. Kaneko, H.; Tohsaki, A.; Hosaka, A. Thermodynamics of quantum ultra-cold neutron gas under gravity of the earth, 2012. Available online: http://arxiv.org/abs/1205.0378 (accessed on 12 December 2012).
7. Roman, F.L.; White, J.A.; Velasco, S. Microcanonical single-particle distributions for an ideal gas in a gravitational field. *Eur. J. Phys.* 1995.
8. Thermodynamics. Available online: http://en.wikipedia.org/wiki/Thermodynamics (accessed on 12 December 20122).
9. Landsberg, P.T.; Dunning-Davies, J.; Polland, D. Entropy of a column of gas under gravitation. *Am. J. Phys.* 1994, *62*, 712–717.
10. Leinonen, R.; Raesaenen, E.; Asikainen, M.; Hirvonen, P.E. Students' pre-knowledge as a guideline in the teaching of introductory thermal physics at university. *Eur. J. Phys.* 2009, *30*, 593–604.
11. Louis-Martinez, D.J. Classical relativistic ideal gas in thermodynamic equilibrium in a uniformly accelerated reference frame. *Class. Quantum Gravity* **2011**, *28*, 035004.
12. Greiner, W.; Neise, L.; Stöcker, H. *Thermodynamics and Statistical Mechanics*; Springer-Verlag: New York, NY, USA, 1997.

CHAPTER 2

A New Entropy Function to Analyze Isentropic Processes of Ideal Gases with Variable Specific Heats

Yunus A. Çengel [1] and Mehmet Kanoglu [2]

[1] Department of Mechanical Engineering, University of Nevada, Reno, NV 89557, USA

[2] Department of Mechanical Engineering, Alanya Alaaddin Keykubat University, 07425 Antalya, Turkey

ABSTRACT

A new entropy function s^+ is defined in terms of the existing entropy function $s°$ and temperature as $s^+ = s° - R \ln T$ to facilitate the analysis of isentropic processes of ideal gases with variable specific heats. The function s^+ also makes it possible to calculate the entropy changes of ideal gases during processes when volume information is available instead of pressure information and the variation of specific heats with temperature is to be

Citation: Çengel, Yunus A., and Mehmet Kanoğlu. 2022. "A New Entropy Function to Analyze Isentropic Processes of Ideal Gases with Variable Specific Heats" Entropy 24, no. 6: 746. https://doi.org/10.3390/e24060746.

accounted for. The introduction of the function s^+ eliminates the need to use the dimensionless isentropic functions relative pressure P_r and relative specific volume v_r of ideal gases and to tabulate their values. The P_r and v_r data are often confused with pressure and specific volume, with an adverse effect on the study of the second law of thermodynamics. The new s^+ function nicely complements the existing $s°$ function in entropy change calculations: the former is conveniently used when volume information is given while the latter is used when pressure information is available. Therefore, the introduction of the new entropy function s^+ is expected to make a significant contribution to the thermodynamics education and research by streamlining entropy analysis of ideal gases.

Keywords: thermodynamics, entropy, entropy functions, isentropic process, ideal gases, variable specific heats, second-law analysis

INTRODUCTION

A good understanding of the fundamentals of thermodynamics is an essential part of an engineering education for most majors. Thermodynamics is usually considered a challenging subject by engineering students because of the abstractions involved. Entropy is an essential part of this course and students often struggle with making sense of it and using it in solving practical problems. It takes time to develop an intuitive understanding of the second-law concepts of entropy, entropy generation and isentropic efficiencies since, unlike energy, entropy is something which can be created out of nowhere. As a result, mastering entropy balance on processes is considered more difficult compared to the energy balance.

Entropy change calculations for real substances such as steam and the refrigerants are performed easily and conveniently by relying on the tabulated values of entropy. For ideal gases, we rely on the entropy change relations in the case of constant specific heats, and a combination of relations and tabulated $s°$ data when the variation of specific heats with temperature is to be accounted for. In computerized calculations, entropy changes can be calculated easily by using property relations and built-in functions.

To facilitate the hand calculations for the analysis of isentropic processes of ideal gas with variable specific heats, the dimensionless functions relative pressure P_r and relative specific volume v_r are defined, and their values are listed in the ideal-gas tables. The values of P_r and v_r are determined by the numerical calculation of their defining integrals, and they have no physical meaning. Although the names and symbols imply otherwise, P_r and v_r have to relation to pressure or specific volume. The use of these terms adds confusion and impacts thermodynamic education negatively.

Costa [1] identified and discussed the reasons why thermodynamics is viewed as a difficult subject by engineering students. A survey of college students indicated significant challenges concerning students' understanding of thermodynamic processes, the first and second laws, and entropy [2]. Mulop et al. [3] acknowledged the difficulties of students learning thermodynamics globally and presented a review and analysis of the different approaches to supporting students' learning of thermodynamics. Goyings and Arnas [4] noted some difficulties in teaching the second law and the related concepts of entropy and exergy, and proposed an alternative way of teaching thermodynamics.

Yang et al. [5] proposed some methods to address some of the challenges associated with the heat transfer and thermodynamics education. Other researchers proposed alternative methods to deal with difficulties involved in learning of thermodynamics and entropy. An alternative to the Carnot–Clausius approach for introducing entropy and the second law of thermodynamics is outlined, and entropy is introduced as a nonequilibrium property [6]. Acevedo et al. [7] presented a new educational package based on e-learning concerning engineering thermodynamics processes, combustion, energy, and entropy analysis. Smith [8] illustrated how thermodynamic energy, entropy, and the second law of thermodynamics can be intuitively related to student's experiences. Recently Çengel [9] addressed the confusion caused by the use of entropy and information interchangeably, and attempted to put things into proper perspective.

There are a number of studies in the literature on the significance of using variable specific heats for the analysis of energy conversion systems involving ideal gases. Many studies indicate that the effects of using variable specific heats for the working fluid in the air-standard cycles such as the diesel cycle, the dual cycle, and the Brayton cycle performance are significant and they should be considered in cycle analysis [10,11,12,13,14].

In this paper, we introduce a new entropy function s^+ to be used in entropy analysis of ideal gases when the variation of specific heats with temperature are considered. The new function is used in both the entropy change calculations and isentropic process analysis of ideal gases. The introduction of this new entropy term s^+ complements the existing entropy function s° and eliminates the need to define and use of relative pressure P_r and relative specific volume v_r in the analysis of isentropic processes of ideal gases. First, we present the existing coverage for the entropy change and isentropic processes of ideal gases as found in most thermodynamics textbooks [15,16,17]. Then, we define the new entropy term s^+ and provide a property table for air which includes the s^+ values. We demonstrate the use of s^+ by solving some sample problems using both the existing approach and the new approach. Finally, we discuss the advantages of the new s^+ term with respect to the existing coverage of entropy.

ENTROPY CHANGE OF IDEAL GASES

The coverage presented here is based on Çengel et al. [15], but is also available in most thermodynamics textbooks. The differential change of entropy is expressed in terms of other properties as

$$ds = \frac{du}{T} + \frac{Pdv}{T} \tag{1}$$

$$ds = \frac{dh}{T} - \frac{vdP}{T} \tag{2}$$

where s is entropy, u is internal energy, h is enthalpy, T is temperature, P is pressure, and v is the specific volume. Substituting the property relations $du = c_v\, dT$, $dh = c_p\, dT$, and $P = RT/v$ where R is the gas constant into the equations above and integrating them for a process between states 1 and 2 gives

$$s_2 - s_1 = \int_1^2 c_p(T)\frac{dT}{T} - R\ln\frac{P_2}{P_1} \tag{3}$$

$$s_2 - s_1 = \int_1^2 c_v(T)\frac{dT}{T} + R\ln\frac{v_2}{v_1} \tag{4}$$

The specific heats of ideal gases, with the exception of monatomic gases, depend on temperature, and performing the integrals in the equations above requires the use of $c_v(T)$ and $c_p(T)$ functions, which is not practical for hand calculations. Assuming constant specific heats at some average value for simplicity gives the following approximate relations:

$$s_2 - s_1 = c_{v,\text{avg}}\ln\frac{T_2}{T_1} + R\ln\frac{v_2}{v_1} \tag{5}$$

$$s_2 - s_1 = c_{p,\text{avg}}\ln\frac{T_2}{T_1} - R\ln\frac{P_2}{P_1} \tag{6}$$

For better accuracy, the variation of specific heats with temperature needs be accounted for. In computerized calculations, this can be easily done by using the built-in functions for properties and numerical integration schemes. For hand calculations, it is convenient to perform these integrals once and tabulate the results. For this purpose, it is common to choose absolute zero as the reference temperature and define a function s° as

$$s^\circ = \int_0^T c_p(T)\frac{dT}{T} \qquad (7)$$

Then the integral in Equation (3) is simply $s^\circ_2 - s^\circ_1$, and Equation (3) becomes

$$s_2 - s_1 = s_2^\circ - s_1^\circ - R\ln\frac{P_2}{P_1} \qquad (8)$$

Now, noting that

$c_v = c_p - R$

for ideal gases and utilizing Equation (7), the integral in Equation (4) can be expressed as

$$\begin{aligned} \textstyle\int_1^2 c_v(T)\frac{dT}{T} &= \textstyle\int_1^2 \left[c_p(T) - R\right]\frac{dT}{T} \\ &= s_2^\circ - s_1^\circ - R\ln\tfrac{T_2}{T_1} \\ &= (s_2^\circ - R\ln T_2) - (s_1^\circ - R\ln T_1) \\ &= s_2^+ - s_1^+ \end{aligned}$$

where the function s^+ is defined as

$$s^+ = s^\circ - R\ln T \qquad (9)$$

That is, the value of s^+ of an ideal at a given temperature can be determined by subtracting $R\ln T$ from the value of s° at the same temperature. Then, the entropy change of an ideal gas can also be expressed as

$$s_2 - s_1 = s_2^+ - s_1^+ + R\ln\frac{v_2}{v_1} \qquad (10)$$

This is equivalent to

$$s_2 - s_1 = s_2^\circ - s_1^\circ - R\ln\frac{T_2}{T_1} + R\ln\frac{v_2}{v_1}$$

The definition of the function s^+ and the new entropy change relation in Equation (10) are important additions to the literature since they enable the calculation of entropy change of ideal gases when the volume information is given instead of pressure information. This way, tedious iterations are avoided in hand calculations.

It is clear that $s°$ and s^+ are functions of temperature alone, and their values are zero at absolute zero temperature. The values of s^+ are calculated from its definition

$$s^+ = s° - R\ln T$$

for air at various temperatures and the results are listed in Table 1 together with other properties of air.

Table 1: Ideal-gas properties of air. Source: Çengel et al. [15] supplemented with s^+ values. (With the introduction of s^+, the P_r and v_r columns are no longer needed).

T, K	*h*, kJ/kg	*u*, kJ/kg	P_r	v_r	$s°$, kJ/kg·K	s^+, kJ/kg·K
200	199.97	142.56	0.3363	1707.0	1.29559	−0.22503
220	219.97	156.82	0.4690	1346.0	1.39105	−0.15692
240	240.02	171.13	0.6355	1084.0	1.47824	−0.09470
260	260.09	185.45	0.8405	887.8	1.55848	−0.03744
280	280.13	199.75	1.0889	738.0	1.63279	0.01561
290	290.16	206.91	1.2311	676.1	1.66802	0.04076
300	300.19	214.07	1.3860	621.2	1.70203	0.06504
320	320.29	228.42	1.7375	528.6	1.76690	0.11139
340	340.42	242.82	2.149	454.1	1.82790	0.15499
360	360.58	257.24	2.626	393.4	1.88543	0.19612
380	380.77	271.69	3.176	343.4	1.94001	0.23518
400	400.98	286.16	3.806	301.6	1.99194	0.27239
420	421.26	300.69	4.522	266.6	2.04142	0.30787
440	441.61	315.30	5.332	236.8	2.08870	0.34180
460	462.02	329.97	6.245	211.4	2.13407	0.37441
480	482.49	344.70	7.268	189.5	2.17760	0.40572
500	503.02	359.49	8.411	170.6	2.21952	0.43593
540	544.35	389.34	11.10	139.7	2.29906	0.49338
580	586.04	419.55	14.38	115.7	2.37348	0.54729
600	607.02	434.78	16.28	105.8	2.40902	0.57310
640	649.22	465.50	20.64	88.99	2.47716	0.62272
680	691.82	496.62	25.85	75.50	2.54175	0.66991
700	713.27	512.33	28.80	69.76	2.57277	0.69261

740	756.44	544.02	35.50	59.82	2.63280	0.73669
780	800.03	576.12	43.35	51.64	2.69013	0.77891
800	821.95	592.30	47.75	48.08	2.71787	0.79939
840	866.08	624.95	57.60	41.85	2.77170	0.83921
880	910.56	657.95	68.98	36.61	2.82344	0.87760
900	932.93	674.58	75.29	34.31	2.84856	0.89627
940	977.92	708.08	89.28	30.22	2.89748	0.93271
980	1023.25	741.98	105.2	26.73	2.94468	0.96795
1000	1046.04	758.94	114.0	25.17	2.96770	0.98517
1040	1091.85	793.36	133.3	23.29	3.01260	1.01882
1100	1161.07	845.33	167.1	18.896	3.07732	1.06744
1140	1207.57	880.35	193.1	16.946	3.11883	1.09870
1200	1277.79	933.33	238.0	14.470	3.17888	1.14403
1240	1324.93	968.95	272.3	13.069	3.21751	1.17325
1300	1395.97	1022.82	330.9	11.275	3.27345	1.21563
1340	1443.6	1058.94	375.3	10.247	3.30959	1.24307
1400	1515.42	1113.52	450.5	8.919	3.36200	1.28291
1440	1563.51	1150.13	506.9	8.153	3.39586	1.30868
1500	1635.97	1205.41	601.9	7.152	3.44516	1.34627
1540	1684.51	1242.43	672.8	6.569	3.47712	1.37067
1600	1757.57	1298.30	791.2	5.804	3.52364	1.40622
1640	1806.46	1335.72	878.9	5.355	3.55381	1.42931
1700	1880.1	1392.7	1025	4.761	3.5979	1.46308
1750	1941.6	1439.8	1161	4.328	3.6336	1.49046
1800	2003.3	1487.2	1310	3.994	3.6684	1.51718
1850	2065.3	1534.9	1475	3.601	3.7023	1.54322
1900	2127.4	1582.6	1655	3.295	3.7354	1.56866
1950	2189.7	1630.6	1852	3.022	3.7677	1.59351
2000	2252.1	1678.7	2068	2.776	3.7994	1.61794
2050	2314.6	1726.8	2303	2.555	3.8303	1.64175
2100	2377.7	1775.3	2559	2.356	3.8605	1.66504
2150	2440.3	1823.8	2837	2.175	3.8901	1.68788
2200	2503.2	1872.4	3138	2.012	3.9191	1.71029
2250	2566.4	1921.3	3464	1.864	3.9474	1.73214

It is noted that, unlike internal energy and enthalpy, the entropy of an ideal gas varies with specific volume or pressure as well as the temperature.

Therefore, entropy cannot be tabulated as a function of temperature alone. The s° and s^+ values in the tables account for the temperature dependence of entropy.

ISENTROPIC PROCESSES OF IDEAL GASES

Several relations for the isentropic processes of ideal gases can be obtained by setting the entropy-change relations equal to zero. When this is done for the case of constant specific heats, we obtain

$Tv^{k-1} = \text{constant}$

$TP^{(1-k)/k} = \text{constant}$

$Pv^{k} = \text{constant}$

Isentropic relations of ideal gases for the case of variable specific heats are obtained by setting Equations (8) and (10) equal to zero, yielding

$$s_2^\circ = s_1^\circ + R \ln \frac{P_2}{P_1} \qquad (11)$$

$$s_2^+ = s_1^+ - R \ln \frac{v_2}{v_1} \qquad (12)$$

These two equations are equivalent, and one can be derived from the other. They vary only in convenience: Equation (11) should be used when pressure information is given for a process and Equation (12) should be used when volume information is given to avoid iteration.

Next, we provide definitions for relative pressure P_r and relative specific volume v_r for the completeness of the treatment. First, we rearrange Equation (11) as

$$\frac{P_2}{P_1} = \exp \frac{s_2^\circ - s_1^\circ}{R}$$

or

$$\frac{P_2}{P_1} = \frac{\exp(s_2^\circ/R)}{\exp(s_1^\circ/R)}$$

The quantity $C_1 \exp(s^\circ/R)$ is defined as the relative pressure P_r where C_1 is a constant with a proper unit. With this definition, the last relation becomes

$$\frac{P_2}{P_1} = \frac{P_{r2}}{P_{r1}} \qquad (13)$$

Note that the relative pressure P_r is a *dimensionless* quantity that is a function of temperature only since $s°$ depends on temperature alone. Therefore, values of P_r can be tabulated against temperature. This is done for air in Table 1. Sometimes, specific volume ratios are given instead of pressure ratios. This is particularly the case when automotive engines are analyzed. In such cases, one needs to work with volume ratios. Therefore, we define another quantity related to specific volume ratios for isentropic processes. This is done by utilizing the ideal-gas relation and Equation (13):

$$\frac{P_1 v_1}{T_1} = \frac{P_2 v_2}{T_2} \rightarrow \frac{v_2}{v_1} = \frac{T_2 P_1}{T_1 P_2} = \frac{T_2 P_{r1}}{T_1 P_{r2}} = \frac{T_2 / P_{r2}}{T_1 / P_{r1}}$$

The quantity C_2T/P_r or $C_2T/\exp(s°/R)$ is a function of temperature only and is defined as relative specific volume v_r where C_2 is a constant with a proper unit. Thus,

$$\frac{v_2}{v_1} = \frac{v_{r2}}{v_{r1}} \qquad (14)$$

Equations (13) and (14) are used for the isentropic processes of ideal gases with variable specific heats. The introduction of the function s^+ renders the definitions of relative pressure P_r and relative specific volume v_r obsolete.

We should note that the relative pressure and relative specific volume are useful when it comes to problems where temperatures are known but pressure or volume needs to be determined. For example, assume we need to calculate final pressure (P_2) during an isentropic process when initial temperature and pressure (T_1 and P_1) and final temperature (T_2) are given. The final pressure can be determined easily using the relation ($P_2/P_1 = P_{r2}/P_{r1}$) after reading P_{r1} and P_{r2} values from Table 1 at the given temperatures T_1 and T_2.

The final pressure can also be determined using the relation [$s°_2 = s°_1 + R\ln(P_2/P_1)$] after reading $s°_1$ and $s°_2$ from Table 1 at T_1 and T_2. This method also involves a single relation but requires more operations due to the gas constant R and the exponential term. This particular case has little practical importance since the exit pressure is normally taken to be the same for the actual and isentropic processes involving work-consuming and work-producing devices such as turbines and compressors.

Next, we solve two example problems using the existing conventional approach as found in most thermodynamics books and the new approach proposed in this paper based on the new entropy term s^+. The first example deals with the calculation of the entropy change of air and the second example deals with the analysis of the ideal Otto cycle that includes two isentropic processes of the air.

EXAMPLE PROBLEM 1 WITH THE EXISTING APPROACH

A 0.75 kg amount of air at 290 K is compressed in a piston-cylinder device from a volume of 0.5 m^3 to 0.2 m^3 and a final temperature of 400 K. Entropy change of air is determined considering variation of specific heats with temperature.

The only relation we have for the entropy change of an ideal gas under variable specific heat assumption is Equation (8):

$$s_2 - s_1 = \overset{\circ}{s_2} - \overset{\circ}{s_1} - R\ln\frac{P_2}{P_1}$$

But we do not have pressure information at the initial and final states. Therefore, we need to first determine pressure values using the ideal-gas relation:

$$P_1V_1 = mRT_1 \rightarrow P_1(0.5\,\text{m}^3) = (0.75\,\text{kg})(0.287\,\text{kJ/kg}\cdot\text{K})(290\,\text{K}) \rightarrow P_1 = 124.8\,\text{kPa}$$
$$P_2V_2 = mRT_2 \rightarrow P_2(0.2\,\text{m}^3) = (0.75\,\text{kg})(0.287\,\text{kJ/kg}\cdot\text{K})(650\,\text{K}) \rightarrow P_2 = 699.6\,\text{kPa}$$

We obtain $s°$ values from Table 1:

$$T_1 = 290\,\text{K} \rightarrow \overset{\circ}{s_1} = 1.66802\,\text{kJ/kg}\cdot\text{K}$$

$$T_2 = 650\,\text{K} \rightarrow \overset{\circ}{s_2} = 2.49364\,\text{kJ/kg}\cdot\text{K}$$

Substituting,

$$\begin{aligned} s_2 - s_1 &= \overset{\circ}{s_2} - \overset{\circ}{s_1} - R\ln\tfrac{P_2}{P_1} \\ &= (2.49364 - 1.66802)\,\text{kJ/kg}\cdot\text{K} - (0.287\,\text{kJ/kg}\cdot\text{K})\ln\tfrac{699.6\,\text{kPa}}{124.8\,\text{kPa}} \\ &= 0.331\,\text{kJ/kg}\cdot\text{K} \end{aligned}$$

The entropy change is

$$\Delta S = m(s_2 - s_1) = (0.75\,\text{kg})(0.331\,\text{kJ/kg}\cdot\text{K}) = 0.248\,\text{kJ/K}$$

EXAMPLE PROBLEM 1 WITH THE NEW APPROACH

Now, we solve the same problem with the new approach by utilizing the s^+ term. Since volume information is given, the entropy change can be determined directly from Equation (10):

$$s_2 - s_1 = s_2^+ - s_1^+ + R \ln \frac{V_2}{V_1}$$

We obtain s^+ values from Table 1:

$$T_1 = 290 \text{ K} \rightarrow s_1^+ = 0.04076 \text{ kJ/kg} \cdot \text{K}$$

$$T_2 = 650 \text{ K} \rightarrow s_2^+ = 0.63475 \text{ kJ/kg} \cdot \text{K}$$

Substituting,

$$\begin{aligned} s_2 - s_1 &= s_2^+ - s_1^+ + R \ln \tfrac{V_2}{V_1} \\ &= (0.63475 - 0.04076) \text{ kJ/kg} \cdot \text{K} + (0.287 \text{ kJ/kg} \cdot \text{K}) \ln \tfrac{0.2 \text{ m}^3}{0.5 \text{ m}^3} \\ &= 0.331 \text{ kJ/kg} \cdot \text{K} \end{aligned}$$

The entropy change is

$$\Delta S = m(s_2 - s_1) = (0.75 \text{ kg})(0.331 \text{ kJ/kg} \cdot \text{K}) = 0.248 \text{ kJ/K}$$

EXAMPLE PROBLEM 2 WITH THE EXISTING APPROACH

An ideal Otto cycle has a compression ratio of 8. At the beginning of the compression process, air is at 95 kPa and 27 °C, and 750 kJ/kg of heat is transferred to air during the constant-volume heat-addition process. Taking into account the variation of specific heats with temperature, the pressure and temperature at the end of the heat-addition process, the network output, and the thermal efficiency are determined.

We solve this problem using the properties in Table 1. We use air-standard assumptions and neglect kinetic and potential energy changes. The gas constant of air is $R = 0.287$ kJ/kg·K. The properties of air are given in Table 1. A *P-v* diagram of the cycle is given in Figure 1.

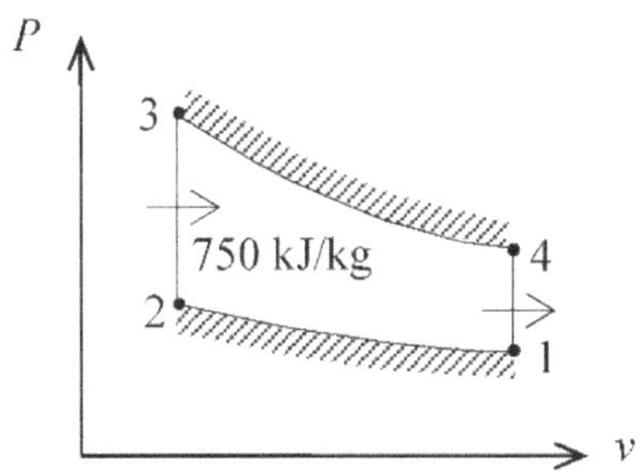

Figure 1: *P-v* diagram of the Otto cycle.

- Process 1-2: isentropic compression

$$T_1 = 300\text{ K} \rightarrow \begin{matrix} u_1 = 214.07\text{ kJ/kg} \\ v_{r1} = 621.2 \end{matrix}$$

$$v_{r2} = \frac{v_2}{v_1} v_{r1} = \frac{1}{r} v_{r1} = \frac{1}{8}(621.2) = 77.65 \rightarrow \begin{matrix} T_2 = 673.1\text{ K} \\ u_2 = 491.2\text{ kJ/kg} \end{matrix}$$

$$\frac{P_2 v_2}{T_2} = \frac{P_1 v_1}{T_1} \rightarrow P_2 = \frac{v_1}{v_2}\frac{T_2}{T_1} P_1 = (8)\left(\frac{673.1\text{ K}}{300\text{ K}}\right)(95\text{ kPa}) = 1705\text{ kPa}$$

- Process 2-3: v = constant heat addition

$$q_{23,\text{in}} = u_3 - u_2 \rightarrow u_3 = u_2 + q_{23,\text{in}} = 491.2 + 750 = 1241.2\text{ kJ/kg} \rightarrow \begin{matrix} T_3 = 1539\text{ K} \\ v_{r3} = 6.588 \end{matrix}$$

$$\frac{P_3 v_3}{T_3} = \frac{P_2 v_2}{T_2} \rightarrow P_3 = \frac{T_3}{T_2} P_2 = \left(\frac{1539\text{ K}}{673.1\text{ K}}\right)(1705\text{ kPa}) = 3898\text{ kPa}$$

- Process 3-4: isentropic expansion

$$v_{r4} = \frac{v_1}{v_2} v_{r3} = r v_{r3} = (8)(6.588) = 52.70 \rightarrow \begin{matrix} T_4 = 774.5\text{ K} \\ u_4 = 571.69\text{ kJ/kg} \end{matrix}$$

- Process 4-1: v = constant heat rejection

$$q_{\text{out}} = u_4 - u_1 = 571.69 - 214.07 = 357.62\text{ kJ/kg}$$

- The network output and the thermal efficiency are

$$w_{\text{net,out}} = q_{\text{in}} - q_{\text{out}} = 750 - 357.62 = 392.4\text{ kJ/kg}$$

$$\eta_{\text{th}} = \frac{w_{\text{net,out}}}{q_{\text{in}}} = \frac{392.4\text{ kJ/kg}}{750\text{ kJ/kg}} = 0.523 = 52.3\%$$

EXAMPLE PROBLEM 2 WITH THE NEW APPROACH

We solve the same Otto cycle problem using the new approach by using the newly defined s^+ term. In the solution we use the values in Table 1.

- Process 1-2: isentropic compression

$$T_1 = 300\text{ K} \rightarrow \begin{matrix} u_1 = 214.07\text{ kJ/kg} \\ s_1^+ = 0.06504\text{ kJ/kg}\cdot\text{K} \end{matrix}$$

$$s_2^+ = s_1^+ - R\ln\frac{v_2}{v_1} = 0.06504 - (0.287)\ln(1/8) = 0.66184\text{ kJ/kg}\cdot\text{K} \rightarrow \begin{matrix} T_2 = 673.1\text{ K} \\ u_2 = 491.2\text{ kJ/kg} \end{matrix}$$

$$\frac{P_2 v_2}{T_2} = \frac{P_1 v_1}{T_1} \rightarrow P_2 = \frac{v_1}{v_2}\frac{T_2}{T_1} P_1 = (8)\left(\frac{673.1\text{ K}}{300\text{ K}}\right)(95\text{ kPa}) = 1705\text{ kPa}$$

- Process 2-3: v = constant heat addition

$$q_{23,\text{in}} = u_3 - u_2 \rightarrow u_3 = u_2 + q_{23,\text{in}} = 491.2 + 750 = 1241.2 \text{ kJ/kg} \rightarrow \begin{array}{l} T_3 = 1539 \text{ K} \\ s_3^+ = 1.37006 \text{ kJ/kg}\cdot\text{K} \end{array}$$

$$\frac{P_3 v_3}{T_3} = \frac{P_2 v_2}{T_2} \rightarrow P_3 = \frac{T_3}{T_2} P_2 = \left(\frac{1539 \text{ K}}{673.1 \text{ K}}\right)(1705 \text{ kPa}) = 3898 \text{ kPa}$$

- Process 3-4: isentropic expansion

$$s_4^+ = s_3^+ - R \ln \frac{v_4}{v_3} = 1.37006 - (0.287)\ln(8) = 0.77326 \text{ kJ/kg}\cdot\text{K} \rightarrow \begin{array}{l} T_4 = 774.5 \text{ K} \\ u_4 = 571.69 \text{ kJ/kg} \end{array}$$

- Process 4-1: v = constant heat rejection

$$q_{\text{out}} = u_4 - u_1 = 571.69 - 214.07 = 357.62 \text{ kJ/kg}$$

$$w_{\text{net,out}} = q_{\text{in}} - q_{\text{out}} = 750 - 357.62 = 392.4 \text{ kJ/kg}$$

- The network output and the thermal efficiency are

$$\eta_{\text{th}} = \frac{w_{\text{net,out}}}{q_{\text{in}}} = \frac{392.4 \text{ kJ/kg}}{750 \text{ kJ/kg}} = 0.523 = 52.3\%$$

DISCUSSION

The solution of the two example problems using the existing approach and the new approach clearly show the convenience of the new approach in entropy analysis of ideal gases with variable specific heat treatment. The advantages of defining the new entropy term s^+ can be explained as follows.

The concept of relative pressure P_r and relative specific volume v_r are eliminated. Entropy is already a challenging subject in thermodynamics and forcing students to learn and use these P_r and v_r functions makes it even more difficult. These functions were introduced a long time ago to allow isentropic process calculations when people did not have electronic calculators and had to use logarithmic tables or slide rulers.

Table 1 is an abbreviated version of air table that exists in most thermodynamics textbooks. It includes one column for the entropy term $s°$ and two more columns for relative pressure P_r and relative specific volume v_r. With the introduction of the s^+ term, a new column for the s^+ term is added but two columns for P_r and v_r are eliminated. With this change, students can perform the same entropy change and isentropic process calculations for air under variable specific heat treatment by dealing with a smaller amount of data in the air table.

In the existing system, for the entropy change of ideal gases under variable specific heat treatment, only one equation (Equation (8)) is available.

If temperature and pressure information is given in the initial and final states (or inlet and exit states), the student can easily calculate entropy change using Equation (8). However, if temperature and volume information is given in the initial and final states, there is no relation available. In this case, the student needs to first find pressure information at the initial and final states and then use Equation (8) to calculate the entropy change (see Example Problem 1 using the existing approach). With the introduction of the s^+ term and its inclusion in the air table, the student can directly use Equation (10) to determine entropy change, as shown in the solution of Example Problem 1 using the new approach.

In the existing system, for the isentropic processes of ideal gases under variable specific heat treatment, three equations (Equations (11), (13) and (14)) are available. Equations (11) and (13) are equivalent as they are both used when pressure information is given. In the new approach, both the relative pressure P_r and the relative specific volume v_r are eliminated and their values in the air table are removed. With the definition of the s^+ term and the addition of its values in the air table, a student only needs two equations (Equations (11) and (12)) to solve isentropic process problems. Equation (11) is used when pressure information is given and Equation (12) is used when volume information is given.

The entropy terms s° and s^+ have some physical sense but not P_r and v_r functions. The change in s° represents the temperature-dependent part of the entropy change when the pressure information is available (Equation (8)). Similarly, the change in s^+ represents the temperature-dependent part of the entropy change when the volume information is available (Equation (10)). The P_r and v_r functions are dimensionless quantities but the s° and s^+ terms have the same unit as entropy.

An additional advantage of eliminating the relative pressure P_r and relative specific volume v_r is that these terms will no longer be confused with pressure and specific volume as well as the reduced pressure P_R and pseudo-reduced specific volume v_R terms used in the compressibility factor calculations.

CONCLUSIONS

The definition of a new entropy term s^+ and its inclusion in ideal-gas tables along with the existing term s° allows a convenient way of calculating entropy change of ideal gases when variation of specific heats with temperature is accounted for. It also allows the calculation of final state properties when isentropic processes of ideal gases are considered. As a result, the need for the definitions of relative pressure P_r and relative specific volume v_r is eliminated and the values of P_r and v_r are removed from ideal-gas tables. Entropy is viewed as a challenging subject by most students of thermodynamics. Thus, the introduction of the new entropy term s^+ is expected to provide considerable convenience and ease in the study of entropy and the second law analysis of energy systems.

ACKNOWLEDGMENTS

The contribution of Tuğberk Hakan Çetin in the preparation of the new air table is greatly appreciated.

Funding Statement

This research received no external funding.

Author Contributions

Conceptualization, Y.A.Ç.; methodology, Y.A.Ç. and M.K.; validation, Y.A.Ç. and M.K.; formal analysis, Y.A.Ç. and M.K.; investigation, Y.A.Ç. and M.K.; resources, Y.A.Ç. and M.K.; data curation, Y.A.Ç. and M.K.; writing—original draft preparation, Y.A.Ç. and M.K.; writing—review and editing, Y.A.Ç. and M.K.; supervision, Y.A.Ç. All authors have read and agreed to the published version of the manuscript.

Conflicts of Interest

The authors declare no conflict of interest.

Footnotes

Publisher's Note: MDPI stays neutral with regard to jurisdictional claims in published maps and institutional affiliations.

REFERENCES

1. Costa V.A.F. Some Considerations and Reflections on Engineering Thermodynamics Teaching and Learning; Proceedings of the 2018 3rd International Conference of The Portuguese Society For Engineering Education (CISPEE); Aveiro, Portugal. 27–29 June 2018.
2. Brown B., Singh C. Development and validation of a conceptual survey instrument to evaluate students' understanding of thermodynamics. *Phys. Rev. Phys. Educ. Res.* 2021;17:010104. doi: 10.1103/PhysRevPhysEducRes.17.010104.
3. Mulop N., Yusof K.M., Tasir Z. A Review on Enhancing the Teaching and Learning of Thermodynamics. *Procedia Soc. Behav. Sci.* 2012;56:703–712. doi: 10.1016/j.sbspro.2012.09.706.
4. Goyings R., Arnas O. *Turbo Expo: Power for Land, Sea, and Air.* Volume 3. Amer Society of Mechanical; New York, NY, USA: 2012. Exergy Versus Energy Analysis: Advantages And Disadvantages In Teaching of Thermodynamics; pp. 457–461.
5. Yang D., Streveler R., Miller R.L., Senocak I., Slotta J. Using schema training to facilitate students' understanding of challenging engineering concepts in heat transfer and thermodynamics. *J. Eng. Educ.* 2020;109:743–759. doi: 10.1002/jee.20360.
6. Struchtrup H. Entropy and the Second Law of Thermodynamics—The Nonequilibrium Perspective. *Entropy.* 2020;22:793. doi: 10.3390/e22070793.
7. Acevedo J.G., Ochoa G.V., Obregon L.G. Development of a new educational package based on e-learning to study engineering thermodynamics process: Combustion, energy and entropy analysis. *Heliyon.* 2020;6:e04269. doi: 10.1016/j.heliyon.2020.e04269.
8. Smith J.L. Teaching the Entropy Concepts of Thermodynamics. In: Beretta G.P., Ghoniem A.F., Hatsopoulos G.N., editors. *AIP Conference Proceedings.* Volume 1033. American Institute of Physics; Cambridge, MA, USA: 2007. pp. 265–270. Meeting The Entropy Challenge: An International Thermodynamics Symposium Held in Honor and Memory of Joseph H Keenan.
9. Çengel Y. On Entropy, Information, and Conservation of Information. *Entropy.* 2021;23:779. doi: 10.3390/e23060779.
10. Wang C., Chen L., Ge Y., Sun F. Comparison of air-standard rectangular cycles with different specific heat models. *Appl. Therm.*

Eng. 2016;109:507–513. doi: 10.1016/j.applthermaleng.2016.08.112.

11. Ebrahimi R. Effects of variable specific heat ratio of working fluid on performance of endoreversible Diesel cycle. *J. Energy Inst.* 2010;83:1–5. doi: 10.1179/014426009X12519696923821.
12. Chen L., Ge Y., Sun F., Wu C. Effects of heat transfer, friction and variable specific heats of working fluid on performance of an irreversible dual cycle. *Energy Convers. Manag.* 2006;47:3224–3234. doi: 10.1016/j.enconman.2006.02.016.
13. Ge Y., Chen L., Sun F., Wu C. Thermodynamic simulation of performance of an Otto cycle with heat transfer and variable specific heats of working fluid. *Int. J. Therm. Sci.* 2005;44:506–511. doi: 10.1016/j.ijthermalsci.2004.10.001.
14. Yanbing L., Yuewu H., Xiaojuan Y. *Advanced Materials Research.* Volume 354–355. Trans Tech Publications Ltd.; Bäch, Switzerland: 2012. The Effects of Variable Specific Heats of Working Fluid on the Performance of Irreversible Reciprocating Brayton Cycle; pp. 1305–1310.
15. Çengel Y.A., Boles M.A., Kanoğlu M. *Thermodynamics. An Engineering Approach.* 9th ed. McGraw-Hill; New York, NY, USA: 2019.
16. Moran M.J., Shapiro H.N., Boettner D.D., Bailey M.B. *Fundamentals of Engineering Thermodynamics.* 9th ed. Wiley; New York, NY, USA: 2018.
17. Borgnakke C., Sonntag R.E. *Fundamentals of Thermodynamics.* 10th ed. Wiley; New York, NY, USA: 2019.

CHAPTER 3

Understanding Atmospheric Behaviour in Terms of Entropy: A Review of Applications of the Second Law of Thermodynamics to Meteorology

Ying Liu, Chongjian Liu and Donghai Wang

State Key Laboratory of Severe Weather, Chinese Academy of Meteorological Sciences, Beijing 100081, China

ABSTRACT

The concept of entropy and its relevant principles, mainly the principle of maximum entropy production (MEP), the effect of negative entropy flow (NEF) on the organization of atmospheric systems and the principle of the

Citation: Liu, Ying, Chongjian Liu, and Donghai Wang. 2011. "Understanding Atmospheric Behaviour in Terms of Entropy: A Review of Applications of the Second Law of Thermodynamics to Meteorology" Entropy 13, no. 1: 211-240. https://doi.org/10.3390/e13010211.

Second Law of thermodynamics, as well as their applications to atmospheric sciences, are reviewed. Some formulations of sub-grid processes such as diffusion parameterization schemes in computational geophysical fluid dynamics that can be improved based on full-irreversibility are also discussed, although they have not yet been systematically subjected to scrutiny from the perspective of the entropy budgets. A comparative investigation shows that the principle of MEP applies to the entropy production of macroscopic fluxes and determines the most probable state, that is, a system may choose a development meta-stable trajectory with a smaller production since entropy production behavior involves many specific dynamical and thermodynamic processes in the atmosphere and the extremal principles only provide a general insight into the overall configuration of the atmosphere. In contrast to the principle of MEP, the analysis of NEF is able to provide a new insight into the mechanism responsible for the evolution of a weather system as well as a new approach to predicting its track and intensity trend.

Keywords: entropy production' entropy flow' second law of thermodynamics' meteorology' review

INTRODUCTION

The atmosphere has been likened to a giant thermodynamic engine in which disorganized energy is transformed into the organized kinetic energy of the winds while the general circulation of the atmosphere can be regarded as simply being driven by temperature differences between the polar and equatorial regions. As mentioned by Sadi Carnot, "We must attribute to heat the great movements that we observe all about us on the Earth. Heat is the cause of currents in the atmosphere, of the rising motion of clouds, of the falling of rain and of other atmospheric phenomena ..." [**1**,**2**,**3**,**4**,**5**].

The notions of entropy and its production in equilibrium and nonequilibrium processes not only form the basis of modern thermodynamics and statistical physics, but have always been at the core of various ideological discussions concerned with the evolution of the world, the course of time, *etc.* These issues were raised by many outstanding scientists, including Clausius, Boltzmann, Gibbs and Onsager. As a result, today we have thousands of books, reviews and papers dedicated to properties of the entropy in different systems [**4**,**6**,**7**].

Some concepts and theories in modern thermodynamics may be relevant and perhaps crucial to enable a deeper understanding of atmospheric dynamics. Recently, more attention has been paid to thermodynamics and statistical mechanics in the atmospheric sciences. In addition to the large number of journal articles [**8**,**9**,**10**,**11**,**12**,**13**,**14**], several monographs have been published in succession [**15**,**16**,**17**,**18**].

As known to all, entropy has a different definition and is measured in a different way for the different fields. For example, in classical physics, the entropy of a physical system is proportional to the quantity of energy no longer available to do physical work. Entropy is central to the Second Law of thermodynamics, which states that in an isolated system any activity increases the entropy; in quantum mechanics, von Neumann entropy extends the notion of entropy to quantum systems by means of the density matrix. In probability theory, the entropy of a random variable measures the uncertainty about the value that might be assumed by the variable; in information theory, the compression entropy of a message (e.g., a computer file) quantifies the information content carried by the message in terms of the best lossless compression rate, in the theory of dynamical systems, entropy quantifies the exponential complexity of a dynamical system or the average flow of information per unit of time; in sociology, entropy is the natural decay of structure (such as law, organization, and convention) in a social system; and in the common sense, entropy means disorder or chaos. By saying that entropy is a measure of disorder we mean that the higher the entropy the greater the disorder, which are applied to the cases: when it is regarded as a measure of the unavailability of a system's energy to do work; as a parameter representing the state of disorder of a system at the atomic, ionic, or molecular level; as a measure of disorder in the universe or simply as a measure of disorder in any dynamical system [**19**,**20**].

Truly in thermodynamics, entropy is commonly associated with the amount of order, disorder, and/or chaos in a thermodynamic system. This stems from Rudolf Clausius' assertion [**21**] that any thermodynamic process always "admits to being reduced to the alteration in some way or another of the arrangement of the constituent parts of the working body" and that internal work associated with these alterations is quantified energetically by a measure of "entropy" change, according to the following differential expression:

$$\int \frac{\delta Q}{T} \geq 0$$

In the years to follow, Ludwig Boltzmann translated these "alterations" into that of a probabilistic view of order and disorder in gas phase molecular systems [**22**]. In 1864, Boltzmann came across Maxwell's paper and was so inspired by it that he spent much of his long and distinguished life developing the subject further. Later, Boltzmann, in efforts to develop a kinetic theory for the behavior of a gas, applied the laws of probability to Maxwell's and Clausius' molecular interpretation of entropy so to begin to interpret entropy in terms of order and disorder. Similarly, in 1882 Hermann von Helmholtz used the word "Unordnung" (disorder) to describe entropy [**22**].

Entropy and disorder also have associations with equilibrium [**23,24**]. Technically, for example, entropy, from this perspective, is defined as a thermodynamic property which serves as a measure of how close a system is to equilibrium—that is, to perfect internal disorder [**23**]. Likewise, the value of the entropy of a distribution of atoms and molecules in a thermodynamic system is a measure of the disorder in the arrangements of its particles [**25**]. In a stretched out piece of rubber, for example, the arrangement of the molecules of its structure has an "ordered" distribution and has zero entropy, while the "disordered" kinky distribution of the atoms and molecules in the rubber in the non-stretched state has positive entropy. Similarly, in a gas, the order is perfect and the measure of entropy of the system has its lowest value when all the molecules are in one place, whereas when more points are occupied the gas is all the more disorderly and the measure of the entropy of the system has its largest value [**25**].

Finally, it is needed to mention that Shannon entropy has been related by physicist Léon Brillouin to a concept sometimes called negentropy early in 1960s. In his 1962 book *Science and Information Theory* [**26**], Brillouin described the Negentropy Principle of Information or NPI, the gist of which is that acquiring information about a system's microstates is associated with a decrease in entropy (work is needed to extract information, erasure leads to increase in thermodynamic entropy). There is no violation of the Second Law of thermodynamics, according to Brillouin, since a reduction in any local system's thermodynamic entropy results in an increase in thermodynamic entropy elsewhere. Negentropy was considered as controversial because its earlier understanding can yield Carnot efficiency higher than one. In 2003 Weiss and Weiss came on a background of psychometric data and theoretical considerations to the conclusion that the negentropy principle underlies the information processing by the brain [**27**]. In 2008 this has been empirically confirmed by a group of neurobiologists [**28**]. In 2009, Mahulikar and Herwig redefined thermodynamic negentropy as the specific entropy deficit of the dynamically ordered sub-system relative to its surroundings [**29**]. This definition enabled the formulation of the *Negentropy Principle*, which is mathematically shown to follow from the Second Law of Thermodynamics, during order existence. In this paper we address only the entropy measures related to meteorology, which include thermodynamic entropy, the entropy of statistical mechanics (Boltzmann entropy or Gibbs entropy), and information entropy (Shannon entropy). In doing so, we understand that there exists an intrinsic connection between Boltzmann entropy and classical thermodynamic entropy as well as between Gibbs and Shannon entropy.

It should be pointed out emphatically that the importance of entropy theory lies not only in the entropy concept itself, but also in the extremal principles developed therein. Glansdorff and Prigogine wrote "Dissipative structures have a quite different [from equilibrium structures] status: they are formed and maintained through the effect of exchange of energy and matter in non-equilibrium conditions" [**30**]. They were referring to the dissipation function of Rayleigh [**31**] that was used also by Onsager [**32,33**]. Glansdorff and Prigogine concluded that at a stable steady state, the dissipation function was minimum [**30**]. 10 years later, Sawada, postulating a principle of largest amount of entropy increment per unit time [**34**], cites work in fluid mechanics by Malkus and Veronis [**35**] as having "proven a principle of maximum heat current, which in turn is a maximum entropy production for a given boundary condition". On the other hand, using the information theoretical formalism of Jaynes, Dewar claimed to have derived a principle of maximum entropy production applying to nonlinear stationary states [**36,37**]. With a background of previous literature, a number of authors [**38**] cited Dewar's purported proof. In particular, Beretta, in an attempt to unify quantum mechanics and thermodynamics [**39,40,41**], proposed a generalization of the time-dependent Schrödinger equation of quantum mechanics, that postulates and implements at the single particle level the principle of maximum entropy production (even though, at that time, he named it "steepest entropy ascent dynamics"). Beretta has also extended his steepest-entropy-ascent nonlinear dynamical equation to model the time evolution of non-equilibrium probability distributions in discrete, non-quantum contexts [**42**].

As a result, this paper will first focus more attention to the works related the maximum entropy production principle (MEPP) and its applications to meteorology. On the other hand, since entropy production is only one component of the whole entropy budget another important component called entropy flow in the entropy balance equation will be intensively addressed here as well. Recently the applications at this direction show an increasing trend owing largely to enlightening by the theory of dissipative structures created by Prigogine, e.g., [**30**].

In fact, entropy flow is a core concept in the modern theory of dissipative structures just as entropy is in classical thermodynamics and statistical physics. According to the Second Law of thermodynamics, an isolated system will evolve spontaneously into equilibrium with maximum entropy in which the order or organization of the system is at the minimum [**43,44,45**]. As a result, negative entropy is very important for a system to keep itself far from its equilibrium, which is also true in biological systems (for a biological individual, equilibrium means death [**46**] as reflected in the statement that life's existence depends on its continuous gain of 'negentropy' from its surroundings [**47**]).

It is noticed that there has been little research based directly on the Second Law of thermodynamics itself in atmospheric sciences. As well-known to all, this law is usually expressed in terms of an inequation, which makes its applications inconvenient, but some progress has nevertheless been reached in this field.

This review paper is arranged as follows: a number of major concepts of entropy used in meteorology are presented in the next section; in **Section 3** starting from the entropy balance equation one introduces briefly the issues regarding entropy budgets and isentropic analyses as well as isentropic coordinates employed in meteorological numerical models; **Section 4** and **Section 5** are devoted to several research results concerning entropy production and entropy flow, respectively; **Section 6** discusses applications of the principle of the Second Law to meteorology; and some discussions and conclusion are given in the last section.

MAJOR CONCEPTS OF ENTROPY USED FREQUENTLY IN METEOROLOGY

Thermodynamic Entropy

The entropy of a system is defined as a summation of "heat supplied" divided by its "temperature" [**48**]. If a certain small amount of heat Q is supplied quasi-statically to a system with an absolute temperature of T, then the entropy of the system will increase by:

$$dS=\delta Q/T \qquad (1)$$

where S is the entropy of the system, d represents an infinitesimal small change of a state function, and represents that of a path function. Heat can be supplied by conduction, convection, or radiation. The entropy of the system will increase by equation (1), no matter which way we may choose. When we extract the heat from the system, the entropy of the system will decrease by the same amount. Thus the entropy of a diabatic system, which exchanges heat with its surrounding system, can either increase or decrease, depending on the direction of the heat exchange. This is not a violation of the Second Law of thermodynamics since the entropy increase in the surrounding system is larger. The Second Law (the law of entropy increase) is valid for a whole (isolated) system. When we sum up all the changes of entropy of interacting subsystems, the total change must be nonnegative. This is a statement of the Second Law of thermodynamics.

The Entropy of Statistical Mechanics

In statistical mechanics, the Boltzmann entropy is an approximation to the normal Gibbs entropy which refers to the statistical formulation of entropy for an ideal gas system of particles as found in Willard Gibbs' 1902 book *Elementary Principles in Statistical Mechanics* [**49**].

The Boltzmann entropy is obtained if one assumes one can treat all the component particles of a thermodynamic system as statistically independent. The probability distribution of the system as a whole then factorises into the product of N separate identical terms, one term for each particle; and the Gibbs entropy simplifies to the Boltzmann entropy:

$$S_B = -N\kappa_B \sum_i p_i \log p_i$$

where the summation is taken over each possible state in the 6-dimensional phase space of a *single* particle (rather than the $6N$-dimensional phase space of the system as a whole) and p_i is the probability of a microstate with k_B as the Boltzmann constant.This reflects the original statistical entropy function introduced by Ludwig Boltzmann in 1872 [**50**]. For the special case of an ideal gas it exactly corresponds to the proper thermodynamic entropy mentioned in **Section 2.1** above.

Information Entropy

In information theory, entropy is a measure of the uncertainty associated with a random variable. The term by itself in this context usually refers to the Shannon entropy (Shannon information entropy) [**51,52**], which quantifies, in the sense of an expected value, the information contained in a message, usually in units such as bits.

The entropy H of a discrete random variable X with possible values $\{x_1, ..., x_n\}$ is:

$H(X)=E(I(X))$

where E is the expected value, and I is the information content of X. $I(X)$ is itself a random variable. If p denotes the probability mass function of X then the entropy can explicitly be written as:

$$H(X) = \sum_{i=1}^{n} p(x_i) I(x_i) = -\sum_{i=1}^{n} p(x_i) \log_b p(x_i)$$

where b is the base of the logarithm used.

By the way, the inspiration for adopting the word *entropy* in information theory came from the close resemblance between Shannon's formula and a very similar formula from thermodynamics. In statistical thermodynamics, the most general formula for the thermodynamic entropy S of a thermodynamic system is the Gibbs entropy:

$$S = -\kappa_B \sum p_i \ln p_i$$

which was defined by J. Willard Gibbs in 1878 [**53**] after earlier work by Boltzmann [**50**]. Note that Gibbs entropy also translates over almost unchanged into the world of quantum physics to give the von Neumann entropy introduced by John von Neumann in 1927 [**19**]:

$$S = -\kappa_B Tr(\rho \ln \rho)$$

where ρ is the density matrix of the quantum mechanical system and Tr is the trace.

Consider a discrete probability distribution among m mutually exclusive propositions. The most informative distribution would occur when one of the propositions was known to be true. In that case, the information entropy would be equal to zero. The least informative distribution would occur when there is no reason to favor any one of the propositions over the others. In that case, the only reasonable probability distribution would be uniform, and then the information entropy would be equal to its maximum possible value, log m. The information entropy can therefore be seen as a numerical measure which describes how uninformative a particular probability distribution is, ranging from zero (completely informative) to log m (completely uninformative).

The principle was first expounded by E.T. Jaynes in two papers in 1957 [**54,55**] where he emphasized a natural correspondence between statistical mechanics and information theory. In particular, Jaynes offered a new and very general rationale why the Gibbsian method of statistical mechanics works. He argued that the entropy of statistical mechanics, and the information entropy of information theory, are principally the same thing. Consequently, statistical mechanics should be seen just as a particular application of a general tool of logical inference and information theory.

As a matter of fact, the explanation of these results (the fluctuation theorem, maximum entropy production and self-organized criticality as in the classical thermodynamics) on general information theoretic grounds underlines their relevance to a broad class of stationary, non-equilibrium systems. In turn, the accumulating empirical evidence for these results lends support to Jaynes' formalism (*i.e.*, Jaynes' information theory formalism of statistical mechanics) as a common predictive framework for equilibrium and non-equilibrium statistical mechanics [**56**].

ON ENTROPY BALANCE

According to classical thermodynamics, the entropy change in an open system consists of two contributory parts: one is entropy flow that is caused by the positive or negative entropy exchange with its surroundings; and, the other is positive definite entropy production owing to irreversible processes within the system. This can be described by the entropy balance equation which is derived from the Gibbs relation and local equilibrium assumption and, for entropy per unit volume, *S*, is of the form:

$$\frac{\partial S}{\partial t} = -div\bar{J}_S + \sigma \qquad (2)$$

where denotes the entropy production, and the "*div*"-term is the entropy flow with $\partial\partial$ denoting the partial differentiation [**45**,**57**]. In the following sections (**Section 4** and **Section 5**) discussions on the classical thermodynamic entropy production and entropy flow are just based on equation (2) above.

In physics, a variety of conservative laws for physical quantity such as mass, momentum and energy have been proved. However, entropy is a very special quantity to which there is no corresponding conservation law. With the local equilibrium assumption employed an entropy balance relation can be derived as mentioned above. As known to all, for an isolated system, its entropy will either increase spontaneously if it is apart from equilibrium or keep unchanged if it is at equilibrium. For an open system, even if the entropy inside the system decreases owing to negative entropy flow into the system from its surroundings a greater system that comprises the original system and its surroundings will experience an entropy-increasing process [**45**].

Indeed, according to [**58**], in the long-term mean the outgoing flux of longwave radiation to space equals the incoming flux of solar energy almost exactly, but the corresponding entropy fluxes are quite different. In fact, the entropy exported by the outgoing longwave radiation is found to be about 22 times larger than the entropy associated with the incoming solar radiation.

At the same time, Peixoto *et al.* claim that the net export of entropy to space by the climate system must be equal to the internal production of entropy by the irreversible processes that occur within the system [**58**]. As a result, the total entropy for the whole climate system should be an invariant.

Usually, the entropy S is expressed as $C_p ln\theta$ with C_p being the specific heat at constant pressure and θ the potential temperature that relates temperature T by:

$$\theta = T(P_0 / P)^{R/C_p}$$

where R is the gas constant for dry air and p_0 reference pressure that is usually taken as 1000 *hPa*. In meteorological applications, not only is the total entropy for the whole climate system regarded as being an invariant but also entropy is assumed to be unchanged in many cases.

In fact, terminology like "isentropic process" is extensively accepted and widely applied in the atmospheric sciences and thus the isentropic analysis or isentropic coordinates become an important tool in diagnosis analysis and numerical weather model development. In the preliminary stage (in early 1950s) the relevant studies were mainly focused on trajectory analysis on isentropic surfaces [**59**] and the isentropic coordinates were not introduced untill 1970s [**60,61,62,63**]. After that some alternative coordinate schemes are suggested, including the (transformed) isentropic coordinates [**64,65,66,67,68,69,70,71**], (hybrid) isentropic-sigma coordinates [**72,73,74,75,76**], combining terrain-following and isentropic coordinates, e.g., [77], and, isentropic semi-geostrophic coordinates [**78**] as well as hybrid isentropic-terrain-following coordinates, e.g., [**79**], employed in either regional or global models. Recently, in addition to the moist isentropic coordinates, an adaptive, hybrid-isentropic vertical coordinate was used in global atmospheric modeling 80,81]. Indeed the continuous improvement of the isentropic vertical coordinate favors the further understanding of the nature of atmospheric motion and the mechanism behind it. On the other hand, Arakawa and Lamb have previously emphasized that global conservation of *lnθ* is important for climate modeling [**82**]. This suggests that if entropy itself were the thermodynamic property and isentropic coordinates are utilized, not only will global conservation of *lnθ* be assured, but also local aphysical sources of entropy by numerical mixing would simply be precluded. However, as Sommerfeld states, "in spite of their not being real, isentropic processes are most important in thermodynamics because definite equations can only be obtained by considering reversible changes" [**83**]. Later Johnson *et al.* further discussed numerical uncertainties in the simulation of reversible isentropic processes and the relevant issues related to entropy conservation [**84**,**85**]. Here, we do not intend to go into their detailed conclusions but just recall the fact that the isentropic motion assumption requires entropy conservation, which can only be valid if the motion is both adiabatic and frictionless, along with the classical thermodynamic entropy definition itself requires the process (the atmospheric motion in our context) to be experienced at equilibrium (that's why the local equilibrium assumption is adopted). As a result, due caution should be exercised when the isentropic approach is employed.

ENTROPY PRODUCTION

From the entropy balance equation (2) above, local changes in entropy cover two parts: one of them is entropy flow and the other is entropy production. We will specifically discuss, in this section, the entropy production that has been more intensively addressed in the field of atmospheric sciences.

Let us first review Paltridge's work in 1975 in which he suggested a well-know relevant principle, the principle of maximum entropy production (MEP) [**86**,**87**]. Indeed, according to Tuck, "On the macroscopic level, the way of MEP has been pioneered by a meteorologist" [**86**,**88**]. Initially Paltridge used the terminology "minimum entropy exchange" [**86**], but after that, he used the current terminology "MEP" to describe the same thing [**88**,**89**,**90**]. This point is clarified in the review by Ozawa *et al.* [**4**].

Studies on the MEP principle (MEPP) originated from meteorology but covered various fields, including physics, chemistry, astronomy, geoscience and life science as well as social sciences, and some reviews have been made on the MEPP recently e.g., [**4**,**7**]. From the basic concepts, Ozawa *et al.* [**4**] introduced first a number of significant works on global climate including Paltridge's, and then discussed the relevant results in the fields such as climates on other planets and mantle convection in planets as well as a simple theory of turbulence that was recently shown to be consistent with the hypothesis of MEP [**91**]. Martyushev and Seleznev summarized and analyzed the studies dealing with MEPP and, also starting from the thermodynamic and statistical basis of the principle (including the relationship of MEPP with the Second Law of thermodynamics and Prigogine's principle), presented various existing applications of the principle in the analysis of nonequilibrium systems, including climate systems [7].

Early in the 1970s Paltridge proposed that Earth's climate structure might be explained from a hypothesis of MEP [**86,92**]. If correct, this proposal would be of crucial importance to future climate research because it provides the hitherto missing global constraint of the Second Law of thermodynamics [**93**].

As pointed out [**86**], the most probable macroscopic flux configuration did be very close to real observations. **Figure 1** shows a schematic of energy transport processes in a planetary system composed of the Earth and the Sun. Shortwave radiation emitted from the Sun with a brightness temperature of about 5,760 K is absorbed by the Earth, mainly in the equatorial region. This energy is transported poleward through direct motions of the atmosphere and oceans (the general circulation).

The energy is finally reemitted to space via longwave radiations from the atmosphere at a temperature of about 252 K, from clouds at 259 K and from the Earth's surface at 288 K. Thus there is a flow of energy from the hot Sun to cold space through the Earth. In the Earth's system the energy is transported from the warm equatorial region to the cool polar regions by the atmosphere and oceans.

Then, according to Carnot, a part of the heat energy is converted into potential energy which is the source of the kinetic energy of the atmosphere and oceans. In this respect, the Earth's system can be regarded as a heat engine operating between thermal reservoirs with different temperatures (equator and poles) [4]. On the other hand, the atmospheric system, especially its troposphere, can be regarded as a heat engine operating between two thermal reservoirs that is the surface at a higher temperature and the tropopause at a lower temperature. That seems to explain why severe weather phenomena, including severe convection, occur in the bottom sphere of the Earth atmosphere—the troposphere. The determination of the strength of the circulation, and hence the rate of heat transport, constitutes a fundamental problem in thermodynamics of the general circulation, e.g., [94].

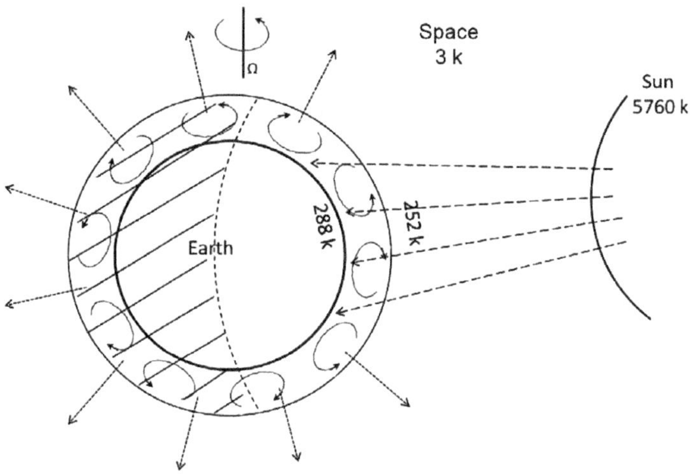

Figure 1. A schematic of energy transport processes in the planetary system of the Earth, the Sun and space. The Earth receives the shortwave radiation from the hot Sun and emits longwave radiation into space. The atmosphere and oceans act as a fluid system that transports heat from the hot region to cold regions via general circulation.

As mentioned by Ozawa *et al.* [**4**], Lorenz suspected that the Earth's atmosphere operates in such a manner as to generate available potential energy at a possible maximum rate [**95**]. The available potential energy is defined as the amount of potential energy that can be converted into kinetic energy. Independently, Paltridge suggested [**86,92**] that the mean state of the present climate is reproducible as a state with a maximum rate of entropy production due to horizontal heat transport in the atmosphere and oceans [**4**]. **Figure 2** shows such an example [**86**]. Without considering the detailed dynamics of the system, the predicted distributions (air temperature, cloud amount, and meridional heat transport) show remarkable agreement with observations [**4**]. Later on, several researchers investigated Paltridge's work and obtained essentially the same result [**96,97,98,99,100,101,102,103**].

Recently an information theoretical formulation of nonequilibrium statistical mechanics (MaxEnt, *i.e.*, the constrained maximization of Shannon information entropy) is used to derive the fluctuation theorem (FT) concerning the probability of Second Law violating phase-space paths [**104**]. Dewar presented a more rigorous and general mathematical derivation of MEP from MaxEnt, and the relationship between MEP and the FT is clarified thereby.

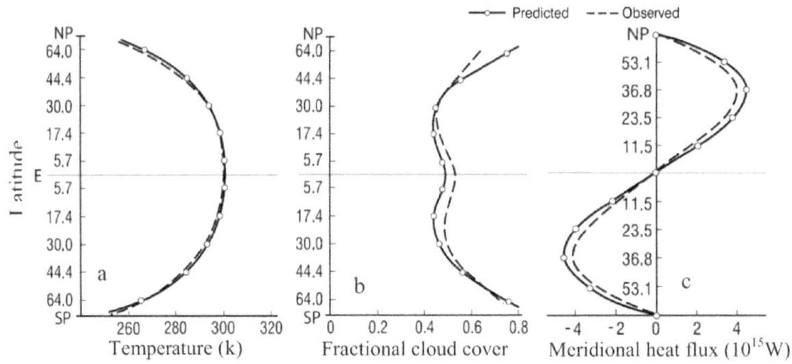

Figure 2. Latitudinal distributions of (a) mean air temperature, (b) cloud cover, and (c) meridional heat transport in the Earth. Solid line curves indicate those predicted with the constraint of maximum entropy production (equation (9)), and dashed lines indicate those observed. (Reproduced with minor changes based on [**4**])

Specifically, it is shown that the FT allows a general orthogonality property of maximum information entropy to be extended to entropy production itself, from which MEP then follows. The new derivation highlights MEP and the FT as generic properties of MaxEnt probability distributions involving anti-symmetric constraints, independently of any physical interpretation [**105**]. Physically, MEP applies to the entropy production of those macroscopic fluxes that are free to vary under the imposed constraints, and corresponds to selection of the most probable macroscopic flux configuration. On the other hand, here we would like to mention that, as Grinstein and Linsker showed, "the question of the existence of possible extremal principles (and in particular, of MaxEP) that might apply to far-from-equilibrium regimes (having non-linear constitutive relations) has not been settled" [**106,107,108**]. We will go into further discussions on it in the **Section** 7 below.

By the way, rooted in information theory, the maximum entropy method (MEM) seeks to extract as much information from a measurement as is justified by the data's signal-to-noise ratio [**109**]. It has been used successfully in a variety of fields, including radio astronomy, NMR, fluorescence, and ligand rebinding [**110,111,112,113,114,115,116**]. The method is by no means restricted to conventional imaging problems, but the word 'image' will be used to refer to any function to be extracted from the data.

Although there is no consensus on the MEM numerous works have been directed at the principle of MEP as well as the MEM and their applications to meteorology, e.g., [**117,118,119,120,121,122**] henceforth. Despite the different details most of them are focused on the climate modeling based on the MEPP, subgrid parameterization scheme and numerical restriction from the entropy budget caused substaintially by entropy production. In addition to the exact thermodynamic origin, some of them are related to the extreme sensitivity of a climate model's temperature response to aphysical entropy sources introduced by spurious numerical dispersion diffusion, Gibbs oscillations, parameterizations, and other factors, see also [**123,124,125,126,127,128,129,130,131,132,133,134,135,136,137,138,139**]. The results in combination with Carathéordary's statement of the Second Law reveal in the strict sense that the presence of positive definite aphysical sources of entropy in a climate model precludes the simulation of unbiased distributions of the additional heat and temperature [**140,141,142**].

Finally, we would like to particularly review the works concerning the potential mechanism for MEP. For instance, in addition to a regulation mechanism by turbulent fluctuations suggested by Paltridge [**86,88,143**], Ozawa *et al.* discussed qualitatively a mechanism by which a turbulent fluid system adjust itself to a state of maximum generation of available potential energy or, equivalently, MEP [**4**]. As a matter of fact, it is obvious from the point of view of information theory that, by choosing to use the distribution with the maximum entropy allowed by our information, the argument goes, we are choosing the most uninformative distribution possible. To choose a distribution with lower entropy would be to assume information we do not possess; to choose one with a higher entropy would violate the constraints of the information we do possess. Thus the maximum entropy distribution is the only reasonable distribution. By the way, as mentioned in the theory of dissipative structures [**30,144**], the stability of a multi-body system is able to be judged in terms of excess entropy production that is the second order variation of entropy, ^2s, where s is specific entropy with ^{2} as the second order variational operator. Liu and Tao suggested a new stability criterion based on the Liapunov second method with the excess entropy production ^2s as a generalized Liapunov function by which the stability characteristics of a series of weather conditions are revealed [**145**]. On the other hand, Pujol and Llebot used a second differential of the entropy as a criterion for the stability in low-dimensional climate models [**146**].

ENTROPY FLOW

As pointed out in **Section 3** above, the "*div*"-term in equation (2) describes the entropy flow entering into a system through its boundaries from the surroundings. The Earth system exchanges radiation with space, which drives and maintains almost all the processes within the Earth system. As shown in **Figure 1**, the essence of the Earth's radiation exchange with space is that the high-energy photons with low entropy from the Sun enter the Earth system and the low-energy photons with high entropy are emitted from the Earth system to space, see also [**147**]. In general, the Earth system as a whole can be thought of as a multi-body system that is closed to nonradiation material exchange and open to radiation exchange with space. For such a complex system, the net entropy flux resulting from the radiation exchange constrains the Earth system's internal entropy production rate, a fundamental measure of the overall activities within the Earth system, including oceanic, atmospheric, and biological processes [**148**].

Because of the entropy in a system determining the order of the system to some extent, the evolution of a specific atmospheric system should be related with entropy flow.

In 1900 Bénard investigated a thin layer of fluid at rest originally, with a free surface at a lower temperature T_1, heated from its bottom at a higher temperature T_2 (see **Figure 3**), and noticed that a rather regular cellular pattern of hexagonal convective cells was abruptly organised when the temperature difference ($T_2 - T_1$) reaching the value of the threshold of primary instability [**30,44,45,149**], suggesting that every system that obtains heat at a higher temperature but loses heat at a lower temperature will experience net negative entropy flow since in this case the entropy exchange (δse) of the system with its environment or entropy flow entering into the system through its boundaries is as follows:

$$\delta s_e = \frac{Q_2}{T_2} - \frac{Q_1}{T_1} = \frac{T_1 - T_2}{T_1 T_2} Q < 0 \tag{3}$$

where Q_1 and Q_2 are the heat fluxes through the top and bottom of the layer of fluid, respectively, and Q is the flux when the system becomes steady so that $Q = Q_1 = Q_2$. Later, in the 1940s, Schrödinger stated in his monograph "What Is Life?" that life's existence depends on its continuous gain of 'negentropy' from its surroundings [**105**]. This implies that negative entropy flow is something very significant for a system, whether it is living or nonliving [**30,44,45,47,150,151**]. That is, negative entropy flow will cause a system initially at equilibrium or even at rest to be organized, or lead a system already at non-equilibrium to a state further from equilibrium (*i.e.*, cause it to strengthen).

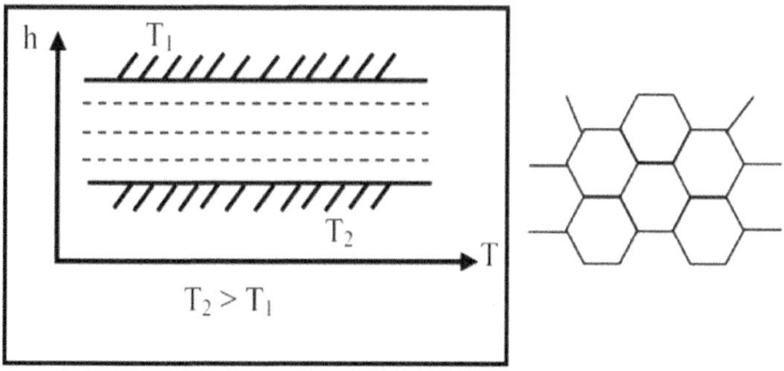

Figure 3. The schematic diagram for Bénard convection as the prototype of self-organization.

Next we will discuss the effect of entropy flow to the organization of tropical atmospheric systems. Recently Liu *et al.* used the entropy flow formula (the "*div*"-term in equation (2) above), as a logical component of the entropy balance equation [**30,44,45**], and calculated the three-dimensional instantaneous entropy flow field of a number of the tropical cyclones [**1,2,152,153,154,155,156**]. The entropy flow vector $J{\rightharpoonup}s$ in $-div \quad J{\rightharpoonup}s$ that is just the "*div*"-term in equation (2) or δse mentioned above can be written as:

$$\vec{J}_s = S\vec{V} + \frac{1}{T}\vec{J}_q - \sum_k \frac{\mu_k}{T}\vec{J}_k \qquad (4)$$

where $J{\rightharpoonup}q$, T, μk and $J{\rightharpoonup}k$ are the heat flow, temperature, chemical potential and diffusive flow for component *k*, respectively, e.g., [**156**]. It is necessary to modify the entropy flow formula so as to make it be suitable for the computation of an atmospheric system such as hurricane. As a result, the entropy flow formula finally becomes:

$$\begin{aligned}
-div\vec{J}_s = & -\frac{\partial\rho su}{\partial x} - \frac{\partial\rho sv}{\partial y} - \rho\frac{\partial\rho su}{\partial p}\left(\frac{\partial\phi}{\partial x}\right) - \rho\frac{\partial\rho sv}{\partial p}\left(\frac{\partial\phi}{\partial y}\right) - \rho\frac{\partial s\omega}{\partial p} \\
& + \frac{\tilde{\lambda}\rho}{T}\{\frac{\partial^2 T}{\partial x^2} + \frac{\partial^2 T}{\partial y^2} + 2\rho\frac{\partial\phi}{\partial x}\frac{\partial^2 T}{\partial p\partial x} + 2\rho\frac{\partial\phi}{\partial y}\frac{\partial^2 T}{\partial p\partial y} + \rho^2[(\frac{\partial\phi}{\partial x})^2 + (\frac{\partial\phi}{\partial y})^2]\frac{\partial^2 T}{\partial p^2} \\
& + \rho(\frac{\partial^2\phi}{\partial x^2} + \frac{\partial^2\phi}{\partial y^2} + \rho\frac{\partial\phi}{\partial x}\frac{\partial^2\phi}{\partial p\partial x} + \rho\frac{\partial\phi}{\partial y}\frac{\partial^2\phi}{\partial p\partial y})\frac{\partial T}{\partial p} + \rho^2 g^2\frac{\partial^2 T}{\partial p^2} \\
& + (\frac{1}{\rho}\frac{\partial T}{\partial x} + \frac{\partial\phi}{\partial x}\frac{\partial T}{\partial p})(\frac{\partial\rho}{\partial x} + \rho\frac{\partial\rho}{\partial p}\frac{\partial\phi}{\partial x} - \frac{\rho}{T}\frac{\partial T}{\partial x} - \frac{\rho^2}{T}\frac{\partial T}{\partial p}\frac{\partial\phi}{\partial x}) \\
& + (\frac{1}{\rho}\frac{\partial T}{\partial y} + \frac{\partial\phi}{\partial y}\frac{\partial T}{\partial p})(\frac{\partial\rho}{\partial y} + \rho\frac{\partial\rho}{\partial p}\frac{\partial\phi}{\partial y} - \frac{\rho}{T}\frac{\partial T}{\partial y} - \frac{\rho^2}{T}\frac{\partial T}{\partial p}\frac{\partial\phi}{\partial y}) + \frac{\rho^2 g^2}{T}(\frac{\partial T}{\partial p})^2\}
\end{aligned} \qquad (5)$$

where *u* and *v* are the velocities in the *x*- and *y*-direction at constant pressure layers, respectively; ω the vertical velocity in *p*-coordinates; ϕ the geopotential height with $\phi = gz$ in which g is the gravitational acceleration; $\tilde{\lambda}$ the thermal conductivity; and s the specific entropy [**156**]. Equation (5) is the fundamental expression for computing entropy flow employed.

Here, we will introduce the computational results for one of the most typical cases—Hurricane Katrina (2005) that is extraordinarily powerful and deadly with a total damage cost of approximately $81 billion being roughly double that of Hurricane Andrew (1992) [**157**] and the total number of fatalities over 1800 and whose track is shown in **Figure 4** [**156,158,159**] and then discuss the relationship of entropy flow with the development of the hurricane.

Let's first have a look at changes in the total entropy flow of Hurricane Katrina based on the National Centers for Environmental Prediction/ National Center for Atmospheric Research (NCEP/NCAR) 1° × 1° (latitude-longitude) resolution reanalysis data (**Figure 5**) [**160**].

Figure 5 shows changes, at 6-hourly intervals, in the accumulative total entropy flow for Katrina as well as the corresponding maximum sustained wind speed representative of the intensity of the hurricane. It is seen from **Figure 5** that there exists a general trend of anti-correlation between these two curves with the phase in the total entropy flow preceding that in the maximum wind speed.

For example, Katrina experienced more and more increasing net negative entropy flow before 00:00 UTC 28 August 2005 and, which was followed closely by its continuous intensification with very rapid strengthening occurring on the morning of 28 August.

On the other hand, the values of negative entropy flow for Katrina decreased distinctly and even turned to positive entropy flow after 00:00 UTC 28 August, which was followed by the gradual weakening of Katrina from its peak intensity of 150 *kt* later at 18:00 UTC 28 August.

Katrina experienced a dramatic decrease of negative entropy flow and a successive turn to the stage of positive entropy flow during the period from 18:00 UTC 28 to 06:00 UTC 29 August when the very rapid weakening of Katrina followed accordingly 6 hours later starting with 12:00 UTC 29 August. As a result, the total entropy flow of a hurricane may contain some significant information conducive to its intensity forecast.

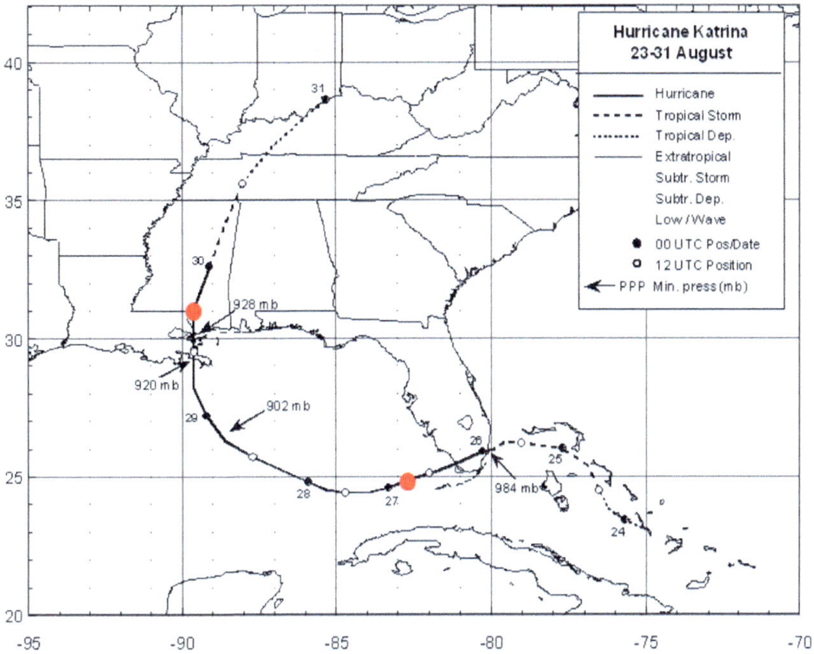

Figure 4. Best track positions for Hurricane Katrina, 23–30 August 2005. The red spot marks are for the corresponding position of the hurricane centres to that shown in **Figure 6** (Reproduced with minor changes based on [**158**]).

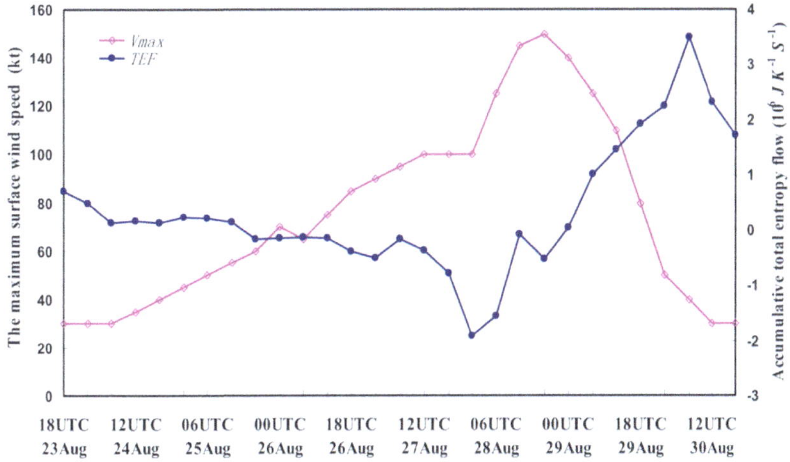

Figure 5. The evolution of total entropy flow (*TEF*, in $10^6\ J\ K^{-1}\ s^{-1}$) for Hurricane Katrina at 6-hourly intervals, with changes in the maximum sustained wind speed ($V_{max,}$ in *kt*) as a reference.

In addition, it is found that the asymmetries of the entropy flow pattern around Katrina's centre in the lower troposphere (e.g., at 850 hPa as are shown in **Figure 6** for the two times of 18 UTC 26 and 28 August which are marked also with the red spot on the Katrina's track in **Figure 4** for easy reference) contained some significant information on hurricane track prediction. In fact in [**156**], there are four points (corresponding to four times of 18:00 UTC 26, 18:00 UTC 28, 18:00 UTC 29 and 00:00 UTC 30 August, respectively) along the Katrina track chosen and discussed to make the finding that Katrina (2005) tends to move towards the side of its centre with stronger negative entropy flows more robust and convincible, since in that way all the four potential moving directions, east-, south-, west- and north-wards, are covered. As stated in [**156**], the finding is the same for all of these four points along the track chosen. As a matter of fact, it occurs throughout the hurricane lifecycle, and is also valid for the other severe tropical storms (see also e.g., [**1,2**] in our MS).

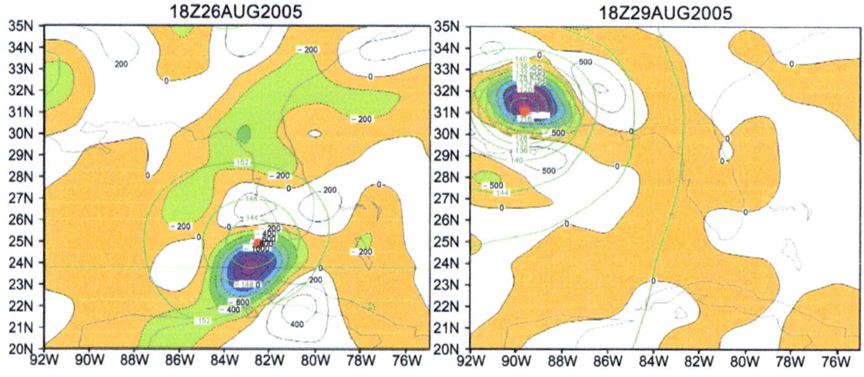

Figure 6. The entropy flow fields at 850 hPa at 18:00 UTC 26 (left panel) and at 18:00 UTC 29 (right one) August 2005. The red spot stands for the corresponding position of the hurricane centre's to the respective time. Green lines are for contours at 850 hPa and black ones are for isopleths of entropy flow with the areas of negative value shaded.

These results imply that classical thermodynamic entropy may serve as an order parameter for an atmospheric system and that entropy flow analysis might provide a new insight into the mechanism responsible for the life cycle of the system. In view of that such severe tropical storms tend to move towards the side of its centre with stronger negative entropy flows, the entropy flow patterns in the lower levels (e.g., 850 hPa) might act as an indicator of the hurricane track. This is understandable from the theory of dissipative structures [**161**]: negative entropy flow entering into an open system will potentially enable the system to depart further from equilibrium or cause it to strengthen, and so here a region of stronger negative entropy flow might become a likely area that the hurricane will approach.

To sum up, it might be argued that negative entropy flow is something very positive which can be regarded as a kind of forces to drive a system to a state further from equilibrium. Next we will give some examples in the Earth's atmosphere.

An estimation of the entropy budget of the atmosphere shows that the entropy flow to the Earth's atmosphere is negative [**162**], which recall spontaneously the fact that the troposphere of the Earth atmosphere experiences a double negative entropy flow process because it gets heat at the tropics with higher temperature but loses heat at both poles with lower temperature horizontally, and also it gets heat at the surface of the Earth with higher temperature but loses heat at the tropopause with lower temperature vertically. This might contribute to the causes, as implied by the Bénard convection shown in **Figure 3**, why the worldwide extreme weather/climate events are getting more and more frequent [**163**,**164**,**165**,**166**] (see also **Figure 7** and **Figure 8** which are based on the data in [**153**], showing the increasing numbers of the severe weather events such as tornados, hails, and thunderstorms with high winds, in the United States, including Wisconsin which is one of the states in the US with the least number of devastating weather disasters in its history, between 1950 and 2003 with a dramatic rise in the numbers after 1990).

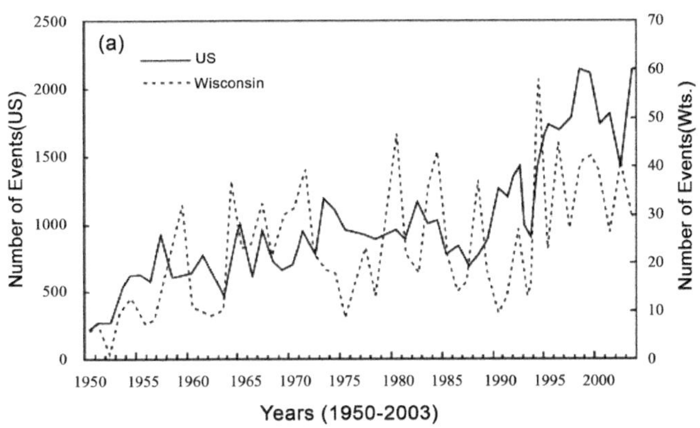

Figure 7. The number of tornados in the United States/Wisconsin between 1950 and 2003.

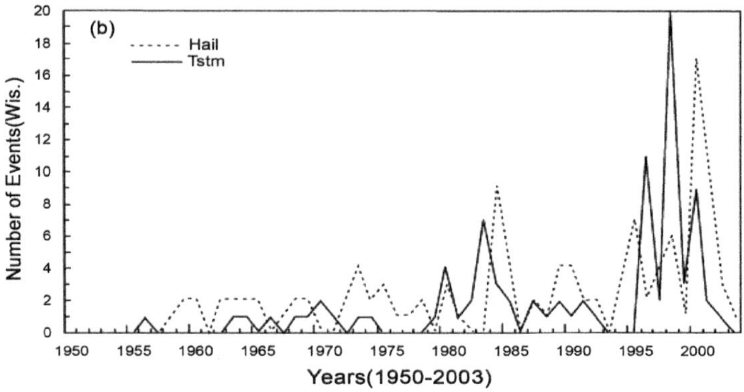

Figure 8. The numbers of thunderstorms, high winds with speed of at least 75 knots and hails with size of at least 2.5 inches in Wisconsin, between 1950 and 2003.

APPLICATIONS OF THE PRINCIPLE OF THE SECOND LAW

So far there has been little research in meteorology based directly on the Second Law of thermodynamics. A plausible cause might be attributed to the fact that this law is expressed in the form of an inequality rather than an equation, which makes its use restricted.

Paltridge proposed Earth's climate structure could be explained via a hypothesis of MEP [**86,89**] which provides global constraint on the Second Law of thermodynamics [**167**]. However, the fact that the Earth's climate system transports heat as does a system in a MEP state does not prove that the Earth's climate system is necessarily seeking such a state [**4**]. The constraint of the Second Law on the thermodynamic behavior of the atmosphere is quite different from that as seen in a "common" physical law such as Newton's Second Law and Energy Conservation law.

Among the studies of the Second Law and its applications, some of them proposed a certain relevant thermodynamic parameter or approximation in a dynamical model that should satisfy or be fully consistent with the Second Law of thermodynamics and some even suggested a generalized Second Law of thermodynamics, including the Second Law of Eddy Thermodynamics [**168**].

Next we show some recent results with the emphases on illustrating how the principle of the Second Law of thermodynamics is employed to improve the dynamic framework of a general circulation model (and thus its simulation accuracy) and the intensity forecasts of a hurricane through a full-irreversability restraint applied on the diffusion scheme in a meso-scale numerical model. as Also we try to understand the potential mechanism responsible for climatic changes in the tropical cyclone number over the western North Pacific in an environment associated with global warming [**169,170,171,172,173,174,175**].

From the very beginning of numerical modeling of atmospheric motions, it was noticed that various mixing/diffusion parameterizations in a model are questionable in the sense that such schemes of parameterization are non-monotonic, which violates the Second Law of thermodynamics, and, dissipative heating associated with vertical momentum diffusion is often included, but in a way that is inconsistent with the Second Law of thermodynamics [**172,176,177,178,179,180,181,182,183**]. As a result, a number of new schemes are suggested based on the principle of the Second Law of thermodynamics. For example, the forecast accuracy of a numerical weather prediction model is improved by emulating physical dissipation rather than using the original diffusion scheme.

The ability of the new physics-based scheme to improve model accuracy is demonstrated first via the case of the one-dimensional viscous Burgers equation that can be written as:

$$\frac{\partial u}{\partial t}+u\frac{\partial u}{\partial x}=\frac{1}{R_e}\frac{\partial^2 u}{\partial x^2} \tag{6}$$

where $u \equiv u(x, t)$ is the flow velocity and R_e the Renolds number. In the case of the initial condition taken as the linear distribution: $u = -x$ at $t = 0$, $-\infty < x < \infty$ with the boundary condition as:

$$u=\begin{cases}1 \; at \; x=-\infty \\ -1 \; at \; x=\infty\end{cases}$$

there exists a special solution: $u = -th[(R_e/2)x]$, which is a steady solution containing no time.

Figure 9 shows that the final solution with the original diffusion scheme has a Gibbs oscillation [**184**] near the "shock wave" where the velocity gradient augments suddenly. On the other hand, the numerical solution consistent with the analytic one is reached when the scheme is changed to the physics-based scheme that meets the restriction of the Second Law of thermodynamics. Using such a scheme, a series of numerical simulations, including the well-known 1998 successive torrential rains along the Yangtze River valley and 365 continuous 24-h integrations for 2005–06 improved with decreased root-mean-square errors and more realistic forecasts in all of the simulations [**176,177,178,179**].

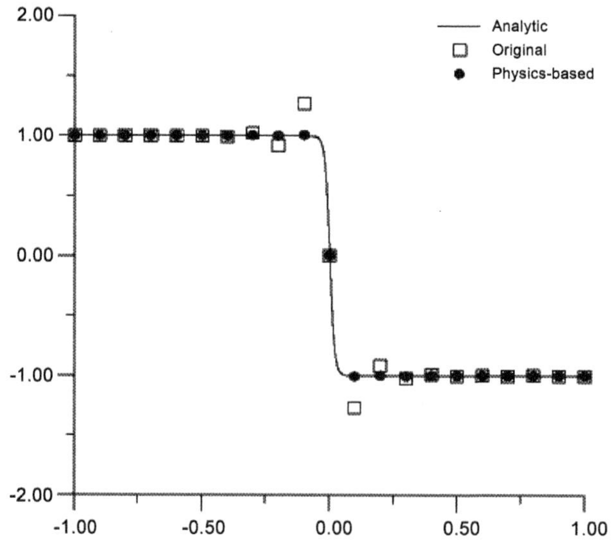

Figure 9. Comparison of analytic solution of the one-dimensional viscous Burger's equation that runs to a steady flow with the numerical solutions between the original and physics-based diffusion schemes used.

Figure 10 shows the intensity forecasts of a hurricane, Hurricane Andrew (1992), improved through a full-irreversability restraint applied on the diffusion scheme in a meso-scale numerical model (the fifth-generation Pennsylvania State University-National Center for Atmospheric Research Mesoscale Model (MM5) [**185**] where a fourth order scheme originally-used created upgradient mass or heat transport, which is spurious and violates the principle of the Second Law of thermodynamics that controls the evolutionary direction of a multi-body system like the atmosphere [**182**]. It is seen from **Figure 10** that the surface maximum sustained wind velocity forecasts based on the new scheme that incorporated the Second Law are distinctly closer to observations than those based on the old one, especially for the period before 54 h. A global spectral model has also been improved by incorporating the second law of thermodynamics in a slightly different way from the above [**186**].

In the end of this section, one would like to mention an application of the Second Law to typhoon climate, in which Liu *et al.* proposed a potential mechanism responsible for changes in the tropical cyclone number during the recent 60 years or so over the western North Pacific in an environment associated with global warming based on the principle of this Law [**187**].

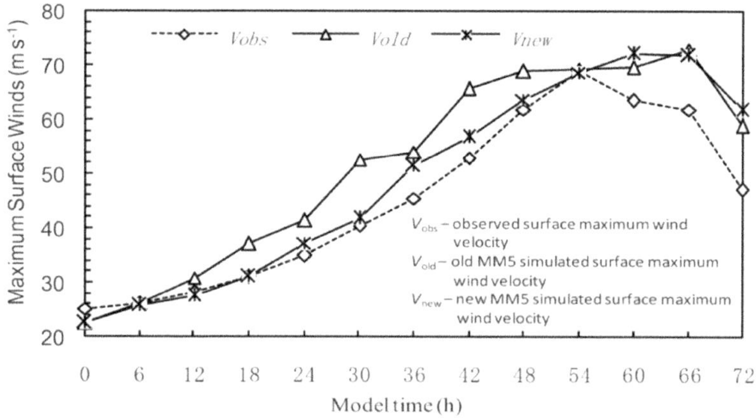

Figure 10. Comparison of the surface maximum sustained wind velocity between the different schemes with observations.

DISCUSSION AND CONCLUSIONS

The Second Law of Thermodynamics is one of the most fundamental principles in thermodynamics that controls the evolutionary direction of a multi-body such as the Earth's atmosphere. Entropy is a crucial physical concept originated from this law. As a result, applications of entropy and the Second Law to atmospheric sciences are significantly important. For the sake of convenience, this paper introduces the works under the subtitles of Entropy balance, Entropy production, Entropy flow, and Applications of the principle of the Second Law, respectively.

Since the 1950s a series of studies of atmospheric dynamics and thermodynamics into the entropy theory and its applications with stress put on entropy production have been made. However, till now the studies on entropy production and its applications are still very preliminary. Early in 1980 Nicolis and Nicolis discussed Paltridge's work and commented that the behaviour of entropy production is far from simple and universal [**188**]. On the other hand, entropy production behaviour involves many specific dynamical and thermodynamic processes in the atmosphere, the extremal principles can only provide a general insight into the overall configuration of the atmosphere. More relevant works need to be done in this direction, see also [**182**]. In fact, later Lucia offers a proof of a theorem for the eventual steady state of a constrained open system: "The principle of maximum irreversible entropy: The irreversible entropy reaches its maximum at the stability" [**189**]. What is more, Lucia's proof is limited to the theory of so-called "rational thermodynamics" in which there is not necessarily an empirically verifiable definition of entropy, let alone entropy production. Sometimes the MEPP mentioned above is used by many researchers who guide themselves by their intuition and are often unaware of the general formalism supporting the validity of MEPP. The point is that MEPP determines just the most probable state and a system may choose a development trajectory with a smaller production, but it will be meta-stable.

In nature, there is much room for improving the entropy theory, including the extremal principles of entropy or its production, as well as the Second Law itself. For example, as Romps mentioned, questions like "Can the entropy budget place stringent bounds on the intensity of storms?" have not yet been addressed, but given the importance of wind-generated dissipation to those living in the atmosphere's boundary layer, these questions certainly deserve attention [**175**]. Recently emphasis has been given to the entropy principle in computational fluid dynamics, although within a different context. Several questions and avenues of inquiry are raised. Specifically, while there are many different formulations of sub-grid processes in the literature on fluid turbulence, they have not yet been systematically subjected to scrutiny from the perspective of the entropy budgets, among which, e.g., the first-order down-gradient Smagorinsky formulation for the sub-grid transport of dry static energy clearly fails to comply with expectations that sub-grid heat transport be a net source of entropy or dry entropy. This would merit further research to assess existing turbulence schemes for their ability to produce entropy sources of the correct sign and magnitude [**175**]. It needs to be pointed out that the studies mentioned in **Section 5** in this paper, on the forecast accuracy of a meso-scale numerical weather prediction model improved by emulating physical dissipation as suggested by the Second Law of thermodynamics, are done in a different way in which the newly-constructed diffusion scheme with a full-irreversability restraint is applied based directly on the principle of the Second Law.

It is noticed that the importance of the entropy budget in relation to its structure, its exchange, and its sources and sinks has received almost no attention in both medium-range weather prediction and climate modeling.

The studies on the effect of negative entropy flow on the self-organization of the atmospheric systems, discussed in **Section 4** in this paper, provide a new insight into the mechanism responsible for the evolution of a weather system as well as a new tool to predict its track and intensity change trend. However, the entropy balance equation and the entropy formulism suitable for the atmospheric applications remain to be further improved e.g., the components of the specific entropy remain to be extended so as to cover all of the water phases.

Lastly, one would like to mention that a variety of computable entropy measures (e.g., block entropy, Kolmogorov entropy, Kolmogorov-Sinai entropy, T complexity, Ledrappier-Young Entropy, Pesin Entropy, Topological entropy, von Neumann entropy, Black hole entropy, and approximate entropy) have been proposed in various subjects [**19**,**20**,**36**,**37**]. The theory of entropy has been developing itself and, its applications to and the proposition of a definition of entropy perfect for meteorology both certainly deserve attention.

ACKNOWLEDGEMENTS

This research is jointly supported by the National Natural Science Foundation of China (Grant Nos. 41075048 and 40633016) and the Basic Research Project of the State Key Laboratory of Severe Weather, Chinese Academy of Meteorological Sciences (2008LaSWZI01).

REFERENCES

1. Liu, C.; Xu, H.; Liu, Y. Linking entropy flow with typhoon evolution: a case-study. *J. Phys. D Appl. Phys.* **2007**, *40*, 6694–6704.
2. Liu, Y.; Liu, C. On the entropy flow properties of a severe tropical storm. *Appl. Phys. Lett.* **2007**, *91*(1).
3. Romps, D.M. The dry-entropy budget of a moist atmosphere. *J. Atmos. Sci.* **2008**, *65*, 3779–3799.
4. Ozawa, H.; Ohmura, A.; Lorenz, R.D.; Pujol, T. The second law of thermodynamics and the global climate system: A review of the maximum entropy production principle. *Rev. Geophys.* **2003**, *41*, 1018.
5. Carnot, S. *Reflexions on the Motive Power of Fire*; Manchester Univ. Press: Manchester, UK, 1986; p. 230.
6. Verkley, W.T.M.; Lynch, P. Energy and enstrophy spectra of geostrophic turbulent flows derived from a maximum entropy principle. *J. Atmos. Sci.* **2009**, *66*(8), 2216–2236.
7. Martyushev, L.M.; Seleznevb, V.D. Maximum entropy production principle in physics, chemistry and biology. *Phys. Rep.* **2006**, *426*, 1–45.
8. Ozawa, H.; Ohmura, A. Thermodynamics of a global-mean state of the atmosphere—A state of maximum entropy increase. *J. Climate* **1997**, *10*, 441–445.
9. Duane, G.S.; Curry, J.A. Entropy of a connecting water-air system and the interpretation of cloud morphogenesis. *Quart. J. Roy. Meteorol. Soc.* **1997**, *123*, 605–629.
10. Barsugli, J.J.; Battisti, D.S. The basic effects of atmosphere-ocean thermal coupling on midlatitude variability. *J. Atmos. Sci.* **1998**, *55*, 477–493.
11. Egger, J. Numerical generation of entropies. *Mon. Wea. Rev.* **1999**, *127*, 2211–2216.
12. Nicolis, C. Irreversible thermodynamics of a simple atmospheric model. *Int. J. Bifurcation Chaos* **2002**, *12*, 2557–2566.
13. Fraedrich, K.; Blender, R. Scaling of atmosphere and ocean temperature correlations in observations and climate models. *Phys. Rev. Lett.* **2003**, *90*, 108501.
14. Gade, H.G.; Gustafson, K. Application of classical thermodynamics principle to study of oceanic overturning circulation. *Tellus* **2004**, *56*,

371–387.

15. Curry, J.A. *Thermodynamics of Atmospheres and Oceans*; Academic Press: San Diego, CA, USA, 1999.
16. Bohren, C.F.; Albrecht, B.A. *Atmospheric Thermodynamics*; Oxford University Press: London, UK, 1999.
17. Tsonis, A.A. *An Introduction to Atmospheric Thermodynamics*; Cambridge University Press: Cambridge, UK, 2002.
18. Zdunkowski, W.; Bott, A. *Thermodynamics of the Atmosphere*; Cambridge University Press: Cambridge, UK, 2004.
19. Downarowicz, T. Entropy. *Scholarpedia* **2007**, *2*(11), 3901. Available online: **http://www.scholarpedia.org/article/Entropy** (accessed on 30 September 2010).
20. Balasis, G.; Daglis, I.A.; Papadimitriou, C.; Kalimeri, M.; Anastasiadis, A.; Eftaxias, K. Investigating dynamical complexity in the magnetosphere using various entropy measures. *J. Geophys. Res.* **2009**, *114*, A00D06.
21. Clausius, R. Ueber verschiedene für die Anwendung bequeme Formen der Hauptgleichungen der mechanischen Wärmetheorie. *Ann. Phys. Chem.* **1865**, *125*, 353–400.
22. Anderson, G. *Thermodynamics of Natural Systems*; Cambridge University Press: Cambridge, UK, 2005; p. 105.
23. Wikipedia. Entropy (order and disorder). Available online: **http://en.wikipedia.org/wiki/Entropy_(order_and_disorder)** (accessed on 2 December 2010).
24. Landsberg, P.T. Is equilibrium always an entropy maximum? *J. Stat. Physics.* **1984**, *35*, 159–69.
25. Greven, A.; Keller, G.; Warnercke, G. *Entropy—Princeton Series in Applied Mathematics*; Princeton University Press: Princeton, NJ, USA, 2003; ISBN 0-691-11338-6. 2003.
26. Wikipedia. Entropy in thermodynamics and information theory. Available online: **http://en.wikipedia.org/wiki/Entropy_in_thermodynamics_and_information_theory** (accessed on 2 December 2010).
27. Weiss, V.; Weiss, H. The golden mean as clock cycle of brain waves. *Chaos Soliton. Fractal.* **2003**, *18*, 643–652.
28. Roopun, A.K.; Kramer, M.A.; Carracedo, L.M.; Kaiser, M.; Davies,

C.H.; Traub, R.D.; Kopell, N.J.; Whittington, M.A. Temporal interactions between cortical rhythms. *Front. Neurosci.* **2008**, *2*, 145–154.

29. Mahulikar, S.P.; Herwig, H. Exact thermodynamic principles for dynamic order existence and evolution in chaos. *Chaos Soliton. Fractal.* **2009**, *41*(4), 1939–1948.
30. Glansdorff, P.; Prigogine, I. *Thermodynamic Theory of Structure, Stability and Fluctuations*; Wiley Interscience: New York, NY, USA, 1971.
31. Strutt, J.W.; Rayleigh, L. Some general theorems relating to vibrations. *Proc. London Math. Soc.* **1873**, *4*, 357–368.
32. Onsager, L. Reciprocal relations in irreversible processes. I. *Phys. Rev.* **1931**, *37*, 405–426.
33. Onsager, L. Reciprocal relations in irreversible processes. II. *Phys. Rev.* **1931**, *38*, 2265–2279.
34. Sawada, Y. A thermodynamic variational principle in nonlinear non-equilibrium phenomena. *Prog. Theor. Phys.* **1981**, *66*, 68–76.
35. Malkus, W.V.R.; Veronis, G. Finite amplitude cellular convection. *J. Fluid Mech.* **1958**, *4*(3), 225–260.
36. Dewar, R. Information theory explanation of the fluctuation theorem, maximum entropy production and self-organized criticality in non-equilibrium stationary states. *J. Phys. A: Math. Gen.* **2003**, *36*, 63.
37. Dewar, R.C. Maximum entropy production and the fluctuation theorem. *J. Phys. A: Math. Gen.* **2005**, *38*, L371.
38. Kleidon, A.; Lorenz, R.D. Nonequilibrium thermodynamics and the production of entropy. In *Life, Earth, and Beyond*; Springer-Verlag: Berlin, Germany, 2005.
39. Beretta, G.P. Steepest entropy ascent in quantum thermodynamics. *Lect. Notes Phys.* **1986**, *278*, 441.
40. Beretta, G.P. Nonlinear quantum evolution equations to model irreversible adiabatic relaxation with maximal entropy production and other nonunitary processes. *Rep. Math. Phys.* **2009**, *64*, 139–168.
41. Maddox, J. Uniting mechanics and statistics. *Nature* **1985**, *316*, 11.
42. Beretta, G.P. Modeling non-equilibrium dynamics of a discrete probability distribution: General rate equation for maximal entropy generation in a maximum-entropy landscape with time-dependent

constraints. *Entropy* **2008**, *10*, 160–182.

43. Prigogine, I. Thermodynamics of irreversible processes. *Bull. Class. Sci. Acad. Roy. Belg.* **1945**, *31*, 600–621.
44. Prigogine, I. *Introduction to Thermodynamics of Irreversible Processes*; Charles C Thomas Publisher: Springfield, IL, USA, 1955.
45. De Groot, S.R.; Mazur, P. *Non-equilibrium Thermodynamics*; North-Holland: Amsterdam, The Netherlands, 1962.
46. Katchalsky, A.; Curran, P.F. *Non-equilibrium Thermodynamics in Biophysics*; Harvard University Press: Cambridge, MA, USA, 1965.
47. Schrödinger, E. *What Is Life?* Cambridge University Press: Cambridge, UK, 1992; pp. 19–25.
48. Clausius, R. Ueber verschiedene für die Anwendung bequeme formen der Hauptgleichungen der mechanischen Wärmetheorie. *Ann. Phys. Chem.* **1865**, *125*, 53–400.
49. Gibbs, J.W. *Elementary Principles in Statistical Mechanics: Developed with Especial Reference to the Rational Foundation of Thermodynamics*; Kessinger Publishing's Photocopy Edition; Kessinger Publishing: Whitefish, MT, USA, 1902.
50. Boltzmann, L. Further studies on the thermal equilibrium of gas molecules. In *Wisssenschaftliche Abhandlungen*; Hasenohrl, F., Ed.; J.A. Barth: Leipzig, Germany, 1909; Volume 1, pp. 316–402.
51. Shannon, C.E. A mathematical theory of communication. *Bell Syst. Tech. J.* **1948**, *27*(7), 379–423.
52. Shannon, C.E. A mathematical theory of communication. *Bell Syst. Tech. J.* **1948**, *27*(10), 623–656.
53. Gibbs, W.J. On the equilibrium of heterogeneous substances. *Transactions of the Connecticut Academy* **1876**. III, October 1875–May 1876, 108–248, May 1877–July 1878, 343–524.
54. Jaynes, E.T. Information theory and statistical mechanics. *Phys. Rev.* **1957**, *106*(4), 620–630.
55. Jaynes, E.T. Information theory and statistical mechanics II. *Phys. Rev.* **1957**, *108*(2), 171–190.
56. Dewar, R. Information theory explanation of the fluctuation theorem, maximum entropy production and self-organized criticality in non-equilibrium stationary states. *J. Phys. A Math. Gen.* **2003**, *36*(3), 634–641.

57. Ruelle, D.P. Extending the definition of entropy to nonequilibrium steady states. *Proc. Natl. Acad. Sci. U. S. A.* **2003**, *100*(6), 3054–3058.
58. Peixoto, J.P.; Oort, A.H.; de Almeida, M.; Tome, A. Entropy budget of the atmosphere. *J. Geophys. Res.* **1991**, *96*, 10981–10988.
59. Carlstead, E.M. A study of constant absolute vorticity trajectories on isentropic surfaces. *J. Meteorol.* **1953**, *10*(5), 356–361.
60. Uccellini, L.W.; Johnson, D.R.; Schlesinger, R.E. An isentropic and sigma coordinate hybrid numerical model: Model development and some initial tests. *J. Atmos. Sci.* **1979**, *36*(3), 390–414.
61. Shapiro, M.A. Simulation of Upper-level frontogenesis with a 20-level isentropic coordinate primitive equation model. *Mon. Weather Rev.* **1975**, *103*(7), 591–604.
62. Bleck, R. Short-range prediction in isentropic coordinates with filtered and unfiltered numerical models. *Mon. Weather Rev.* **1974**, *102*(12), 813–829.
63. Anthens, R.A. A numerical model of the slowly varying tropical cyclone in isentropic coordinates. *Mon. Weather Rev.* **1971**, *99*(8), 617–635.
64. Yang, H.; Olaguer, E.; Tung, K.K. Simulation of the present-day atmospheric ozone, odd nitrogen, chlorine and other species using a coupled 2-D model in isentropic coordinates. *J. Atmos. Sci.* **1991**, *48*(3), 442–471.
65. Yueh-Jiuan, G.; Hsu, A.A. Numerical modeling of the atmosphere with an isentropic vertical coordinate. *Mon. Weather Rev.* **1990**, *118*(10), 1933–1959.
66. Trevisan, A.; Ferranti, L.; Malguzzi, P. Further developments of normal mode theory of lee cyclogenesis: Isentropic coordinate model. *J. Atmos. Sci.* **1988**, *45*(24), 3880–3888.
67. Black, T.L. A Comparison of Key Forecast variables derived from isentropic and sigma coordinate regional models. *Mon. Weather Rev.* **1987**, *115*(12), 3097–3114.
68. Merrill, J.T.; Bleck, R.; Boudra, D. Techniques of lagrangian trajectory analysis in isentropic coordinates. *Mon. Weather Rev.* **1986**, *114*(3), 571–581.
69. Andrews, D.G. A Finite-amplitude eliassen-palm theorem in isentropic coordinates. *J. Atmos. Sci.* **1983**, *40*(8), 1877–1883.
70. Ka, K.T. On the two-dimensional transport of stratospheric trace gases in isentropic coordinates. *J. Atmos. Sci.* **1982**, *39*(10), 2330–2355.

71. Brankovic, C. A transformed isentropic coordinate and its use in an atmospheric model. *Mon. Weather Rev.* **1981**, *109*(9), 2029–2039.

72. Reames, F.M.; Zapotocny, T.H. Inert trace constituent transport in sigma and hybrid isentropic-sigma models. Part II: Twelve Semi-Lagrangian Algorithms. *Mon. Weather Rev.* **1999**, *127*(2), 188–200.

73. Reames, F.M.; Zapotocny, T.H. Inert trace constituent transport in sigma and hybrid isentropic-sigma models. Part I: Nine advection algorithms. *Mon. Weather Rev.* **1999**, *127*(2), 173–187.

74. Zapotocny, T.H.; Johnson, D.R.; Reames, F.M. A comparison of regional isentropic-sigma and sigma model simulations of the January 1979 Chicago blizzard. *Mon. Weather Rev.* **1993**, *121*(7), 2115–2136.

75. Johnson, D.R.; Zapotocny, T.H.; Reames, F.M.; Wolf, B.J.; Bradley, P.R. A comparison of simulated precipitation by hybrid isentropic-sigma and sigma models. *Mon. Weather Rev.* **1993**, *121*(7), 2088–2114.

76. Pierce, R.B.; Reames, F.M.; Zapotocny, T.H.; Johnson, D.R.; Wolf, B.J. Numerical investigations with a hybrid isentropic-sigma model. Part I: Normal-mode characteristics. *J. Atmos. Sci.* **1991**, *48*(18), 2005–2024.

77. Bleck, R.; Benjamin, S.G. Regional weather prediction with a model combining terrain-following and isentropic coordinates. Part I: Model description. *Mon. Weather Rev.* **1993**, *121*(6), 1770–1785.

78. Buzzi, A.; Trevisan, A.; Salustri, G. Internal frontogenesis: A two-dimensional model in isentropic, Semi-Geostrophic coordinates. *Mon. Weather Rev.* **1981**, *109*(5), 1053–1060.

79. Benjamin, S.G.; Grell, G.A.; Brown, J.M.; Smirnova, T.G.; Bleck, R. Mesoscale weather prediction with the RUC hybrid isentropic-terrain-following coordinate model. *Mon. Weather Rev.* **2004**, *132*(2), 473–494.

80. Bleck, R; Benjamin, S.; Lee, J.; MacDonald, A.E. On the use of an adaptive, hybrid-isentropic vertical coordinate in global atmospheric modeling. *Mon. Weather Rev.* **2010**, *138*(6), 2188–2210.

81. Pauluis, O.; Czaja, A.; Korty, R. The global atmospheric circulation in moist isentropic coordinates. *J. Climate* **2010**, *23*(11), 3077–3093.

82. Arakawa, A.; Lamb, V.R. *Computational Design of the Basic Dynamical Process of the UCLA General Circulation Model. Methods in Computational Physics*; Academic Press: New York, NY, USA, 1977; pp. 173–265.

83. Sommerfeld, A. *Mechanics of Deformable Bodies—Lectures on*

Theoretical Physics Volume II; Academic Press: New York, NY, USA, 1964; Kuerti, G., Translator.

84. Johnson, D.R.; Lenzen, A.J.; Zapotocny, T.H.; Schaack, T.K. Numerical uncertainties in the simulation of reversible isentropic processes and entropy conservation. *J. Climate* **2000**, *13*(21), 3860–3884.
85. Johnson, D.R.; Lenzen, A.J.; Zapotocny, T.H.; Schaack, T.K. Numerical uncertainties in simulation of reversible isentropic processes and entropy conservation: Part II. *J. Climate* **2002**, *15*(14), 1777–1804.
86. Paltridge, G.W. Global dynamics and climate—A system of minimum entropy exchange. *Quart. J. Roy. Meteorol. Soc.* **1975**, *101*, 475–484.
87. Wikipedia. Non-equilibrium_thermodynamics. Available online: **http://en.wikipedia.org/wiki/Non-equilibrium_thermodynamics** (accessed on 2 December 2010).
88. Tuck, A.F. *Atmospheric Turbulence: A Molecular Dynamics Perspective*; Oxford University Press: London, UK, 2008; p. 33.
89. Paltridge, G.W. A physical basis for a maximum of thermodynamic dissipation of the climate system. *Quart. J. Roy. Meteorol. Soc.* **2001**, *127*, 305–313.
90. Platridge, G.W. The steady-state format of global climate. *Quart. J. Roy. Meteorol. Soc.* **1978**, *104*, 927–945.
91. Paltridge, G.W. Climate and thermodynamic systems of maximum dissipation. *Nature* **1979**, *279*, 630–631.
92. Ozawa, H.; Shimokawa, S.; Sakuma, H. Thermodynamics of fluid turbulence: A unified approach to the maximum transport properties. *Phys. Rev. E* **2001**, *64*, 026303.
93. Goody, R.M. Sources and sinks of climate entropy. *Quart. J. Roy. Meteorol. Soc.* **2000**, *126*, 1953–1970.
94. Lorenz, E.N. *The Nature and Theory of the General Circulation of the Atmosphere*; World Meteorological Organization: Geneva, Switzerland, 1967; p. 161.
95. Lorenz, E.N. Generation of available potential energy and the intensity of the general circulation. In *Dynamics of Climate*; Pfeffer, R.L., Ed.; Pergamon: Tarrytown, NY, USA, 1960; pp. 86–92.
96. Grassl, H. The climate at maximum entropy production by meridional atmospheric and oceanic heat fluxes. *Quart. J. Roy. Meteorol. Soc.* **1981**, *107*, 153–166.

97. Shutts, G.J. Maximum entropy production states in quasigeostrophic dynamical models. *Quart. J. Roy. Meteorol. Soc.* **1981**, *107*, 503–520.
98. Mobbs, S.D. Extremal principles for global climate models. *Quart. J. Roy. Meteorol. Soc.* **1982**, *108*, 535–550.
99. Noda, A.; Tokioka, T. Climates at minima of the entropy exchange rate. *J. Meteorol. Soc. Jpn.* **1983**, *61*, 894–908.
100. Sohn, B.J.; Smith, E.A. Energy transports by ocean and atmosphere based on an entropy extremum principle, part I, Zonal averaged transports. *J. Climate* **1993**, *6*, 886–899.
101. Sohn, B.J.; Smith, E.A. Energy transports by ocean and atmosphere based on an entropy extremum principle, part II, Two-dimensional transports. *Meteorol. Atmos. Phys.* **1994**, *53*, 61–75.
102. Ozawa, H.; Ohmura, A. Thermodynamics of a global mean state of the atmosphere—A state of maximum entropy increase. *J. Climate* **1997**, *10*, 441–445.
103. Pujol, T.; Llebot, J.E. Extremal principle of entropy production in the climate system. *Quart. J. Roy. Meteorol. Soc.* **1999**, *125*, 79–90.
104. Dewar, R. Information theory explanation of the fluctuation theorem, maximum entropy production and self-organized criticality in non-equilibrium stationary states. *J. Phys. A Math. Gen.* **2003**, *36*, 631–641.
105. Dewar, R.C. Maximum entropy production and the fluctuation theorem. *J. Phys. A Math. Gen.* **2005**, *38*, L371–L381.
106. Grinstein, G.; Linsker, R. Comments on a derivation and application of the 'maximum entropy production' principle. *J. Phys. A Math. Gen.* **2007**, *40*, 9717–9720.
107. Dewar, R. Information theory explanation of the fluctuation theorem, maximum entropy production and self-organized criticality in non-equilibrium stationary states. *J. Phys. A Math. Gen.* **2003**, *36*, 63.
108. Dewar, R.C. Maximum entropy production and the fluctuation theorem. *J. Phys. A Math. Gen.* **2005**, *38*, L371.
109. Center for Molecular Modeling (CMM). Available online: **http://cmm.cit.nih.gov/maxent/letsgo.html** (accessed on 30 September 2010).
110. Cornwell, T. J.; Evans, K.F. A simple maximum entropy deconvolution algorithm. *Astron. Astrophys.* **1985**, *143*, 77–83.
111. Wernecke, S.J.; D'Addario, L.R. Maximum entropy image reconstruction. *IEEE Trans. Comput.* **1976**, *C-26*, 351–364.

112. Gull, S.F.; Daniell, G. Image reconstruction from noisy and incomplete data. *Nature* **1978**, *272*, 686–690.

113. Jaynes, E.T. The ratinable of maximum entropy methods. *Proc. IEEE* **1982**, *70*, 939–952.

114. Frieden, B.R. Restoring with maximum likelihood and maximum entropy. *J. Opt. Soc. Am.* **1972**, *62*, 511–518.

115. Narayan, R.; Nityananda, R. *Maximum Entropy—Flexibility versus Fundamentalism, in Indirect Imaging*; Roberts, J.A., Ed.; Cambridge University Press: Cambridge, UK, 1984; pp. 281–290.

116. Narayan, R.; Nityananda, R. Maximum entropy image restoration in astronomy. *Ann. Rev. Astron. Astrophys.* **1986**, *24*, 127–170.

117. Ou, H.-W. Possible Bounds on the earth's surface temperature: From the perspective of a conceptual Global-Mean model. *J. Climate* **2001**, *14*(13), 2976–2988.

118. Nye, T.; Yamamoto, T.; Trevorrow, M. Measurements of the directional spectra of shallow water waves using the maximum entropy principle and a single ocean bottom seismometer. *J. Atmos. Ocean. Technol.* **1990**, *7*(5), 781–791.

119. Steinwagner, J.; Schwarz, G.; Hilgers, S. Use of a maximum entropy method as a regularization technique during the retrieval of trace gas profiles from limb sounding measurements. *J. Atmos. Ocean. Technol.* **2006**, *23*(12), 1657–1667.

120. Kivman, G.A.; Kurapov, A.L.; Guessen, A.V. An entropy approach to tuning weights and smoothing in the generalized inversion. *J. Atmos. Ocean. Technol.* **2001**, *18*(2), 266–276.

121. Bernard, Urban. Humidity profile retrieval using a maximum entropy principle. *Mon. Weather Rev.* **1995**, *123*(4), 1185–1196.

122. Verkley, W.T.M.; Lynch, P. Energy and enstrophy spectra of geostrophic turbulent flows derived from a maximum entropy principle. *J. Atmos. Sci.* **2009**, *66*(8), 2216–2236.

123. Elsner, J.B.; Kara, A.B.; Owens, M.A. Fluctuations in North Atlantic hurricane frequency. *J. Climate* **1999**, *12*(2), 427–437.

124. Sultan, B.; Janicot, S.; Correia, C. Medium lead-time predictability of intraseasonal variability of rainfall in West Africa. *Weather Forecast.* **2009**, *24*(3), 767–784.

125. Kane, R.P. Quasi-biennial and quasi-triennial oscillations in the summer monsoon rainfall of the meteorological subdivisions of India. *Mon. Weather Rev.* **1995**, *123*(4), 1178–1184.
126. Christensen, R.A.; Eilbert, R.F. Seasonal precipitation forecasting with a 6–7 month lead time in the Pacific Northwest using an information theoretic model. *Mon. Weather Rev.* **1985**, *113*(4), 502–519.
127. Simpson, J. Use of the gamma distribution in single-cloud rainfall analysis. *Mon. Weather Rev.* **1972**, *100*(4), 309–312.
128. Mo, K.C. Adaptive filtering and prediction of intraseasonal oscillations. *Mon. Weather Rev.* **2001**, *129*(4), 802–817.
129. Bocquet, M.; Pires, C.A.; Wu, L. Beyond Gaussian statistical modeling in geophysical data assimilation. *Mon. Weather Rev.* **2010**, *138*(8), 2997–3023.
130. Prieto, R.; Schubert, W.H. Analytical predictions for zonally symmetric equilibrium states of the stratospheric polar vortex. *J. Atmos. Sci.* **2001**, *58*(18), 2709–2728.
131. Schubert, W.H.; Montgomery, M.T.; Taft, R.K.; Guinn, T.A.; Fulton, S.R.; Kossin, J.P.; Edwards, J.P. Polygonal eyewalls, asymmetric eye contraction, and potential vorticity mixing in hurricanes. *J. Atmos. Sci.* **1999**, *56*(9), 1197–1223.
132. Liu, Y.G.; Hallett, J. On size distributions of cloud droplets growing by condensation: A new conceptual model. *J. Atmos. Sci.* **1998**, *55*(4), 527–536.
133. Frederiksen, J.S.; Dix, M.R.; Kepert, S.M. Systematic energy errors and the tendency toward canonical equilibrium in atmospheric circulation models. *J. Atmos. Sci.* **1996**, *53*(6), 887–904.
134. Tai, K.-S.; Ogura, Y. An observational study of easterly waves over the eastern pacific in the northern summer using FGGE data. *J. Atmos. Sci.* **1987**, *44*(2), 339–361.
135. Hayashi, Y.; Golder, D.G. The seasonal variation of tropical transient planetary waves appearing in a GFDL general circulation model. *J. Atmos. Sci.* **1980**, *37*(4), 705–716.
136. Cadet, D.; Teitelbaum, H. Observational evidence of internal inertia-gravity waves in the tropical stratosphere. *J. Atmos. Sci.* **1979**, *36*(5), 892–907.
137. Hayashi, Y.; Golder, D.G. The generation of equatorial transient planetary waves: control experiments with a GFDL general circulation model. *J. Atmos. Sci.* **1978**, *35*(11), 2068–2082.

138. Schönwiese, C-D. Moving spectral variance and coherence analysis and some applications on long air temperature series. *J. Clim. Appl. Meteorol.* **1987**, *26*(12), 1723–1730.
139. Tapiador, F.J.; Kidd, C.; Levizzani, V.; Marzano, F.S. A neural networks-based fusion technique to estimate half-hourly rainfall estimates at 0.1° resolution from satellite passive microwave and infrared data. *J. Appl. Meteorol.* **2004**, *43*(4), 576–594.
140. Johnson, D.R. "General coldness of climate models" and the second law: Implications for modeling the earth system. *J. Climate* **1997**, *10*(11), 2826–2846.
141. Adams, D.K.; Rennó, N.O. Thermodynamic efficiencies of an idealized global climate model. *Clim. Dynam.* **2005**, *25*, 801–813.
142. Johnson, D.R. General coldness of climate models and the second law: Implications for modeling the earth system. *J. Climate* **1997**, *10*, 2826–2846.
143. Paltridge, G.W. Thermodynamic dissipation and the global climate system. *Quart. J. Roy. Meteorol. Soc.* **1981**, *107*, 531–547.
144. Nicolis, G.; Prigogine, I. *Self-organization in Non-equilibrium Systems*; John Wiley & Sons Inc.: New York, NY, USA, 1977.
145. Liu, C. The generalized Liapunovian stability for the atmospheric systems. *Acta Meteorologica Sinica* **1988**, *2*(1), 30–40.
146. Pujol, T.; Llebot, J.E. Second differential of the entropy as a criterion for the stability in low-dimensional climate models. *Quart. J. Roy. Meteorol. Soc.* **1999**, *125*, 91–106.
147. Stephens, G.L.; O'Brien, D.M. Entropy and climate. I: ERBE observations of the entropy production. *Quart. J. Roy. Meteorol. Soc.* **1993**, *119*, 121–152.
148. Wu, W.; Liu, Y. Radiation entropy flux and entropy production of the Earth system. *Rev. Geophys.* **2010**, *48*, RG2003.
149. Eckert, K.; Bestehorn, M.; Thess, A. Square cells in surface-tension-driven Bénard convection: experiment and theory. *J. Fluid Mech.* **1998**, *356*, 155–197.
150. Olby, R. Schrodinger's problem: What is life? *J. Hist. Biol.* **1971**, *4*, 119–148.
151. Katchalsky, A.; Curran, P.F. *Non-Equilibrium Thermodynamics in Biophysics*; Harvard University Press: Cambridge, MA, USA, 1965.

152. Liu, C.; Liu, Y. Negative entropy flow and its effect on the organization of synoptic-scale severe atmospheric systems. *Geophys. Res. Lett.* **2004**, *31*, L01108.

153. Xu, H.; Liu, C. Entropy flow properties of a typhoon as simulated by a meso-scale model. *EPL.* **2008**, *83*, 18001.

154. Liu, Y.; Liu, C. Entropy flow and the evolution if a storm. *Entropy* **2008**, *10*, 430–440.

155. Liu, Y.; Liu, C. Negative entropy flow and the life-cycle of a severe tropical storm. *Atmos. Res.* **2009**, *9*, 39–43.

156. Liu, C.; Luo, Z.; Liu, Y.; Yu, H.; Zhou, X.; Wang, D.; Ma, L.; Xu, H. Implication of entropy flow for the development of a system as suggested by the life cycle of a hurricane. *Mod. Phys. Lett. B* **2010**, *24*(16), 1747–1757.

157. Liu, Y.; Zhang, D.L.; Yau, M.K. A multiscale numerical study of Hurricane Andrew (1992), Part I: Explicit simulation and verification. *Mon. Wea. Rev.* **1997**, *125*, 3073–3093.

158. Knabb, R.D.; Rhome, J.R.; Brown, D.P. Tropical cyclone 1 report: Hurricane Katrina (23–30 August 2005). Available online: **http://www.nhc.noaa.gov/pdf/TCR-AL122005_Katrina.pdf** (accessed on 30 September 2010).

159. McTaggart-Cowan, R.; Bosart, L.F.; Gyakum, J. R.; Atallah, E. H. Hurricane Katrina (2005). Part I: Complex lifecycle of an intense tropical cyclone. *Mon. Wea. Rev.* **2007**, *135*, 3905–3926.

160. E. Kalnay, E.; Kanamitsu, M.; Kistler, R.; Collins, W.; *et al.* The NCEP/NCAR 40-year reanalysis project. *Bull. Amer. Meteor. Soc.* **1996**, *77*(3), 437–471.

161. Nitschke, K.; Bestehorn, M.; Thess, A. Square cells in surface-tension-driven Bénard convection. *J. Fluid Mech.* **1998**, *356*, 155–197.

162. Peixoto, J.P.; Oort, A.H.; Almeida, M.D.; Tomé, A. Entropy budget of the atmosphere. *J. Geophys. Res.* **1991**, *96*, 10,981–10,988.

163. Easterling, D.R.; Evana, J.L.; Grosman, P.Ya.; Karl, T.R.; Kunkel, K.E.; Ambenje, P. Observed variability and trends in extreme climate. *Bull. Am. Metor. Soc.* **2000**, *81*, 417–425.

164. Karl, T.R.; Kukla, G.; Razuvayev, V.N.; Changery, M.J.; Quayle, R.G.; Heim, R.R., Jr.; Easterling, D.R.; Fu, C.B. Global warming: Evidence for asymmetric diurnal temperature change. *Geophys. Res. Lett.* **1991**, *18*, 2253–2256.

165. Manton, M.J.; Eral, J. Trend in extreme daily rainfall and temperature in southeast Asia and the South Pacific: 1961–1998. *Int. J. Climatol.* **2001**, *21*, 269–284.
166. Wang, S.W.; Gong, D.Y. Enhancement of the warming trend in China. *Geophys. Res. Lett.* **2000**, *27*(16), 2581–2584.
167. Goody, R. Maximum entropy production in climate theory. *J. Atmos. Sci.* **2007**, *64*(7), 2735–2739.
168. Dryden, W.A. On the flux of eddy energy in small-scall turbulent motion. *J. Meteorol.* **1956**, *13*(5), 433–447.
169. Blackadar, A. Extension of the laws of thermodynamics to turbulent systems. *J. Meteorol.* **1955**, *12*(2), 165–175.
170. Pauluis, O. Thermodynamic consistency of the anelastic approximation for a moist atmosphere. *J. Atmos. Sci.* **2008**, *65*(8), 2719–2729.
171. Weaver, A. J.; Eby, M. On the numerical implementation of advection schemes for use in conjunction with various mixing parameterizations in the GFDL ocean model. *J. Phys. Oceanogr.* **1997**, *27*(2), 369–377.
172. Wulf, O.R.; Davis, L., Jr. On the efficiency of the engine driving the atmosphere circulation. *J. Meteorol.* **1952**, *9*(2), 80–82.
173. Burkhardt, U.; Becker, E. A consistent diffusion-dissipation parameterization in the ECHAM climate model. *Mon. Weather Rev.* **2006**, *134*(4), 1194–1204.
174. Kleeman, R. Measuring dynamical prediction utility using relative entropy. *J. Atmos. Sci.* **2002**, *59*(13), 2057–2072.
175. Romps, D.M. The dry-entropy budget of a moist atmosphere. *J. Atmos. Sci.* **2008**, *65*(12), 3779–3799.
176. Liu, C.; Liu, Y.; Kang, H.W. A new technique of physical dissipation and its application to a mesoscale numerical weather prediction model. *Sci. China Ser. D* **2002**, *45*(9), 769–780.
177. Liu, Y.; Liu, C.; Kang, H.W. The modeling improvement of hurricane Andrew by the technique of irreversible thermodynamic operators. *Acta Meteorologica Sinica* **2002**, *16*(4), 441–450.
178. Liu, Y.; Liu, C.; Xu, H.; Zhao, Y.M. A numerical study of sensitivity of the physical dissipative technique to precipitation parameterization in a mesoscale model. *Acta Meteorologica Sinica* **2005**, *19*(1), 67–75.

179. Liu, C.; Zhao, L.N.; Liu, Y. Physical dissipative technique and its application to the MM4. *Acta Meteorologica Sinica* **2002**, *59*(5), 516–524.

180. Liu, C.; Liu, Y.; Xu, H. A physics-based diffusion scheme for numerical models. *Geophys. Res. Lett.* **2006**, *33*, L12805.

181. Liu, C.; Liu, Y.; Xu, H. A new diffusion scheme for numerical models based on full irreversibility. *Weather Forecast.* **2009**, *24*, 595–600.

182. Liu, C.; Liu, Y.; Luo, Z. Improving numerical simulation of a hurricane by irreversible thermodynamic operators. *Atmos. Res.* **2009**, *93*, 44–49.

183. Shutts, G.J. Maximum entropy production states in quasi-geostrophic dynamical models. *Quart. J. Roy. Meteorol. Soc.* **1981**, *107*, 503–520.

184. Pielke, R.A. *Mesoscale Meteorological Modeling*; Academic Press: New York, NY, USA, 1984; p. 599.

185. Grell, G.A.; Dudhia, J.; Stauffer, D.R. *A Description of the Fifth-generation Penn State/NCAR Mesoscale Model (MM5)*; National Center for Atmospheric Research: Boulder, CO, USA, 1995; p. 138.

186. Liu, C.; Liu, Y. An attempt at improving a global spectral model by incorporating the second law of thermodynamics. *Geophys. Res. Lett.* **2005**, *32*, L03806.

187. Liu, C.; Ma, L.; Liu, Y.; Luo, Z.; Lei, X.; Zhou, X.; Wang, D.; Xu, H. Linking tropical cyclone frequency over the western north pacific with sea surface temperatures. In *Hurricanes and Climate Change*; Elsner, J.B., Hodges, R.E., Malmstadt, J.C., Scheitlin, K.N., Eds.; Springer: New York, NY, USA, 2010; Volume 2.

188. Nicolis, G.; Nicolis, C. On the entropy balance of the earth-atmosphere system. *Quart. J. Roy. Meteorol. Soc.* **1980**, *125*, 1859–1878.

189. Lucia, U. Irreversible entropy variation and the problem of the trend to equilibrium. *Physica A* **2007**, *376*, 289–292.

CHAPTER 4

Mechanics and Thermodynamics of a New Minimal Model of the Atmosphere

Gabriele Vissio[1,2], Valerio Lucarini[3,4]

[1] CEN, Meteorological Institute, University of Hamburg, Hamburg, Germany

[2] Institute of Geosciences and Earth Resources (IGG) - National Research Council (CNR), Turin, Italy

[3] Department of Mathematics and Statistics, University of Reading, Reading, UK

[4] Centre for the Mathematics of Planet Earth, University of Reading, Reading, UK

Citation: Vissio, G., Lucarini, V. "Mechanics and thermodynamics of a new minimal model of the atmosphere". Eur. Phys. J. Plus 135, 807 (2020). https://doi.org/10.1140/epjp/s13360-020-00814-w.

ABSTRACT

The understanding of the fundamental properties of the climate system has long benefitted from the use of simple numerical models able to parsimoniously represent the essential ingredients of its processes. Here, we introduce a new model for the atmosphere that is constructed by supplementing the now-classic Lorenz '96 one-dimensional lattice model with temperature-like variables. The model features an energy cycle that allows for energy to be converted between the kinetic form and the potential form and for introducing a notion of efficiency. The model's evolution is controlled by two contributions—a quasi-symplectic and a gradient one, which resemble (yet not conforming to) a metriplectic structure. After investigating the linear stability of the symmetric fixed point, we perform a systematic parametric investigation that allows us to define regions in the parameters space where at steady-state stationary, quasi-periodic, and chaotic motions are realised, and study how the terms responsible for defining the energy budget of the system depend on the external forcing injecting energy in the kinetic and in the potential energy reservoirs. Finally, we find preliminary evidence that the model features extensive chaos. We also introduce a more complex version of the model that is able to accommodate for multiscale dynamics and that features an energy cycle that more closely mimics the one of the Earth's atmosphere.

INTRODUCTION

The climate is a nonequilibrium system whose dynamics is primarily driven by the uneven absorption of solar radiation, which is mainly absorbed near the surface and in the tropical latitudes, rather than aloft and in the mid-high latitudes, respectively. The system reacts to such an inhomogeneity in the local energy input through a complex set of instabilities and feedbacks affecting its dynamical processes and thermodynamic and radiative fluxes. Such processes lead to an overall reduction in the temperature gradients inside the system and allow for the establishment of approximate steady-state conditions [1, 2].

An example of the re-equilibration mechanism can be described as follows. The large scale energy transport, which tends to reduce the temperature difference between low and high latitudes, is mainly performed by atmospheric disturbances in the form of synoptic and (to a lesser extent) planetary eddies which are, in turn, fuelled by the baroclinic conversion of available potential energy into kinetic energy, which is then dissipated by friction. In turn, the presence of large low-high latitudes temperature differences is responsible for the existence of a reservoir of available potential energy, which is continuously replenished thanks to the dishomogeneity of the radiative energy budget across the globe [1]. This is the core of the celebrated Lorenz energy cycle, which provides a powerful representation of the climate as an engine [3]. The thermodynamic viewpoint on the climate allows to define its efficiency and irreversibility [4,5,6,7,8,9].

The reconstruction, interpretation, and analysis of observative data (now facilitated by the recent advances in data science); analytical tools borrowed (mostly) from mathematics and physics; and numerical simulations all contribute to our understanding of the climate. This task is exceedingly demanding since the system features nontrivial variability on a vast range of temporal and spatial scales and, furthermore, our ability to observe it has changed enormously over time. Additionally, the presence of periodic as well as irregular fluctuations in the boundary conditions does not allow the climate to reach an exact steady state [10, 11].

One of the features of the numerical investigation of the climate system is the reliance on hierarchies of models. In other terms, climate phenomena are investigated using a full range of models going from low dimensional ones to state-of-the-art Earth system models, able to represent with higher precision many aspects of the climate. See a discussion of the meaning and of the use of hierarchies of climate models in [11,12,13,14]. It is important to remark that, given the multiscale nature of the climate system, the heterogeneity of its subdomains, and the number of the active physical, chemical, and biological processes, the endeavour of constructing a model able to directly simulate all of them appears as a Sisyphean task, whilst, instead, the parametrisation of the effect of the unresolved scales on those that are explicitly simulated is an essential component of any reasonable model of the Earth system [15,16,17].

Low-order models have played and still play today a very important role for improving our understanding of the geophysical flows. Apart from the landmark 3-dimensional model developed by Lorenz in 1963 [18] starting from the truncation of the equations describing the Rayleigh–Benard convection introduced by Saltzman [19], simple models have been key to scientific advances in oceanography [20,21,22], dynamical meteorology [23,24,25], climate dynamics [26,27,28,29], turbulence [30,31,32,33], and convection [34, 35], among others. Additionally, low-order models supplemented by stochastic forcings have also provided the backbone of the stochastic theory of climate [36,37,38]. Aside from sheer mathematical-related curiosity, many of these models were created in order to shed some light on specific problems by using a physically meaningful benchmark tool that is easier to analyse mathematically and faster to simulate numerically with respect to more complex models.

The Lorenz '96 model

Of special relevance for the present study is the now-celebrated Lorenz '96 model [25, 39] (hereafter L96), whose structure is briefly recapitulated below. The model consists of a lattice of N gridpoints, whose state is described by a variable. The model has periodic boundary conditions, so that it can loosely be interpreted as describing the properties of the atmosphere along a latitudinal circle. The model features in an extremely simplified—almost metaphorical—way the main processes of the atmosphere: forcing, dissipation, and advection.

Two versions of the model have been proposed: a one-level version, where the dynamics takes place in a limited range of spatial and temporal scales, and a two-level version, where the lattice is augmented in order to describe dynamical processes on smaller spatial and temporal scales. The two-level version of the L96 model has the especially attractive property that the time-scale separation between the fast and the slow variables can be controlled by modulating one parameter.

The L96 model has rapidly gained relevance among geoscientists, physicists, and applied mathematicians, as it has become a benchmark testbed for parametrisations [40,41,42,43,44,45,46], for studying extreme events [47,48,49,50], for developing data assimilation schemes [51,52,53,54], for developing ensemble forecasting techniques [55,56,57], for studying the properties of Lyapunov exponents and covariant Lyapunov vectors [58,59,60,61], for developing and testing ideas in nonequilibrium statistical mechanics [62,63,64,65,66], and for investigating bifurcations [67,68,69,70,71,72,73]. By looking at these references, the reader can find a very thorough analysis of the properties of the L96 model.

The one-level L96 model can be written as:

$$\frac{\mathrm{d}X_k}{\mathrm{d}t} = X_{k-1}(X_{k+1} - X_{k-2}) - \gamma X_k + F, \tag{1}$$

with boundary conditions:

$$X_{k-N} = X_{k+N} = X_k. \tag{2}$$

k = 1,...., N is the index of the gridpoints defining the lattice, the nonlinear term defines a nontrivial process of advection, F is an external forcing, and γ (usually taken with unitary value) modulates the dissipation. In the unforced and inviscid regime—i.e. setting F = γ = 0, the energy of the system, expressed as the sum of the squares of the variables, is conserved:

$$\frac{\mathrm{d}E}{\mathrm{d}t} = \frac{\mathrm{d}}{\mathrm{d}t}\sum_{k=1}^{N}\frac{X_k^2}{2} = \sum_{k=1}^{N} X_k \frac{\mathrm{d}X_k}{\mathrm{d}t} = 0. \tag{3}$$

If γ = 1 and N >> 1, the model's attractor is a fixed point for $0 \leq F \leq 8/9$. As F is increased, the fixed point $X_1 = X_2 = \cdots = X_K = F$ loses stability as the system undergoes bifurcations leading to a quasi-periodic behaviour for moderate values of F and chaotic behaviour for $F \geq 5.0$ [39]. This is only a rough description of the complexity of the bifurcations taking place in the L96 model as F is changed: as discussed in detail in [70–72], the properties of the system depend on N in a very nontrivial way in the regime of moderate forcing. In the regime of strong forcing and developed turbulence, instead, some sort of universality emerges, as the L96 model is strongly chaotic when $F \geq 8$ and its properties are extensive with respect the number of nodes N [65]. By and large, the mechanism of instability of the L96 model boils down to exchanges of energy between the symmetric state and the perturbations away from it, as particularly clear in the case of the linear instability analysis [39]. Nonetheless, the L96 model clearly features only one form of energy, which we may refer to as kinetic.

This paper

In this paper, we propose an extension of the L96 model whereby a second variable is attached to each gridpoint, representing, metaphorically, the local thermodynamical properties, which are advected by the dynamical variables of the L96 model, and undergo forcing and diffusion. The model proposed here features a meaningful definition of energy that includes the kinetic part already present in the L96 model plus a potential part associated to the fluctuations of the *temperature* in the domain. The fundamental advantage of the new model proposed here and analysed in detail in Sect. 2 is that it features an energy cycle that allows for conversion between the kinetic and potential forms of energy. Conceptually, this mirrors the change one has going from one a one-layer quasi-geostrophic model, where only barotropic processes are possible, to a two-layer quasi-geostrophic model, which, instead, features a coupling between dynamical and thermodynamic processes via baroclinic conversion [74]. Note that, the Lorenz '63 model [18] and, more completely so, its extensions to higher order modes [75] features a nontrivial energetics where exchanges take place between potential and kinetic energy [76]. The energy of the system defines the symplectic component that contributes—together with the metric one—to defining the evolution of the model [47].

The rest of the paper is structured as follows. Section 2 provides a thorough introduction to the one-level version of the model presented here. The evolution equations are presented together with the rationale on which the model is based. Additionally, a detailed analysis of its mechanical and thermodynamic properties is carried out. Finally, the linear stability analysis of the symmetric fixed point is also described. In Sect. 3, we present the results of a large set of numerical integrations of the model, aimed at exploring its properties in a rather vast range of values of two main parameters, which control the input of energy in the kinetic and potential form. We discuss the thermodynamics of the model in terms of mean values and the fluctuations of the main terms describing the energetics of the model. Such a physical characterisation of the model is complemented by the analysis of how the first Lyapunov exponent depends on the two considered parameters, in order to be able to separate the regions where the asymptotic dynamics of the system takes place in a regular vs in a strange attractor [67, 77]. We will discover a nontrivial interplay between the two sources of forcings applied to the model. We then perform a preliminary analysis for assessing to what extent the system obeys extensive chaos. In Sect. 4, we summarise the main features of the model and the results obtained so far, and propose future lines of investigations. As in the case of original L96 model, the model introduced here can be formulated in a two-level fashion—see Appendix A—with nontrivial couplings among different levels and variables and with a fairly sophisticated energetics, which is conceptually rather similar to the one described by the Lorenz energy cycle in the atmosphere. The analysis of the properties of the two-level model will not be performed in this paper and will be the subject of future studies.

MODEL FORMULATION AND PROPERTIES

We want to extend the standard L96 model presented in the introduction by adding a second set of variables for all the gridpoints k = 1,..., N. The goal is to construct a toy model able to describe in a very simple yet conceptually correct way the interaction between dynamical and thermodynamical processes of the atmosphere. The evolution equations of the model we propose in this contribution are the following:

$$\frac{dX_k}{dt} = X_{k-1}(X_{k+1} - X_{k-2}) - \alpha\theta_k - \gamma X_k + F, \tag{4}$$

$$\frac{d\theta_k}{dt} = X_{k+1}\theta_{k+2} - X_{k-1}\theta_{k-2} + \alpha X_k - \gamma\theta_k + G, \tag{5}$$

with k = 1,..., N. The variable θ_k can be loosely interpreted as temperature at the grid-point k. The boundary conditions are defined as

$$X_{k-N} = X_{k+N} = X_k,$$
$$\theta_{k-N} = \theta_{k+N} = \theta_k, \tag{6}$$

The variable θk undergoes a constant forcing G, a linear dissipation term, and a nonlinear term representing, loosely speaking, the advection performed by the X variables. Additionally, θk and Xk are linearly coupled through a term proportional to α. The purpose of this coupling is to represent, in a very simplified way, the effect of correlated thermal and dynamical fluctuations on the dynamics, which allow for an exchange between kinetic and potential energy associated with thermal fluctuations, as discussed below. The introduction of a term proportional to α is the only—yet important—modification in the dynamics of the X variables for this model as compared to the classical L96 model, see Eq. (1). In what follows, we consider F, G, $\alpha, \gamma \geq 0$.

The coupling between the X and the θ variables is constructed in such a way that in the unforced and inviscid limit (F = G = γ = 0) the total energy of the system

$$E = K + P = \sum_{k=1}^{N} \left(\frac{X_k^2}{2} + \frac{\theta_k^2}{2} \right)$$

given by the sum of its kinetic and potential components, is conserved:

$$\frac{dE}{dt} = \frac{dK}{dt} + \frac{dP}{dt} = \sum_{k=1}^{N} X_k \frac{dX_k}{dt} + \sum_{k=1}^{N} \theta_k \frac{d\theta_k}{dt} = 0, \tag{7}$$

whereas, in general, *K* and *P* are not separately conserved. The quadratic functional form of the potential energy is inspired by the fact that the available potential energy in the global circulation of the atmosphere is approximately proportional to the variance of the temperature fluctuations [1, 3, 78]. The dynamical role of the function *E* is explored in the next section.

Mechanics

By definition, the time derivative of any smooth observable Ψ ($X_1,\ldots, X_K$, $\theta_1,\ldots,\theta_k$) is obtained by applying the generator of the Koopman operator to Ψ, as follows:

$$\frac{d\Psi}{dt} = \mathscr{L}[\Psi].$$

We will now show that linear operator L can be written as the sum of a contribution coming from a symplectic (indeed, quasi-symplectic, for the reasons detailed at the end of this section) term and a contribution coming from a gradient term. Indeed, we can write:

$$\frac{d\Psi}{dt} = \mathscr{L}[\Psi] = \{\Psi, E\} + \langle\Psi, \Gamma\rangle \tag{8}$$

where {A, B} = − {B, A} is a suitably defined Poisson bracket for the functions A and B, whilst <A, B>=<B, A> gives the gradient contribution. The evolution Eqs. (4)–(5) are obtained by setting $\Psi = X_i$ and $\Psi = \theta_i$, i = 1,..., N, respectively

We have that $\langle A, B\rangle = (\partial_{X_i} A)(\partial_{X_i} B) + (\partial_{\theta_i} A)(\partial_{\theta_i} B)$, where we use the Einstein convention for the indices. The function Γ defining the gradient contribution to the dynamics is:

$$\Gamma = -\gamma E + F\sum_{k=1}^{N} X_k + G\sum_{k=1}^{N}\theta_k = -\gamma E + F\Xi + G\Theta, \tag{9}$$

where $\Xi = \sum_{k=1}^{N} X_k$ and $\Theta = \sum_{k=1}^{N}\theta_k$. it is clear that such a component describes the irreversible dynamics as it vanishes in the unforced, inviscid limit F = G = γ = 0.

We then discuss the symplectic term associated with the Poisson bracket. We have that

$$\begin{aligned}\{A, B\} = &(\partial_{X_i} A) J_{ij} (\partial_{X_j} B) + (\partial_{\theta_i} A) Y_{ij} (\partial_{\theta_j} B)\\ &+ (\partial_{X_i} A) L_{ij} (\partial_{\theta_j} B) + (\partial_{\theta_i} A) M_{ij} (\partial_{X_j} B)\end{aligned} \tag{10}$$

where

$$\begin{aligned}J_{ij} &= X_{i-1}\delta_{i+1,j} - X_{i-1}\delta_{i-2,j}\\ Y_{ij} &= X_{i+1}\delta_{i+2,j} - X_{i-1}\delta_{i-2,j}\\ L_{ij} &= -\alpha\delta_{i,j}\\ M_{ij} &= \alpha\delta_{i,j}\end{aligned} \tag{11}$$

It is clear that the energy E is the generator of time translations according to the symplectic contribution and the antisymmetry of the Poisson brackets enforces the corresponding conservation law already discussed in Eq. 7.

We remark that, in the inviscid and unforced limit the system is not Hamiltonian because the Poisson brackets do not fulfil the Jacobi identity {A,{B,C}}+{C,{A, B}}+{B,{C, A}} = 0.. Because of this and of the fact that {Γ,E}≠0, the system given in Eqs. (4)–(5) is not metriplectic, i.e. the standard generalisation of Hamiltonian system to the dissipative case. [79]. Note that, the dynamics of dissipative fluids is, instead, metriplectic [80], and so is the dynamics of the (extended) Lorenz '63 model, which is in fact derived from the Rayleigh-Bénard equations through systematic modal truncation [47]. The lack of an underlying Hamiltonian skeleton confirms the well-known fact that the L96 model cannot be easily related to any model of fluid flows.

Thermodynamics

Using Eqs. (8)–(9), we obtain the time evolution of the energy of the system:

$$\frac{\mathrm{d}E}{\mathrm{d}t} = \langle \Psi, \Gamma \rangle = -2\gamma E + F\Xi + G\Theta = \Gamma - \gamma E$$

which implies that, at steady state $2\gamma\bar{E} = F\bar{\Xi} + G\bar{\Theta}$, where $\bar{\Psi}$ is the long term average of the quantity Ψ.

We next analyse the separate budget of the kinetic and potential energy. By inserting K and P in Eq. (8), we obtain:

$$\frac{\mathrm{d}K}{\mathrm{d}t} = \sum_{k=1}^{N} X_k \frac{\mathrm{d}X_k}{\mathrm{d}t} = I_{\mathrm{K}} - D_{\mathrm{K}} + C_{\mathrm{P,K}} \quad I_{\mathrm{K}} = F\Xi, \quad D_{\mathrm{K}} = 2\gamma K,$$

$$C_{\mathrm{P,K}} = -\sum_{k=1}^{N} (\alpha X_k \theta_k), \tag{12}$$

$$\frac{\mathrm{d}P}{\mathrm{d}t} = \sum_{k=1}^{N} \theta_k \frac{\mathrm{d}\theta_k}{\mathrm{d}t} = I_{\mathrm{P}} - D_{\mathrm{P}} - C_{\mathrm{P,K}} \quad I_{\mathrm{P}} = G\Theta, \quad D_{\mathrm{P}} = 2\gamma P, \tag{13}$$

where I_K (I_P) is the rate of input of kinetic (potential) energy, D_K (D_P) is the dissipation rate of kinetic (potential) energy, and $C_{P,K}$ is the conversion rate from potential to kinetic energy. We remark that the input and dissipation of energy in either kinetic or potential form is due by the metric component of the dynamics. Instead, the conversion of energy between the potential and kinetic form is controlled by the Poisson brackets given in Eqs. (10)–(11). Nonetheless, the components J and Y of the Poisson brackets, which describe advection, do not give any net contribution. Equations (12)–(13) describe the energetics of the model presented in this work, which is represented by the diagram shown in Fig. 1. One can draw a parallel between the energetics of this model and the Lorenz energy cycle of the atmosphere, where, as well known, the input of energy comes almost entirely through the potential energy channel via baroclinic forcing associated with the differential heating of low versus high latitude regions [1, 3]. The two-level version of the model introduced in this paper features an energetics that is conceptually closer to the one of the true atmosphere because it is able to describe energy cascades across scales on top of energy conversion processes, see "Appendix A".

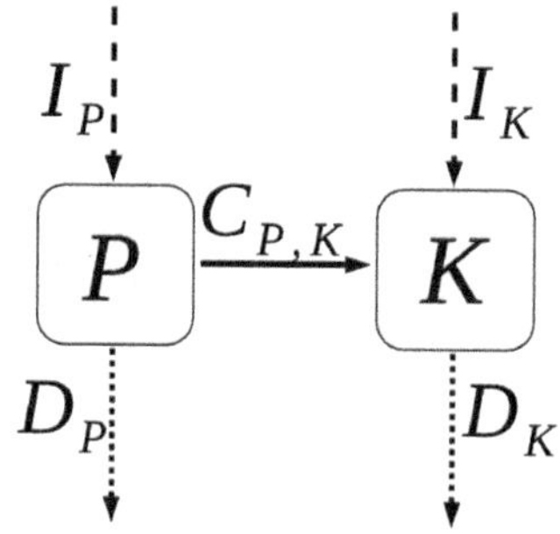

Figure 1: Energetics of the model presented in Eqs. (4)–(5). We indicate the fluxes of energy in and out of the reservoirs of kinetic (K) and potential (P) energy. Dashed lines represent input of energy; dotted lines represent energy dissipation; and solid lines represent energy conversion terms. The direction of the arrows indicates a positive energy flux. See text for details

At steady-state conditions, one has $2\gamma\bar{K} = F\bar{\Xi} + \bar{C}$ and $2\gamma\bar{P} = G\bar{\Theta} - \bar{C}$, which relate the size of the reservoirs of kinetic and potential energy to intensity of the acting forcings and energy exchange. We can also introduce a notion of efficiency of this model $\eta = \bar{C}/(F\bar{\Xi} + G\bar{\Theta}) = \bar{C}/(2\gamma\bar{E})$, which relates the amount of energy exchanged between the two reservoirs of energy to the total energy input. Since $X_k^2 + \theta_k^2 \geq 2|X_k\theta_k|\ \forall k$, we have that $2E \geq |C|/\alpha$. Therefore, $|\eta| \leq \alpha/\gamma$, which provides a constraint on the efficiency of the system. Note that, η is positive if, on the average, energy is converted from potential to kinetic, and negative otherwise.

Note that, $K \geq 0$ and $P \geq 0$ by definition. As a result, if $G,\gamma > 0$ and $F = 0$, one has $\bar{C} \geq 0$ (on the average we have an energy flux from potential to kinetic),1 whereas if $F,\gamma > 0$ and $G = 0$, $\bar{C} \leq 0$ (on the average, we have an energy flux from kinetic to potential). If $F = G = 0$, instead, we have that $\bar{K} = \bar{P} = \bar{E} = 0$, where the origin is a stable fixed point for the system.

Linear Stability Analysis

We investigate the linear stability of the system analysed here around the fixed point corresponding to the symmetric solution $X_j = X_k$ = const, 1,..., j, k,... N and $\theta_j = \theta_k$ = const, 1,..., j, k,... N . By plugging this ansatz in Eqs. (4)–(5), one gets:

$$X_k = \tilde{X} = \frac{\gamma F - \alpha G}{\gamma^2 + \alpha^2} \forall k, \tag{14}$$

$$\theta_k = \tilde{\theta} = \frac{\alpha F + \gamma G}{\gamma^2 + \alpha^2} \forall k. \tag{15}$$

Taking inspiration from [39],[Footnote2] we then investigate the linear stability of this solution by substituting $X_k = \tilde{X} + A\exp(\sigma t)\exp(\mathrm{i}\kappa k - \mathrm{i}\omega t)$ and $\theta_k = \tilde{\theta} + B\exp(\sigma t)\exp(\mathrm{i}\kappa k - \mathrm{i}\omega t)$ in Eqs. (4)–(5), where $\tilde{X}$ and $\tilde{\theta}$ have been defined in Eqs. (14) and (15), respectively; *A* and *B* are complex constants, σ is a real number defining the growth rate (if positive) of the amplitude of the wave, whilst κ is the wavenumber and ω is the angular frequency of the wave. Neglecting terms that are quadratic in the wave amplitude, one obtains:

$$(\sigma - \mathrm{i}\omega)A = \tilde{X}A(\exp(\mathrm{i}\kappa) - \exp(-2\mathrm{i}\kappa)) - \alpha B - \gamma A \tag{16}$$

$$(\sigma - \mathrm{i}\omega)B = 2\mathrm{i}\tilde{X}B\sin(2\kappa) + 2\mathrm{i}\tilde{\theta}A\sin(\kappa) + \alpha A - \gamma B, \tag{17}$$

We exclude the trivial solution A = B = 0 and, thanks to linearity, we set A = 1 (only the ratio b = B/A is indeed relevant). We separate real and imaginary part in the previous equations and obtain:

$$\sigma = \tilde{X}(\cos(\kappa) - \cos(2\kappa)) - \alpha\mathbf{Re}\{b\} - \gamma \tag{18}$$

$$\omega = -\tilde{X}(\sin(\kappa) + \sin(2\kappa)) + \alpha\mathbf{Im}\{b\} \tag{19}$$

$$\sigma\mathbf{Re}\{b\} + \omega\mathbf{Im}\{b\} = -2\tilde{X}\mathbf{Im}\{b\}\sin(2\kappa) + \alpha - \gamma\mathbf{Re}\{b\}, \tag{20}$$

$$\sigma\mathbf{Im}\{b\} - \omega\mathbf{Re}\{b\} = 2\tilde{X}\mathbf{Re}\{b\}\sin(2\kappa) + 2\tilde{\theta}\sin(\kappa) - \gamma\mathbf{Im}\{b\}, \tag{21}$$

where Re{x} and Im{x} are the real and imaginary part of the complex number x, respectively. The conditions leading to the bifurcation associated with the loss of stability of the fixed point given in Eqs. (14)–(15) can be derived by setting σ = 0 in Eqs. (18)–(21) and finding ω, κ, Re{b}, and Im{b} and a function of the parameters F, G, α, and γ . In the case ω ≠ 0, the onset of the neutral wave corresponds to a Hopf bifurcation

Solving the previous Eqs. (18)–(21) and finding the expression of σ and ω as a function of κ and of the parameters F, G, α, and γ gives the dispersion relation of the waves. Additionally, obtaining the real and imaginary part of b allows for understanding the relative amplitude of the waves in the X and θ variables.

Note that, the linear stability analysis of the L96 model can be obtained by setting $\alpha = 0$, $\gamma = 1$ in Eq. (16) and neglecting, instead, the θ variables. One then recovers the result first presented in [39] and discussed in greater detail in [70]. In what follows, we consider $F \geq 0$; an analysis of the dynamics occurring for $F < 0$ has been presented in [72]. It is possible to derive the minimal value of F such that the fixed point of the system loses stability and, correspondingly, to obtain the wavelength and frequency of the emerging neutral wave. One finds that, taking a continuum approximation ($N \rightarrow \infty$), the neutral wave is realised when $F = F_{crit} = 8/9$, where the critical wavenumber is $\kappa = \kappa_{crit} = arc_{cos}(1/4)$, and the critical frequency is $\omega_{crit} = -F_{crit}(\sin(\kappa_{crit}) + \sin(2\kappa_{crit})) \approx -1.29$. If one assumes that the gridpoints are arranged like along a latitudinal circle where the longitude increases with the index of the gridpoints (note that, the periodic boundary conditions of the system impose a toroidal topology), we have that the crest of the neutral wave moves westward, because the phase velocity $vp = \omega_{crit}/\kappa_{crit} =\approx -0.98$ is negative. Instead, the group velocity $vg = \partial\omega_{crit}/\partial\kappa|_{\kappa=\kappa crit} = -F_{crit}(\cos(\kappa_{crit}) + 2\cos(2\kappa_{crit})) \approx 1.33 > 0$ so that the wave packets have an eastward propagation.

As a result of the presence of the coupling between the X and θ variables, it is hard to find for the model introduced in this paper an explicit expression for the conditions supporting the presence of a neutral wave, also if one takes the special cases where one between F and G vanishes. A simple solution is instead found if one takes $\gamma = \alpha = 1$ and $F = G$, which implies $X^{\sim} = 0$ and $\theta^{\sim} = F$. One then obtains the following results when imposing $\sigma = 0$ and taking the continuum approximation: $Re\{b\}=-1$, $\omega = Im\{b\} = \sqrt{2}$, $\kappa = arc_{sin}(\sqrt{2}/F)$. This indicates that $F_{crit} = \sqrt{2}$, and $\kappa_{crit} = \pi/2$ (corresponding to a critical wavelength of 4), and $\omega_{crit} = \sqrt{2}$. Therefore, the phase velocity of the neutral wave $vp = \omega_{crit}/\kappa_{crit} = 2\sqrt{2}/\pi$ is positive, corresponding to an eastward motion of the wave crests. Since $\omega_{crit} = F_{crit}\sin(\kappa_{crit})$, we have $vg = \partial\omega_{crit}/\partial\kappa_{|\kappa=\kappa crit} = Fc_{rit}(\cos(\kappa_{crit})) = 0$, implying no net motion of the wave packets.

RESULTS

Many are the possible scientific questions one can address regarding the model introduced above. Building on the large literature on the L96 model discussed in the introduction, and taking into account the extra features of the current model, we can mention the following lines of investigation:

- Analysis of the bifurcations leading the system from fixed point to a periodic and quasi-periodic behaviour to a chaotic regime as the forcing is increased;
- Systematic investigation of the predictability of the system—e.g. analysis of the finite-time and asymptotic Lyapunov exponents and the corresponding covariant Lyapunov vectors as a function of the two forcing parameters F and G;
- Systematic investigation of the energetics of the system as a function of the two forcing parameters F and G;
- Analysis of the signal propagation through waves in the quasi-periodic and weakly chaotic regime;
- Definition of asymptotic scaling laws for the properties if the system for large values of F and G;
- Detection and analysis of chaos extensivity as the number of gridpoints $N\rightarrow\infty$;
- Extension of the model to multiple scales and analysis of dynamics and of the energetics of scale-to-scale interaction.

Obviously, it is impossible to address with a high level of detail all these aspects in the present paper.

Rather than focusing on one or few aspects among those above, since this is the first time this model is proposed to the scientific community, we will present some preliminary results that address partially each of the points above, in the hope of stimulating a reader into going in greater detail. Further studies by the authors that focus specifically in some of the aspects mentioned above will be reported elsewhere.

All the numerical integrations presented below are performed using a Dormand–Prince method with adaptive time step and a spin time of 100 time units, with runs of 1000 time units.

We make use of the Python module JiTCODE [81], an extension of SciPy's ODE that allows to numerically simulate ordinary differential equations, computing quantities of interest as Lyapunov exponents as well. All results have been double checked and confirmed using the MATLAB function ode45 where integrations are performed using the 4th order Runge–Kutta integrator with adaptive time step [82].

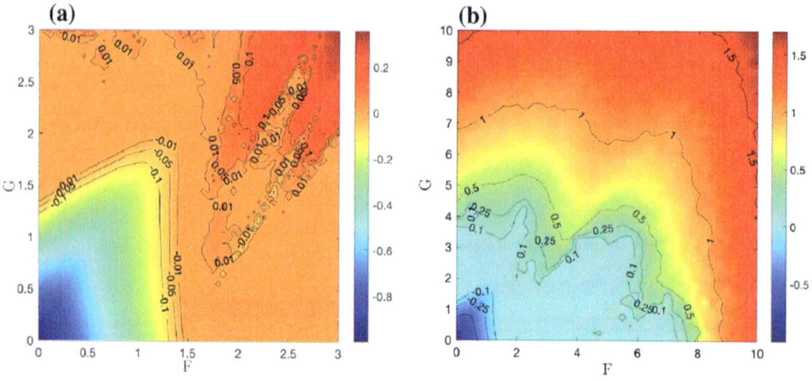

Figure 2: First Lyapunov exponent of the one-level model with N = 36 grid-points as a function of F and G. a Detail for $0 \leq F, G \leq 3$. b Full range for $0 \leq F, G \leq 10$. In all simulations, $\gamma = \alpha = 1$.

Transition to Chaos and Predictability

A simple yet fundamentally correct way to characterise at qualitative level the dynamical properties of a system is to investigate to what extent its evolution is sensitive to its initial conditions. Roughly speaking, the first Lyapunov exponent of a system measures the asymptotic rate of growth or decay of the distance between two orbits which are initialised in the attractor of the system at infinitesimal distance from each other [83]. Similarly, one can define the sum of the first p Lyapunov exponents as defining the asymptotic average rate of growth or decay of the p-volume defined by p + 1 orbits that are initialised in the attractor of the system at infinitesimal distance from each other [77]. Indeed, for a Q dimensional continuous time dynamical system, it is possible to compute Q Lyapunov exponents $\lambda_1,\ldots,\lambda_Q$, where the customary ordering is such that $\lambda_j \leq \lambda_k$ if $k \leq j$ of the [84].

If $\lambda_1 < 0$ the attractor is a fixed point, whilst if $\lambda_1 = 0$ the attractor is periodic or quasi-periodic. Finally, the presence of a positive first Lyapunov exponent is a significant evidence that the system is chaotic, and the value of such exponent determines quantitatively the rapidity with which two nearby trajectories diverge from each other. In this case, one has that there is at least one $\lambda_j = 0$, $j > 1$, which corresponds to the direction of the flow [77,83].

Figure 2 shows the estimate of λ_1 for N = 36 as a function of F and G in the range $0 \leq F, G \leq 10$. We remark that the Python JiTCODE module allows for the computation of the full spectrum of Lyapunov exponents using the algorithm proposed in [84]. The system has a negative λ1 for small values of the forcings, as expected, see Fig. 2a. We remark that if G = 0 $\lambda_1 \leq 0$ for $F \leq 1.4$, whereas for the L96 model $F_{crit} = 8/9$, indicating that presence of a mechanism of energy transfer between kinetic and potential energy and the presence of a new channel of dissipation (for potential energy) leads to higher stability for the system. We observe that λ1 depends in a very nontrivial way on both F and G, as the system's behaviour depends delicately on how the energy is injected into it, because the dynamics of the X and θ variables is, in fact, quite distinct. It is extremely different to force the system through kinetic or the potential energy channel. We also observe that the theoretical prediction of Fcrit = √2 for F = G agrees with what shown in Fig. 2a.

Many other interesting features appear. If we increase F from zero to 3 whilst keeping G = 1.7, the asymptotic dynamics of the system changes first from quasi-periodic to a fixed point, then again to quasi-periodic, which then alternates with chaotic behaviour. Indeed, one can observe two complex tongue-like structures in Fig. 2a for $F \geq 1.7$, $G \geq 0.5$, which indicate the presence of a very nontrivial set of bifurcations for that regions of the parameters' space, defining the transition between the quasi-periodic behaviour—the light orange region—ad the chaotic regime—the dark orange and red region in Fig. 2a.

Zooming out towards a larger range of values for *F* and *G* the intuitive argument that increasing either *F* or *G* makes the system less predictable becomes more robust, even though there are regions where a destructive interference is clear (in terms of reduced values of λ_1) between the two forms of forcing, compare the two troughs near the diagonal in Fig. 2b.

We remark that it is reasonable to expect that, as in the case of the L96 model [70,71,72], in the regime of moderate forcing the position and nature of the bifurcations will depend delicately on the number of nodes *N*, so that one should expect modifications especially in Fig. 2a when performing simulations for a value of *N* other than 36 considered here. Instead, as shown in Sect. 3.4, one finds some indication of universality associated with the continuum limit $N \rightarrow \infty$ when sufficiently strong forcing is considered.

Energetics

It is useful to investigate the long-term average of the terms in Eqs. (12)–(13) as a function of F and G, see Fig. 3. The lack of equivalence between applying forcing to the X vs to the θ variables is extremely clear by looking at the conversion term (panel e). C^{-} is positive for large values of G and moderate values of F, and negative vice versa. The absolute value of C^{-}

increases with F (G) if G (F) is kept constant. The zero isoline strongly deviates from the diagonal and indicates that if F = G there is a net transfer of energy from kinetic to potential. The zero isoline of C^{-} coincides with the ridge in the value of λ1 shown in Fig 2b, indicating that the condition of no net energy exchange between the two reservoirs of energy corresponds to a state where instabilities are rather strong. The zero-isoline of the efficiency η (see panel f), by definition, coincides with the one of C^{-}. The absolute value of the efficiency grows with the asymmetry of the forcing and peaks for moderate intensity of either F or G, suggesting—see Sect. 3.5—that the energy conversion becomes less efficient decreases when stronger forcings are considered.

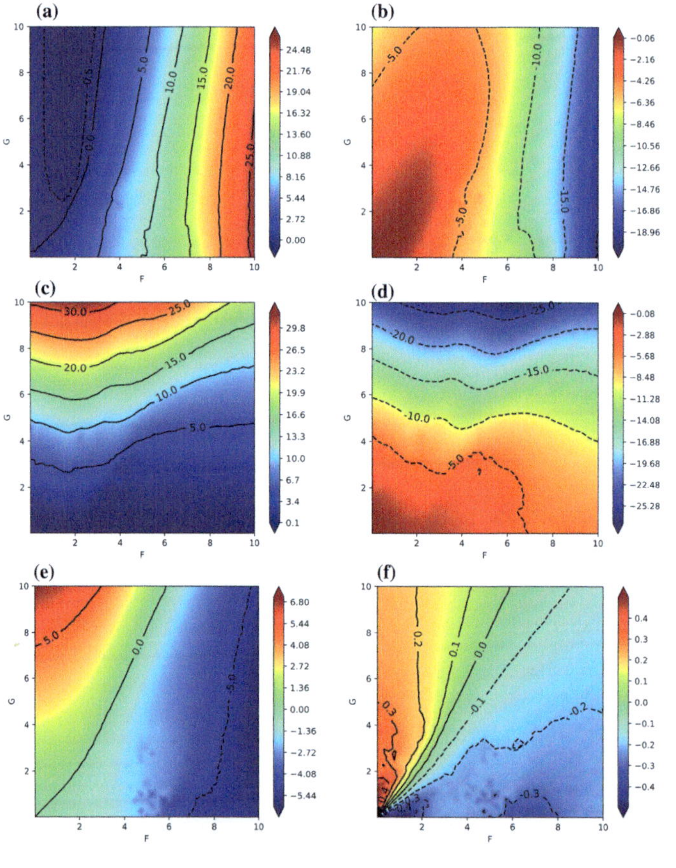

Figure 3: **a** $F\frac{1}{N}\bar{\Xi} = F\frac{1}{N}\sum_{k=1}^{N} X_k$; **b** $-2\gamma\frac{1}{N}\bar{K} = -\gamma\frac{1}{N}\sum_{k=1}^{N} X_k^2$; **c** $G\frac{1}{N}\bar{\Theta} = G\frac{1}{N}\sum_{k=1}^{N} \theta_k$; **d** $-2\frac{1}{N}\gamma\bar{P} = -\gamma\frac{1}{N}\sum_{k=1}^{N} \theta_k^2$; **e** $\frac{1}{N}\bar{C} = -\frac{1}{N}\sum_{k=1}^{N} (\alpha X_k \theta_k)$; **f** $\eta = \bar{C}/(2\gamma\bar{E})$ as a function of F and G. In all simulations $\gamma = \alpha = 1$.

The behaviour of the other thermodynamical quantities is somehow unsurprising, as we have that both input and dissipation of kinetic (potential) energy increase with F (G). We remark that, once again, the response of the system to the two individual forcings is quantitatively different. It should be noted that, when one considers $F \leq 2$, for $G \geq 2$ one has that the net input of kinetic energy is negative (with the dissipation of kinetic energy, being, by definition, positive). This indicates a very nontrivial impact of the thermodynamic variables on the dynamical ones, which are the only ones performing advection. As a result, there is an additional mechanism of energy loss for the system, whilst all the energy input takes place through the potential energy channel. Instead, when considering low values of G, the potential energy input is always positive—yet small.

Waves Amidst Chaos

We highlight some qualitative features of the dynamics of the model that indicate the presence of wave-like structures amidst chaos in the regime of moderate forcing. Figure 4 shows some examples of evolution of the system of the system in the case of N = 36 sectors with F = 10, G = 0 (panel a); F = 10, G = 10 (panel b); and F = 0, G = 10 (panel c). We are using a Hovmöller-type diagramme [85], where time is on the vertical axis and the variables X_k and θ_k, $k = 1, \ldots, 36$ are on the horizontal axis. This diagramme is particularly well-suited for appreciating wave-like structures, as it is easy to visualise wave crests. If the forcing on the θ variables is switched off, the X variables behave similarly to the case of the L96 model, where, amidst chaos, the clear signature of a westward propagating phase velocity can be found, as already observed in [39] and recently discussed in [73]. As can be guessed from the evolution equations, the θ variables feature weaker variability and similar pattern of the wave crests, as they are advected by the X variables and receive energy from them. The situations is qualitatively similar when both the X and θ variables are forced, but, quite naturally, the fluctuations of the θ variables are stronger than in the previous case. Note that in the case analysed here of F = G = 10, the wave crests travel in the opposite direction with respect to what we have found for the neutral wave emerging for $F = G = \sqrt{2}$, see Sect. 2.3. Therefore, the presence of a turbulent background radically changes the kinematics of the waves. If, instead, the forcing acts on the θ variables only, the wave crests have a much less clear direction of propagation, both for the θ and for the X variables, where the latter feature a much lower variability, as expected. In other terms, the setup where F vanishes is characterised by absolute instability, with little or no advection of anomalies, whereas the other two cases are characterised by convective instability, where anomalies are spatially advected [86].

We will further discuss in the following sections in more quantitative terms the differences emerging when forcing the X variables only, the θ variables only, or all variables.

Chaos Extensivity

Ruelle [87] proposed that systems with short-range interaction can feature extensive chaos, because large domains can be hierarchically partitioned into smaller, weakly interacting subdomains with similar properties. One way to test whether chaos extensivity is to analyse the finite-size scaling of the Lyapunov exponents. Specifically, one plots the obtained spectrum of Lyapunov exponents for different values of system size Q (in our case, Q = 2N) against the rescaled index x = (j − 1/2)/Q and tests whether a universal curve is obtained in the limit of large values of Q [65,88]. We remark that chaos extensivity implies that the ratio between the Kaplan-Yorke dimension of the attractor [89], also referred to as Lyapunov dimension [67], and Q tends to a constant as Q → ∞.

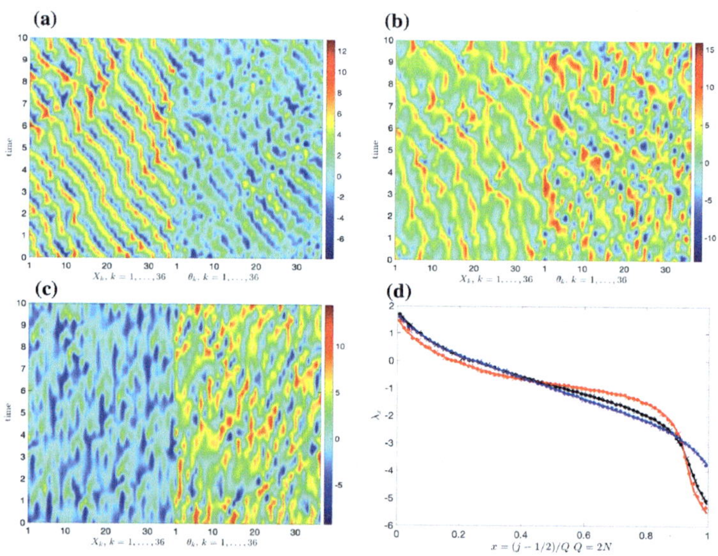

Figure 4: Slice of 10 time units of the evolution of the system with N=36 for F =10, G=0 (a); F=10, G=10 (b); and F=0, G=10 (c). Time is on the vertical axis. On the horizontal axis, the first 36 variables are the X_k, k=1,...,36,

followed by the variables θ_k, k=1,…,36. d Extensivity of the system—spectrum of Lyapunov exponents as a function of the rescaled index $x=(j-1/2)/(2N)$ for F=10, G=0 (red); F=10, G=10 (black); F=0, G=10 (blue). Solid lines: N=72; dots: N=36; crosses: N=18. In all simulations, $\alpha=\gamma=1$

In order to prove convincingly the extensive nature of chaos in the system analysed here, one should test such property for all values of F and G. This is beyond the scope of this paper. Yet, preliminary results do confirm extensivity for the three reference cases F = 10, G = 0; F = G = 10; F = 0, G = 10 shown in Panels (a)–(c) of Fig. 4. Indeed, we have here performed simulations with N = 18, N = 36, and N = 72 and, as shown in Fig. 4d, the Lyapunov exponents spectra seem to collapse to universal curves as N grows. Indeed, the Lyapunov spectra plotted against their respective rescaled indices can hardly be visually distinguished. This is especially encouraging in view of the clear evidence for chaos extensivity in the L96 model [60,65].

Scaling Laws for Strong Forcings

As thoroughly analysed in [65], in the one-layer L96 model the average energy per unit site scales to a very high degree of approximation as $F^{1.33}$ for large values of *F*. The origin of such a scaling law is still unknown. We report some preliminary results of scaling laws obtained for the current model in some special configurations of parameters. We have performed long integrations (1000 time units) at steady state for three set of experiments:

1. $F = 2^j$, $j = 3, \ldots, 14$; $G = 0$
2. $F = G = 2^j$, $j = 3, \ldots, 14$;
3. $G = 2^j$, $j = 3, \ldots, 14$; $F = 0$

which correspond to applying a forcing of increasing strength on the X variables only, on both the X and the θ variables, or on the θ variables only, respectively.

These are regimes of forcing where, see the case of the L96 model [65], one might expect that chaos extensivity applies with a very good approximation, see Sect. 3.4. We obtain the following approximate asymptotic scaling laws, which are rather accurate when F and/or G are larger than 256:

1. $\bar{K}, \bar{E} \propto F^{1.33}$, $\bar{\Xi} \propto F^{0.33}$, $\bar{\Theta} \propto F^{-0.28}$, $\bar{P} = -\bar{C}/2 \propto F^{0.71}$, $\lambda_1 \propto F^{0.66}$;
2. $\bar{K}, \bar{E}, \bar{P} \propto F^{1.33}$, $\bar{P} \approx 0.7\bar{K}$, $\bar{C} < 0$, $|\bar{C}| \propto F^{0.70}$, $\bar{\Xi}, \bar{\Theta} \propto F^{0.33}$, $\bar{\Theta} \approx 0.7\bar{\Xi}$, $\lambda_1 \propto F^{0.66}$;
3. $\bar{P}, \bar{E} \propto G^{1.50}$, $\bar{K} = \bar{C}/2 \propto G^{1.00}$, $\bar{\Theta} \propto G^{0.50}$, $\bar{\Xi} \propto G^{0.00}$, $\lambda_1 \propto G^{0.50}$;

where the uncertainty is 0.01 for all the numbers above. As clear from these scaling relations, and in agreement with what one could intuitively guess by looking at Fig. 4, it is rather different to force the system through the X or the θ variables, and the interplay between the two reservoirs of energy is far from trivial. If the forcing is applied to only one set of variables, the energy cycle is more enhanced, ceteris paribus, when the θ variables undergo the forcing. Indeed, the reservoir of total energy and the conversion C^{-} of energy between the two forms are larger than for corresponding case of forcing applied uniquely to the X variables. The behaviour of the quantities $\bar{\Xi}$ and $\bar{\Theta}$ is also extremely different in the two cases, implying a qualitatively different way the forcing impacts the spatially coherent fluctuations of the variables. If F = G, the ratio of the average size of the two reservoirs of energy is a constant, with K^{-} being larger that P^{-} (and, correspondingly, $\bar{\Xi}$ being larger than $\bar{\Theta}$), and the average flux of energy goes from kinetic to potential. We remind that the dynamics of the F = G case is characterised by convective instability, similarly to the case where G = 0, compare Fig. 3a and b. The reason why the forcing through the kinetic channel dominates in the special case of F = G is still unclear and should be further investigated.

In all cases, the amount of energy that is converted between the two forms becomes a negligible fraction of the total incoming energy in the limit of large forcing. In other terms, the efficiency of the model $\eta = \bar{C}/(2\gamma\bar{E})$ tends to zero if either *F* or *G* tend to infinity, even if if C^{-} tends to infinity. It is then unsurprising that when considering the limit of large F, regardless of whether G is also increased, we obtain for the X variables results that are in agreement with what featured by the one-layer L96, compare with [65]. At this regard, a useful piece of information is obtained by looking at the properties of λ1 in the large forcing limit. One obtains that in scenarios 1 and 2, $\lambda_1 \propto F^{0.66}$, which is again in excellent agreement with what obtained for the L96 model (including the pre-exponential factor). Scenarios 1) and 2) seem like featuring a rather similar dynamics, the main difference between the two being the strength of the fluctuations of the θ variables; compare with panels a) and b) of Fig. 4.

The growth of λ_1 with G is slower in the case F is set to zero, where we have absolute instability. We can gain a qualitative understanding of the different impact on λ1 of changes in the value of G vs F by comparing panels a), b) and c) of Fig. 4, which nonetheless describe weaker regimes of forcing (what follows stands also in the case of stronger applied forcing).

CONCLUSIONS

Simple and conceptual models have proved extremely useful for better understanding the dynamics of climate as a whole as well as of its individual components. Indeed, their usefulness spans from being the testbeds for developing new methods in terms of data analysis, data assimulation, and model testing; to supporting the definition of new metrics for testing more complex models; to providing valuable insights in the basic active physical mechanisms and most prominent mathematical features.

The model presented in this paper goes in this direction and has been constructed in order to provide a new layer of physical complexity to the L96 model by adding a new variable to each gridpoint of the model. This variable can be loosely interpreted as a local temperature and allows for the establishment of a complex energetics for the system, encompassing energy input, output, and conversion. Two forms of energy are present in the system, a kinetic one and potential one. We are also able to introduce a notion of efficiency, which is useful for studying the conversion of energy from one form to the other one. The energetics of the model is reminiscent of the one of the real atmosphere. Extending previous analyses, we have provided a fairly complete analysis of the mechanics of the new model by separating a quasi-symplectic and a metric component to its dynamical structure. The energy of the system is used to construct the antisymmetric evolution operator, whose corresponding brackets are not true Poisson brackets because they do not obey the Jacobi identity; hence, the symplectic structure is not complete.

We have then performed a preliminary analysis of some of the key aspects of the new model by investigating how its properties change as a function of the two parameters that control the input of kinetic and potential energy. We have studied, in a special case, the Hopf bifurcation leading to the onset of the neutral wave from the fixed point solution. The interplay between the two forcings is extremely nontrivial in the weak forcing regime, where much needs to be explored regarding the transition from fixed point to quasi-periodic to chaotic asymptotic states, and one expects that the structure and position of the bifurcations might depend delicately on the number of modes included in the system, similarly to the case of the L96 model. When considering regimes associated with stronger forcing, the system exhibits extensive chaos, even if there is clear evidence of wave-like structures emerging in the context of an overall strongly chaotic flow. Understanding the interplay between ordered wave-like structures and turbulence seems of great interest. The system reacts differently depending on how we force it. The nature of the flow is impacted because absolute vs convective instability dominate if we force the system through the potential energy vs kinetic energy channel, respectively. The mechanism of energy conversion makes sure that also the variables that are not directly forced feature nontrivial variability. If the strength of the forcing is the same in the two channels, the kinetic energy channel ends up being more efficient: the dynamics is characterised by convective instability, and, on the average, energy is transferred from the kinetic to the potential form. The reason for this behaviour is still unclear. Similarly to the case of the L96 model, it is possible to obtain accurate power laws describing how some of the fundamental dynamical and thermodynamical properties of the system scale with the forcing parameters in the limit of very strong forcings.

The analysis presented here is only a first step in the direction of better understanding the properties of this model, which we believe has the potential of being of great interest for investigations in areas like statistical physics, nonlinear dynamics, data assimilation, mechanics, model reduction techniques, and extreme events.

Finally, again along the lines of the L96 model, we have introduced a two-level version—see Appendix A—of the model, which allows for studying multi-scale dynamics and which features an energetics that resembles, conceptually, the one of the atmosphere, where the Lorenz energy cycle describes succinctly the input and output of energy in the kinetic and potential form as well as the conversion between the two forms and between energy compartments at small vs large scales. The study of the properties of this model, which is *a fortiori* extremely promising in the fields above, will be carried out in a future work.

Notes

1. Note that, this is, to a very good approximation, what applies to the climate system as a whole, because the geophysical fluids do not receive any input of mechanical energy, apart from the very small lunar and solar tidal forcing.
2. See also the rather sophisticated analysis of the stability of the L96 model presented in [70,71,72].

TWO-LEVEL NEW MODEL

Here, we propose an extension of the model able to represent multiscale dynamics and energy exchanges across scales. A detailed investigation of the properties of this model will be discussed elsewhere. We present the evolution equation and discuss the energetics of the model. Mimicking the structure of the classical two-layer L96 model, we define the following evolution equations:

$$\frac{dX_k}{dt} = X_{k-1}(X_{k+1} - X_{k-2}) - \alpha\theta_k - \gamma X_k + F - \frac{hc}{b}\sum_{j=1}^{J} Y_{j,k}, \tag{22}$$

$$\frac{dY_{j,k}}{dt} = cbY_{j+1,k}(Y_{j-1,k} - Y_{j+2,k}) - \alpha c\phi_{j,k} - \gamma cY_{j,k} + f + \frac{hc}{b}X_k, \tag{23}$$

$$\frac{d\theta_k}{dt} = X_{k+1}\theta_{k+2} - X_{k-1}\theta_{k-2} + \alpha X_k - \gamma\theta_k + G - \frac{hc}{b}\sum_{j=1}^{J}\phi_{j,k}, \tag{24}$$

$$\frac{d\phi_{j,k}}{dt} = cb(Y_{j-1,k}\phi_{j-2,k} - Y_{j+1,k}\phi_{j+2,k}) + \alpha cY_{j,k} - \gamma c\phi_{j,k} + g + \frac{hc}{b}\theta_k, \tag{25}$$

with k = 1,..., N; j = 1,..., J , where $Y_{j,k}$'s are j small-scale variables coupled with X_k , similarly to the two-level L96 model, and $\varphi_{j,k}$'s are j small-scale variables similarly coupled with θ_k . Additionally, the variables $Y_{j,k}$ and $\varphi_{j,k}$ are also mutually coupled. The constant hc/b determines the strength of the coupling between variables at different scale, whilst c defines the time scale separation between the two levels and b controls the relative amplitude of the fluctuations between large and small scales. Finally, f (g) defines the forcing on the variables $Y_{j,k}$ ($\varphi_{j,k}$). The boundary conditions are the following:

$$X_{k-N} = X_{k+N} = X_k,\quad Y_{j,k-N} = Y_{j,k+N} = Y_{j,k},\quad Y_{j-J,k} = Y_{j,k-1},\quad Y_{j+J,k} = Y_{j,k+1},$$
$$\theta_{k-N} = \theta_{k+N} = \theta_k,\quad \phi_{j,k-N} = \phi_{j,k+N} = \phi_{j,k},\quad \phi_{j-J,k} = \phi_{j,k-1},\quad \phi_{j+J,k} = \phi_{j,k+1}. \tag{26}$$

Note that choosing α = 0 and neglecting the θ and φ variables, we obtain the classic two-layer L96 model.

Below we show the equations for the fluxes of energy for X and θ alongside with those for Y and φ, as we have done in Sect. 2 for the one-level model.

We define $K_L = \sum_{k=1}^{N} X_k^2/2$ and $K_S = \sum_{j=1}^{J}\sum_{k=1}^{N} Y_{j,k}^2/2$ as the kinetic energy at large and small scales, respectively. Similarly, we define $P_L = \sum_{k=1}^{N}\theta_k^2/2$, , and $\sum_{j=1}^{J}\sum_{k=1}^{N}\phi_{j,k}^2/2$, as the potential energy at large and small scales, respectively. One derives the following equations for the various reservoirs of energy:

$$\frac{\mathrm{d}}{\mathrm{d}t}K_\mathrm{L} = I_{K_\mathrm{L}} - C_{K_\mathrm{L},K_\mathrm{S}} + C_{P_\mathrm{L},K_\mathrm{L}} - D_{K_\mathrm{L}} \tag{27}$$

$$\frac{\mathrm{d}}{\mathrm{d}t}K_\mathrm{S} = I_{K_\mathrm{S}} + C_{K_\mathrm{L},K_\mathrm{S}} + C_{P_\mathrm{S},K_\mathrm{S}} - D_{K_\mathrm{S}} \tag{28}$$

$$\frac{\mathrm{d}}{\mathrm{d}t}P_\mathrm{L} = I_{P_\mathrm{L}} - C_{P_\mathrm{L},P_\mathrm{S}} - C_{P_\mathrm{L},K_\mathrm{L}} - D_{P_\mathrm{L}} \tag{29}$$

$$\frac{\mathrm{d}}{\mathrm{d}t}P_\mathrm{S} = I_{P_\mathrm{S}} + C_{P_\mathrm{L},P_\mathrm{S}} - C_{P_\mathrm{S},K_\mathrm{S}} - D_{P_\mathrm{S}} \tag{30}$$

where

$$\begin{aligned}
&I_{K_\mathrm{L}} = F\sum_{k=1}^{N} X_k \quad I_{K_\mathrm{S}} = f\sum_{j=1}^{J}\sum_{k=1}^{N} Y_{j,k} \quad D_{K_\mathrm{L}} = 2\gamma K_\mathrm{L} \quad D_{K_\mathrm{S}} = 2\gamma c K_\mathrm{S} \\
&I_{P_\mathrm{L}} = G\sum_{k=1}^{N} \theta_k \quad I_{P_\mathrm{S}} = g\sum_{j=1}^{J}\sum_{k=1}^{N} \phi_{j,k} \quad D_{P_\mathrm{L}} = 2\gamma P_\mathrm{L} \quad D_{P_\mathrm{S}} = 2\gamma c P_\mathrm{S} \\
&C_{K_\mathrm{L},K_\mathrm{S}} = \frac{hc}{b}\sum_{k=1}^{N}\sum_{j=1}^{J} X_k Y_{j,k} \quad C_{P_\mathrm{L},P_\mathrm{S}} = \frac{hc}{b}\sum_{k=1}^{N}\sum_{j=1}^{J} \theta_k \phi_{j,k} \\
&C_{P_\mathrm{L},K_\mathrm{L}} = -\alpha\sum_{k=1}^{N} X_k\theta_k \quad C_{P_\mathrm{S},K_\mathrm{S}} = -\alpha c\sum_{k=1}^{N}\sum_{j=1}^{J} Y_{j,k}\phi_{j,k}
\end{aligned} \tag{31}$$

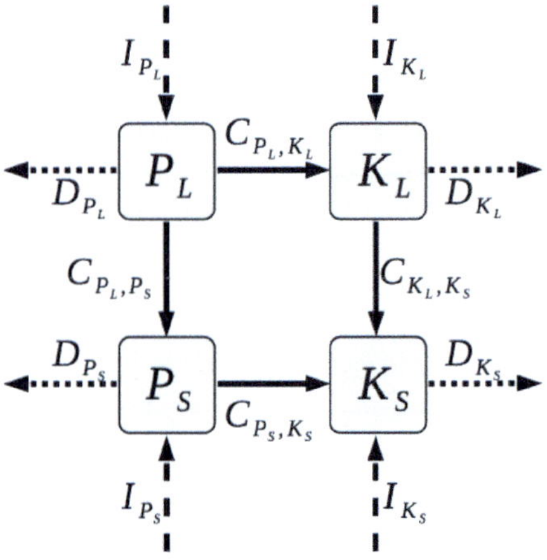

Figure 5: Diagram of the Lorenz energy cycle of the two-level model. We indicate the fluxes of energy between large (L) and small (S) scales and between kinetic (K) and potential (P) energy. Dashed lines represent input of energy; dotted lines represent energy dissipation; and solid lines represent energy conversion terms.

The energy cycle of this model is depicted in Fig. 5 and is closely reminiscent of the Lorenz energy cycle of the atmosphere. Note that, if we add Eqs. (27) with (28), on the one hand, and Eqs. (29) with (30), on the other hand, we derive the energy budget for the total kinetic and total potential energy, respectively. Instead, if we add Eqs. (27) with (29), on the one hand, and Eqs. (28) with (30), on the other hand, we derive the budgets for the total energy at large and small scale, respectively.

ACKNOWLEDGEMENTS

GV wishes to thank C. Franzke for several useful discussions. VL wishes to thank R. Blender, K. Fraedrich, G. Gallavotti, and A. Speranza for the many inspiring conversations on the topics covered in this paper. VL acknowledges the support provided by the EU Horizon 2020 project TiPES (Grant No. 820970) and by the EPSRC project "Applied Nonautonomous Dynamical Systems: Theory, Methods and Examples" (Grant No. EP/T018178/1) The two authors have equally contributed to this paper.

REFERENCES

1. J.P. Peixoto, A.H. Oort, Physics of Climate (American Institute of Physics, New York, 1992)
2. V. Lucarini, R. Blender, C. Herbert, F. Ragone, S. Pascale, Mathematical and physical ideas for climate science. Rev. Geophys. 52, 809–859 (2014)
3. E.N. Lorenz, The Nature and Theory of the General Circulation of the Atmosphere, vol. 218 (World Meteorological Organization, Geneva, 1967)
4. O. Pauluis, Sources and sinks of available potential energy in a moist atmosphere. J. Atmos. Sci. 64, 2627–2641 (2007)
5. V. Lucarini, Thermodynamic efficiency and entropy production in the climate system. Phys. Rev. E 80, 021118 (2009)
6. M.H.P. Ambaum, Thermal Physics of the Atmosphere (Wiley, New York, 2010)
7. V. Lucarini, K. Fraedrich, F. Lunkeit, Thermodynamics of climate change: generalized sensitivities. Atmos. Chem. Phys. 10(20), 9729–9737 (2010)
8. F. Laliberté, J. Zika, L. Mudryk, P.J. Kushner, J. Kjellsson, K. Döös, Constrained work output of the moist atmospheric heat engine in a warming climate. Science 347(6221), 540–543 (2015)
9. V. Lembo, F. Lunkeit, V. Lucarini, Thediato (v1.0)—a new diagnostic tool for water, energy and entropy budgets in climate models. Geosci. Model Dev. 12(8), 3805–3834 (2019)
10. M. Ghil, A mathematical theory of climate sensitivity or, How to deal with both anthropogenic forcing and natural variability?, in Climate Change: Multidecadal and Beyond, vol. 6, ed. by C.-P. Chang, M. Ghil, M. Latif, J.M. Wallace (World Scientific Publishing Co., Singapore, 2015), pp. 31–52
11. M. Ghil, V. Lucarini, The physics of climate variability and climate change. Rev. Mod. Phys. 92, 035002 (2020)
12. S.H. Schneider, R.E. Dickinson, Climate modeling. Rev. Geophys. 12(3), 447–493 (1974)
13. I.M. Held, The gap between simulation and understanding in climate modeling. Bull. Am. Meteorol. Soc. 86, 1609–1614 (2005)
14. V. Lucarini, Modeling complexity: the case of climate science, in

Models, Simulations, and the Reduction of Complexity, ed. by U. Gohde, S. Hartmann, J.H. Wolf (De Gruyter, Berlin, 2013), pp. 229–254

15. T.N. Palmer, P.D. Williams, Introduction. Stochastic physics and climate modelling. Philos. Trans. Ser. A Math. Phys. Eng. Sci. 366(1875), 2421–2427 (2008)
16. C.L.E. Franzke, T.J. O'Kane, J. Berner, P.D. Williams, V. Lucarini, Stochastic climate theory and modeling. Wiley Interdiscip. Rev. Clim. Change 6(1), 63–78 (2015)
17. J. Berner, U. Achatz, L. Batté, L. Bengtsson, A. De La Cámara, H.M. Christensen, M. Colangeli, D.R.B. Coleman, D. Crommelin, S.I. Dolaptchiev, C.L.E. Franzke, P. Friederichs, P. Imkeller, H. Järvinen, S. Juricke, V. Kitsios, F. Lott, V. Lucarini, S. Mahajan, T.N. Palmer, C. Penland, M. Sakradzija, J.-S. Von Storch, A. Weisheimer, M. Weniger, P.D. Williams, J.-I. Yano, Stochastic parameterization: towards anew view of weather and climate models. Bull. Am. Meteorol. Soc. 98(3), 565–588 (2017)
18. E.N. Lorenz, Deterministic nonperiodic flow. J. Atmos. Sci. 20, 130–141 (1963)
19. B. Saltzman, Finite amplitude free convection as an initial value problem—I. J. Atmos. Sci. 19, 329–341 (1962)
20. H. Stommel, Thermohaline convection with two stable regimes of flow. Tellus 2, 244–230 (1961)
21. G. Veronis, An analysis of the wind-driven ocean circulation with a limited number of Fourier components. J. Atmos. Sci. 20, 577–593 (1963)
22. C. Rooth, Hydrology and ocean circulation. Prog. Oceanogr. 11, 131–149 (1982)
23. J.G. Charney, J.G. DeVore, Multiple flow equilibria in the atmosphere and blocking. J. Atmos. Sci. 36, 1205–1216 (1979)
24. E.N. Lorenz, Irregularity: a fundamental property of the atmosphere. Tellus A Dyn. Meteorol. Oceanogr. 36, 98–110 (1984)
25. E.N. Lorenz, Predictability—a problem partly solved, in Predictability of Weather and Climate, ed. by T. Palmer, R. Hagedorn (Cambridge University Press, Cambridge, 1996), pp. 40–58
26. M.I. Budyko, The effect of solar radiation variations on the climate of the earth. Tellus 21, 611–619 (1969)

27. W.D. Sellers, A global climatic model based on the energy balance of the earth atmosphere. J. Appl. Meteorol. 8, 392–400 (1969)
28. M. Ghil, Climate stability for a Sellers-type model. J. Atmos. Sci. 33, 3–20 (1976)
29. K. Fraedrich, Catastrophes and resilience of a zero-dimensional climate system with ice-albedo and greenhouse feedback. Q. J. R. Meteorol. Soc. 105(443), 147–167 (1979)
30. E.B. Gledzer, System of hydrodynamic type admitting two quadratic integrals of motion. Dokl. Akad. Nauk SSSR 209, 1046–1048 (1973)
31. L. Biferale, A. Lambert, R. Lima, G. Paladin, Transition to chaos in a shell model of turbulence. Phys. D 80, 105–119 (1995)
32. V.S. L'vov, E. Podivilov, A. Pomyalov, I. Procaccia, D. Vandembroucq, Improved shell model of turbulence. Phys. Rev. E 58, 1811 (1998)
33. L. Biferale, Shell models of energy cascade in turbulence. Annu. Rev. Fluid Mech. 35, 441–468 (2003)
34. A. Brandenburg, Energy spectra in a model for convective turbulence. Phys. Rev. Lett. 69, 605–608 (1992)
35. J. Mingshun, L. Shida, Scaling behavior of velocity and temperature in a shell model for thermal convective turbulence. Phys. Rev. E 56, 441–446 (1997)
36. K. Hasselmann, Stochastic climate models, part I: theory. Tellus 28(6), 473–485 (1976)
37. B. Saltzman, Dynamical Paleoclimatology: Generalized Theory of Global Climate Change (Academic Press, New York, 2001)
38. P. Imkeller, J.S. von Storch, Stochastic Climate Models (Birkhauser, Basel, 2001)
39. E.N. Lorenz, Designing chaotic models. J. Atmos. Sci. 62(5), 1574–1587 (2005)
40. D. Orrell, Model error and predictability over different timescales in the Lorenz '96 systems. J. Atmos. Sci. 60(17), 2219–2228 (2003)
41. D.S. Wilks, Effects of stochastic parametrizations in the Lorenz '96 system. Q. J. R. Meteorol. Soc. 131(606), 389–407 (2005)
42. R. Abramov, A simple stochastic parameterization for reduced models of multiscale dynamics. Fluids 1(1), 2 (2016)

43. H.M. Arnold, I.M. Moroz, T.N. Palmer, Stochastic parametrizations and model uncertainty in the Lorenz system. Philos. Trans. R. Soc. A Math. Phys. Eng. Sci. 371(1991), 20110479 (2013)
44. G. Vissio, V. Lucarini, A proof of concept for scale-adaptive parametrizations: the case of the Lorenz '96 model. Q. J. R. Meteorol. Soc. 144, 63–75 (2018)
45. G. Vissio, Statistical mechanical methods for parametrization in geophysical fluid dynamics, in Reports on Earth System Science, Vol. 212 (2018) 46. A. Chattopadhyay, P. Hassanzadeh, D. Subramanian, Data-driven predictions of a multiscale Lorenz 96 chaotic system using machine-learning methods: reservoir computing, artificial neural network, and long short-term memory network. Nonlinear Process. Geophys. 27(3), 373–389 (2020)
46. R. Blender, V. Lucarini, Nambu representation of an extended Lorenz model with viscous heating. Phys. D 243(1), 86–91 (2013)
47. T. Bódai, Extreme Value Analysis in Dynamical Systems: Two Case Studies (Cambridge University Press, Cambridge, 2017), pp. 392–429
48. A.E. Sterk, D.L. van Kekem, Predictability of extreme waves in the Lorenz-96 model near intermittency and quasi-periodicity. Complexity 2017, 9419024 (2017)
49. G. Hu, T. Bódai, V. Lucarini, Effects of stochastic parametrization on extreme value statistics. Chaos Interdiscip. J. Nonlinear Sci. 29(8), 083102 (2019)
50. A. Trevisan, F. Uboldi, Assimilation of standard and targeted observations within the unstable subspace of the observation—analysis—forecast cycle system. J. Atmos. Sci. 61, 103–113 (2004)
51. A. Trevisan, D. Isidoro, O. Talagrand, Four-dimensional variational assimilation in the unstable subspace and the optimal subspace dimension. Q. J. R. Meteorol. Soc. 136, 487–496 (2010)
52. G. Hu, C. Franzke, Data assimilation in a multi-scale model. Math. Clim. Weather Forecast. 3, 118–139 (2017)
53. J. Brajard, A. Carrassi, M. Bocquet, L. Bertino, Combining data assimilation and machine learning to emulate a dynamical model from sparse and noisy observations: a case study with the Lorenz 96 model.
54. J. Comput. Sci. 44, 101171 (2020)
55. D.S. Wilks, Comparison of ensemble-MOS methods in the Lorenz '96 setting. Meteorol. Appl. 13(3), 243–256 (2006)

56. W. Duan, Z. Huo, An approach to generating mutually independent initial perturbations for ensemble forecasts: orthogonal conditional nonlinear optimal perturbations. J. Atmos. Sci. 73(3), 997–1014 (2016)
57. N. Le Carrer, P.L. Green, A possibilistic interpretation of ensemble forecasts: experiments on the imperfect Lorenz 96 system. Adv. Sci. Res. 17, 39–45 (2020)
58. D. Pazó, I.G. Szendro, J.M. López, M.A. Rodríguez, Structure of characteristic Lyapunov vectors in spatiotemporal chaos. Phys. Rev. E 78, 016209 (2008)
59. S. Hallerberg, D. Pazó, J.M. López, M.A. Rodríguez, Logarithmic bred vectors in spatiotemporal chaos: structure and growth. Phys. Rev. E Stat. Nonlinear Soft Matter Phys. 81(6), 1–8 (2010)
60. A. Karimi, M.R. Paul, Extensive chaos in the Lorenz-96 model. Chaos Interdiscip. J. Nonlinear Sci. 20, 043105 (2010)
61. M. Carlu, F. Ginelli, V. Lucarini, A. Politi, Lyapunov analysis of multiscale dynamics: the slow bundle of the two-scale Lorenz 96 model. Nonlinear Process. Geophys. 26(2), 73–89 (2019)
62. R.V. Abramov, A.J. Majda, New approximations and tests of linear fluctuation-response for chaotic nonlinear forced-dissipative dynamical systems. J. Nonlinear Sci. 18(3), 303–341 (2008)
63. V. Lucarini, S. Sarno, A statistical mechanical approach for the computation of the climatic response to general forcings. Nonlinear Process. Geophys. 18(1), 7–28 (2011)
64. V. Lucarini, Stochastic perturbations to dynamical systems: a response theory approach. J. Stat. Phys. 146(4), 774–786 (2012)
65. G. Gallavotti, V. Lucarini, Equivalence of non-equilibrium ensembles and representation of friction in turbulent flows : the Lorenz 96 model. J. Stat. Phys. 156, 1027–1065 (2014)
66. R.V. Abramov, Leading order response of statistical averages of a dynamical system to small stochastic perturbations. J. Stat. Phys. 166(6), 1483–1508 (2017)
67. E. Ott, Chaos in Dynamical Systems (Cambridge University Press, Cambridge, 1993)
68. H. Broer, C. Simó, R. Vitolo, Bifurcations and strange attractors in the Lorenz-84 climate model with seasonal forcing. Nonlinearity 15,

1205–1267 (2002)
69. D. Orrell, L.A. Smith, Visualising bifurcations in high dimensional systems: the spectral bifurcation diagram. Int. J. Bifurcat. Chaos 13(10), 3015–3027 (2003)
70. D.L. van Kekem, A.E. Sterk, Travelling waves and their bifurcations in the Lorenz-96 model. Phys. D 367, 38–60 (2018)
71. D.L. van Kekem, A.E. Sterk, Wave propagation in the Lorenz-96 model. Nonlinear Process. Geophys. 25(2), 301–314 (2018)
72. D.L. van Kekem, A.E. Sterk, Symmetries in the Lorenz-96 model. Int. J. Bifurcat. Chaos 29(01), 1950008 (2019)
73. J. Kerin, H. Engler, On the Lorenz '96 model and some generalizations (2020)
74. J.R. Holton, An Introduction to Dynamic Meteorology. International Geophysics Series, 4th edn. (Elsevier Academic Press, Burlington, 2004)
75. Z.-M. Chen, W.G. Price, On the relation between Rayleigh–Bénard convection and Lorenz system. Chaos Solitons Fractals 28(2), 571–578 (2006)
76. V. Lucarini, K. Fraedrich, Symmetry breaking, mixing, instability, and low-frequency variability in a minimal Lorenz-like system. Phys. Rev. E 80, 026313 (2009)
77. J. Eckmann, D. Ruelle, Ergodic theory of chaos and strange attractors. Rev. Mod. Phys. 57(4), 617–656 (1985)
78. R. Grotjahn, Global Atmospheric Circulations: Observations and Theories (Oxford University Press, Oxford, 1993)
79. A.N. Kaufman, Dissipative hamiltonian systems: a unifying principle. Phys. Lett. A 100(8), 419–422 (1984)
80. M. Grmela, Bracket formulation of dissipative fluid mechanics equations. Phys. Lett. A 102(8), 355–358 (1984)
81. G. Ansmann, Efficiently and easily integrating differential equations with JiTCODE, JiTCDDE, and JiTCSDE. Chaos Interdiscip. J. Nonlinear Sci. 28(4), 043116 (2018)
82. L.F. Shampine, M.W. Reichelt, The MATLAB ODE Suite. SIAM J. Sci. Comput. 18(1), 1–22 (1997)
83. S. Strogatz, Nonlinear Dynamics and Chaos: With Applications to Physics, Biology, Chemistry, and Engineering, 2nd edn. (Westview Press, Boulder, 2014)

84. G. Benettin, L. Galgani, A. Giorgilli, J.M. Strelcyn, Lyapunov characteristic exponents for smooth dynamical systems and for hamiltonian systems; a method for computing all of them. Meccanica 15, 21–30 (1980)
85. E. Hovmöller, The trough-and-ridge diagram. Tellus 1(2), 62–66 (1949)
86. P. Huerre, P.A. Monkewitz, Local and global instabilities in spatially developing flows. Annu. Rev. Fluid Mech. 22(1), 473–537 (1990)
87. D. Ruelle, Large volume limit of the distribution of characteristic exponents in turbulence. Commun. Math. Phys. 87(2), 287–302 (1982)
88. K.A. Takeuchi, H. Chaté, F. Ginelli, A. Politi, A. Torcini, Extensive and subextensive chaos in globally coupled dynamical systems. Phys. Rev. Lett. 107, 124101 (2011)
89. J.L. Kaplan, J.A. Yorke, Chaotic behavior of multidimensional difference equations, in Functional Differential Equations and Approximation of Fixed Points, pp. 204–227 (1979)

SECTION 2
EFFECTS OF GRAVITY

CHAPTER 5

The Effects of Gravity on the Climate and Circulation of a Terrestrial Planet

Stephen I. Thomson, Geoffrey K. Vallis

College of Engineering, Mathematics and Physical Sciences, University of Exeter, UK

ABSTRACT

The climate and circulation of a terrestrial planet are governed by, among other things, the distance to its host star, its size, rotation rate, obliquity, atmospheric composition and gravity. Here we explore the effects of the last of these, the Newtonian gravitational acceleration, on its atmosphere and climate. We first demonstrate that, if the atmosphere obeys the hydrostatic primitive equations, which are a very good approximation for most terrestrial atmospheres, and if the radiative forcing is unaltered, changes in

Citation: Thomson, SI, Vallis, GK. "The effects of gravity on the climate and circulation of a terrestrial planet". Q J R Meteorol Soc. 2019; 145: 2627–2640. https://doi.org/10.1002/qj.3582.

gravity have no effect at all on the circulation except for a vertical rescaling. That is to say, the effects of gravity may be completely scaled away and the circulation is unaltered. However, if the atmosphere contains a dilute condensible that is radiatively active, such as water or methane, then an increase in gravity will generally lead to a cooling of the planet because the total path length of the condensible will be reduced as gravity increases, leading to a reduction in the greenhouse effect. Furthermore, the specific humidity will decrease, leading to changes in the moist adiabatic lapse rate, in the Equator-to-Pole heat transport, and in the surface energy balance because of changes in the sensible and latent fluxes. These effects are all demonstrated both by theoretical arguments and by numerical simulations with moist and dry general circulation models.

Keywords: gravitational acceleration, planetary atmosphere, primitive equations, radiative forcing, greenhouse effect, general circulation models

INTRODUCTION

The climate of a terrestrial planet depends on an almost uncountable number of factors, including the distance to its host star, the nature of that host star, the size and rotation rate of the planet, the atmospheric composition and many other factors. The variety of planetary climates is large, and there is and can be no single theory of planetary climate, nor is there a planetary analogue of the Hertzsprung–Russell diagram showing how the luminosity of stars varies with their effective temperature. However, this is not to say that we cannot apply general physical principles to atmospheric circulation and planetary climate. Thus, for example, Read (2011) and Wang *et al.* (2018) describe how various non-dimensional parameters describe the general circulation of a large class of planetary atmospheres, Kaspi and Showman (2015) illustrate how the planetary circulation patterns vary over a wide range of orbital parameters, and Pierrehumbert (2010) applies building blocks based on elementary physical principles to construct a plentiful panoply of planetary climates.

As regards planetary circulation, among the most studied parameters are the planetary radius and rotation rate; these combine to give the external Rossby number that is one of the single most influential parameters on planetary circulation. Atmospheric composition obviously plays a key radiative role in determining the surface temperature, especially if the composition gives rise to a greenhouse effect, or an anti-greenhouse effect, and the changing composition of Earth's atmosphere is obviously of current interest. Less well studied is the effect of gravity, here meaning the Newtonian gravitational acceleration as measured at the planetary surface. One expects that a planet with a higher gravity than another, but otherwise the same, would have a thinner (meaning less extended) atmosphere with a higher surface density, but the effects on the circulation and temperature are less clear. The matter was partially investigated by Kaspi and Showman (2015) and Kilic *et al.* (2017), but their model set-ups were very different and their results too incompatible to compare, with the former using a simplified general circulation model (GCM) without many Earth-like effects, such as the radiative effect of water vapour, and the latter fixing their surface temperatures independent of gravity. In this paper we revisit the issue, looking at it both as a problem in geophysical fluid dynamics and a problem in planetary climate.

We first, in Section 2, examine how the adiabatic equations of motion, both the primitive equations and the full Navier–Stokes equations, scale with gravity. We find that in the primitive equations the effects of gravity can be completely scaled out of the problem and that, if the diabatic forcing is sufficiently simple, the circulation is unaltered. This invariance is broken both by non-hydrostatic effects and by having a non-shallow atmosphere, but in many planetary atmospheres these effects will be small, although not always negligible (Mayne *et al.* 2019). In Section 3 we describe how changes in gravity lead to non-negligible changes in moisture content. We then explore the effects of these changes using some idealized numerical simulations: first, in Section 4, we describe the radiative effects of those changes, and then in Sections 5–7 we explore the dynamical effects of the changes in specific humidity, In Section 8 we look at the role of gravity with a more complete model, and in Section 9 we provide our conclusions.

INVARIANCE OF THE EQUATIONS OF MOTION

The momentum equations in the primitive equations on the sphere may be written, in standard notation, as (Vallis, 2019)

$$\frac{\mathrm{D}u}{\mathrm{D}t} - 2\Omega v \sin\vartheta + \frac{uv\tan\vartheta}{a} = -\frac{1}{\rho a\cos\vartheta}\frac{\partial p}{\partial\lambda}, \tag{1}$$

$$\frac{\mathrm{D}v}{\mathrm{D}t} + 2\Omega u \sin\vartheta + \frac{u^2\tan\vartheta}{a} = -\frac{1}{\rho a}\frac{\partial p}{\partial\vartheta}, \tag{2}$$

$$\frac{\partial p}{\partial z} = -\rho g. \tag{3}$$

The mass continuity and adiabatic thermodynamic equations are, respectively,

$$\frac{\mathrm{D}\rho}{\mathrm{D}t} + \rho\nabla \cdot \mathbf{v} = 0, \tag{4}$$

and

$$c_v \frac{\mathrm{D}T}{\mathrm{D}t} + \frac{p}{\rho}\nabla \cdot \mathbf{v} = 0, \qquad \text{or} \qquad \frac{\mathrm{D}\theta}{\mathrm{D}t} = 0, \tag{5}$$

where $\mathbf{v}$ is the three-dimensional velocity, λ is longitude, ϑ is latitude, T and θ are temperature and potential temperature, and the other notation is quite standard. These equations remain invariant under the following transformation:

$$\begin{gathered} g \to \alpha g, \qquad p \to \gamma p, \qquad \rho \to \gamma \rho, \qquad (T,\theta) \to (T,\theta), \\ t \to t, \qquad (x,y) \to (x,y), \qquad z \to z/\alpha, \\ (u,v) \to (u,v), \qquad w \to w/\alpha. \end{gathered} \tag{6}$$

Here α represents a change in the surface gravity, and the appropriate scaling of other quantities, and γ represents a change in the atmospheric surface pressure and appropriate scaling of other quantities. The assumption that pressure p and density ρ both scale like γ is based on assuming the atmosphere is an ideal gas. This scaling would not be valid for an non-ideal gas, such as a Van der Waals gas, or a liquid such as sea water.

If we substitute Equation 6 into Equations 1–5 then all the factors of α and γ cancel and the equations are unchanged, as was noted in section 2.8 of Frierson (2005). In the special case where $\alpha = \gamma$, meaning that we scale gravity and surface pressure by the same amount, equivalent to changing gravity and keeping the atmospheric mass the same, the equations remain unchanged, as was noted by Vallis (2019). If $\gamma \neq \alpha$ then the mass of the atmosphere changes. Given the invariance of the unforced equations themselves, it is a simple matter to confirm that all quantities of dynamical interest, such as the deformation radius, $L_d \equiv NH/f$ and the Eady growth rate, $0.31\,U/L_d$, remain invariant.

The gravitational invariance does not hold in the full Navier–Stokes equations on the sphere. The full momentum equations are

$$\frac{\mathrm{D}u}{\mathrm{D}t} - \left(2\Omega + \frac{u}{r\cos\vartheta}\right)(v\sin\vartheta - w\cos\vartheta) = -\frac{1}{\rho r\cos\vartheta}\frac{\partial p}{\partial \lambda}, \tag{7}$$

$$\frac{\mathrm{D}v}{\mathrm{D}t} + \frac{wv}{r} + \left(2\Omega + \frac{u}{r\cos\vartheta}\right)u\sin\vartheta = -\frac{1}{\rho r}\frac{\partial p}{\partial \vartheta}, \tag{8}$$

$$\frac{\mathrm{D}w}{\mathrm{D}t} - \frac{u^2+v^2}{r} - 2\Omega u\cos\vartheta = -\frac{1}{\rho}\frac{\partial p}{\partial r} - g. \tag{9}$$

The additional metric terms in the horizontal momentum equations, for example uw/r, and the vertical acceleration term $\mathrm{D}w/\mathrm{D}t$ in the vertical equations, are not invariant with respect to the gravitational transformation, as they depend on α. Interestingly, however, the equations *are* invariant with respect to changes in γ. That is, the full momentum equations are invariant to changing the atmospheric mass, equivalent to changing γ alone.

The importance of the metric terms depends on the ratio of the thickness of the atmosphere to the radius of the planet and this is quite small in most terrestrial atmospheres. In the solar system, Titan perhaps comes closest to refuting this statement; Titan has a radius is 2576 km , a scale height of about 20 km , a tropopause at about 40 km and a stratopause at about 300 km , still only 12% of the planetary radius. Non-hydrostatic motion within an atmosphere also violates the invariance, as is implicit in the "hypo-hydrostatic" rescaling of Garner *et al.* (2007). Finally, terms on the right-hand side of the thermodynamic equation might also violate the invariance, as we consider later.

Simulations with Primitive Equations and Newtonian Relaxation

To demonstrate how the invariance manifests itself in practice, we perform simulations with a dry dynamical core obeying the primitive equations, using the Isca framework (Vallis *et al.* 2018). The forcing is that of Held and Suarez (1994), which is a Newtonian relaxation back to a specified temperature that is a function of pressure and latitude. We perform an integration with the standard value of gravity (9.8 m s^{-2}) and one with double that value, keeping the total mass of the atmosphere constant in the two integrations (i.e. setting $\gamma = \alpha$). Figure 1 shows the temperature field in the two integrations and, as is evident, they are identical (in their early stages they are bit-wise identical, but numerical artifacts mean that their final state is not). The velocity and pressure fields (not shown) are also identical. If the fields were plotted in height co-ordinates, then the case with doubled gravity would appear as flatter (with $z \rightarrow z/2$), but this has no dynamical effect in the primitive equations. It is important to note that the same invariance of dry-dynamical core solution can be found if $\gamma \neq \alpha$, i.e. the atmospheric mass is changed when gravity is changed, as discussed in section 2.8 of Frierson (2005), and in agreement with our analysis of the primitive equations above.

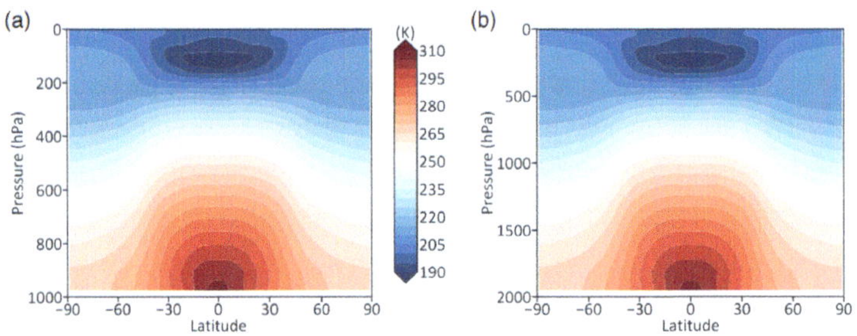

Figure 1: (a) The zonal-mean temperature as a function of pressure and latitude in a Held–Suarez run with normal Earth gravity. (b) The same but with twice Earth gravity.

The invariance discussed above does not necessarily hold if we add more realistic forcing to the thermodynamic equation, and in particular if the radiative forcing is sensitive to gravity, as suggested by the dimensional analysis of such scenarios in Frierson (2005) and Koll and Abbot (2015). Although the effects we describe below can be quite subtle, those due to changes in the moisture content, or any other radiatively active condensible, are more clear. We investigate some of these effects in the following sections.

CHANGE IN MOISTURE CONTENT

There are two distinct changes that a condensible may bring about as gravity changes, one due to its radiative properties if it is a greenhouse gas and the other due to the release of latent heat when it condenses. For specificity we deal with water vapour, and assume that the condensible is dilute (meaning the fraction of the condensible is small). Both of the effects arise because the fraction of condensible, relative to the rest of the atmosphere, will diminish if gravity increases, as we discuss below. The overall temperature of the planet's surface will then diminish as gravity increases (since water vapour is a potent greenhouse gas), and the dynamical effects of condensation (for example, in setting the saturated adiabatic lapse rate) will also diminish.

Total Water Vapour Content

The total water vapour content of a column of atmosphere, W, is given by

$$W = \int_0^\infty \frac{e}{R_v T} \, dz, \tag{10}$$

where e is the vapour pressure of the water vapour, R_v is its specific gas constant and T is the temperature. Suppose that we increase the gravity of a planet by a factor α, without initially changing the temperature. The vapour pressure is a strong function of temperature, and at fixed relative humidity is only a function of temperature (since saturation vapour pressure is a function only of temperature in an ideal gas). If gravity increases by a factor α, then the lowest-order effect is for temperature to fall with height by a factor α more rapidly than before, following the scaling of Section 2. That is, at any given height z the value of e will be lower than before, and the total water content of the atmosphere will fall roughly by a factor α. Changes in relative humidity can quantitatively alter this conclusion, but unless relative humidity also changes by a factor α, which is in most circumstances very unlikely, that change will be small.

To see the above argument another way, we transform Equation 10 into an integral over pressure and, using the hydrostatic and ideal gas relations, obtain

$$W = -\frac{1}{g} \int_0^\infty e \frac{R_d}{R_v} \, d\log(p/p_{surf}), \tag{11}$$

where R_d is the specific gas constant for air, p_{surf} is the surface pressure, and the limits of the integral are the same even as g and p_{surf} change. Now, as noted above, e is determined largely by temperature, and the value of temperature at any given value of log (p/ p_{surf}) is, to lowest order, unaltered (as in Figure 1). Thus, the integrand is unaltered by the transformation, but the factor of $1/g$ outside the integral indicates that the total water content will scale by a factor of $1/\alpha$. Of course, once the water content changes, the temperature changes because of radiative effects, which causes the water content to change again, so the effect is not a simple one. Nonetheless, the most basic effect that can be expected is that if gravity increases, water vapour content will fall. Since water vapour is a potent greenhouse gas, temperature will also fall. It is important to note that this change in total water vapour mass is related to a change in α only, and not a change in γ. The implication of this is that changing the atmospheric mass without changing gravity will not lead to a change in water-vapour mass (under the assumptions made above that changes to temperature and relative humidity are small).

Specific Humidity

In addition to changes in the total water vapour content, the specific humidity, q, will fall as surface pressure increases (regardless of how this happens) and this can have an important dynamical effect. The specific humidity is defined as the ratio of the mass of water vapour to the total mass of air and in terms of pressures it may be written

$$q = \frac{\epsilon e}{p - e(1-\epsilon)} \approx \frac{\epsilon e}{p}, \qquad (12)$$

where ϵ is the ratio of the molar mass of water vapour to that of dry air and the approximation giving $\epsilon e/p$ holds for a dilute atmosphere. Since pressure scales with γ but e does not (it does not depend on γ or α at lowest order) we expect that the specific humidity will fall as surface pressure increases, scaling roughly as $1/\gamma$. (Note that vp = He_{sat} where H is relative humidity and *e*sat is the saturation vapour pressure, a function only of temperature. As with the argument for total water vapour content, unless H changes, e will not change as γ or α changes.)

The consequence of this is that the hydrology cycle will weaken as surface pressure falls, essentially because the condensation will have a smaller effect on a denser atmosphere. That is, if there is a change in specific humidity of Δq then the temperature change is given by

$$c_p \Delta T = -L\Delta q, \qquad (13)$$

so that ΔT is smaller as q falls.

Thus, in summary, an increase in gravity and/or surface pressure has two somewhat distinct effects on the condensible (with the opposite effect for a decrease in gravity and/or surface pressure). First, the total amount of condensible decreases, roughly in proportion to the increase in gravity, because of the reduced scale height of the temperature field. One effect of this is to reduce the greenhouse effect of the condensible and so make the atmosphere cooler. Second, the specific humidity falls in a dilute atmosphere, not primarily because of the cooling of the atmosphere, but because of the increase in total pressure of the atmosphere and the approximate constancy of the vapour pressure, e, as given by Equation 12. This effect will be further amplified by the cooling of the planet because of the reduced greenhouse effect, but does not depend upon it. The main consequences of this are that the hydrology cycle will weaken and, concomitantly, the magnitude of the saturated adiabatic lapse rate will increase as it approaches the dry adiabatic lapse rate. In the sections that follow, we first explore and quantify the radiative effect, and then the dynamical effects of the weaker hydrology cycle; in both cases we use idealized radiative transfer schemes to isolate the effects. In Section 8 we use a more accurate radiative transfer scheme to see how the effects work together. In all of the following experiments, we choose to set $\gamma = \alpha$, meaning that when we change the gravity, we change the surface pressure by the same amount in order to keep the atmospheric mass the same between experiments. From now on we therefore drop the use of γ and consider changes in both pressure and gravity as being proportional to α.

In what follows, we choose to vary α between 0.8 and 2, equivalent to varying g between 7.85 and 19.62 $\mathrm{m\,s^{-2}}$. We deliberately do not explore values of α beyond this range as we do not want to push our model far outside of an Earth-like regime. Pushing far beyond this range would invalidate our assumptions that, for example, the atmospheric composition stays the same accross a range of α values. In reality, planets with very high gravity values would retain lighter elements like hydrogen and helium during their formation and evolution, meaning that changes in gas properties with gravity ought to be accounted for. Therefore, in order to retain the simplicity of our investigation, we limit our α range to stay close to 1.

RADIATIVE EFFECTS

As noted above, water vapour is a greenhouse gas so that increasing gravity, and thereby reducing total water vapour content, will lead to a cooling of the atmosphere. We illustrate this effect by a set of integrations with a moist GCM, again using the Isca framework. In all of the following experiments, the gravitational acceleration is changed, and the model's mean surface pressure is prescribed to change like α, representing a constant atmospheric mass between experiments. We configure Isca to use a grey radiative transfer with an optical depth prescription that depends on the atmospheric specific humidity, q. We follow Byrne and O'Gorman (2013), except that we change the parameter a=0.1627 so that the time-averaged surface temperatures are similar to that achieved with a complex radiative transfer code with a surface albedo of 0.3, as discussed in Vallis *et al.* (2018). In other respects the model is similar to that described in Frierson *et al.* (2006), except that virtual temperature effects are included, with a slab-ocean surface and time-constant insolation that represents the annual mean of Earth's incoming radiation. For simplicity we omit the seasonal cycle using instead a time-constant insolation profile which well approximates annual mean insolation on Earth.

The zonal-mean surface temperature profiles under different gravitational accelerations are shown in Figure 2a. Additionally, global-average values of W are plotted against $1/\alpha$ in Figure 2b, and the response of the zonal-mean atmospheric temperature to doubling gravity is shown in Figure 2c. The latter is presented in so-called "sigma" coordinates, where $=p/p_{surf}$, allowing the difference between Earth gravity and twice-Earth gravity to be presented on one plot. In Figure 2b, the increase in W with increasing $1/\alpha$ is consistent with expectations for the decrease in W with increasing gravity. The slope is, however, different from a simple $1/\alpha$ dependence, owing to the increase in gravity and the concomitant decrease in temperature, both of which act to decrease W for increasing α. Alongside a decrease in W with increasing α, the associated decrease in long-wave optical depth and subsequent surface cooling is evident in Figure 2a.

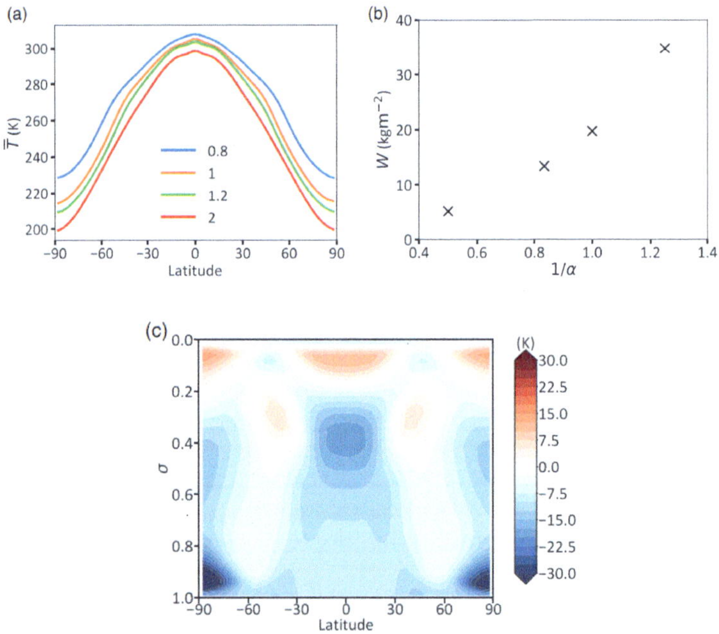

Figure 2: (a) The zonal-mean surface temperature against latitude in an aqua-planet run with optical depths like Byrne and O'Gorman (2013). (b) The time- and area-averaged value of total water-vapour content W plotted against $1/\alpha$. (c) The zonal-mean atmospheric temperatures in a twice-Earth-gravity run minus the same in a normal-Earth-gravity run, plotted versus σ coordinates.

SPECIFIC HUMIDITY EFFECTS

In addition to the overall cooling, the structure of the cooling has a distinctive pattern, as can be seen in Figure 2c. Two effects are particularly noticeable: an enhanced cooling in both the tropical upper troposphere and near the surface at high latitudes (a "polar amplification"). In this section we determine the mechanisms determining this structure, with more details in sections following.

These effects are essentially the inverse of a global-warming response, illustrated for example in figure 6 of Vallis *et al.* (2015), and the mechanisms are similar (but inverted), and are due to the changes in specific humidity. To isolate the effect, we perform the same set of experiments as those described above using a radiative scheme with a fixed optical depth (and so one that does not depend on water-vapour amount), as in Frierson *et al.* (2006). We thereby eliminate the overall global cooling effect.

The profiles of zonal-mean surface temperature in these runs are shown in Figure 3a. It is clear from comparison between this figure and Figure 2a that removing the water-vapour–optical-depth feedback has altered the response to changing gravity considerably. Without the long-wave optical depth feedback, the twice gravity profile is now warmer in the Tropics and colder at the Poles than its Earth-gravity equivalent, unlike the response seen with the feedback.

The increase in total column water vapour, W, with increasing $1/\alpha$ shown in panel Figure 3b is also present, but scaling is closer to $1/\alpha$ than in Figure 2b, owing to the lack of global cooling in the newer experiments. The atmospheric temperature response is also different, as seen by comparing Figure 2c with Figure 3c; the enhanced low-level cooling over the Poles is not as conspicuous, and this is because this cooling relies in part on a direct radiative effect not present in the runs with fixed optical depth.

However, the enhanced upper-level cooling in the Tropics is still present, and this is due to changes in saturated adiabatic lapse rate. As q diminishes, then the saturated adiabatic lapse rate increases in magnitude, so that in the Tropics upper levels cool preferentially, as can be seen in both Figure 2c and Figure 3c. Changes in the q profiles in the latter experiments are shown in Figure 3d. We explore the lapse-rate effect further in Section 7.

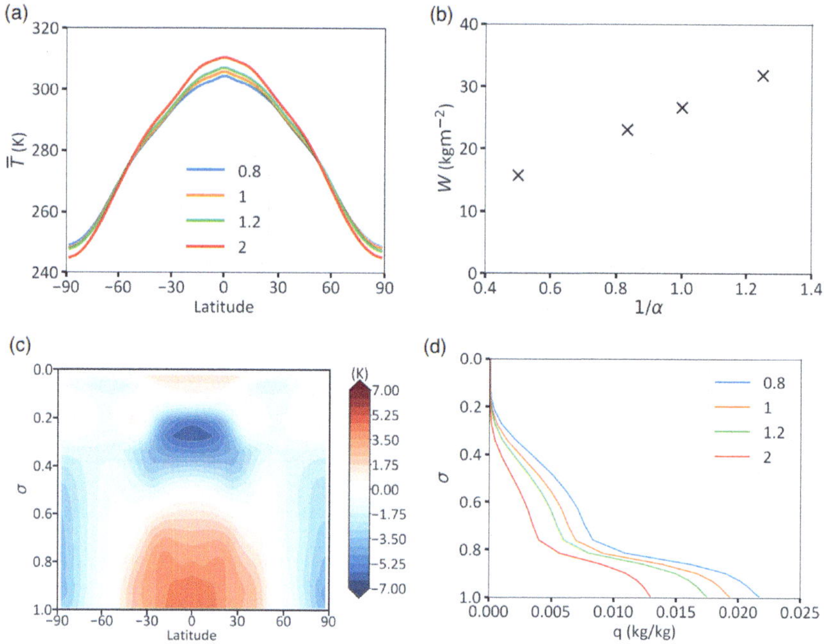

Figure 3: (a) The zonal-mean surface temperature against latitude in an aqua-planet run with optical depths as in Frierson *et al.* (2006). (b) The time- and area-averaged value of W plotted against $1/\alpha$. (c) The zonal-mean atmospheric temperatures in a twice-Earth-gravity run minus the same in a normal-Earth-gravity run, plotted versus σ coordinates. (d) Vertical profiles of q averaged between 10°S and 10°N.

Interestingly, it is found that the poleward moist-static-energy flux or 'heat transport' (not shown) changes very little between the experiments with varying gravity. The moist-static energy flux is given by

$$[vMSE] = \frac{-1}{g}\int_{p_{\text{surf}}}^{0}\int_{0}^{2\pi} v\,(C_pT + gz + L_vq)\,\mathrm{d}\lambda\,\mathrm{d}p, \tag{14}$$

where MSE is the moist static energy, v is the meridional velocity, λ is longitude and the square brackets denote a vertical mass-weighted integral. Despite the consistency in the overall transport, the balance of terms in this equation does change. Specifically increasing gravity decreases q, increases v and also changes the temperature structure, thereby affecting all the terms. The lack of change in overall transport is consistent with results found in an idealized GCM by Frierson *et al.* (2007) and references therein. However, unchanged overall transport when changing parameters is certainly not always the case (e.g. Schneider *et al.* 2010). The changes in temperature structure, then, are not primarily caused by changes in overall heat transport. Rather, further investigation indicates that these changes have two main causes, namely changes in the fluxes from the surface to the atmosphere, and (as previously noted) changes in the tropical lapse rates. We now discuss each of these in turn.

SURFACE-FLUX EFFECTS

A Vertical-Flux-Based Argument

Consider now the effects of heat, momentum and moisture exchange between the surface and the lower atmosphere. The effects on the atmosphere can be written as the vertical gradient of upward eddy fluxes of the relevant quantity:

$$\frac{\partial T}{\partial t} = \ldots - \frac{\partial}{\partial z}\left(\overline{T'w'}\right), \qquad (15a)$$

$$\frac{\partial u}{\partial t} = \ldots - \frac{\partial}{\partial z}\left(\overline{u'w'}\right), \qquad (15b)$$

$$\frac{\partial q}{\partial t} = \ldots - \frac{\partial}{\partial z}\left(\overline{q'w'}\right), \qquad (15c)$$

with the overbars representing a mean over some area, and the primes being a departure from that mean – these are subgrid-scale quantities in a GCM.

If we consider how each of these flux terms scale with a change in gravity, an obvious difference between them is that T and u do not scale simply with α, but q scales like $1/\alpha$, as discussed in Section 3.2. In addition, the fluxes themselves will vary with gravity, as we now show. If we were to scale the various terms in Equation (15) using Equation 6, continuing to use $\gamma = \alpha$, then all factors of α cancel and it might appear that the surface fluxes are unaltered. However, this scaling is unwarranted because the fluxes are non-hydrostatic. We would expect z to still scale like $1/\alpha$, as described in Section 2, but in the boundary layer the turbulence is essentially isotropic, meaning that w' scales like u', which does *not* scale with α. This suggests that the tendencies of the vertical flux terms will scale like α for u and T. The tendencies from the q term do not scale with α, but if we account for the factor of $1/\alpha$ on the left-hand side, then the effect of the vertical flux on the scaled q also scales like α. The same conclusions can be drawn if we formulate the boundary-layer fluxes as diffusion terms of the form

$$\frac{\partial T}{\partial t} = \dots \frac{\partial}{\partial z}\left(\kappa \frac{\partial T}{\partial z}\right), \qquad (16)$$

and similarly for u and q. We expect the eddy diffusivity, κ, to scale like an eddy velocity multiplied by a vertical length-scale, $l' w'$. Now, w' does not scale with α (since $w' \sim u'$), whereas l' scales like $1/\alpha$, and so κ itself scales like $1/\alpha$. The right-hand side of Equation 16 then scales like α, as before.

Surface Flux Implementation in Isca

The surface fluxes in Isca, as in most GCMs, are parametrized with bulk-aerodynamic laws, but these obey the same scalings as above as we will show. First consider the simplest case of a neutrally stable boundary layer over a smooth surface. Here, it is common to take $\kappa = K_{vk} u_* z$ (e.g. Kraus, 1972, equation (5.15)) where $K_{vk} \approx 0.4$ is the von Kármán constant and u_* is the turbulent velocity, which does not scale with α. Thus, $\kappa \sim 1/\alpha$, and the effects of the boundary-layer flux convergence scales like α.

The surface fluxes in the model are generalizations of this and are described by bulk-aerodynamic formulae with coefficients determined by Monin–Obukhov scaling, namely

$$S = \rho_{\rm atm} C_p C(z, z_{\rm sens}) |\mathbf{V_a}| \, (\theta_{\rm surf} - \theta_{\rm atm}), \qquad (17a)$$

$$Q = \rho_{\rm atm} C(z, z_{\rm moist}) |\mathbf{V_a}| \, (q_{\rm surf} - q_{\rm atm}), \qquad (17b)$$

$$\tau = \rho_{\rm atm} C(z, z_{\rm stress}) |\mathbf{V_a}| \mathbf{V_a}, \qquad (17c)$$

where S is the sensible heat flux out of the surface, Q is the latent heat flux out of the surface, and τ is the stress exerted by the surface on the atmosphere. In these formulae $\rho_{\rm atm}$ is the atmospheric density at the lowest model level, V_a is the horizontal wind velocity on the lowest model level, C_p is the heat capacity of dry air, $\theta_{\rm surf}$ and $\theta_{\rm atm}$ are the surface and lowest-model-level potential temperatures, respectively, $q_{\rm surf}$ and $q_{\rm atm}$ are the surface and lowest-model-level specific humidities, respectively, where $q_{\rm surf}$ is the saturated specific humidity at the temperature of the surface. $C(z, z_{\rm rough})$ is a function of the stability of the boundary layer as calculated by Monin–Obukhov similarity theory, where z is the height on the lowest model level and $z_{\rm rough}$ is the roughness length appropriate for each quantity. (In the simulations we take $z_{\rm sens} = z_{\rm moist} = z_{\rm stress} = z_{\rm rough} = 3.21 \times 10^{-5}$m, and scale these values like $1/\alpha$ with changing gravity). The acceleration provided by τ in the momentum equations is $\rho^{-1} \partial \tau / \partial z$. Because τ scales like α in Equation (17c), and the factors associated with $\rho^{-1}\partial/\partial z$ cancel out, the momentum tendency scales like α. This is equivalent to having an eddy diffusivity κ that scales like $\sim 1/\alpha$, as before, and this holds for velocity, temperature and specific humidity. The difference between surface latent heat fluxes and temperature fluxes lies solely in the different ways that T and q scale under a change in gravity, not in their effective eddy diffusivities. Specifically, the latent heat flux given by (17b) remains invariant under a change in gravity, whereas the sensible heat flux scales like α, because ρ scales like α. These changes in surface fluxes affect the surface (mixed-layer) temperature which obeys an equation of the form

$$C_{\text{surf}}\frac{\partial T_{\text{surf}}}{\partial t} = SW - LW - S - Q, \tag{18}$$

where C_{surf} is the mixed-layer›s heat capacity, T_{surf} is the surface temperature, SW is the net short-wave flux into the surface, LW is the net long-wave cooling of the surface. The gravitational acceleration is not explicitly present on the left-hand side, which will go to 0 in a steady-state regardless, or in SW or LW, and changes in g will have no effect at lowest order on these terms. As discussed above, Q has no dependence on α, but S increases like α. Thus, under an increase in gravity we expect latent heat fluxes to play a relatively smaller role in the heat balance of the surface layer.

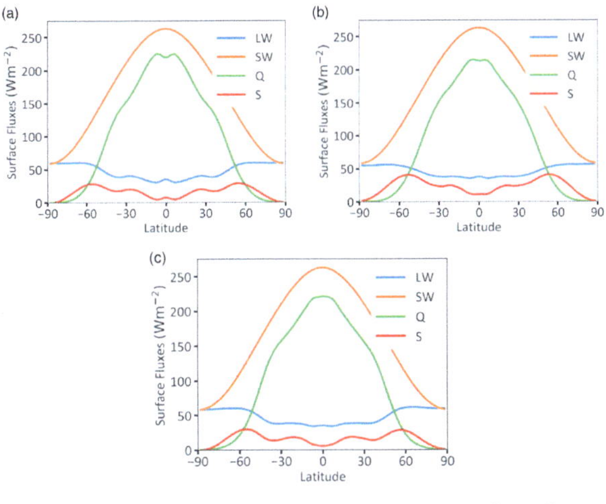

Figure 4: The terms in the surface-temperature equation for a simulation (a) with Earth-gravity and (b) with twice-Earth gravity. LW and SW are the long- and short-wave fluxes, Q is the latent heat flux and S the sensible heat flux. (c) is as (b), but from the twice-Earth gravity case in the scaled-surface-flux experiments described in Section 6.3.

To see these various effects, Figure 4 shows the time-averaged terms on the RHS of Equation 18 in a case with normal-Earth gravity in (a) and twice-Earth gravity in (b).

The short-wave heating of the surface is the same in both cases, as our radiation scheme has a fixed solar absorption, but there are notable changes in the other flux components. The sensible heat fluxes, S, increase everywhere because of the changes in the atmospheric density, as is predicted by the simple scaling. In the polar regions, where the latent cooling of the surface is small, the increase in S necessitates a decrease in the long-wave cooling of the surface in order to maintain a balance (in the surface heat budget) with the short-wave heating. This decrease in long-wave cooling is achieved by a decrease in both the atmospheric and surface temperatures. This is consistent with the polar cooling seen in Figure 3.

In the Tropics, the latent heat fluxes decrease slightly with increased gravity. The scaling result is that they stay constant, and the decrease arises because $|V_a|$ decreases with increased gravity. This is because of the increased surface stress τ due to the increased atmospheric density, which leads to weaker near-surface winds. The decrease in tropical latent heat fluxes is partly offset by the increase in tropical sensible heat fluxes, but the sum of the two decreases when gravity is doubled, necessitating an increase in long-wave cooling, which is provided by way of surface and the lower tropical atmosphere warming. It is clear that the changes in the surface energy budget are consistent with the temperature changes seen in Figure 3.

Note that in the experiments described above, the contribution of $C(z,z_{rough})$ does not change significantly with gravity. This is partly due to our scaling of z_{rough} like $1/\alpha$, but is also a reflection that the stability of the boundary layer does not change significantly with gravity.

A Scaling Experiment

In order to isolate the influence of the changes in surface fluxes with changed gravity, we conduct an experiment where ρ_{atm} in the S and τ formulae is divided by α, so that S and τ no longer scale proportionally with gravity. We leave the ρ_{atm} in Q as it is, so that none of the three fluxes then scale with gravity. The results of this experiment are shown in Figure 5.

Comparing the scaled-flux results in Figure 5 with the unscaled-flux results in Figure 3, it is clear that the polar cooling apparent in the unscaled experiment is no longer present in the scaled experiment. Analysis of the surface energy budget in this case, shown for the twice-Earth-gravity case in Figure 4c, is consistent with the polar cooling in the unscaled case being caused by a change in the sensible heat fluxes. The tropical latent heat fluxes do change by a small amount in the scaled experiments, but by a smaller amount than in the unscaled experiments.

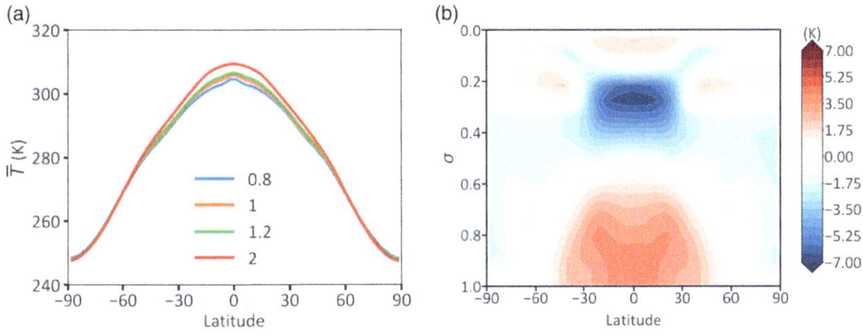

Figure 5: (a) shows the zonal-mean surface temperature against latitude in an aquaplanet run with optical depths as in Frierson *et al.* (2006), but also with scaled S and τ so that they do not scale with gravity. (b) shows the zonal-mean atmospheric temperatures in a twice-Earth-gravity run minus the same in a normal-Earth-gravity run, plotted versus coordinates.

This is consistent with part of this change being due to the changed surface winds in the unscaled experiment. It is also clear, particularly from a comparison of Figure 3c with Figure 5b that the changes in tropical temperatures are not caused by the scaling of surface fluxes with gravity.

LAPSE-RATE CHANGES

To investigate the changes in tropical temperatures associated with changing the gravitational constant, we turn to analysing changes in the tropical lapse rates. On Earth, the tropical lapse rate remains close to the moist-adiabatic lapse rate, which is calculated based on the condensation of ascending saturated parcels. In climate change projections, it is well-known that a significant component of the tropical temperature changes are caused by changes in the tropical lapse rate and the tropopause height (e.g. Vallis *et al.* 2015). Warming (associated with increased greenhouse gases), leads to a higher moisture content and a decreased magnitude of the moist-adiabatic lapse rate, and so more warming in the upper tropospheric regions than at the surface. In addition, because the atmosphere remains in radiative balance with the incoming short-wave radiation, the temperature of the tropopause (which in a grey atmosphere is directly related to the emission temperature) stays constant. A corollary of this is that the tropopause height increases with global warming. Although the argument is only exact for grey radiation, similar effects are seen in GCMs with full radiation schemes.

The same process, but in reverse, is operating in our experiments with increased gravity that include water vapour–optical depth feedback, as seen in Figure 2. Here, increased gravity decreases surface moisture, and so decreases long-wave optical depth, leading to a colder surface. The decreased surface moisture increases the magnitude of the moist-adiabatic lapse rate and the upper troposphere cools more than the surface. The tropopause height decreases under increased gravity, to maintain a constant outgoing long-wave radiation.

The experiments with no water vapour–optical depth feedback do not get a surface cooling under increased gravity (Figures 3 and 5). However, the specific humidity does decrease considerably (Figure 3d). Consequently the magnitude of the moist adiabatic lapse rate increases, leading to a cooler upper troposphere and a lower tropopause.

To quantify these notions we construct simplified tropical temperature profiles using the following assumptions.

- The stratosphere is optically thin and in radiative balance such that it has a constant temperature equal to the emission temperature.
- Radiative transfer is grey in the infrared, with a surface optical depth of τ_s.
- The lapse rate, Γ_r, is a constant in height.

The tropopause height can then be calculated according to the following equation, from Vallis *et al.* (2015).

$$H_{\text{trop}} = \frac{1}{16\Gamma_r}\left(CT_{\text{trop}} + \sqrt{C^2T_{\text{trop}}^2 + 32\Gamma_r\tau_s H_a T_{\text{trop}}}\right). \tag{19}$$

Here H_{trop} is the height of the tropopause, Γ_r is a representative lapse rate, Γ_r, H_a is the scale height of the atmospheric absorber, and T_{trop} is the tropopause temperature, which can be approximated via the incoming solar radiation assuming that the stratosphere is optically thin. The tropospheric temperature profile, $T(z)$, can then be calculated using

$$T(z) = T_{\text{trop}} - \Gamma_r(q_r)(H_{\text{trop}} - z), \tag{20}$$

where z is the height above the surface and q_r is a representative specific humidity. A slight extension to this formalism in useful, in which we continue to use Equation 19 for H_{trop}, but the vertical profile for temperature is constructed with a vertically varying lapse rate, Γ_s, which can be taken to be the saturated adiabatic lapse rate. The temperature profile is then given by

$$T(z) = T_{\text{trop}} - \Gamma_s\{q_s(z), T(z)\}(H_{\text{trop}} - z), \tag{21}$$

where $q_s(z)$ is the saturation specific humidity at height z. In our calculations of H_{trop} we take $\Gamma_r = \Gamma_s$ calculated using T=255 K and q_s calculated at a representative pressure of =0.3, making Γ_r close to 6 α K/km, which is a representative value for Earth. Using Equations 21 and 19, we construct representative temperature profiles for the range of α values used in our experiments, which are shown in Figure 6a. These artificial profiles show that, under an increase in gravity, Γ_s increases, the tropopause height drops, and surface temperatures rise, with temperatures in the upper troposphere falling.

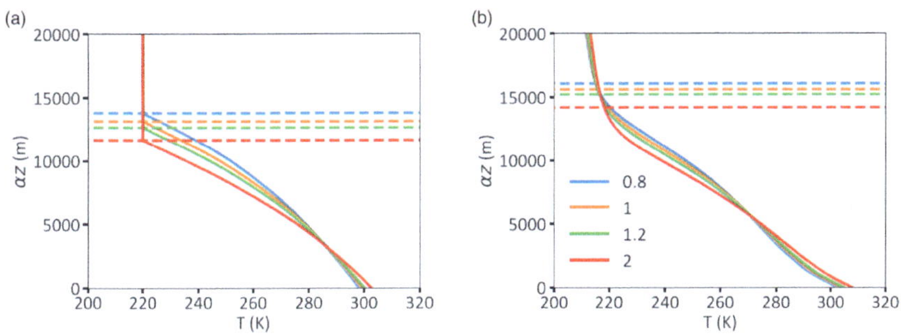

Figure 6: (a) shows representative temperature profiles constructed using Equation 21 and related assumptions for the multiples of Earth gravity used in other experiments. The horizontal dashed lines are the tropopause heights for the various gravity values, and the solid lines are the temperature profiles. The z axis has been scaled by α so that the profiles can be compared on the same height scale. (b) shows temperature profiles from the surface-flux-scaled experiments, thereby isolating the lapse-rate effects. These profiles have been averaged between 10°S and 10°N, and have their tropopause heights calculated from the 2 K/km threshold definition.

In Figure 6b we show time- and latitude-averaged vertical temperature profiles from our grey-radiation experiments with scaled surface fluxes, as shown in Figure 5. These are qualitatively similar to those in Figure 6a, verifying that effects included in our artificial profiles, i.e. changes in the tropical lapse rate and tropopause height, are sufficient to explain the temperature changes seen in Figures 3 and 5. In contrast to the above, let us also consider artificial profiles in regions far from saturation, where $\Gamma \approx \Gamma_d = g/$

c_p, so that $T(z) = T_{trop} - \Gamma_d(H_{trop} - z)$. This dry adiabatic lapse rate increases like α, so at a given value of z/α the atmospheric temperatures would be the same independent of changes in gravity. (This result is actually demanded by the fact that the dry equations are invariant with respect to changes in α, with $z \rightarrow z/\alpha$.) Therefore the lapse-rate effects seen in Figure 6 should not be present in regions that are far from saturation, hence why the lapse-rate changes are only apparent in the Tropics in Figures 3 and 5.

RESPONSES WITH REALISTIC RADIATIVE TRANSFER

In the above sections we isolated various effects using idealized models. We now use a more realistic model to explore their combined effect, and in particular we use the SOCRATES radiative transfer code (Edwards and Slingo, 1996; Manners *et al.*, 2017). SOCRATES is a highly flexible radiative transfer code that has been used extensively in operational UK Met Office models and in the study of exoplanetary atmospheres (e.g. Amundsen *et al.* 2016). Here we use SOCRATES with 12 long-wave bands and 21 short-wave bands, and run it without ozone absorption in the stratosphere for ease of comparison across different gravity values. We also run without a seasonal cycle, instead forcing the model with the incoming short-wave profile used in the grey-radiation experiments.

The zonal-mean surface temperatures in these experiments are shown in Figure 7a. The Earth-like climate is colder with SOCRATES than it was with the grey radiation schemes of Frierson *et al.* (2006) shown in Figures 3 and 5, and is somewhat similar to the temperatures in Figure 2. This is also reflected in the total water-vapour amounts shown in Figure 7b, which are lower with SOCRATES than they were with the grey radiation schemes. The difference in temperature in the Earth-like cases is likely because of the increased absorption of short-wave radiation in SOCRATES compared with the grey schemes, and the inclusion of the well-known spectral window for long-wave cooling in SOCRATES .

Despite the mean-state differences, a number of familiar features are apparent in the atmospheric temperature response to a doubling of gravity shown in Figure 7c. The cooling of the surface outside the Tropics is consistent with a decrease in long-wave optical depth due to decreased column moisture.

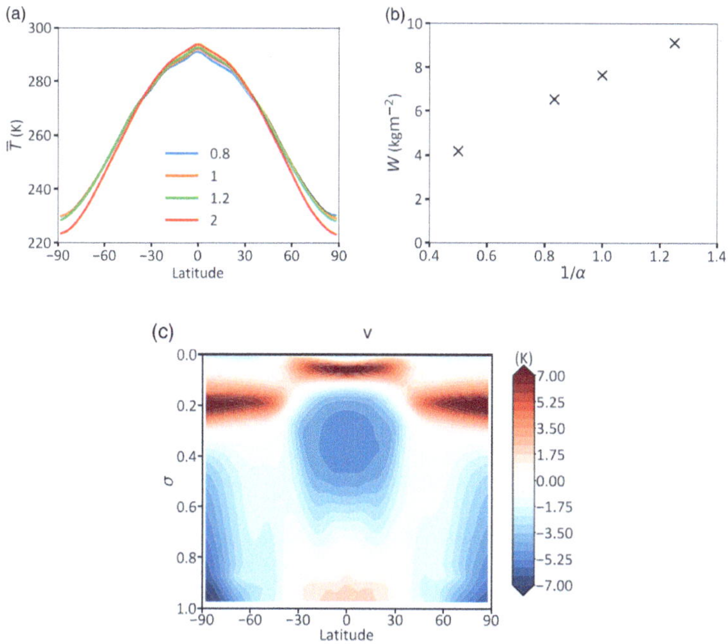

Figure 7: (a) The zonal-mean surface temperature against latitude in an aqua-planet run with the SOCRATES radiation scheme. (b) The time- and area-averaged value of W plotted against $1/\alpha$. (c) The zonal-mean atmospheric temperatures in a twice-Earth-gravity run minus the same in a normal-Earth-gravity run, plotted versus σ coordinates.

In the tropical regions we see warming at the surface and in the lower troposphere with cooling aloft, consistent with an increase in the saturated adaibatic lapse rate changes, but with an additional cooling due to the longwave optical depth decrease. In contrast with Figure 2, the decrease in longwave optical depth is not enough to offset the surface heating from the lapse rate changes, meaning that a realistic combination of these effects is not quite the same as an inverse of a climate-change response.

The Equator-to-Pole heat transport decreases in these experiments (not shown), consistent with this increase in Equator-to-Pole temperature gradient. This result was not found in the grey radiation runs. The lack of transport change with grey radiation may well be a special case for the grey radiation prescription and parameters that are used, consistent with the contrasting transport changes found with grey radiation in Frierson *et al.* (2007) and Schneider *et al.* (2010). The latent and sensible heat fluxes changes with SOCRATES are broadly similar to those shown in Figure 4, with a diminished role for latent heat fluxes under increased gravity. One notable difference is that the sensible heat fluxes are (with SOCRATES) negative in the polar regions, because the atmosphere is warmer than the surface. As a result, the increase in the magnitude of the sensible heat fluxes, due to the α scaling described above, necessitates an increase in surface long-wave cooling in the polar regions (whereas previously it gave rise to a decrease). Finally, we note that an increase in gravity gives rise to an increase in pressure broadening of the spectral lines in the radiative transfer, an effect only included in our runs with SOCRATES . This broadening is related to the absolute atmospheric pressure i.e. p not p/p_{surf}, so that a higher gravity gives rise to higher pressure and more broadening. This turns out to be a small effect compared with the reduction of water vapour condensible, and we do not describe the results. However, it may play a more important role in a dry atmosphere.

CONCLUSIONS

In this paper we have investigated the response of terrestrial atmospheres to a change in the gravitational acceleration at the planet's surface. The full Navier–Stokes momentum equations in spherical geometry do have a dependence on gravity, but these dependence is usually small in a terrestrial atmosphere, consistent with the (normally very good) approximations used to derive the primitive equations. The adiabatic primitive equations remain invariant under a transformation where gravity is changed by a factor α if the vertical co-ordinate z is scaled by a factor $1/\alpha$. In addition, both the primitive and full momentum equations are invariant under a transformation where pressure and density are both scaled by a factor γ, as would happen for an ideal gas with a change in atmospheric mass. Any changes found due to changes in gravity must then (in a primitive equation atmosphere) arise from thermodynamical and radiative aspects of the planet's atmosphere, and their interaction with the dynamics, rather than the dynamics alone.

The effects of a change in gravity on an Earth-like atmosphere with a constant atmospheric mass arise from two main phenomena:

1. A change in the total column water vapour under gravity, arising from a change in the atmospheric scale height combined with the scaling invariance of the temperature field. Thus, in a higher-gravity planet, the atmosphere has a smaller vertical extent and less total water vapour. Since water vapour is a potent greenhouse gas, this effect leads to an overall cooling of the atmosphere.

2. A change in the specific humidity, at least in a dilute atmosphere in which the condensible is a small fraction of the total atmosphere. In such an atmosphere $q \approx e/p$, and since p scales with gravity while e does not, an increase of gravity leads to a general reduction in specific humidity. Thus, an increase in gravity leads to a reduction in the effects of condensation, with the following main effects:
    - Changes in specific humidity lead to changes in the saturated adiabatic lapse rate, which is the dominant factor determining lapse rate in the Tropics. An increase in gravity leads to warming near the surface and cooling aloft. This effect is very robust across all experiments and parameters.
    - Changes in surface sensible heat flux, surface stress, and latent heat fluxes from the surface, with the first two scaling like α but the third not. A reduction in q at higher gravity thus leads to a reduction in the relative importance of the latent heat flux compared to sensible heat flux. The effects of this are rather complicated, and lead to different amounts of long-wave cooling and polar cooling under higher gravity, differing quantitatively across experiments with different radiation schemes.
    - Changes in the relative components of the meridional energy flux. A reduction in q at higher gravity leads to a smaller meridional latent heat flux, but in many experiments this is compensated by an increase in the sensible heat flux. We do not ascribe a universality to this result.

In addition, changes in the pressure broadening of spectral lines due to changes in atmospheric pressure have a small effect. In our experiments this effect is much smaller than the change in greenhouse effect due to changes in the amount of condensible, but in a dry atmosphere the effect would be apparent.

The balance between the above effects will determine the overall response, and that balance is determined by the properties of the atmosphere and condensible. In this paper we have focussed on an Earth-like planet, but a condensible with a smaller latent heat content than water, but a larger effect on the long-wave optical depth, would make the radiative effects more important than the condensation effects. In our experiments we have assumed that the atmospheric mass stays constant when changing gravity, meaning that the surface pressure scales with α. If we were to split the above effects found when changing gravity into those that are due to changes in gravity alone, and those due to changes in surface pressure alone, then we may ascribe the results as follows. The change in the total column water vapour is due to a change in gravity alone, for it does not depend on the atmospheric mass. The change in the specific humidity is caused by a change in surface pressure alone, and could be accomplished by a change in gravity or a change in atmospheric mass. The subsequent effects on the atmosphere can therefore also be partly categorized in this way, with changes in the saturated lapse rates, and changes in the Equator-to-Pole energy transports being due to changes in surface pressure, and the change in long-wave optical depth being caused by a change in gravity. The changes in surface fluxes are slightly more ambiguous, as the scaling with α of the vertical flux formulation in Equation (15) is caused by changes in gravity, but the latent-heat flux is dependent on both gravity and surface pressure. The model›s surface energy balance, as determined by the bulk-aerodynamic formulae in Equation (17), is similarly dependent on both factors, and further investigation of this is left to future work. In thinking about such a division of effects, it is important to note that changing the surface pressure by changing the atmospheric mass, rather than by changing the gravity, comes with the additional effect of changing the atmospheric heat capacity. This leads more significant changes in the Equator-to-Pole temperature gradient than are found in the present work, and thus more significant changes in the atmospheric circulation, as discussed by Chemke and Kaspi (2017).

In a non-dilute atmosphere – that is, one in which the condensible is not a minor constituent – the effects we describe above would be different again since the approximation leading to Equation 12 is no longer valid and the relative amount of the condensible would not necessarily change with gravity.

However, assuming that the amount of condensible is determined primarily by the planetary temperature, a reduction in scale height of the atmosphere with increased gravity would still lead to a smaller total amount of condensible, and (if the condensible is a greenhouse gas) to a cooler planet (and then still less condensible). Evidently, the properties of any condensible species are key in setting the atmospheric temperature structure, its circulation, and ultimately its habitability, for any given planet or exoplanet.

ACKNOWLEDGEMENTS

SIT is supported by the Natural Environment Research Council (grant number NE/M006123/1), and GKV acknowledges support from the Royal Society (Wolfson Foundation), the Leverhulme Trust and the Newton Fund. The authors wish to thank Bob Beare for useful discussions, and Mark Hammond and James Manners for their help with setting up SOCRATES . The authors also wish to thank Dorian Abbot and an anonymous reviewer, who helped to improve the manuscript. The authors declare no conflict of interest. Documentation and code required to run the Isca model framework can be found at https://execlim.github.io/IscaWebsite/ and https://github.com/ExeClim/Isca (accessed 11 June 2019).

REFERENCES

1. Amundsen, D.S., Mayne, N.J., Baraffe, I., Manners, J., Tremblin, P., Drummond, B., Smith, C., Acreman, D.M. and Homeier, D. (2016) The UK Met Office global circulation model with a sophisticated radiation scheme applied to the hot Jupiter HD 209458b. *Astronomy & Astrophysics*, **595**(A36). https://doi.org/10.1051/0004-6361/201629183
2. Byrne, M.P. and O'Gorman, P.A. (2013) Land–ocean warming contrast over a wide range of climates: convective quasi-equilibrium theory and idealized simulations. *Journal of Climate*, **26**, 4000–4016
3. Chemke, R. and Kaspi, Y. (2017) Dynamics of massive atmospheres. *The Astrophysical Journal*, **845**(1). https://doi.org/10.3847/1538-4357/aa7742
4. Edwards, J.M. and Slingo, A. (1996) Studies with a flexible new radiation code. I: choosing a configuration for a large-scale model. *Quarterly Journal of the Royal Meteorological Society*, **122**, 689–719
5. Frierson, D.M.W. (2005). Studies of the general circulation of the atmosphere with a simplified moist general circulation model. Ph.D. thesis, Princeton University, Princeton, NJ.
6. Frierson, D.M.W., Held, I.M. and Zurita-Gotor, P. (2006) A gray radiation aquaplanet moist GCM. Part 1: static stability and eddy scales. *Journal of the Atmospheric Sciences*, **63**, 2548–2566
7. Frierson, D.M.W., Held, I.M. and Zurita-Gotor, P. (2007) A gray radiation aquaplanet moist GCM. Part 2: energy transports in altered climates. *Journal of the Atmospheric Sciences*, **64**, 1680–1693
8. Garner, S.T., Frierson, D.M.W., Held, I.M., Pauluis, O.M. and Vallis, G.K. (2007) Resolving convection in a global hypohydrostatic model. *Journal of the Atmospheric Sciences*, **64**, 2076–2090
9. Held, I.M. and Suarez, M.J. (1994) A proposal for the intercomparison of the dynamical cores of atmospheric general circulation models. *Bulletin of the American Meteorological Society*, **75**, 1825–1830
10. Kaspi, Y. and Showman, A.P. (2015) Atmospheric dynamics of terrestrial exoplanets over a wide range of orbital and atmospheric parameters. *The Astrophysical Journal*, **804**(60). https://doi.org/10.1088/0004-637X/804/1/60
11. Kilic, C., Raible, C.C., Stocker, T.F. and Kirk, E. (2017) Impact of variations of gravitational acceleration on the general circulation of the planetary atmosphere. *Planetary and Space Science*, **135**, 1–16. https://

doi.org/10.1016/j.pss.2016.11.001

12. Koll, D.D.B. and Abbot, D.S. (2015) Deciphering thermal phase curves of dry, tidally locked terrestrial planets. *The Astrophysical Journal*, **802**(1). https://doi.org/10.1088/0004-637X/802/1/21
13. Kraus, E.B. (1972) *Atmosphere–Ocean Interaction*. Oxford University Press, Oxford, UK.
14. Manners, J., Edwards, J.M., Hill, P. and Thelen, J.-C. (2017). SOCRATES technical guide: Suite Of Community Radiative Transfer codes based on Edwards and Slingo. Technical Report, Met Office, Exeter, UK.
15. Mayne, N.J., Drummond, B., Debras, F., Jaupart, E., Manners, J., Boutle, I.A., Baraffe, I. and Kohary, K. (2019) The limits of the primitive equations of dynamics for warm, slowly rotating small Neptunes and super Earths. *The Astrophysical Journal*, **871**. https://doi.org/10.3847/1538-4357/aaf6e9
16. Pierrehumbert, R.T. (2010) *Principles of Planetary Climate*. Cambridge University Press, Cambridge, UK.
17. Read, P.L. (2011) Dynamics and circulation regimes of terrestrial planets. *Planetary and Space Science*, **59**, 900–914
18. Schneider, T., O'Gorman, P.A. and Levine, X.J. (2010) Water vapor and the dynamics of climate changes. *Reviews of Geophysics*, **48**(3). https://doi.org/10.1029/2009RG000302
19. Vallis, G.K. (2019) *Essentials of Atmospheric and Oceanic Dynamics*. Cambridge University Press, Cambridge, UK.
20. Vallis, G.K., Zurita-Gotor, P., Cairns, C. and Kidston, J. (2015) The response of the large-scale structure of the atmosphere to global warming. *Quarterly Journal of the Royal Meteorological Society*, **141**, 1479–1501
21. Vallis, G.K., Colyer, G., Geen, R., Gerber, E., Jucker, M., Maher, P., Paterson, A., Pietschnig, M., Penn, J. and Thomson, S.I. (2018) Isca, v1.0: A framework for the global modelling of the atmospheres of Earth and other planets at varying levels of complexity. *Geoscientific Model Development*, **11**, 843–859
22. Wang, Y., Read, P.L., Tabataba-Vakili, F. and Young, R.M.B. (2018) Comparative terrestrial atmospheric circulation regimes in simplified global circulation models. Part I: from cyclostrophic super-rotation to geostrophic turbulence. *Quarterly Journal of the Royal Meteorological Society*, **144**, 2537–2557

CHAPTER 6

Gravity is Vertical in Geophysical Fluid Dynamics

Andrew L. Stewart & James C. McWilliams

Department of Atmospheric and Oceanic Sciences, University of California, Los Angeles, CA, USA.

INTRODUCTION

Earth is a rotating, gravitating planet, with space-time variations in these properties that are small compared to their average values. In analyzing Earth's atmospheric and oceanic dynamics, it is often assumed that the gravitational acceleration g and the rotation rate Ω are constants and that the

Citation: Stewart, A.L., McWilliams, J.C. "Gravity is vertical in Geophysical Fluid Dynamics". Sci Rep 12, 6029 (2022). https://doi.org/10.1038/s41598-022-10023-3.

shape of the planet is a sphere, or at least a spheroid of rotation about the polar axis that bulges outward at the Equator.

None of these simplifications is fully accurate. Nevertheless, a common theoretical practice is to act as if they are true, with the proviso that the fluid dynamical equations are fundamentally expressed in geopotential coordinates, with the vertical direction z^ defined as parallel to the spatial gradient of the combined gravitational-rotational potential, $-\nabla\Phi$, and with orthogonal horizontal coordinates, e.g., in the directions of longitude and latitude[1]. The advantages of this practice are to hide the complexity of the true gravity and rotation in a simple coordinate frame and, in particular, to be able to express the widespread approximate hydrostatic force balance between pressure gradient and density in only the vertical direction.

Chu[2] asserts that the scientific community has hitherto neglected a potentially important, "horizontal" component of Earth's gravitational acceleration. This is a misrepresentation of the community-standard formulation of the equations of Geophysical Fluid Dynamics (GFD), discussed in the preceding paragraph.

The author claims instead that the widespread practice is to describe spatial variations on Earth in an absolute spherical coordinate system. In such a coordinate system the elevations of geopotential surfaces vary as functions of latitude and longitude, and thus there is a horizontal component of the gravitational acceleration.

The author asserts that this component of gravity has simply been neglected in previous studies of the ocean and atmosphere. In fact, the implicit use of geopotential coordinates in the standard formulation of the equations of GFD ensures that there is no horizontal component of gravity.

Below we provide a brief overview of the history of geopotential coordinates in GFD, to demonstrate the extent to which Chu[2] misrepresents this standard practice.

We then explicitly consider the formulation of the equations describing ocean circulation in an absolute spherical coordinate system, as suggested by Chu[2]. We show that even in such a coordinate system Chu[2] vastly overestimates the importance of the "horizontal gravity" terms. Finally, we show that subsequently published findings by Chu[3],[4], built upon the "horizontal gravity" formulation of Chu[2], erroneously attribute modifications of the flow in oceanic and atmospheric Ekman layers to the effects of "horizontal gravity".

HISTORICAL USE OF GEOPOTENTIAL COORDINATES TO DESCRIBE GEOPHYSICAL FLOWS

The practice of using geopotential coordinates in Geophysical Fluid Dynamics has deep roots. It may have arisen in the context of Earth's tides, and it is certainly implicit in the Laplace tidal equation[5] (likely with overlap with work by Lagrange[6],[7]; see review by Craik[8]). The predictions of these equations have been validated extensively against observations[9]. At least by the time of Lamb[10] (Sec. 213), it was recognized that the appropriate coordinate frame was relative to a resting free surface under the influence of both gravity and Earth's rotation (i.e. centrifugal force), which we can equate to a "geopotential" surface, Φ= constant. Explicitly he says "denote by *z* the altitude measured outwards along a normal, of any point on this surface", i.e. in the direction of $-\nabla\Phi$. He also declared that this surface "is a surface of revolution about the polar axis, but the ellipticity will not in the first instance be taken to be small". We cannot discern whether Lamb recognized that this ellipsoid was partly due to the non-uniformity of the gravity field, as well as to the centrifugal force. We believe there was no observational evidence about the geographical variation of $\vec{g} = -\nabla\Phi$ at that time.

The tides are more generally discussed by Hendershott and Munk[9] and by Hendershott[11]. In the latter is a statement that "it is convenient to represent the tide-generating potential by its horizontal and time variation over some near-sea-level equipotential (the geoid) of the gravitational potential due to the earth's shape, internal mass distribution, and rotation". Obviously, the variation of g was known by then. Modern measurements[1] have shown that the radius of an iso-potential surface near Earth's sea surface varies by O(100) m, which is small compared to the mean radius of ≈ 6368 km.

Phillips[12] writes equations for large-scale flows as a template for modern global atmospheric and oceanic models that explicitly implicate $-\nabla\Phi$ as the direction of the vertical coordinate and he refers to Φ as the geoid that "is not exactly a spherical surface of revolution but undulates slightly above and below the reference spheroid. The latter is a theoretically and empirically defined ellipsoid of revolution whose shape and mass is such that [it] ...gives a good approximation to observed gravity." He also identifies a geographic latitude as the angle between g and the equatorial plane, with longitude the third orthogonal coordinate direction. He even discusses a formal curvilinear coordinate representation in terms of the reference spheroid, though he soon shifts to spherical coordinates. Therefore, the basis for preferring geopotential coordinates, in which the gravitational force acts vertically, is well established in GFD. In this coordinate system gravity acts only in the vertical direction, so there is no "horizontal gravity" missing from previous GFD studies. Chu[2] presents equations of motion in an alternative, absolute spherical coordinate system, in which the elevations of geopotentials vary with latitude and longitude, and thus there is a horizontal component of the gravitational acceleration. However, this formulation is much less convenient for oceanic or atmospheric models due to the large (O(100m)) undulations of the ocean surface associated with the marine geoid.

NEGLIGIBLE "HORIZONTAL GRAVITY" IN AN ABSOLUTE SPHERICAL COORDINATE SYSTEM

Chu[2] derives a formulation of the Boussinesq equations of motion in an absolute spherical coordinate system. They compare the "horizontal gravity" terms that emerge in this coordinate system with the magnitude of the Coriolis force at the ocean surface, as derived from Ocean Surface Current Analysis Real-time (OSCAR) surface currents. The Coriolis force is particularly relevant because its balance with the horizontal gradient of the sea surface elevation (relative to the geoid) determines the horizontal, geostrophic surface current. The author finds that the horizontal gravity force is typically an order of magnitude larger than the Coriolis force. The author concludes that existing formulations of ocean dynamics are missing a key contribution to the horizontal momentum balance because they do not account for "horizontal gravity".

As noted above, existing formulations of ocean dynamics are, in fact, posed in geopotential coordinates, so there is no "horizontal gravity" force that needs to be accounted for. If one insists on formulating the equations of motion in an absolute spherical coordinate system then "horizontal gravity" terms do indeed appear. However, below we show that the comparison performed by Chu[2] grossly overestimates the importance of this "horizontal gravity" force. This is because the "horizontal gravity" force is largely compensated by the horizontal component of the hydrostatic pressure, which reduces its effective strength by three orders of magnitude.

The horizontal momentum equation (Eq. (22) of Chu[2]) is

$$\underbrace{\rho_0 \frac{\mathrm{D}\boldsymbol{U}}{\mathrm{D}t}}_{\text{Advection}} + \underbrace{\rho_0 f \hat{\boldsymbol{z}} \times \boldsymbol{U}}_{\text{Coriolis}} + \underbrace{\nabla_h p}_{\text{Pressure gradient}} = \underbrace{\rho \nabla_h V}_{\text{Horizontal gravity}} + \underbrace{\rho_0 \boldsymbol{F},}_{\text{External forcing}} \tag{1}$$

and the vertical momentum equation (Eq. (26) of Chu[2]) is

$$\frac{\partial p}{\partial z} = \rho \frac{\partial V}{\partial z}. \tag{2}$$

Here ρ0 is the reference density, ρ is the spatially-varying density, *f* is the Coriolis parameter, U is the horizontal velocity vector, *p* is the pressure, ∇h denotes the horizontal gradient operator, and F denotes additional sources and sinks of momentum. Here it is assumed that the tilt of geopotential surfaces relative to spheroidal surfaces is small, such that the conventional assumption of a small vertical-to-horizontal aspect ratio holds[13]. The gravitational *V* is given approximately by (Eq. (16) of Chu[2])

$$V \approx g_0(N - z), \tag{3}$$

where g_0 is the reference gravitational acceleration and *N* is the horizontally-varying elevation of a reference geopotential surface that everywhere lies below the sea surface.

Chu[2] compares the "horizontal gravity" and Coriolis terms in ([1]), and finds the former to exceed the latter by an order of magnitude for typical ocean surface currents. However, this comparison overlooks an almost complete compensation between the "horizontal gravity" term and the horizontal pressure gradient term. Noting that $\partial V/\partial z \approx -g_0$ from ([3]), we integrate ([2]) downward from the sea surface (z=S) to obtain

$$p = -g_0 \int_{z'=S}^{z'=z} \rho \, dz'. \tag{4}$$

We can then write the combined horizontal pressure gradient and "horizontal gravity" terms in ([1]) as

$$\rho_0 \frac{DU}{Dt} + \rho_0 f \hat{z} \times U - \rho_0 F = -\nabla_h p + \rho \nabla V = g_0 \int_{z'=S}^{z'=z} \nabla_h \rho \, dz' - \rho_0 g_0 \nabla_h S + g_0 \rho \nabla_h N$$
$$= \underbrace{g_0 \int_{z'=S}^{z'=z} \nabla_h \rho \, dz'}_{\text{Baroclinic pressure gradient}} \underbrace{-\rho_0 g_0 \nabla_h (S - N)}_{\text{Surface pressure gradient}} + \underbrace{g_0 (\rho - \rho_0) \nabla_h N}_{\text{Horizontal gravity anomaly}}, \tag{5}$$

where $\rho_0 = \rho|z=S$ is the surface density. In Eq. ([5]), the surface pressure gradient is now expressed in terms of the surface relative to the geopotential surface. Eq. ([5]) is identical to the expression for the horizontal pressure gradient that would be obtained in conventional geopotential coordinates[14], plus an additional "horizontal gravity anomaly" term. Thus the appropriate comparison to assess the importance of "horizontal gravity" in this formulation of the equations of motion is to compare the "horizontal gravity anomaly" and Coriolis terms in ([5]). At the surface z=S the "horizontal gravity anomaly" term is zero by construction because ρ=ρ0. In the subsurface, while the "horizontal gravity anomaly" term in ([5]) is non-zero, it is approximately three orders of magnitude smaller than the "horizontal gravity" term in ([1]). Based on the finding of Chu[2] that the "horizontal gravity" term is an order of magnitude larger than the Coriolis force, it is reasonable to expect that the "horizontal gravity anomaly" is typically one to two orders of magnitude smaller than the Coriolis force.

In summary, contrary to Chu's claims[2], the conventional formulation of ocean dynamical equations is posed implicitly in geopotential coordinates. In this coordinate system there is no horizontal component of the gravitational force, and thus "horizontal gravity" does not exist. If one were to formulate the equations of motion in an absolute spherical coordinate system, with spatially-varying geopotential elevations, then these equations should indeed include a term due to "horizontal gravity" (c.f. Eq. [1]). However the effect of this term is almost completely compensated by the horizontal gradient of the hydrostatic pressure, with the residual "horizontal gravity anomaly" term being typically three orders of magnitude smaller. Consequently, "horizontal gravity" would likely have a negligible impact on ocean circulation even in a model formulated in absolute spherical coordinates.

"HORIZONTAL GRAVITY" DOES NOT MODIFY THE OCEANIC NOR ATMOSPHERIC EKMAN LAYERS

This section discusses findings reported in publications by Chu[3],[4], which build directly upon the work of Chu[2]. The articles examine the influence of "horizontal gravity" on flows in the oceanic and atmospheric Ekman layers. As noted above, existing formulations of ocean dynamics are, in fact, posed in geopotential coordinates, so there is no "horizontal gravity" force that needs to be accounted for in these Ekman layers. However, if one insists on formulating the equations of motion in an absolute spherical coordinate system then "horizontal gravity" terms do indeed appear.

In both articles the author concludes that "horizontal gravity" dramatically alters the flow in the Ekman layer, and thus should be accounted for in models of the ocean and atmosphere. This would be a remarkable result, given that it results from only a slight change in the coordinate system (from geopotential coordinates to absolute spherical coordinates). However, it is misleading to include "horizontal gravity" in the momentum balance of the Ekman layer. Instead, "horizontal gravity" and the horizontal pressure gradient jointly define both the resting hydrostatic state and the geostrophic flow that the Ekman layer solution should approach far from the atmosphere-ocean interface. This adjustment leaves the classical Ekman layer solution unmodified by the switch from geopotential to absolute spherical coordinates. This argument is explained in more detail in the Appendix.

In summary, contrary to the claims of Chu[3] and Chu[4], the conventional formulation of GFD in geopotential coordinates eliminates any horizontal component of the gravitational force. Thus, under this formulation "horizontal gravity" does not exist and has no impact on the flow in the oceanic nor atmospheric Ekman layers. If one formulates the equations of motion in an absolute spherical coordinate system, with spatially-varying geopotential elevations, then a term due to "horizontal gravity" does appear. However, it is misleading to label the flows resulting from this force as a component of the Ekman flow; rather, the combination of the horizontal pressure gradient and "horizontal gravity" forces define the hydrostatic resting state and the geostrophic flow outside the Ekman layer, and the ageostrophic Ekman transport should be defined as the deviation from this geostrophic flow. Under this decomposition the Ekman flow in absolute spherical coordinates is identical to the Ekman flow in conventional geopotential coordinates.

DISCUSSION

In summary, we find Chu's advocacy of horizontal gravity to be erroneously presented and, more importantly, inapposite even if correctly done. The foundational ideas of geostrophic, hydrostatic balance and Ekman boundary layers remain intact. The value of satellite altimetry for estimating large- and meso-scale oceanic surface geostrophic currents is preserved (to paraphrase Walter Munk, "the greatest experiment ever done in physical oceanography").

SUPPLEMENTARY INFORMATION

Appendix

"Horizontal gravity" does not modify oceanic Ekman transport

The analysis of Chu[1] starts from the Boussinesq momentum equation, posed in an absolute spherical coordinate system in the limit of small Rossby number (Eq. 17 of Chu[1]):

$$\underbrace{\rho_0 f\hat{\mathbf{z}}\times \mathbf{U}}_{\text{Coriolis}} + \underbrace{\nabla_3 p}_{\text{Pressure gradient}} \approx \underbrace{\rho \mathbf{g}}_{\text{Gravity}} + \underbrace{\rho_0 \mathbf{F}}_{\text{External forcing}} . \quad (1)$$

Here $\rho 0$ is the reference density, ρ is the spatially-varying density, f is the Coriolis parameter, U is the horizontal velocity vector, p is the pressure, ∇_3 denotes the three-dimensional gradient operator, and F denotes additional sources and sinks of momentum. Note that the gravity vector g has both horizontal and vertical components, g = $\mathbf{g} = \mathbf{g}_h + g_z\hat{\mathbf{z}}$. Following Chu[1] , we write these as

$$\mathbf{g}_h \approx g_0 \nabla_h N, \quad (2a)$$

$$g_z = -g_0, \quad (2b)$$

where N is the elevation of the marine geoid, g_0 is the reference gravitational acceleration, and ∇h denotes the horizontal gradient operator.

The central issue with the analysis of Chu[1] lies in the identification of a "geostrophic" component that balances the horizontal pressure gradient via the Coriolis force:

$$\rho_0 f\hat{\mathbf{z}}\times \mathbf{U}_g^{\mathrm{C21}} = -\nabla_h p. \quad (3)$$

The remaining component of the flow balances the horizontal component of the gravity vector and the external forcing via the Coriolis force, and is referred to as the "ageostrophic" or "Ekman" flow:

$$\rho_0 f\hat{\mathbf{z}}\times \mathbf{U}_E^{\mathrm{C21}} = \rho \mathbf{g}_h + \rho_0 \mathbf{F}. \quad (4)$$

Here the superscript ^{C21} denotes definitions made by Chu[1] . We argue that this is an inappropriate decomposition, and that in fact the horizontal component of gravity should be included in the "geostrophic" flow, rather than the "ageostrophic"/"Ekman" flow.

The basis for this argument is that: (i) the horizontal component of the gravity vector is almost entirely compensated by the horizontal component of the hydrostatic pressure, so this separation produces components of the flow that largely cancel one another;

(ii) the residual between the horizontal pressure gradient force and horizontal gravity produces horizontal flows throughout the water column, rather than just in the surface boundary layer; (iii) the formulation of Chu1 implies that even in the absence of wind forcing F = 0, there is still an ageostrophic Ekman transport driven by the "horizontal gravity", which contradicts established theory and measurements of the wind-driven ocean surface boundary layer[2]. Thus, both the horizontal pressure gradient and the "horizontal gravity" jointly determine the geostrophic flow external to the Ekman layer.

In support of this argument, first write the density as a perturbation relative to the constant reference density,

$$\rho = \rho_0 + \tilde{\rho}. \tag{5}$$

Next, note that the gravity term in (1) corresponding to the reference density ρ_0 can be written as an exact gradient,

$$\rho \boldsymbol{g} = \rho_0 \boldsymbol{g} + \tilde{\rho} \boldsymbol{g} = \nabla_3 \Phi_0 + \tilde{\rho} \boldsymbol{g}, \tag{6}$$

where the reference gravitational potential is given by

$$\Phi_0 = \rho_0 g_0 (N - z). \tag{7}$$

We can therefore define a reference hydrostatic pressure p_0 that exactly balances the component of gravity associated with the reference density, i.e.

$$p = p_0 + \tilde{p}, \qquad \nabla_3 p_0 = \nabla_3 \Phi_0. \tag{8}$$

Substituting (8) into the horizontal component of (1) yields

$$\underbrace{\rho_0 f \hat{\boldsymbol{z}} \times \boldsymbol{U}}_{\text{Coriolis}} \approx \underbrace{-\nabla_h \tilde{p}}_{\text{Pressure gradient}} + \underbrace{\tilde{\rho} g_0 \nabla_h N}_{\text{Horizontal gravity}} + \underbrace{\rho_0 \boldsymbol{F}}_{\text{External forcing}} . \tag{9}$$

This form of the horizontal momentum equation largely removes the issue of compensation between the horizontal pressure gradient force and the horizontal gravity force, which would otherwise greatly exaggerate the dynamical influence of the "horizontal gravity term".

We now consider the horizontal flows produced by the action of each of the terms on the right-hand side of (9). Following Chu[1], we consider an external forcing F resulting from vertical eddy momentum fluxes,

$$\boldsymbol{F} = \frac{\partial}{\partial z}\left(K \frac{\partial \boldsymbol{U}}{\partial z}\right), \tag{10}$$

where K is the eddy viscosity. The surface boundary condition is

$$\rho_0 K \frac{\partial \boldsymbol{U}}{\partial z} = \boldsymbol{\tau} \quad \text{at} \quad z = 0, \tag{11}$$

where τ is the surface wind stress, and we assume that the eddy viscosity becomes negligibly small outside the surface boundary layer,

$$K \to 0 \qquad \text{for} \qquad z < -H_{\mathrm{bl}}, \tag{12}$$

where H_{bl} denotes the thickness of the surface boundary layer. By integrating (9) from the surface to an arbitrary elevation $z < -H_{bl}$, we can then write an expression for the transport above that elevation:

$$\underbrace{\int_z^0 \boldsymbol{U}\,\mathrm{d}z'}_{\text{Total transport}} = \underbrace{\frac{1}{\rho_0 f}\int_z^0 \hat{\mathbf{z}}\times\nabla_h\tilde{p}\,\mathrm{d}z'}_{\text{Pressure gradient}} + \underbrace{\frac{1}{f}\int_z^0 b\hat{\mathbf{z}}\times\nabla_h N\,\mathrm{d}z'}_{\text{Horizontal gravity}} \underbrace{-\frac{1}{\rho_0 f}\hat{\mathbf{z}}\times\boldsymbol{\tau}}_{\text{Ekman transport}}. \tag{13}$$

Here $b = -g\rho^{\sim}/\rho_0$ is the buoyancy. The last term in (13) is the familiar expression for the Ekman transport[3] ; this component of the transport is confined entirely to the surface boundary layer. In contrast, both the pressure gradient and "horizontal gravity" components of the transport in (13) vary, in general, throughout the water column. Thus the pressure gradient and "horizontal gravity" set the far-field flow, below the Ekman layer, and the external forcing term in (9) drives the ageostrophic, Ekman layer flow.

Furthermore, arbitrary changes in the "horizontal gravity" term may be imposed simply by changing the reference density ρ_0. To be precise, consider a change in the reference density from ρ_0 to $\rho^\star{}_0 = \rho_0 + \delta\rho_0$, where $\delta\rho_0 \ll \rho_0$ and $(\rho - \rho_0) \ll \rho0$. The buoyancy becomes

$$b^\star = -g(\rho - \rho_0^\star)/\rho_0^\star \approx b + \delta b_0, \qquad \delta b_0 = g\delta\rho_0/\rho_0^\star, \tag{14}$$

where we have approximated $\rho^\star{}_0 \approx \rho_0$ in the denominator. Thus (13) becomes, after some manipulations,

$$\underbrace{\int_z^0 \boldsymbol{U}\,\mathrm{d}z'}_{\text{Total transport}} \approx \underbrace{\frac{1}{\rho_0 f}\int_z^0 \hat{\mathbf{z}}\times\nabla_h\tilde{p}^\star\,\mathrm{d}z'}_{\text{Pressure gradient}} + \underbrace{\frac{1}{f}\int_z^0 b\hat{\mathbf{z}}\times\nabla_h N\,\mathrm{d}z' - z\frac{\delta b_0}{f}\hat{\mathbf{z}}\times\nabla_h N}_{\text{Horizontal gravity}} \underbrace{-\frac{1}{\rho_0 f}\hat{\mathbf{z}}\times\boldsymbol{\tau}}_{\text{Ekman transport}}. \tag{15}$$

Here $p^{\sim}{}_\star$ is the perturbation pressure, defined relative to a reference pressure $\rho^\star{}_0$ that hydrostatically balances the reference gravitational potential $\Phi^\star{}_0 = \rho^\star{}_0 g_0 (N - z)$ (c.f. Eq. (8)). Eq. (15) implies that an arbitrary change in the reference density leads to a vertically-uniform addition to the "horizontal gravity"-driven component of the flow, and thus a vertically-integrated transport that increases linearly with depth. This implies that the "horizontal gravity"-driven component of the flow is ill-defined, and thus that analyzing this flow in isolation, or as part of the 'Ekman' transport (as done by Chu[1]) is misleading.

To reconcile the apparent arbitrariness in the definition of the "horizontal gravity"-driven component of the flow in Eq. (15), we note that the perturbation pressure gradient may be written, using (8), as

$$\nabla_h \tilde{p} = \nabla_h \tilde{p}^\star + \delta\rho_0 g_0 \nabla_h N. \tag{16}$$

Substituting Eq. (16) into Eq. (15) we obtain, after some manipulations,

$$\underbrace{\int_z^0 \boldsymbol{U}\,\mathrm{d}z'}_{\text{Total transport}} \approx \underbrace{\frac{1}{\rho_0 f}\int_z^0 \hat{\mathbf{z}}\times\nabla_h\tilde{p}\,\mathrm{d}z' + z\frac{\delta b_0}{f}\hat{\mathbf{z}}\times\nabla_h N}_{\text{Pressure gradient}} + \underbrace{\frac{1}{f}\int_z^0 b\hat{\mathbf{z}}\times\nabla_h N\,\mathrm{d}z' - z\frac{\delta b_0}{f}\hat{\mathbf{z}}\times\nabla_h N}_{\text{Horizontal gravity}} \underbrace{-\frac{1}{\rho_0 f}\hat{\mathbf{z}}\times\boldsymbol{\tau}}_{\text{Ekman transport}}. \tag{17}$$

Thus the shift in the reference density produces equal and opposite perturbations to the pressure gradient-driven and "horizontal gravity"-driven components of the flow. Simplifying Eq. (17) by cancelling these terms recovers Eq. (13). This highlights the close dynamical link between the pressure gradient and "horizontal gravity" terms, and indicates that they should be considered to set the geostrophic component of the flow in tandem.

Based on these considerations, we argue that in an absolute spherical coordinate system, the appropriate decomposition of the flow would be to relate the geostrophic flow to the residual between the horizontal pressure gradient and "horizontal gravity",

$$\rho_0 f\hat{\mathbf{z}}\times\boldsymbol{U}_g = -\nabla_h p + \rho\boldsymbol{g}_h. \tag{18}$$

It follows that the "ageostrophic" or "Ekman" flow is

$$\rho_0 f\hat{\mathbf{z}}\times\boldsymbol{U}_E = \rho_0\boldsymbol{F}. \tag{19}$$

Thus the "Ekman" flow is unchanged by the shift to absolute spherical coordinates.

"Horizontal gravity" does not modify atmospheric Ekman pumping

The analysis of Chu[4] starts from the same momentum equation as that of Chu[1] (Eq. 19a of Chu[4]), but now describes flow in the atmosphere:

$$\underbrace{\rho_0 f\hat{\mathbf{z}}\times\boldsymbol{U}}_{\text{Coriolis}} + \underbrace{\nabla_3 p}_{\text{Pressure gradient}} \approx \underbrace{\rho\boldsymbol{g}}_{\text{Gravity}} + \underbrace{\rho_0\boldsymbol{F}}_{\text{External forcing}}. \tag{21}$$

Here ρ_0 is the reference density, ρ is the spatially-varying density, f is the Coriolis parameter, U is the horizontal velocity vector, p is the pressure, ∇_3 denotes the three-dimensional gradient operator, and F denotes additional sources and sinks of momentum. Note that the gravity vector g has both horizontal and vertical components, $g = gh + gz\ z\hat{}$. The analysis of Chu4 proceeds by identifying a "geostrophic" component of the flow that balances the horizontal pressure gradient via the Coriolis force:

$$\rho_0 f \hat{\mathbf{z}} \times \boldsymbol{U}_g^{C21} = -\nabla_h p. \tag{22}$$

The remaining component of the flow balances the horizontal component of the gravity vector and the external forcing via the Coriolis force, and is referred to as the "ageostrophic" or "Ekman" flow:

$$\rho_0 f \hat{\mathbf{z}} \times \left(\boldsymbol{U} - \boldsymbol{U}_g^{C21}\right) = \rho \boldsymbol{g}_h + \rho_0 \boldsymbol{F}. \tag{23}$$

Here the superscript ^{C21} denotes definitions made by1 . The author then solves (22) for the vertical structure of the "ageostrophic" flow, $\left(\boldsymbol{U} - \boldsymbol{U}_g^{C21}\right)$, throughout the atmospheric boundary layer, under the canonical assumption that the external forcing F takes the form of a vertical Laplacian operator with a constant eddy viscosity[3] .

This decomposition, from which all of the analysis of Chu[4] is derived, is critically flawed. The "horizontal gravity" force, which appears in (20) only because an absolute spherical coordinate system is used, should contribute to the definition of the geostrophic flow, rather than the ageostrophic/ Ekman flow. Far above the atmospheric Ekman layer, i.e. for $z \gg DE$, the eddy vertical viscosity may be neglected and thus $F = 0$. Thus the far-field horizontal velocity is given by

$$\rho_0 f \hat{\mathbf{z}} \times \boldsymbol{U} = -\nabla_h p + \rho \boldsymbol{g}_h \quad \text{for} \quad z \gg D_E. \tag{24}$$

The canoncial formulation of the atmospheric Ekman layer3 describes the behavior of a viscous boundary layer in response to an externally-imposed, geostrophic flow. In the absolute spherical coordinate system used by Chu[4] the externally-imposed flow is given by (23). Therefore, in this coordinate system the appropriate decomposition would be to define the geostrophic flow as

$$\rho_0 f \hat{\mathbf{z}} \times \boldsymbol{U}_g = -\nabla_h p + \rho \boldsymbol{g}_h, \tag{25}$$

and the "ageostrophic" or "Ekman" flow as

$$\rho_0 f \hat{\mathbf{z}} \times \left(\boldsymbol{U} - \boldsymbol{U}_g\right) = \rho_0 \boldsymbol{F}. \tag{26}$$

Thus the "Ekman" flow and pumping are unchanged by the shift to absolute spherical coordinates.

References

1. Chu, P. C. True gravity in ocean dynamics. Part 1: Ekman transport. Dyn. Atmospheres Ocean. 96, 101268 (2021).
2. Svensson, U. The structure of the turbulent Ekman layer. Tellus 31, 340–350 (1979).
3. Pedlosky, J. Geophysical fluid dynamics (Springer Science & Business Media, 2013).
4. Chu, P. C. True gravity in atmospheric ekman layer dynamics. J. Geophys. Res. Atmospheres 126, e2021JD035293 (2021)

ACKNOWLEDGEMENTS

We appreciate extensive discussions with colleagues in coming to understand the contents of these papers.

Author contributions

A.L.S. and J.C.McW. wrote and reviewed the manuscript.

Competing interests

The authors declare no competing interests.

Footnotes

Publisher's note

Springer Nature remains neutral with regard to jurisdictional claims in published maps and institutional affiliations.

REFERENCES

1. Wunsch C, Stammer D. Satellite altimetry, the marine geoid, and the oceanic general circulation. *Annu. Rev. Earth Planet. Sci.* 1998;**26**:219–253. doi: 10.1146/annurev.earth.26.1.219.
2. 2. Chu PC. Ocean dynamic equations with the real gravity. *Sci. Rep.* 2021;**11**:1–10. doi: 10.1038/s41598-020-79139-8. [PMC free article] [PubMed] Retracted
3. Chu, P. C. True gravity in ocean dynamics. Part 1: Ekman transport. *Dyn. Atmos. Oceans***96**, 101268 (2021).
4. Chu, P. C. True gravity in atmospheric Ekman layer dynamics. *J. Geophys. Res. Atmos.***126**, e2021JD035293 (2021).
5. de Laplace, P.-S. M. Suite des récherches sur plusieurs points du système du monde (XX.V–XXVII). *Mem. Présentés Divers Savans Acad. R. Sci. Inst. France*, 525–552 (1776).
6. Lagrange, J.-L. Mémoire sur la théorie du mouvement des fluides. *Nouv. Mem. Acad. Berlin*, p. 196 (1781).
7. Lagrange, J.-L. Sur la manière de rectifier deux endroits des principes de newton relatifs à la propagation du son et au mouvement des ondes. *Nouv. Mém. Acad. Berlin***5**, 591–609 (1786).
8. Craik ADD. The origins of water wave theory. *Annu. Rev. Fluid Mech.* 2004;**36**:1–28. doi: 10.1146/annurev.fluid.36.050802.122118.
9. Hendershott M, Munk W. Tides. *Annu. Rev. Fluid Mech.* 1970;**2**:205–224. doi: 10.1146/annurev.fl.02.010170.001225.
10. Lamb H. *Hydrodynamics*. Dover Publications; 1932.
11. Hendershott, M. Long waves and ocean tides (1980).
12. Phillips, N. A. Principles of large scale numerical weather prediction. In *Dynamic Meteorology* 1–96 (Springer, 1973).
13. Pedlosky, J. *Geophysical Fluid Dynamics* (Springer Science & Business Media, 2013).
14. Vallis GK. *Atmospheric and Oceanic Fluid Dynamics*. Cambridge University Press; 2017.

CHAPTER 7

Convective Cells of Internal Gravity Waves in the Earth's Atmosphere with Finite Temperature Gradient

O. Onishchenko[1,2], O. Pokhotelov[1], and V. Fedun[3]

[1]Institute of Physics of the Earth, 123995, 10 B. Gruzinskaya str., Moscow, Russia
[2]Space Research Institute, 117997, 84/32 Profsoyuznaya str., Moscow, Russia
[3]Dept. of Automatic Control and Systems Engineering, The University of Sheffield, Mappin Street, Sheffield, S1 3JD, UK

ABSTRACT

In this paper, we have investigated vortex structures (e.g. convective cells) of internal gravity waves (IGWs) in the earth's atmosphere with a finite vertical temperature gradient. A closed system of nonlinear equations for these waves and the condition for existence of solitary convective cells are obtained. In the atmosphere layers where the temperature decreases with height, the presence of IGW convective cells is shown. The typical parameters of such structures in the earth's atmosphere are discussed.

Citation: Onishchenko, O., Pokhotelov, O., and Fedun, V.: "Convective cells of internal gravity waves in the earth's atmosphere with finite temperature gradient", Ann. Geophys., 31, 459–462, https://doi.org/10.5194/angeo-31-459-2013, 2013.

Keywords. Meteorology and atmospheric dynamics (Middle atmosphere dynamics)

INTRODUCTION

The study of the lithosphere and neutral atmosphere/ionosphere coupling is the fundamental problem of geophysics. According to modern concepts the mechanism of lithosphere–ionosphere interaction is closely related to the internal gravity waves (IGWs). In the atmosphere of the earth, the internal gravity waves are generated by various terrestrial processes. For example, thunderstorms, volcanic eruptions, tsunami, typhoons, zonal winds, and earthquakes can be a strong sources of such waves (see e.g. Kazimirovsky et al., 2003; Occhipinti and Lognonne′, 2008; Varney et al., 2009; Grimalsky et al., 2010; Rozhnoi et al., 2012, for details). It was shown (see e.g. Hocking, 1996; Fritts and Alexander, 2003; Liu and Gardner, 2004; Lu et al., 2009) that, in an atmosphere in which the density decreases with height, the internal gravity waves propagate to high altitudes with increasing of their amplitude and, therefore, can perturb the ionosphere. Satellite and ground-based electromagnetic sounding of the neutral atmosphere and lower ionosphere has shown a strong connection between terrestrial catastrophes and perturbations of these layers. Therefore, the problem of forecasting and monitoring of terrestrial events by electromagnetic sensing is closely linked with the problem of IGW propagation.

Generation of nonlinear structures may limit the growth of IGW perturbations. The parametric generation of zonal structures by small-scale weakly turbulence of IGW in the atmosphere has been investigated previously by Horton et al. (2008) and Onishchenko and Pokhotelov (2012). Generated zonal wind is a self-regulating system of shear flows, significantly weakening the transport processes in the vertical direction. This mechanism provides an effective channel of energy transfer from the small-scale turbulence region to the region of IGW global convective motions and plays an important role in the regularization of atmospheric turbulence. Such a process is the paradigm of the inverse cascade in turbulent two-dimensional theory of anisotropic turbulence as a result of the formation of a regular large-scale structure from the small-scale chaos.

Nonlinear IGWs can also exist in the form of the vortex structures (see Stenflo, 1987, 1990, 1991; Stenflo and Shukla, 2009, for example). In these works, authors have shown that horizontal velocity of vortices should be of the order or greater than the speed of sound.

The finite vertical gradient of temperature determines dynamics of the vertical transport and convective stability of the atmosphere. The stratified atmosphere is stable, if the square of height-averaged Brunt–Vais ¨ al¨ a frequency ¨ ω_g, i.e.

$$\omega_g^2 = g\left(\frac{\gamma - 1}{\gamma H} + \frac{1}{T}\frac{dT}{dz}\right) \qquad (1)$$

is greater than zero. Here, g is the gravitational acceleration, d/dz is the derivative in height, γ is the ratio of specific heats, H is the local scale of height of the atmosphere and T is an unperturbed temperature. Usually in the earth's atmosphere ω^2_g is positive. But, due to time dependence of temperature gradient it may be negative in some layers. In the regions where the temperature decreases with height and the square of height-averaged Brunt–Vais ¨ al¨ a frequency is negative, the powerful vertical convective transport occurs. The condition $\omega^2_g < 0$ corresponds to the absolute instability of IGWs; i.e. frequencies of waves are imaginary.

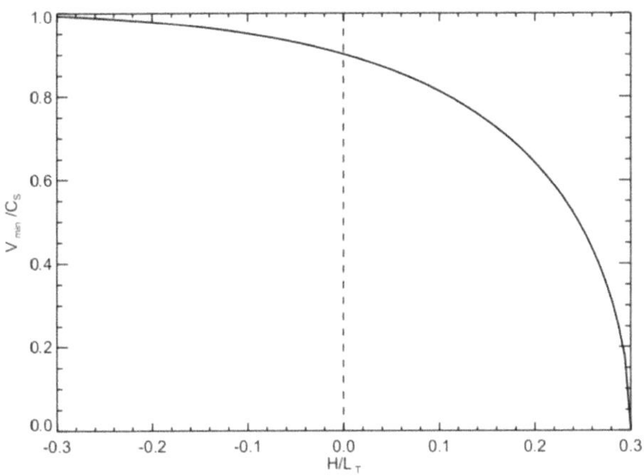

Figure 1. The normalized minimum speed of vortices (v_{min}/c_S) as a function of the dimensionless parameter H/L_T .

Nonlinear IGWs in the atmosphere with a finite temperature gradient were investigated previously by Jovanovic et al. ′ (2001, 2002). Authors have shown that many different types of vortex structures can exist in the atmosphere layers where the condition $(d^2\rho_0/dz^2)^{1/2} \gg d\rho_0/dz$ is satisfied. Here, ρ_0 is the unperturbed density.

In this paper we present a new theoretical study of the nonlinear vortex structures (i.e. convective cells) of IGWs in the earth's atmosphere layers where density and temperature vary in height according to the exponential law. The vertical temperature gradient is taken into account, and the vortex structures have been studied not only in the stably stratified atmosphere, but also in the case of unstable stratified atmosphere.

The paper is structured as follows. In Sect. 2 we derive the close system of the nonlinear equations for IGWs. Section 3 is devoted to the stationary vortex solutions and criteria for existence of such structures. Our discussion and conclusions are found in Sect. 4.

THE REDUCED EQUATIONS FOR THE FINITE AMPLITUDE IGWS

In the current study we neglect the influence of dissipative processes, e.g. viscosity, thermal conductivity, heat flow from outside, friction, etc. We begin with the hydrodynamic equations in the local Cartesian coordinate system (x, y,z), where the axis z coincides with the vertical. To investigate the non linear dynamics of low-frequency IGWs, we use conservation equation of momentum

$$\rho\frac{d\boldsymbol{u}}{dt}+\nabla p-\rho\boldsymbol{g}=0 \qquad (2)$$

and equation for transport of potential temperature,

$$\frac{d\theta}{dt}=0. \qquad (3)$$

Here $\theta = p^{1/\gamma}/\rho$ is the potential temperature; ρ and p are the density and pressure, respectively; $d/dt = \partial/\partial t + u \cdot \nabla$ is the Euler (convective) time derivative; u is the velocity of matter; $g = -g\hat{z}$ is the gravity acceleration and $\hat{z}$ is the unit vector of the local Cartesian coordinate system (x, y,z) which is directed along the vertical. To describe the two-dimensional motion in the gravitational waves, assume that u = (u,0,w), where $u = -\partial\psi/\partial z$ and $w = \partial\psi/\partial x$. Here, ψ (t, x,z) is the stream function.

Assuming a weak perturbation, the density ρ and pressure p can be written as follows: $\rho = \rho_0(z) + \tilde{\rho}(t, x,z)$, $p = p0(z) + \tilde{p}(t, x,z)$, where ρ_0 and p_0 are the unperturbed density and pressure, with $\tilde{\rho}$ and $\tilde{p}$ the corresponding disturbances, $|\tilde{\rho}| << \rho_0$ and $|\tilde{p}| << p_0$. As a result, from Eqs. (2) and (3) one obtains a system of two nonlinearly coupled equations:

$$\frac{\partial}{\partial t}\left(\nabla^2\psi + \frac{dln\left(\rho_0\right)}{dz}\frac{d\psi}{dz}\right) + J\left(\psi, \nabla^2\psi\right)$$
$$= -\frac{\partial \chi}{\partial x} + \frac{1}{\rho_0^2}J\left(\tilde{\rho}, \tilde{p}\right) \tag{4}$$

and

$$\frac{\partial \chi}{\partial t} - \omega_g^2\frac{\partial \psi}{\partial x} + \{\psi, \tilde{\rho}\} = 0. \tag{5}$$

Here J (a, b) = $(\partial a/\partial x)\partial b/\partial z - (\partial a/\partial z)\partial b/\partial x$ is the Jacobian, $\nabla^2 = \partial^2/\partial x^2 + \partial^2/\partial z2$, and $\chi = g\rho/\tilde{\rho}_0$ is the normalized density perturbation. In an unperturbed atmosphere the pressure decreases exponentially with height; i.e. $p_0(z - z_0) = p_0(z_0)\exp[-(z - z_0)/H]$, where $H = c_S^2/\gamma g$ is the scale height, and $c_S = (\gamma p_0/\rho_0)^{1/2}$ is the speed of sound. Next, let us define the unperturbed temperature as a function of height as $T(z - z_0) = T(z_0)\exp[-(z - z_0)/L_T]$, where L_T is the characteristic scale of the vertical temperature gradient. Note that, in the layers of the atmosphere where the temperature decreases with height, $L_T > 0$; in an inversion layer, $L_T < 0$. According to the ideal gas law, i.e. $p/\rho T$ = constant, we obtain the expression for the unperturbed density,

$$\rho_0(z - z_0) = \rho_0(z_0)\,exp\left[-(z - z_0)/L_\rho\right],$$

and the relation between the characteristic scale of the vertical density gradient ($L\rho$), the scale height (H) and the characteristic scale of the vertical temperature gradient (L_T):

$$1/L_\rho = 1/H - 1/L_T.$$

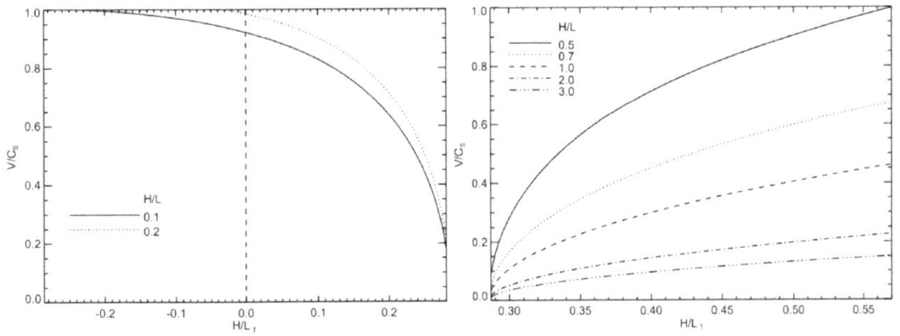

Figure 2. The normalized horizontal speed of vortex (v/c_S) as a function of the dimensionless parameter H/L and H/L_T. The vortex speeds have been plotted by assuming the applicable parameters for the stable and unstable stratified atmospheres (see the left and right panels, respectively).

Next, by using the substitution $\psi = \hat{\psi}(t, x, z)$ exp (z − z_0)/2L_ρ , and $\chi = \hat{\chi}(t, x, z)$ exp (z − z_0)/2L_ρ , Eqs. (4) and (5) can be represented in the form

$$\frac{\partial}{\partial t}\left(\nabla^2\hat{\psi} - \frac{\hat{\psi}}{4L_\rho^2}\right) + J\left(\hat{\psi}, \nabla^2\hat{\psi}\right) = -\frac{\partial\hat{\chi}}{\partial x}, \tag{6}$$

and

$$\frac{\partial\hat{\chi}}{\partial t} - \omega_g^2\frac{\partial\hat{\psi}}{\partial x} + J\left(\hat{\psi}, \hat{\rho}\right) = 0. \tag{7}$$

This closed system of Eqs. (6) and (7) can be used for numerical simulations of the nonlinear dynamics of IGWs in the earth's atmosphere with a finite vertical temperature gradient. In the limit of an isothermal atmosphere, i.e. if L_T >> H, the system of Eqs. (6) and (7) coincides with the system of equations studied previously by Horton et al. (2008); Onishchenko and Pokhotelov (2012); Kaladze and Tsamalashvili (1997). From Eqs. (4) and (5) follows the law of conservation of wave energy ∂E/∂t = 0, where

$$E = \frac{1}{2}\int\rho_0\left[(\nabla\psi)^2 + \frac{\psi^2}{4L_\rho^2} + \frac{\chi^2}{\omega_g^2}\right]dxdz. \tag{8}$$

Equation (8) shows that, in the case of an unstable stratified atmosphere, the energy can be negative.

CONVECTIVE CELLS

To find a stationary vortex solution in a coordinate system that is moving with the velocity v along the x-axis, let us introduce the new variable η = x −vt. Under this coordinate transformation the system of the two Eqs. (6) and (7) can be reduced to a single equation:

$$J_1\left(\nabla^2\hat{\psi} - \Lambda\hat{\psi}, \hat{\psi} - vz\right) = 0, \tag{9}$$

where J1 (a, b) = (∂a/∂η)∂b/∂z − (∂a/∂z)∂b/∂η and

$$\Lambda = \frac{1}{4L_\rho^2} - \frac{\omega_g^2}{v^2}. \tag{10}$$

A similar equation has been studied analytically and numerically; see e.g. Onishchenko et al. (2008) and references therein. It has been shown that the solution of Eq. (9) can be found in the form of dipole vortices. In the outer region of the vortex condition,

$$\nabla^2 \hat{\psi} - \Lambda \hat{\psi} = 0 \tag{11}$$

is satisfied. Solitary vortices in the atmosphere with a finite vertical temperature gradient exist if $\Lambda > 0$. From Eq. (10) one can obtain the relation that connects the characteristic scale of the vortex (convective cell) $L = \Lambda^{-1/2}$ with the typical atmospheric parameters scales L_T, H and the horizontal velocity of the vortex v:

$$\frac{H}{L} = \frac{1}{2}\left[\left(1 - \frac{H}{L_T}\right)^2 - \frac{4c_S^2}{\gamma v^2}\left(\frac{\gamma - 1}{\gamma} - \frac{H}{L_T}\right)\right]^{1/2}. \tag{12}$$

The right term of Eq. (12) should be real. Therefore, in the atmosphere layer with decreasing temperature, the vortex velocity should be limited from below. From here we obtain the minimum possible value of the vortex velocity:

$$\frac{v_{\min}^2}{c_S^2} = \frac{4}{\gamma}\left(\frac{\gamma - 1}{\gamma} - \frac{H}{L_T}\right)\left(1 - \frac{H}{L_T}\right)^{-2}. \tag{13}$$

Figure 1 and the left panel of Fig. 2 show dependence of the vortex parameters vmin and v in a stable stratified atmosphere with the effective value $\gamma = 1.4$ and $H/L_T < 0.28$. In unstable stratified atmosphere layers when $H/L_T > 0.28$, there is no limit on horizontal velocity of vortices and the dependence of normalized vortex velocity on parameters H/LTand H/L shown in the right panel of Fig. 2. One can see from these figures that the vortex structures with spatial scales L of the order of H can exist in the atmosphere with $H/L_T > 0.28$. The right panel of Fig. 2 shows that spatial scale of the vortex structures can be smaller than H. With the growth of the vortex scale, the vortex speed is growing.

CONCLUSIONS

In this paper we studied the vortex structure of IGWs in the earth's atmosphere with a finite vertical temperature gradient. The possibility of the vortices' existence in the thermospheric layers, with the vertical temperature gradient of the order of the dry adiabatic gradient, has been shown. The condition for the minimum velocity of vortices in such an atmosphere has been obtained. We have found that the horizontal velocity of the vortices is significantly less than the speed of sound. We have investigated IGW vortex structures in the case of an unstable stratified atmosphere, i.e. for $\omega_g^2 < 0$. It is shown that in such atmospheric layers (i.e. in the thermosphere) there is no lower limit on the velocity of the vortices.

ACKNOWLEDGEMENTS

This work was supported by the Russian Fund for Basic Research (grant 11-05-00920) and Program 8 of the Russian Academy Sciences. Also, authors are grateful for the partial financial support from the International Space Science Institute (Bern, Switzerland) through the grant "Large-scale vortices and zonal winds in planetary atmospheres/ionospheres: Theory vs. observations"

Guest Editor M. Balikhin thanks two anonymous referees for their help in evaluating this paper.

REFERENCES

1. Fritts, D. and Alexander, M.: Gravity wave dynamics and effects in the middle atmosphere, Rev. Geophysics, 41, 1–64, 2003.
2. Grimalsky, V., Kotsarenko, A., Pulinets, S., Koshevaya, S., and Perez-Enriquez, R.: On the modulation of intensity of Alfven resonances before earthquakes: Observations and model, J. Atmos. Solar-Terr. Phys., 72, 1–6, 2010.
3. Hocking, W.: Dynamical coupling processes between the middle atmosphere and lower ionosphere, J. Atmos. Terr. Phys., 58, 735– 752, 1996.
4. Horton, W., Kaladze, T., Van Dam, J., and Garner, T.: Zonal flow generation by internal gravity waves in the atmosphere, J. Geophys. Res., 113, A08312, doi:10.1029/2007JA012952, 2008.
5. Jovanovic, D., Stenflo, L., and Shukla, P. K.: Acoustic gravity tripolar vortices, Phys. Lett. A, 279, 70–74, 2001.
6. Jovanovic, D., Stenflo, L., and Shukla, P. K.: Acoustic-gravity ´ nonlinear structures, Nonlin. Processes Geophys., 9, 333–339, doi:10.5194/npg-9-333-2002, 2002.
7. Kaladze, T. and Tsamalashvili, L.: Solitary dipole vortices in the Earth's ionosphere, Phys. Lett. A., 232, 269–274, 1997.
8. Kazimirovsky, E., Herraiz, M., and De La Morena, B. A.: Effects on the Ionosphere Due to Phenomena Occurring Below it, Surv. Geophys., 24, 139–184, 2003.
9. Liu, A. and Gardner, C.: Vertical dynamical transport of mesospheric constituents by dissipating gravity waves, J. Atmos. Sol. Terr. Phys., 66, 267–275, 2004.
10. Lu, X. Liu, A. Z., Swenson, G., Li, T., Leblanc, T., and McDermid, I. S.: Gravity wave propagation and dissipation from the stratosphere to the lower thermosphere, J. Geophys. Res., 114, D11101, doi:10.1029/2008JD010112, 2009.
11. Occhipinti, G., Kherani, E. A. and Lognonne, P.: Geomagnetic de- ´ pendence of ionospheric disturbances induced by tsunamigenic internal gravity waves, Geophysical Journal International, 173, 753–765, 2008.
12. Onishchenko, O. and Pokhotelov, O.: Generation of zonal structure by internal gravity waves in the Earths Atmosphere, Doklady Earth Sciences, 445, 845–848, 2012.

13. Onishchenko, O., Pokhotelov, O., and Astafieva, N.: Generation of large-scale eddies and zonal winds in planetary atmospheres, Physics-Uspekhi, 51, 577–589, 2008.
14. Rozhnoi, A., Shalimov, S., Solovieva, M., Levin, B., Hayakawa, M., and Walker, S.: Tsunami-induced phase and amplitude perturbations of subionospheric VLF signals, J. Geophys. Res. (Space Physics), 117, A09313, doi:10.1029/2012JA017761, 2012.
15. Stenflo, L.: Acoustic solitary vortices, Phys. Fluids, 30, 3297–3299, 1987. Stenflo, L.: Acoustic gravity vortices, Phys. Scripta, 41, 641–642, 1990.
16. Stenflo, L.: Acoustic gravity vortices, Z. Naturforsch., 46, 560–560, 1991.
17. Stenflo, L. and Shukla, P.: Nonlinear acoustic-gravity waves, J. Plasma Phys., 75, 841–847, 2009.
18. Varney, R. H., Kelley, M. C., and Kudeki, E.: Observations of electric fields associated with internal gravity waves, J. Geophys. Res. (Space Physics), 114, 2304, doi:10.1029/2008JA013733, 2009.

SECTION 3
THE WATER AND THE VERTICAL STRUCTURE OF MOISTURE IN THE ATMOSPHERE

CHAPTER 8

Water from Air: an Overlooked Source of Moisture in Arid and Semiarid Regions

Theresa A. McHugh[1,2], Ember M. Morrissey[2], Sasha C. Reed[1], Bruce A. Hungate[2,3] & Egbert Schwartz[2,3]

[1] U.S. Geological Survey, Southwest Biological Science Center, Moab, UT, USA.

[2] Center for Ecosystem Science and Society, Northern Arizona University, Flagstaff, AZ, USA.

[3] Department of Biological Sciences, Northern Arizona University, Flagstaff, AZ, USA.

Citation: McHugh, T., Morrissey, E., Reed, S. et al. "Water from air: an overlooked source of moisture in arid and semiarid regions". Sci Rep 5, 13767 (2015). https://doi.org/10.1038/srep13767.

ABSTRACT

Water drives the functioning of Earth's arid and semiarid lands. Drylands can obtain water from sources other than precipitation, yet little is known about how non-rainfall water inputs influence dryland communities and their activity. In particular, water vapor adsorption – movement of atmospheric water vapor into soil when soil air is drier than the overlying air – likely occurs often in drylands, yet its effects on ecosystem processes are not known. By adding ^{18}O-enriched water vapor to the atmosphere of a closed system, we documented the conversion of water vapor to soil liquid water across a temperature range typical of arid ecosystems. This phenomenon rapidly increased soil moisture and stimulated microbial carbon (C) cycling, and the flux of water vapor to soil had a stronger impact than temperature on microbial activity. In a semiarid grassland, we also observed that non-rainfall water inputs stimulated microbial activity and C cycling. Together these data suggest that, during rain-free periods, atmospheric moisture in drylands may significantly contribute to variation in soil water content, thereby influencing ecosystem processes. The simple physical process of adsorption of water vapor to soil particles, forming liquid water, represents an overlooked but potentially important contributor to C cycling in drylands.

In drylands, precipitation is so low relative to demand that the availability of water represents the primary control over biological processes[1]. Accordingly, understanding both the sources of dryland water and how water is accessed is critical for understanding and forecasting arid and semiarid ecosystem function. Drylands cover 41% of Earth's terrestrial surface[2] and represent the planet's largest biome[3]. Thus, improving our understanding of the primary factor controlling dryland function – water availability – is important at the global scale. This is particularly true with regard to the global C cycle as, due to their large spatial extent, dryland soils store nearly twice as much soil organic C as temperate forest soils[4,5]. Recent research also suggests that C turnover rates in semiarid ecosystems are an increasingly dominant driver of inter-annual variability in the global C cycle[6].

This significant role in global biogeochemical cycles and the central influence of water in regulating activity has generated interest in the relationships between precipitation and C cycling in dryland ecosystems[7,8,9,10]. Climate change in the form of altered precipitation timing and amount could alter dryland biota and biogeochemical cycling[11,12]. However, dryland ecosystems obtain water from sources other than precipitation and, in contrast to the many studies exploring the links between precipitation and dryland C cycling, almost nothing is known about how soil water derived from atmospheric water vapor influences dryland communities and their activity, nor about how climate change will affect this water input and its interactions with soil C cycling. This knowledge is particularly relevant as increased temperature results in increased capacity of the atmosphere to hold more water[13,14].

In the absence of precipitation, there are three primary mechanisms for the addition of water to surface soils: fog deposition, dew formation, and water vapor adsorption[15]. Fog occurs only when atmospheric water vapor concentrations reach saturation, an uncommon condition in most drylands (but see reference[16]). Dew is formed when the surface temperature is lower than or equal to the dewpoint temperature, at which time water vapor from the air in contact with the cold soil surface condenses to form dew. While dew has been suggested to be an important source of water in some drylands[17,18,19], there are many systems and seasons where dew formation is likely to be rare[15,20]. In contrast, the formation of liquid water in soil pores resulting from the adsorption of water vapor from air can occur any time the relative humidity in the soil atmosphere is lower than the relative humidity of the overlying air[15] – in other words, when soil air is drier than the overlying atmosphere. Environmental conditions are such that this phenomenon could take place frequently in many dryland regions throughout the world[15], constituting an important link in the water cycle of arid and semiarid regions[21,22,23]. Yet, the limited knowledge we do have about the ecological role of water vapor adsorption – as a liquid water source for vascular plants[22,23] and as a driver of litter decomposition[24,25,26] – comes from coastal regions that experience high relative humidity[20,27]. Strong water limitation in drylands means that soil water inputs from adsorption of atmospheric water vapor could be an important driver of soil biogeochemistry, yet whether the phenomenon is quantitatively significant and ecologically relevant is not known.

To test for the occurrence of water vapor adsorption and examine the conditions under which it occurs, we conducted two sets of laboratory experiments. First, ^{18}O-enriched water vapor was added to the atmosphere of sealed jars containing field dry soil from a semiarid grassland. Temperature and relative humidity were monitored over a three-day period as soils experienced simulated diurnal temperature fluctuations (23 °C for 4 h, 40 °C for 8 h, 23 °C for 4 h, and 10 °C for 8 h) typical for an arid region. Through the addition of ^{18}O-enriched water vapor to the atmosphere (0, 50, 100, or 500 μg fully evaporated into ~1 L headspace), we were able to trace the movement of water vapor into the soil. Addition of water vapor to the atmosphere increased the atmospheric relative humidity (Fig. 1a) and soil gravimetric water content (Fig. 1b). The increase in soil gravimetric water content was due to vapor adsorption, as the soil water ^{18}O atom percent excess (APE) confirmed the transfer of moisture from the atmosphere to the soil (Fig. 1c). Further, the amount of water vapor adsorbed (calculated from ^{18}O APE) was linearly related to the gravimetric moisture content of the soil ($r^2=0.91$, $p<0.00001$).

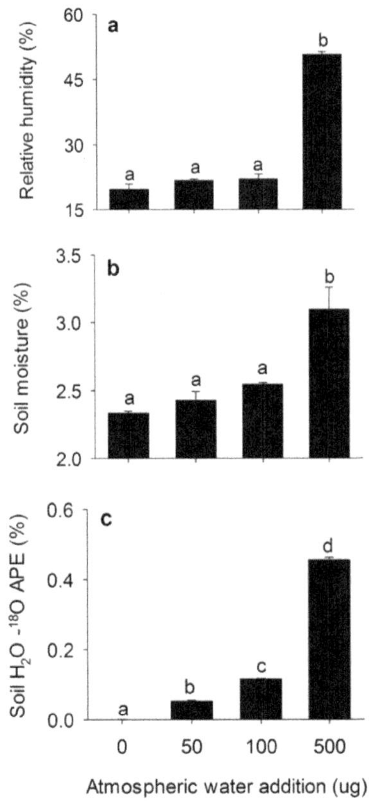

Figure 1: Atmospheric relative humidity (**a**) soil gravimetric moisture content (**b**) and soil water ^{18}O atom percent excess (**c**) in response to atmospheric water vapor addition. Error bars are standard error for means (n=3). Significant differences as determined by one-way ANOVA and Tukey's post-hoc test are indicated using lowercase letters ($\alpha=0.05$).

Across all treatments, 35–50% of the water added to the atmosphere was adsorbed. Water vapor adsorption occurred at atmospheric relative humidity values ranging from 20–60% (Fig. 1a,b), which are consistent with humidity values reported for arid and semiarid lands across the globe[28]. Because soil temperatures were well above dewpoint temperatures, the transfer of water from the atmosphere to the soil could not have occurred through dew.

A second laboratory experiment was performed to examine the influence of temperature on water vapor adsorption and the extent to which it regulates soil microbial activity. Field dry soil was incubated for three days under the simulated diurnal temperature fluctuations previously described, or at a constant temperature of either 23 or 40 °C. Water vapor was added to the atmosphere of treatment jars (500 μg fully evaporated into ~1 L headspace), while experimental controls received no water addition. Water vapor adsorption was confirmed by an increase in the gravimetric water content of the soils over the course of the incubation. Relative to their moisture content at the initiation of the experiment (2.3%), the average moisture content of soils in the water vapor adsorption treatment increased by 65.9% for the simulated diurnal treatment, 63.3% for the 23 °C treatment, and 59.9% for the 40 °C temperature regime. Thus, water vapor adsorption occurred under all temperature regimes, and the concomitant increases in CO_2 efflux (Fig. 2) indicate that vapor adsorption stimulated soil CO_2 production. The effect of water vapor adsorption was significantly greater in soils experiencing simulated diurnal fluctuations ($p<0.0001$, Fig. 2). Overall, temperature explained 23% of the variation in soil CO_2 efflux rates, while water vapor addition explained 47% (two-way ANOVA). These results suggest that temperature may have opposing effects on microbial activity in dryland soils. On one hand, higher temperatures, which increase molecular diffusion rates, will lead to greater respiration activity but also deplete soils of scarce moisture. On the other hand, colder temperatures will allow more water vapor adsorption, thereby increasing soil moisture and microbial habitat in soil. Maximum respiration rates may be observed when soils are warming up early in the morning, as they still contain moisture trapped through water vapor adsorption, yet are warm enough to support high microbial activity. Others have also observed peaks in respiration during the relatively cool morning hours[29], and these results, which challenge commonly held beliefs about the relationship between respiration and temperature, could be explained by the movement of water from air to soil. Future climate conditions could alter the importance of water vapor adsorption in dryland function, not only through changes in precipitation and relative humidity, but also via interactions with temperature.

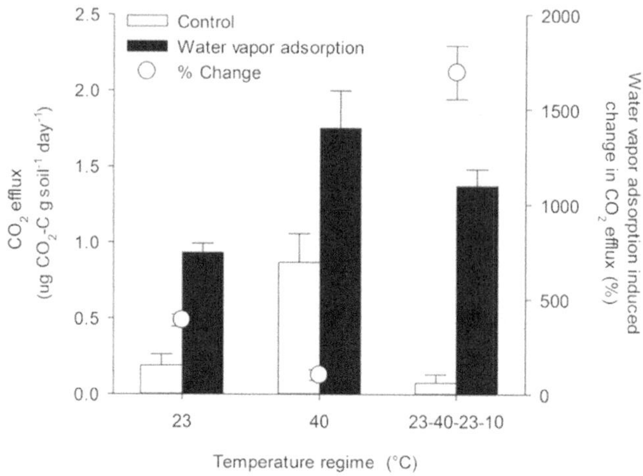

Figure 2: Soil CO_2 **efflux rates across temperature regimes for control and water** vapor adsorption treatment.

Error bars are standard error for means (n=5). The effect of water vapor adsorption (percent change relative to control) is shown with white circles.

We also explored atmospheric water inputs in the field. At a semiarid grassland in northern Arizona, USA (Supplementary Fig. S1), changes in atmospheric relative humidity, soil moisture, and soil respiration were monitored over a 24-h period during the dry season when native grasses were dormant. We observed a strong diurnal pattern in atmospheric relative humidity that was correlated with temperature, and as expected, during the late night and early morning when temperatures were lowest, relative humidity was highest. There was a strong positive correlation between atmospheric relative humidity and soil moisture ($r^2=0.42$, $p<0.001$). Relative humidity varied from 14 to 49%, gravimetric soil moisture content varied from 0.36 to 4.79%, and both were generally higher during the morning hours. These findings are consistent with other studies showing that diurnal fluctuations in water vapor adsorption by soil during the summer season follow fluctuations in atmospheric relative humidity[22,30,31]. The increase in the moisture content of the top five centimeters of soil, in conjunction with no associated changes in the moisture content of deeper soil layers, indicates that the water addition was the result of atmospheric inputs and not simply the redistribution of water within the soil profile. Though the resolution of our temperature measurements did not allow us to determine if dew formation also contributed to the observed increase in moisture and activity, the robust relationship between soil moisture and relative humidity – in combination with the lack of moisture input from other sources – provides strong evidence that atmospheric water was added to soil in this grassland ecosystem.

While we focused our study efforts on bare patches of soil within a grassland ecosystem, it is important to consider the microclimate that develops within the plant canopy[32] and how these conditions impact the addition of non-rainfall water. In a semiarid grassland in southeastern Spain, net water gains from water vapor adsorption were found to be higher in bare soil than in soil under *Stipa tenacissima* tussocks, as vegetation limits fluctuations in atmospheric relative humidity near the soil[30].

In order to test the hypothesis that increased soil moisture from non-rainfall water inputs was the main driver of soil CO_2 efflux in the field (Supplementary Fig. S2), we conducted a path analysis to determine how soil temperature and moisture affected soil respiration. A theory-constrained full model was developed to represent all hypothetical relationships between soil temperature, soil moisture, and soil CO_2 efflux within our system. Using the experimental data, a reduced model was attained by sequentially removing paths in order of largest probability value until all remaining paths were significant (Fig. 3). These results are consistent with a scenario in which low temperatures enable the movement of water vapor from air to soil, and the subsequent increase in soil moisture, in turn, stimulates soil CO_2 efflux rates. This demonstration of increased CO_2 efflux in the laboratory and under field conditions provides strong evidence that atmospheric water influences the activity of soil microorganisms.

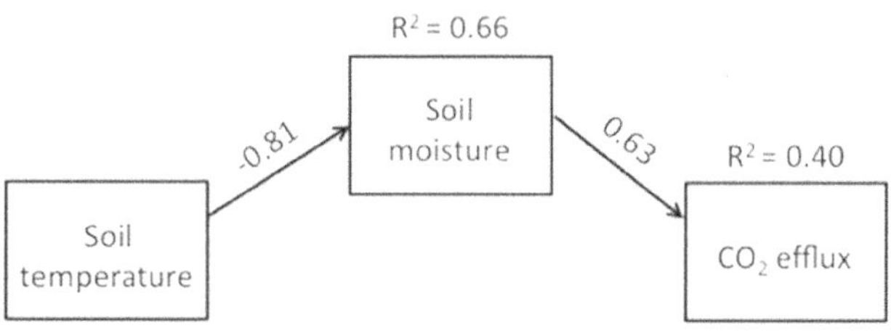

Figure 3: Path diagram displaying the role of soil temperature and soil moisture in regulating soil CO_2 **efflux rates under a reduced model.**

Arrows represent unidirectional causal relationships. The amount of variation that can be explained by the model is indicated by the R^2 values associated with each response variable. Standardized path coefficients (r) associated with each arrow reflect the strength of each relationship ($p < 0.001$ in both cases). The full model also included the direct assessment of temperature effects on CO_2 efflux rates, but since that relationship was not statistically significant, no arrow is shown.

While we know little about the biological significance of atmospheric water vapor on soil microbial activity, it has been well documented that very small moisture inputs can activate ecosystem processes in dryland regions[33,34]. Indeed, small inputs of water can drive dryland community composition and biogeochemical cycles[8,10,12]. Thus, liquid water from the atmosphere has the potential to be a critical source of water for dryland organisms. More regions worldwide are expected to experience reduced rainfall as a result of climate change[35], a scenario where non-rainfall atmospheric water inputs could become an increasingly important source of soil moisture as higher amounts of atmospheric water vapor are predicted globally[13,14]. Although the nature of the interactions among temperature, atmospheric humidity, and soil humidity require further inquiry, our data suggest that these unexplored changes could impact dryland C fluxes. In particular, our laboratory and field results demonstrated that non-rainfall addition of moisture to soil can occur across a large range of relative humidity values, and established that this phenomenon influences soil C cycling in dryland soils. Future research should aim to address the applicability of these observations to global drylands, quantify the role of dew versus water vapor adsorption inputs, and determine the contribution of this water source to daily and seasonal C fluxes.

METHODS

Laboratory Experiments

In the first laboratory experiment, 30 g samples of sieved (2 mm) soil from the grassland were placed in 1000 ml sealed Nalgene jars also containing a hygro-thermometer (Extech Instruments; Nashua, NH) and a small reservoir with either 0, 50, 100, or 500 µg of ^{18}O-enriched water (2 atom%). Triplicate jars for each treatment were then subjected to temperature cycling designed to simulate diurnal temperature fluctuations typical of an arid environment: 4 h at 23 °C, 8 h at 40 °C, 4 h at 23 °C, 8 h at 10 °C, repeated over a 72-h period. Immediately prior to each temperature transition, relative humidity and temperature values were recorded. Gravimetric moisture content was determined for each soil sample at the end of the experiment (105 °C for 24 h). Water was extracted from soil samples using a cryogenic vacuum line[36]. Extracted samples were then analyzed for $\delta^{18}O$ using a G1102-i isotopic H_2O cavity ring down spectrometer (Picarro Inc; Santa Clara, CA). The amount of water vapor adsorbed (H_2O_{WVA}) was calculated using the known isotope compositions of the water added to the atmosphere (2 atom% ^{18}O), the native soil water (~natural abundance: 0.19 atom% ^{18}O), and the isotopic mass balance equation where:

$$H_2O_{WVA} = H_2O_{total}(atom\%_{total} - atom\%_{native})/(atom\%_{WVA} - atom\%_{native}).$$

For the second laboratory experiment, 30 g samples of sieved (2 mm) soil from the grassland were placed in each of 30 sealed Mason jars (946 ml). Half of the jars included a small reservoir with 500 μg of water, while the other half experienced no atmospheric water addition. The jars were then equally divided so that five replicate jars were incubated for 72 h at high (40 °C) or moderate (23 °C) temperature, or the temperature cycling previously described. Gas samples (5 ml) were collected from the headspace of each jar 0, 52, and 72 h after the initiation of the experiment using an air-tight syringe inserted through a septum on the lid of each jar. Headspace samples were analyzed for CO_2 concentration using an infrared gas analyzer model L1-6262 (LI-COR Biosciences; Lincoln, NE). All samples showed a linear increase in CO_2 concentration over time, and production rates (μg CO_2-C g soil^{-1} day^{-1}) were calculated using linear regression (median correlation coefficient=0.92).

Site Description and Field Study

The field site is a semiarid, high-desert grassland near Flagstaff, AZ, USA (35°34'20'N, 111°34'4'W, 1755 m above sea level, 230 mm of rain annually). The dominant plants are perennial grasses (*Bouteloua eriopoda*, *Bouteloua gracilis*, *Sporobulus cryptandrus*, and *Pleuraphis jamesii*) with few shrubs (*Ericameria nauseousa*, *Gutierrezia sarothrae*). Soils are cindery and are classified in the U.S. Department of Agriculture Soil Taxonomic Subgroup of Typic Haplustolls. There is a Campbell Scientific (Logan, UT) weather station located at this field site that collects temperature, moisture, and humidity data in addition to other environmental variables.

In early July of 2013, prior to the arrival of monsoon precipitation and during a period when grasses were physiologically dormant, a 24-hour observational study was conducted in order to characterize diurnal soil moisture fluctuations and CO_2 efflux rates. Soil respiration measurements were taken with a LI-8100A Automated Soil Flux System (LI-COR Biosciences; Lincoln, NE). Ten soil collars (50 cm diameter) were installed in plant interspaces throughout the site, and CO_2 efflux measurements were taken at these locations every hour for 24 h. One soil core (5 cm diameter and depth) was collected every hour during the 24-h period for gravimetric moisture determination. Soil samples were sieved (2 mm) and weighed in the field, then carefully transported back to the laboratory where they were dried (105 °C for 24 h) and reweighed.

Data Analysis

Treatment effects in the laboratory experiments were assessed with one- or two-way ANOVA in JMP Pro 9.0.2 (Cary, NC). Linear regressions were performed in R (http://www.R-project.org). Path analysis was performed with SPSS Amos 22, and the path model was a good fit to the data (Chi square=0.635, p=0.426).

ADDITIONAL INFORMATION

Supplementary Material

Supplementary Figure S1: Semiarid, high elevation grassland where the field study was conducted (a), and close-up of the soil surface during the study period when plants were dormant (b).

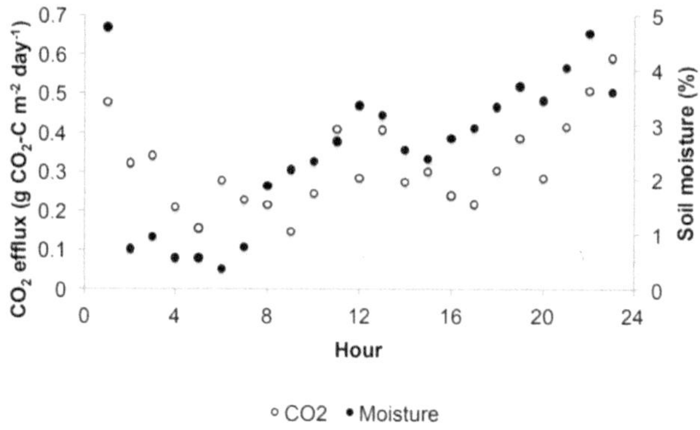

Supplementary Figure S2: Mean CO_2 efflux rates and soil moisture contents in the field during the 24-hour study.

ACKNOWLEDGMENTS

This material is based upon work supported by the National Science Foundation Postdoctoral Research Fellowship in Biology to T.A.M. under Grant No. 1402451. S.C.R. was supported by the U.S. Department of Energy Office of Science, Office of Biological and Environmental Research Terrestrial Ecosystem Sciences Program, under Award Number DE-SC-0008168. Any use of trade, firm, or product names is for descriptive purposes only and does not imply endorsement by the U.S. Government.

Footnotes

Author Contributions T.A.M. designed and conducted the studies and analyses, and led the manuscript preparation. E.M.M. and E.S. contributed to the laboratory experiments and analyses. All authors explored the data and contributed to the writing of this manuscript.

REFERENCES

1. Noy-Meir I. Desert ecosystems: environment and producers. *Annu Rev Ecol Syst* **4**, 25–51 (1973).
2. Reynolds J. F., Maestre F. T., Kemp P. R., Stafford-Smith D. M. & Lambin E. Natural and human dimensions of land degradation in drylands: causes and consequences. In: *Terrestrial Ecosystems in a Changing World* (eds. Canadell D., Pataki D. & Pitelka L. F.) 247–257 (Springer, 2007).
3. Schimel D. S. Drylands in the earth system. *Science* **327**, 418–419 (2010).
4. Lal R. Carbon sequestration in dryland ecosystems. *Env Manage* **33**, 528–544 (2004).
5. Jobbagy E. G. & Jackson R. B. The vertical distribution of soil organic carbon and its relation to climate and vegetation. *Ecol Appl* **10**, 423–436 (2000).
6. Poulter B. *et al.* Contribution of semi-arid ecosystems to interannual variability of the global carbon cycle. *Nature* **509**, 600–603 (2014).
7. Knapp A. K. *et al.* Rainfall variability, carbon cycling, and plant species diversity in a mesic grassland. *Science* **298**, 2202–2205 (2002).
8. Huxman T. E. *et al.* Precipitation pulses and carbon fluxes in semiarid and arid ecosystems. *Oecologia* **141**, 254–268 (2004).
9. Sala O. E., Gherardi L. A., Reichmann L., Jobbagy E. & Peters D. Legacies of precipitation fluctuations on primary production: theory and data synthesis. *Philos Trans R Soc London [Biol]* **367**, 3135–3144 (2012).
10. Collins S. L. *et al.* A multiscale, hierarchical model of pulse dynamics in arid-land ecosystems. *Annu Rev Ecol Evol Syst* **45**, 397–419 (2014).
11. Scott R. L., Jenerette G. J., Potts D. L. & Huxman T. E. Effects of seasonal drought on net carbon dioxide exchange from a woody-plant-encroached semiarid grassland. *J Geophys Res [Biogeosci]* **114**, 2005–2012 (2009).
12. Reed S. C. *et al.* Increased precipitation results in rapid moss mortality and altered fertility in a dryland ecosystem. *Nature Clim Change* **2**, 752–755 (2012).
13. Held I. M. & Soden B. J. Robust responses of the hydrological cycle to global warming. *J Climate* **19**, 5686–5699 (2006).

14. Manabe S. & Stouffer R. J. Sensitivity of a global model to an increase of CO_2 concentration in the atmosphere. *J Geophys Res* **85**, 5529–5554 (1980).
15. Agam N. & Berliner P. R. Dew formation and water vapor adsorption in semi-arid environments – a review. *J Arid Env* **65**, 572–590 (2006).
16. Azua-Bustos A. *et al.* Hypolithic cyanobacteria supported mainly by fog in the coastal range of the Atacama Desert. *Microb Ecol* **61**, 568–581 (2011).
17. Malek E., McCurdy G. & Giles B. Dew contribution to the annual water balance in semi-arid desert valleys. *J Arid Env* **42**, 71–80 (1999).
18. Kidron G. J., Hernstadt I. & Barzilay E. The role of dew as a moisture source for sand microbiotic crusts in the Negev Desert, Israel. *J Arid Env* **52**, 517–533 (2002).
19. Budel B., Bendix J., Bicker F. R. & Allan Green T. G. Dewfall as a water source frequently activates the endolithic cyanobacterial communities in the granites of Taylor Valley, Antarctica. *J Phycol* **44**, 1415–1424 (2008).
20. Agam N. & Berliner P. R. Diurnal water content changes in the bare soil of a coastal desert. *J Hydrometeorol* **5**, 922–933 (2004).
21. Danalatos N. G., Kosmas C. S., Moustakas N. C. & Yassoglou N. Rock fragments: their impact on soil physical properties and biomass production under Mediterranean conditions. *Soil Use Manage* **11**, 121–126 (1995).
22. Kosmas C., Danalatos N. G., Poesen J. & van Wesemael B. The effect of water vapour adsorption on soil moisture content under Mediterranean climatic conditions. *Agr Water Manage* **36**, 157–168 (1998).
23. Kosmas C. *et al.* Parameters affecting water vapor adsorption under semi-arid climatic conditions. *Agr Water Manage* **48**, 61–78 (2001).
24. Newell S. Y., Fallon R. D., Cal Rodriguez R. M. & Groene L. C. Influence of rain, tidal wetting, and relative humidity on release of carbon dioxide by standing-dead salt-marsh plants. *Oecologia* **68**, 73–79 (1985).
25. Keuhn K. A., Steiner D. & Gessner M. O. Diel mineralization patterns of standing-dead plant litter: implications for CO_2 flux from wetlands. *Ecology* **85**, 2504–2518 (2004).

26. Dirks I., Navon Y., Kanas D., Bumbar R. & Grunzweig J. M. Atmospheric water vapor as a driver of litter decomposition in Mediterranean shrubland and grassland during rainless seasons. *Glob Change Biol* **16**, 2799–2812 (2010).
27. Verhoef A., Diaz-Espejo A., Knight J. R., Villagarcia L. & Fernandez J. E. Adsorption of water vapor by bare soil in an olive grove in southern Spain. *J Hydrometeorol* 7, 1011–1027 (2006).
28. New M. G., Hulme M. & Jones P. D. Representing 20th century space-time climate variability: development of a 1961-1990 mean monthly terrestrial climatology. *J Climate* **12**, 829–856 (1999).
29. Wang B. *et al.* Soil moisture modifies the response of soil respiration to temperature in a desert shrub ecosystem. *Biogeosci* **11**, 259–268 (2014).
30. Ramirez D. A., Bellot J., Domingo F., Blasco A. Can water responses in *Stipa tenacissima* L. during the summer season be promoted by non-rainfall water gains in soil? *Plant Soil* **291**, 67–79 (2007).
31. Ucles O., Villagarcia L., Canton Y., Lazaro R., Domingo F. Non-rainfall water inputs are controlled by aspect in a semiarid system. *J Arid Environ* **113**, 43–50 (2014).
32. Jones M. B. Plant microclimate. In: Photosynthesis and Production in a Changing *Environment* (eds. Hall D. O., Scurlock J. M. O., Bolhar-Nordenkampf H. R., Leegood R. C. & Long S. P.) 47–64 (Chapman & Hall, 1993).
33. Cui M. & Caldwell M. M. A large ephemeral release of nitrogen upon wetting of dry soil and corresponding root responses in the field. *Plant Soil* **191**, 291–299 (1997).
34. Austin A. T. *et al.* Water pulses and biogeochemical cycles in arid and semiarid ecosystems. *Oecologia* **141**, 221–235 (2004).
35. Intergovernmental Panel on Climate Change. Synthesis Report. *IPCC, Geneva* (2008).
36. Kerhoulas L. P., Kolb T. E. & Koch G. W. Tree size, stand density, and the source of water used across seasons by ponderosa pine in northern Arizona. *Forest Ecol Manage* **289**, 425–433 (2013).

CHAPTER 9

Observed Behaviours of Precipitable Water Vapour and Precipitation Intensity in Response to Upper Air Profiles Estimated from Surface Air Temperature

Mikiko Fujita [1] **& Tomonori Sato**[2]

[1] Japan Agency for Marine-Earth Science and Technology, 3173-25 Showa-machi, Kanazawa-ku, Yokohama, 236- 0001, Japan.

[2] Faculty of Environmental Earth Science, Hokkaido University, Kita-10, Nishi-5, Sapporo, 060–0810, Japan.

Citation: Fujita, M., Sato, T. "Observed behaviours of precipitable water vapour and precipitation intensity in response to upper air profiles estimated from surface air temperature". Sci Rep 7, 4233 (2017). https://doi.org/10.1038/s41598-017-04443-9.

ABSTRACT

Extremely heavy precipitation affects human society and the natural environment, and its behaviour under a warming climate needs to be elucidated. Recent studies have demonstrated that observed extreme precipitation increases with surface air temperature (SAT) at approximately the Clausius–Clapeyron (CC) rate, suggesting that atmospheric water vapour content can explain the relationship between extreme precipitation and SAT. However, the relationship between atmospheric water vapour content and SAT is poorly understood due to the lack of reliable observations with sufficient spatial and temporal coverage for statistical analyses. Here, we analyse the relationship between atmospheric water vapour content and SAT using precipitable water vapour (PWV) derived from global positioning system satellites. A super-CC rate appears in hourly PWV when the SAT is below 16 °C, whereas the rate decreases at high SAT, which is different from the precipitation-SAT relationship. The effects of upper air temperature and water vapour can consistently explain the super-CC rate of PWV relative to SAT. The difference between moist and dry adiabatic lapse rates increases with SAT, in consequence of more ability to hold water vapour in the free atmosphere under higher SAT conditions; therefore, attainable PWV increases more rapidly than the CC rate as SAT increases.

INTRODUCTION

Many disasters related to extremely heavy precipitation have been reported throughout the worldwide1. A likely increase in the intensity and frequency of precipitation extremes is projected by many atmosphere–ocean coupled GCMs with elevated greenhouse gas concentration scenarios2, 3. This is an issue for policymakers who must respond to a societal demand to ensure resiliency to disasters caused by heavy precipitation events. The efforts of scientists have been aimed at assessing the degree to which heavy precipitation events would be enhanced in a warming climate. Recently, many precipitation products, such as those from rain gauges4, 5, ground-based radar6, space-borne precipitation radar7, and climate models2, have become available for the statistical analyses of heavy precipitation extremes, which requires a high temporal frequency of measurements and a long-term record.

Theoretically based studies have conventionally utilized observed surface air temperature (SAT) for scaling the variation of extreme precipitation intensity, probably because SAT is the most fundamental and reliable meteorological parameter. Relationships between extreme precipitation intensity, which is typically defined as a certain high threshold percentile, and SAT have been investigated in both mid-latitude regions4,5,6,7,8,9,10,11 and tropical regions12, 13.

Recently, as available observational data have accumulated, a consensus has been reached that precipitation intensity can increase more rapidly than the Clausius–Clapeyron (CC) rate if the SAT is above 12–15 °C4, 9 or in convective precipitation6. However, the physical mechanism that satisfactorily explains this super-CC rate remains controversial. To elucidate the relationship between extreme precipitation and SAT, the tropospheric water vapour content and the vertical profile of air temperature must be considered. Radiosonde observation is a traditional direct measurement of vertical atmospheric structure, but its spatial and temporal coverage is insufficient, especially for capturing short-lived extreme precipitation events that may occur on hourly timescales.

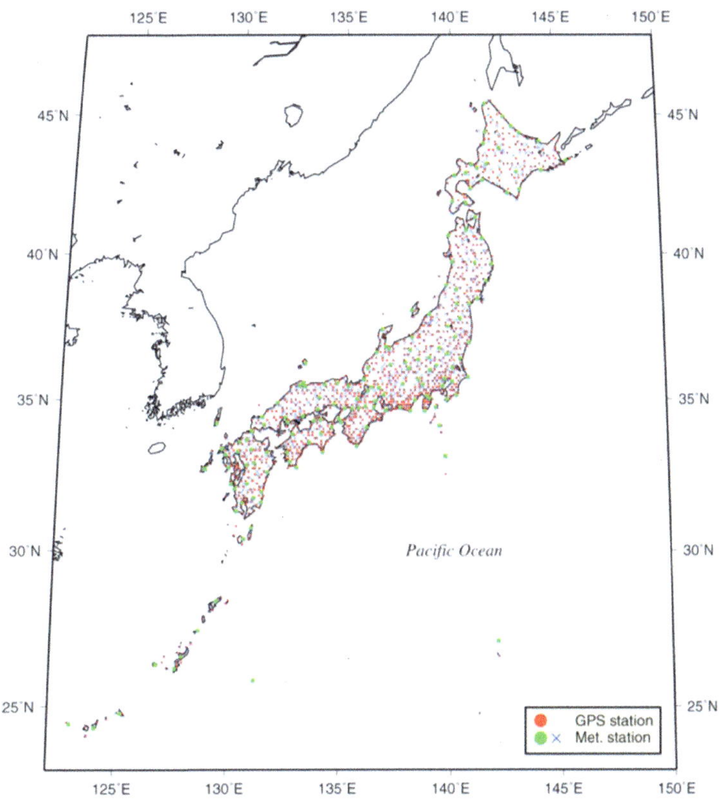

Figure 1: Observational stations. Map of the study area showing GPS and sur-

face meteorological stations in Japan. Red circles indicate GPS stations. Blue crosses and green circles indicate meteorological stations. We use surface pressure, air temperature and relative humidity observed at the green circles and surface air temperature and precipitation observed at the blue crosses. The map was created using GMT Generic Mapping Tools (GMT)27 version 5.1.1 available at http://www.soest.hawaii.edu/gmt/.

In climate models, the atmospheric moisture profile associated with intense precipitation is highly dependent on cumulus and microphysics schemes. Indeed, difficulties in obtaining a dataset of atmospheric water vapour content with high spatial and temporal coverage have been the major impediment to understanding the vertical structure. Here, we use precipitable water vapour (PWV), which is the vertically integrated water vapour mixing ratio, derived from global positioning system (GPS) data archives for the 15 years from 1996 to 2010 to describe a relationship between tropospheric water vapour content and SAT. The GPS-derived PWV data enable sub-hourly estimations of PWV (see Methods) with high spatial density over Japan at ~20 km intervals (Fig. 1)

RESULTS AND DISCUSSION

We analysed all hourly PWV data to determine percentiles for each 2 °C daily SAT bin (Fig. 2a). The 99th percentile PWV increases with super-CC rate, which is larger than the CC rate but less than two times the CC rate, for SAT below 14 °C. Over 14 °C, the extreme PWV increases approximately following CC rate and it become slower. The dependence of extreme PWV (i.e., 99th percentile PWV) on daily SAT differs from that of extreme precipitation4·5·6·7·8·9·10·11, which exhibited a super-CC rate in a similar SAT range over the southern part of Japan, where the climatological SAT is high5· 14. The 90th percentile PWV also increases with super-CC rate for SAT below 16 °C.

Around 18–20 °C, the 90th percentile of PWV increases approximately following the CC rate. For the 50th and 75th percentiles, the super-CC rate appears when the SAT is below 22 °C. Double the CC rate appears when the SAT is between 18 and 24 °C for PWV lower than 25th percentiles, whereas the rate of PWV change becomes slower at all percentiles as SAT increases and tends to decrease at an SAT range higher than 18 °C for 99th percentile and 28 °C for 25th percentile. These features of the PWV-SAT relationship indicate the importance of vertical evolution of the moist layer in the air column, as discussed below.

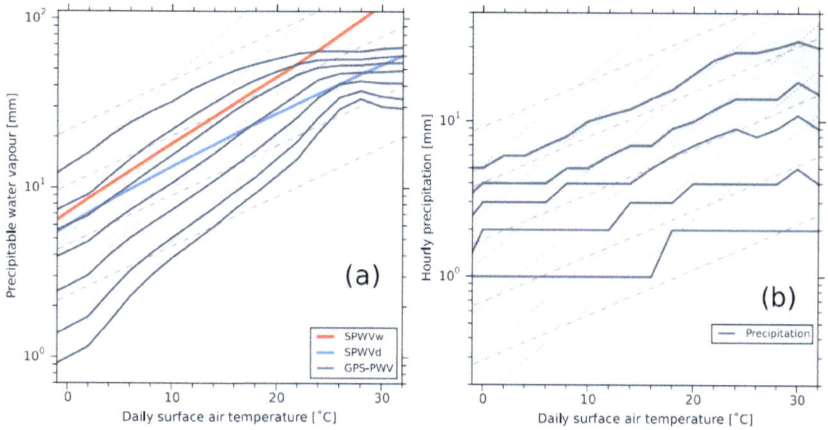

Figure 2: Percentiles of hourly PWV and precipitation on a logarithmic scale as a function of daily surface temperature (SAT). (**a**) Hourly GPS-PWV percentiles. Solid black lines are, from top to bottom, the 99th, 90th, 75th, 50th, 25th, 10th and 5th percentiles of hourly GPS-PWV. Red and blue lines are the estimated $SPWV_w$ and $SPWV_d$ profiles, respectively. (**b**) Hourly precipitation percentiles. Solid black lines are, from top to bottom, the 99th, 90th, 75th, 50th, and 25th percentiles of hourly precipitation. Dashed and dotted grey lines in both panels are the exponential relations corresponding to one and two times the Clausius–Clapeyron relation, respectively. The shaded area, plotted only for 99th and 90th percentiles, indicates the 90% confidence intervals estimated using the bootstrap method (see Methods).

The percentiles of hourly precipitation at the nearest station to the GPS site were also analysed for each 2 °C daily SAT bin for comparison (Fig. 2b). The precipitation data from observations at 808 stations (Fig. 1) with 1 mm intervals were used, and the percentiles were computed from raw data over 1 mm hr^{-1}. The 99th percentile of hourly precipitation increases at a CC-like rate below 16 °C, and it reaches a double-CC rate around 18–22 °C. The percentiles become stable and decrease over 24 °C, although the statistical errors were large. In the 90th, 75th, and 50th percentiles, the hourly precipitation increases at the CC or lower rate. The rates of increase of precipitation are consistent with previous studies of extreme precipitation[4,5,6,7,8,9,10,11], but there are significant differences in the rates of observed PWV, with super-CC even below 16 °C. Additionally, if daily maximum PWV is used, double-CC rates in lower percentiles are also clearly recognized when the SAT exceeds 18 °C (Supplementary Fig. 1). Key questions that are raised include why does PWV increase in super-CC rate with SAT and how does the maximum water vapour capacity in the air column (saturated precipitable water vapour: SPWV) vary with SAT?

In the CC equation, the water vapour capacity of a given volume of air is constrained by the air temperature at that height, not necessarily by the SAT. To interpret the CC rate in observed PWV, we compare it with theoretically obtained SPWV indices, $SPWV_d$ and $SPWV_w$, by assuming two tropospheric temperature profiles corresponding to the dry and wet adiabatic lapse rates (Γ_d and Γ_m), respectively, for an air parcel at ground level with a temperature equal to the SAT (see Methods). The rate of increase in $SPWV_d$ is close to the CC rate of 7% °C^{-1}, confirming that the CC relationship holds between the theoretically derived attainable PWV and SAT for sub-saturation conditions (blue line in Fig. 2a). By contrast, $SPWV_w$ increases more rapidly than the CC rate, reaching 11–12% °C^{-1} (red line in Fig. 2a). This pattern is due to the increasing amount of latent heat energy released per unit SAT rise, similar to the relationship between potential temperature and saturated equivalent potential temperature as their difference becomes larger with increasing air temperature. The GPS-PWV lies above $SPWV_d$ at the 75th, 90th, and 99th percentiles, indicating that there is more water vapour than in a standard sub-saturation atmosphere. Furthermore, the 90th and 99th percentiles PWV are above the $SPWV_w$ for SAT below 22 °C. These findings require a further analysis of the vertical distribution of water vapour. To discuss the vertical distribution of water vapour, the observed GPS-PWV is separated into two parts: water vapour in the boundary layer that is assumed to be constrained by SAT and water vapour in the free atmosphere that reflects deep convective processes. To determine the typical boundary layer height, we adopt the water vapour scale height15 (WSH; see Methods) as an index representing how deeply the near-surface atmosphere is mixed by thermal turbulent eddies. In Fig. 3a, estimated WSHs are plotted for each percentile of GPS-PWV.

The WSH increases with PWV percentiles: the maximum WSH for the 99th percentile reaches 3,500 m at an SAT of 14 °C, whereas the 25th percentile has a maximum WSH of 2,100 m at 26 °C. The maximum WSH tends to appear at higher SAT as the percentile value decreases. For WSHs below the 25th percentile, WSH increases with SAT until 26 °C, whereas it decreases at SAT over 28 °C. At other percentiles, WSH also increases when SAT is below 14–24 °C with a lower rate of increase than at lower percentiles. The WSH decreases with SAT at SAT over 14 °C for 99th percentile and over 24 °C for 50th percentile at a rate similar to that at the lower percentiles.

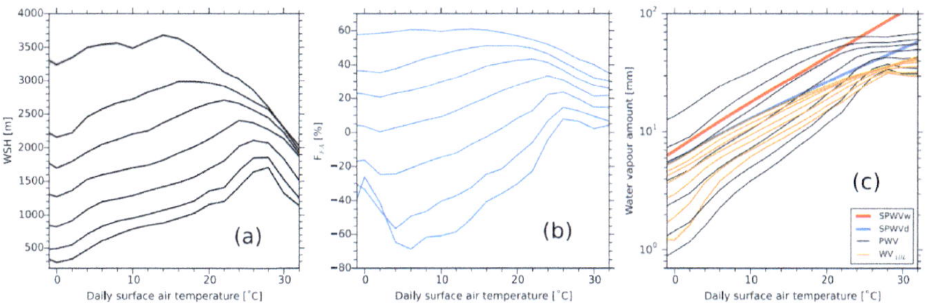

Figure 3: Percentiles of water vapour parameters as a function of SAT. (**a**) WSH. (**b**) Percentage of water vapour amount in the free atmosphere (F_{FA}). (**c**) Water vapour amount in the boundary layer (WV_{ABL}; orange), GPS-PWV (black), $SPWV_w$ (bold red) and $SPWV_d$ (bold blue). Multiple lines of WSH, F_{FA}, and WV_{ABL} represent the values when the 99th, 90th, 75th, 50th, 25th, 10th and 5th percentiles values of GPS-PWV (from top to bottom in the figures) are observed. In panel (**c**), the GPS-PWV percentiles are the same as in Fig. 2a except for the number of stations used. The shaded area indicates the 90% confidence intervals (see Methods). Note the vertical log scale in (**c**).

Based on the WSH, the percentage of water vapour in the free atmosphere (above the WSH) relative to PWV was analysed to explain the super-CC rate in observed PWV. If the saturated atmosphere in the boundary layer is considered, the maximum amount of water vapour within the WSH (WV_{ABL}; see Methods) is determined by the vertical integral of the saturated water vapour mixing ratio for the temperature profile assuming SAT and Γ_d. The WV_{ABL} (orange lines in Fig. 3c) in the higher percentile is very similar to $SPWV_d$ (blue line in Fig. 3c) and the CC rate (see Fig. 2a). The amount of water vapour in the free atmosphere (WV_{FA}) is calculated as the difference between PWV and WV_{ABL}.

The percentage fraction of WV_{FA} relative to PWV (F_{FA}; see Methods) is stably high with positive values at high percentiles (Fig. 3b). Indeed, at the 99th percentile, approximately 50% of PWV is in the free atmosphere when SAT is below 20 °C, indicating a deep moist layer. For lower percentiles, the F_{FA} is negative for low SAT but becomes positive as SAT increases. The change in F_{FA} from negative to positive with increasing SAT suggests the vertical expansion of the moist layer from the near surface into the free atmosphere.

As upward moisture transport is vital for moist convection, the F_{FA} is high for high percentiles, as shown in Fig. 3b. Furthermore, for high SAT, convective precipitation is likely to dominate, which also contributes to the vertical expansion of the moist layer by pushing up lower-level water vapour[6]. For the 99th percentile of precipitation (Fig. 2b), the rate of increase with SAT around 16 °C exceeds the CC rate, reaching the double-CC rate. At SAT around 16 °C, the water vapour in the free atmosphere significantly increases even in the 25th percentile (Fig. 3b). Additionally, the PWV during heavy precipitation events shows a higher increasing rate with SAT (Supplementary Fig. 2).

These results also support the hypothesis that at higher SAT, vertical water vapour transport in convective precipitation systems penetrates into the free atmosphere, charging more water vapour above the boundary layer in addition to the increased water vapour holdings in the boundary layer constrained by SAT. As the surface precipitation intensity depends on both cloud microphysical processes and mesoscale circulations, the precipitation intensity does not always have a linear relationship with PWV. However, the PWV analysis indicates that vertical transport of water vapour is vigorous under high SAT that allows enhanced water vapour in the free atmosphere as condensation heating warms the air temperature. The vertical distributions of air temperature and water vapour mixing ratio are crucial for describing the behaviour of convective systems under different SAT conditions[16].

For an SAT higher than 16 °C, the rate of increase in the 99th percentile GPS-PWV is reduced (Fig. 2a), which is associated with the reduction of WSH (Fig. 3a). High SAT tends to appear in boreal summer, when there is a quasi-stationary anticyclone around the Western North Pacific as part of the subtropical high belt. The surface divergent flow under the high-pressure system reduces PWV because the descending airflow transports low-moisture air downward, which prevents the wetter surface layer from developing vertically. On hot summer days with an SAT higher than approximately 20 °C, the stable atmospheric profile is unfavourable for developing convective precipitation. A similar decrease of precipitation at high SAT was observed in other regions when the dominant synoptic circulation in a certain SAT range did not attract intense precipitation[12,17]. In addition, the PWV is less sensitive to SAT during winter, as the near-surface atmospheric layer tends to be capped by an inversion layer. The study area is influenced by a continental air mass during the winter monsoon that efficiently reduces the GPS-PWV. Therefore, the rapid increase in low-percentile PWV with SAT (Fig. 2a) is caused mainly by a skewed probability distribution with a long tail on the low-PWV side, which likely reflects the dominant synoptic patterns in the study area. These features show that synoptic circulation change is as important as SAT because it can alter the probability distribution of precipitation intensity, as indicated by future projected changes in mean precipitation that will involve decreases in many regions of the world despite increasing SAT18.

In this study, a theoretical analysis has revealed that the rate of SPWV increase with SAT becomes logarithmically steeper at high SAT (red line in Fig. 2) because of the increasing contribution of water vapour condensation heating that warms the upper air and thus increases water vapour content nonlinearly under moist convection. A quantitative evaluation of the source of increased PWV with SAT, i.e., how much of a contribution is derived from horizontal moisture convergence under different seasons or geographical backgrounds, would be an interesting topic for future studies. Although this study is just one example in a humid climate zone, the fundamental idea is applicable to other areas. The effect of latent heating in a cloud system on the vertical temperature profile is a key consideration to answer the general question "How much will PWV increase, and how strong will extreme precipitation be, in a future climate?".

METHODS

GPS Precipitable Water Vapour

The GPS raw signal data from 1,219 stations provided by the GNSS Earth Observation NETwork system (GEONET) of the Geospatial Information Authority of Japan (Fig. 1) were used for the calculation of PWV over the period 1996–2010. The PWV was calculated from zenith tropospheric delay (ZTD), which is a measure of the communication time lag between GPS satellites and the GPS receiver.

The moisture component of ZTD was converted to PWV by removing the effect of dry atmosphere estimated from the surface air pressure19 and SAT20 at meteorological stations near the GPS station. The ZTD was computed by the GPS processing software RTNet21,22, and the coordinates of the GPS sites were estimated monthly to avoid any un-modelled geodetic error23 or site-specific biases, which can affect the observed vertical depth of the atmosphere.

Here, the Saastamoinen troposphere model24 and the global mapping function25 were used. The observation cut-off angle was 10°, and the data-sampling interval was 30 seconds26. Finally, the estimated GPS-PWV was averaged hourly. The GPS-PWV represents the mean PWV in the atmosphere over an inverted conical shape with a base radius of approximately 30 km. The accuracy of the GPS-PWV was verified by a comparison with PWV derived from a direct humidity profile observed by radiosonde, and it was shown to be sufficiently high26. In the present study, the GPW-PWV data were rejected if the surface pressure was below 980 hPa to avoid the observational errors that may occur if the horizontal atmospheric gradient is very large, such as around typhoons.

Statistical Tests

The percentiles of PWV and precipitation were computed from the raw data. The 90% confidence intervals of the percentiles were determined using the bootstrap method.

After 3,000 estimations of percentile values by using random subsamples, 90% confidence intervals were determined based on the assumption that 3,000 estimated values follow a normal distribution. The number of subsamples was set to 10% of all data in each SAT bin. The estimated confidence intervals of PWV are very small and difficult to visually identify in the figures.

Calculation of Theoretical Saturated Precipitable Water Vapour

The theoretical saturated water vapour (SPWV) was estimated as follows. First, vertical air temperature profiles were estimated at 100 m intervals. An air parcel with temperature equal to SAT and pressure of 1,013 hPa at the ground was lifted to 10 km by applying the dry or wet adiabatic lapse rate.

Two SPWV values, $SPWV_d$ and $SPWV_w$, were then obtained for each of the two air temperature profiles by calculating the vertical integrals of the saturated mixing ratio of water vapour assuming the hydrostatic balance. Here, $SPWV_d$ and $SPWV_w$ represent the attainable PWV corresponding to dry and wet adiabatic lapse rates, respectively, and hence are expressed as functions of SAT. We repeated these steps for SAT varying from −2 to 32 °C at 2 °C intervals. The dry adiabatic lapse rate (Γ_d) is a constant (0.98 K 100 m^{-1}), whereas the wet adiabatic lapse rate (Γ_w) varies from approximately 0.35 to 0.98 K 100 m^{-1}. The wet adiabatic lapse rate depends on the saturated water vapour mixing ratio as

$$\Gamma_w = \Gamma_d / \left(1 + \frac{L}{c_p}\frac{dq_{sat}}{dT}\right) \quad (1)$$

where L is the latent heat of condensation of water (2,501 kJ kg^{-1}), c_p is the specific heat of dry air at constant pressure (1,003.5 J kg^{-1} K^{-1}), q_{sat} is the saturated water vapour mixing ratio, and T is air temperature.

Estimation of boundary layer height and water vapour amount in and above the boundary layer

The scale height of water vapour WSH was derived as15.

$$WSH = PWV/(\rho_d q_{sfc}) \quad (2)$$

where q_{sfc} is the water vapour mixing ratio at the surface and ρ_d is the density of dry air (1.2 kg m^{-3}) at the surface. For WSH, we used only 321 GPS stations within 20 km of the meteorological stations where observations of relative humidity and air pressure, required for calculating q_{sfc}, are routinely performed with hourly interval (green circles in Fig. 1). The WSHs (Fig. 3a, black line) corresponding to each PWV percentile were calculated as the arithmetic mean of hourly interval WSHs for the events when the PWV value was within the 90% confidence interval for each percentile (Fig. 2a, shading). The amount of water vapour in the boundary layer (WV_{ABL}) was calculated based on the vertical integral of the saturated water vapour mixing ratio (q_{sat}) between the ground surface and WSH height. Here, we considered the vertical distribution of q_{sat} assuming SAT and Γ_d similarly to the calculation of $SPWV_d$. For this calculation, we also used air pressure observed at ground surface level and that at WSH level assuming the hydrostatic balance. As the PWV can be divided into two components—water vapour in the boundary layer (WV_{ABL}) and water vapour in the free atmosphere (WV_{FA})—the following can be derived:

$$WV_{FA} = PWV - WV_{ABL} \quad (3)$$

The fraction of water vapour in the free atmosphere (F_{FA}) is defined as the percentage of WV_{FA} relative to the PWV, i.e., $F_{FA} = WV_{FA}/PWV * 100$. The negative F_{FA} can occur at the low percentiles of PWV because saturated mixing ratio was used in estimating WV_{ABL}. Therefore, the negative F_{FA} for the low PWV percentiles means that water vapour holding capacity in the boundary layer is larger than PWV ($WV_{ABL} > PWV$), whereas the positive F_{FA} means that the PWV is sufficiently high to exceed the holding capacity of water vapour in the boundary layer ($WV_{ABL} < PWV$), suggesting that the moist layer develops vertically beyond WSH.

SUPPLEMENTARY INFORMATION

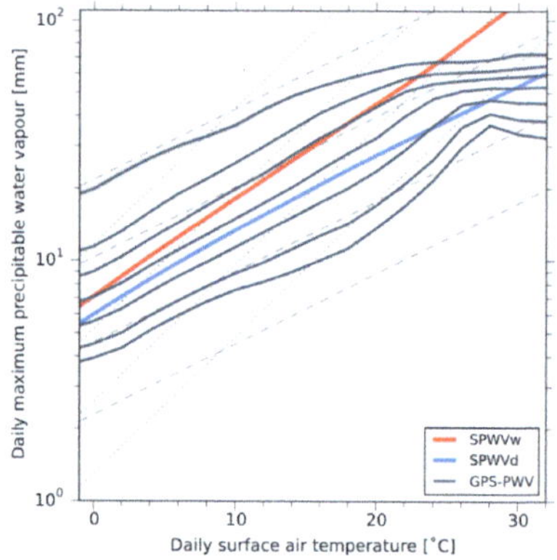

Figure S1: Percentiles of daily maximum of hourly PWV on a logarithmic scale as a function of daily surface temperature (SAT). Solid black lines are, from top to bottom, the 99th, 90th, 75th, 50th, 25th, 10th and 5th percentiles of daily maximum of hourly GPS-PWV. Red and blue lines are the estimated $SPWV_w$ and $SPWV_d$ profiles, respectively. Dashed and dotted grey lines are the exponential relations corresponding to one and two times the Clausius–Clapeyron relation, respectively. The shade area, plotted only for 99th percentiles, indicate 90 % confidence intervals estimated by the bootstrap method (see Methods).

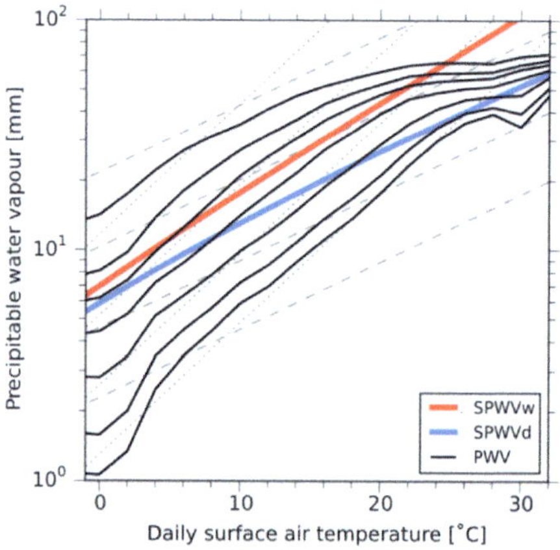

Figure S2: Percentiles of hourly PWV before and after extreme precipitation on a logarithmic scale as a function of daily surface temperature (SAT). Solid black lines are, from top to bottom, the 99th, 90th, 75th, 50th, 25th, 10th and 5th percentiles of hourly GPS-PWV. These GPS-PWV percentiles were calculated from the data when the extreme precipitation over 95th percentile in the each

SAT bin occurred in the nearest meteorological stations (blue crosses in Fig.1) during 6 hours before and after. If multiple extreme hourly precipitation events occur within the 12 hours, we regard them as one event to avoid duplicative count of GPS-PWV data. Red and blue lines are the estimated $SPWV_w$ and $SPWV_d$ profiles, respectively. Dashed and dotted grey lines are the exponential relations corresponding to one and two times the Clausius–Clapeyron relation, respectively. The shade area, plotted only for 99th percentiles, indicate 90 % confidence intervals estimated by the bootstrap method (see Methods).

ACKNOWLEDGEMENTS

The authors acknowledge the Geospatial Information Authority of Japan for providing the GEONET GPS data and the Japan Meteorological Agency for providing the surface meteorological data. M.F. acknowledges support from the Global Environmental Research Fund (RFa11-01, 2RF-1304) of the Ministry of the Environment, Japan. M.F. and T.S. acknowledge the Social Implementation Program on Climate Change Adaptation Technology (SI-CAT) and as well as a Grant-in-Aid for Young Scientists B (Grant No. 15K16316) and A (Grant No. 15H05464), respectively, awarded by the Ministry of Education, Culture, Sports, Science and Technology (MEXT) of Japan.

Contributions

M.F. conceived the project, designed the study and wrote the manuscript. T.S. advised on interpretation and assisted in writing the manuscript.

Competing Interests

The authors declare that they have no competing interests.

REFERENCES

1. Easterling, D. R. *et al.* Climate extremes: observations, modeling, and impacts. *Science* **289**, 2068–2074 (2000).
2. Emori, S. & Brown, S. J. Dynamic and thermodynamic changes in mean and extreme precipitation under changed climate. *Geophys. Res. Lett.* **32** (2005).
3. Trenberth, K. Changes in precipitation with climate change. *Clim. Res* **47**, 123–138 (2011).
4. Lenderink, G. & van Meijgaard, E. V. Increase in hourly precipitation extremes beyond expectations from temperature changes. *Nature Geosci* **1**, 511–514 (2008).
5. Utsumi, N. *et al.* Does higher surface temperature intensify extreme precipitation? *Geophys. Res. Lett.* **38** (2011).
6. Berg, P., Moseley, C. & Haerter, J. O. Strong increase in convective precipitation in response to higher temperatures. *Nature Geosci* **6**, 181–185 (2013).
7. Hamada, A., Takayabu, Y. N., Liu, C. & Zipser, E. J. Weak linkage between the heaviest rainfall and tallest storms. *Nat. Commun.* **6**, 6213 (2015).
8. Haerter, J. O. & Berg, P. Unexpected rise in extreme precipitation caused by a shift in rain type? *Nature Geosci* **2**, 372–373 (2009).
9. Lenderink, G. & van Meijgaard, E. V. Linking increases in hourly precipitation extremes to atmospheric temperature and moisture changes. *Environ. Res. Lett.* **5**, 025208 (2010).
10. Haerter, J. O., Berg, P. & Hagemann, S. C. D. Heavy rain intensity distributions on varying time scales and at different temperatures. *J. Geophys. Res.* **115** (2010).
11. Loriaux, J. M., Lenderink, G., Roode, S. R. D. & Siebesma, A. P. Understanding convective extreme precipitation scaling using observations and an entraining plume model. *J. Atmos. Sci.* **70**, 3641–3655 (2013).
12. Maeda, E. E., Utsumi, N. & Oki, T. Decreasing precipitation extremes at higher temperatures in tropical regions. *Nat. Hazards* **64**, 935–941 (2012).
13. Vittal, H. *et al.* Lack of dependence of Indian summer monsoon rainfall extremes on temperature: An observational evidence. *Sci. Rep.* **6**, 31039 (2016).

14. Fujibe, F. Annual variation of extreme precipitation intensity in Japan: Assessment of the validity of Clausius-Clapeyron scaling in seasonal change. *SOLA* **12**, 106–110 (2016).
15. Kanemaru, K. & Masunaga, H. A satellite study of the relationship between sea surface temperature and column water vapor over tropical and subtropical oceans. *J. Climate* **26**, 4204–4218 (2013).
16. Romps, D. M. Clausius–Clapeyron scaling of CAPE from analytical solutions to RCE. *J. Atmos. Sci.* **73**, 3719–3737 (2016).
17. Berg, P. *et al*. Seasonal characteristics of the relationship between daily precipitation intensity and surface temperature. *J. Geophys. Res.* **114**, D18102 (2009).
18. Chou, C. *et al.* Increase in the range between wet and dry season precipitation. *Nature Geosci* **6**, 263–267 (2013).
19. Elgered, G., Davis, J. L., Herring, T. A. & Shapiro, I. I. Geodesy by radio interferometry: Water vapor radiometry for estimation of the wet delay. *J. Geophys. Res.* **96**(B4), 6541–6555 (1991).
20. Bevis, M. *et al.* GPS meteorology: Remote sensing of atmospheric water vapor using the global positioning system. *J. Geophys. Res.* **97**, 15787 (1992).
21. Rocken, C. *et al.* "Real-time ionospheric and atmospheric corrections for wide area single frequency carrier phase ambiguity resolution," Proceedings of the 19th International Technical Meeting of the Satellite Division of The Institute of Navigation (ION GNSS 2006), Fort Worth, TX, September 2006, pp. 1208-1218.
22. Iwabuchi, T. *et al.* "PPP and network true real-time 30 sec estimation of ZTD in dense and giant regional GPS network and the application of ZTD for nowcasting of heavy rainfall," Proceedings of the 19th International Technical Meeting of the Satellite Division of The Institute of Navigation (ION GNSS 2006), Fort Worth, TX, September 2006, pp. 1902-1909.
23. King, M. A., Watson, C. S., Penna, N. T. & Clarke, P. J. Subdaily signals in GPS observations and their effect at semiannual and annual periods. *Geophys. Res. Lett.* **35** (2008).
24. Saastamoinen, J. Atmospheric correction for the troposphere and stratosphere in radio ranging satellites. *The use of artificial satellites for geodesy 247–251* (2013).

25. Boehm, J., Niell, A., Tregoning, P. & Schuh, H. Global Mapping Function (GMF): A new empirical mapping function based on numerical weather model data. *Geophys. Res. Lett.* **33** (2006).
26. Fujita, M., Wada, A., Iwabuchi, T. & Rocken, C. "GPS Precipitable water vapor dataset for climate science," Proceedings of the 25th International Technical Meeting of The Satellite Division of the Institute of Navigation (ION GNSS 2012), Nashville, TN, September 2012, pp. 3454–3458.
27. Wessel, P., Smith, W. H. F., Scharroo, R., Luis, J. F. & Wobbe, F. Generic Mapping Tools: Improved version released. *EOS Trans. AGU* **94**, 409–410 (2013).

CHAPTER 10

Vertical Structure of Moisture Content Over Europe

Agnieszka Wypych [1] and Bogdan Bochenek[2]

[1] Jagiellonian University, 7 Gronostajowa St., 30-387 Krakow, Poland

[2] Institute of Meteorology and Water Management—National Research Institute, 14 Piotra Borowego St., 30-215 Krakow, Poland

ABSTRACT

The vertical structure of water vapor content in the atmosphere strongly affects the amount of solar radiation reaching the Earth's surface and processes associated with the formation of clouds and atmospheric precipitation. The

Citation: Agnieszka Wypych, Bogdan Bochenek, "Vertical Structure of Moisture Content over Europe", Advances in Meteorology, vol. 2018, Article ID 3940503, 13 pages, 2018. https://doi.org/10.1155/2018/3940503.

purpose of this study was to assess the vertical differentiation of water vapor over Europe on a seasonal basis and also to evaluate the role of atmospheric circulation in changes therein. Daily values of specific humidity (SHUM) for the time period 1981–2015 were obtained from pressure levels available from ECMWF Era-Interim reanalysis data and used in the study. Eight grid points were analyzed in detail. Each point is representative of a region with different moisture conditions. SHUM profiles were then used to identify cases of moisture inversion. Horizontal flux of specific humidity (SHUMF) was analyzed for principal pressure levels that occur in both inversion-type and inversion-free situations. In addition, SHUM and SHUMF anomalies were identified for advection directions. The research results showed the existence of differences in the vertical structure of water vapor content in the troposphere over Europe, and the Northeastern Atlantic and the presence of moisture inversions not only in areas north of 60°N but also in temperate and subtropical zones. Inversions can occur in two different forms—surface-based and elevated. The occurrence of inversions varies with the seasons. The role of atmospheric circulation is observable in the winter and triggers both surpluses and shortages of moisture via the effect of specific pressure system types (significant role of seasonal pressure high) and via advection directions. In addition, there exists a clear difference between the structure of moisture in the atmospheric boundary layer and in the free atmosphere.

INTRODUCTION

The vertical structure of water vapor content in the Earth's atmosphere strongly impacts the amount of solar radiation reaching the surface of the planet and the process of formation of clouds and atmospheric precipitation. In light of this, water vapor is thought to be the most important trace gas in the atmosphere, and mutual feedbacks triggered by water vapor affect climate change the most by increasing its sensitivity. The total water vapor feedback factor is estimated to equal 0.47 [1, 2], although water vapor in the free atmosphere is believed to play a much more important role than that in the planet's boundary layers [2]. Increases in the quantity of water vapor increase the amount of longwave radiation trapped near the surface of the Earth and at the same time lead to an increase in air temperature and exacerbated warming. Given that it is air temperature that determines the pressure of saturated water vapor, the spatial distribution of water vapor and its vertical structure closely depend on temperature changes, which further implies an increased intensity of atmospheric processes.

The complexity of relationships between air temperature and water vapor content affects the radiation balance and water circulation patterns. For these very reasons, studies on the vertical structure of water vapor in the atmosphere, its variability over time, and differentiation across geographic space play a key role in the analysis of climate change including parametrization of climate models and weather forecasting [3, 4]. Studies are conducted first and foremost in areas where water vapor is especially vital due to its complex linkages with local conditions that also affect weather processes on a global scale in subtropical areas and polar areas. Many studies on the vertical structure of water vapor content focus on the key relevance of moisture inversions and their impact on selected processes related to water circulation and energy exchange. Kloesel and Albrecht [5] note the substantial role of the Earth's boundary layer in the regulation of water vapor transfers into the free atmosphere in the tropics, especially via convection, which is driven by low moisture inversions. The importance of studies in tropical areas is observable in the work of Holloway and Neelin [6] in the content of entrainment in convective boundary layers. In addition, Peters and Neelin [7] investigated factors that determine the formation of atmospheric precipitation. Finally, Wagner et al. [8] examined differences in the vertical structure of water vapor over the Atlantic Ocean. A large influx of solar energy along with strong air temperature inversions, cloud cover that is sensitive to multiple factors, and mesoscale circulation determining moisture transport through meridional inflows cause large differences in water vapor content across geographic space and in the vertical profile in polar areas. This subject is examined in multiple studies that emphasize the importance of the vertical structure of water vapor content in energy balances [9, 10]. Increases in the amount of water vapor present with altitude (moisture inversion) imply an increased share of the downward component of longwave radiation in the radiation balance, as shown by Devasthale et al. [11] for the winter season when water vapor from a moisture inversion constitutes more than 50% of the total column water vapor. This yields a significant effect on cloud cover [12]. The frequent occurrence of inversions in the Arctic and Antarctic was confirmed by Tomasi et al. [13], Vihma et al. [14], and Nygård et al. [15, 16] who showed their link with temperature inversions and the occurrence of low clouds.

Most research studies on this subject were conducted based on either regular weather balloon measurements or occasional field measurements. However, they also note the point nature of these measurements and the lack of opportunity to make measurements over areas of sea. One potential way to solve this problem is the use of reanalysis data, as noted by Serreze et al. [17] and Nygård et al. [15, 16]. Brunke et al. [18] compared data from five reanalysis attempts and were able to show that moisture inversions can also occur outside Arctic and Antarctic areas, although much less frequently and with less intensity. This is especially true in stratus regions and temperate regions—mostly in the cooler months of the year. On the contrary, frequencies exceed 80% in the winter, but only 10% in the summer, near the Arctic and Antarctic circles. The former are associated with lower (near the ground) temperature inversions over land areas and often are accompanied by the occurrence of fog [18, 19, 20]. These studies show that given the significance of water vapor content in key atmospheric processes, detailed analysis of the vertical structure of water vapor content in Arctic and Antarctic areas may provide additional information on factors modifying changes in moisture content in the atmosphere. At the same time, it may help in the parametrization of regional climate models (RCMs) and mesometeorological forecasting models.

The purpose of this study was to evaluate the vertical differentiation of water vapor content in the atmosphere over Europe from a seasonal perspective and to examine the role of atmospheric circulation and its impact thereupon. Moisture content was identified at pressure levels up to 300 hPa with a special focus on cases of moisture inversion. Circulation patterns were described via advection directions for air masses including eastward intensity along with the northward moisture flux.

DATA AND METHODS

Three spatial domains were examined: (1) the study area, (2) distinguished moisture regions, and (3) selected grid points. The study area consisted of Europe and the Northeastern Atlantic between 27°W and 45°E as well as 33°N and 73.5°N, with the exception of Greenland (Figure 1). Moisture regions (Figure 1; polygons) were identified by Wypych et al. [21] via the use of three variables describing moisture content, that is, total column water vapor of the entire atmosphere, specific humidity, and relative humidity at selected pressure levels (950 hPa, 850 hPa, 700 hPa, and 500 hPa). Cluster analysis helped identify six regions with different moisture conditions: Northern-Atlantic, Northern-Continental, Mid-Atlantic, Mid-Continental, Subtropical-Atlantic, and Mediterranean. Although the distinction was based on the grouping of the *k*-means method via largest possible differences and largest similarities within each type, internal cohesion tests for the identified moisture regions have shown them to be quite different. In addition, the studied moisture regions differ in the surface area (Figure 1). Hence, detailed analyses were carried out at key grid points (Figure 1). This included an analysis of the vertical structure of moisture conditions and the intensity of moisture flux. Moisture conditions at the studied grid points were similar to areal averages for the studied regions. At the same time, they also are representative of latitudinal (50.25°N) and longitudinal (15.75°W, 20.25°E) cross sections of the study area manifesting the effect of geographic location on moisture conditions. Table 1 shows the location of all the grid points used in the study.

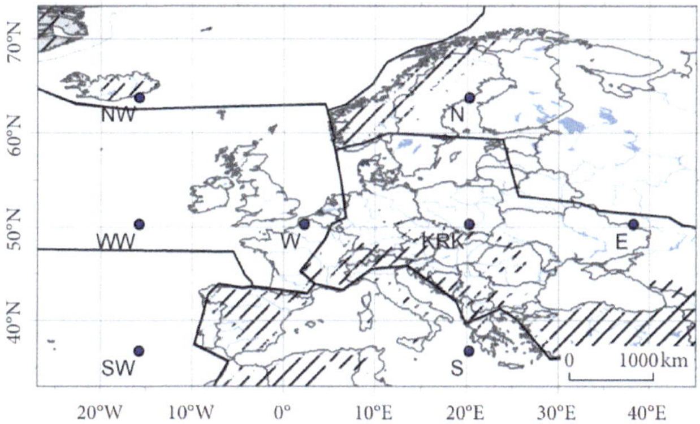

Figure 1: Study area with selected grid point locations representing different humidity regions in Europe (polygons); slanted lines are used to denote areas situated above the average 950 hPa level.

Table 1: Location of selected grid points.

Point ID	Latitude (°N)	Longitude (°)
NW	63.75	15.75 W
WW	50.25	15.75 W
SW	36.75	15.75 W
W	50.25	2.25 E
N	63.75	20.25 E
KRK	50.25	20.25 E
E	50.25	38.25 E
S	36.75	20.25 E

The study uses data from the period 1981–2015 obtained from ECMWF reanalysis data sets (ERA-Interim) at a base horizontal resolution of 0.75° [22]. Daily specific humidity values (q or SHUM) served as the basis for the study and were obtained for 18 pressure levels ranging from 950 hPa to

300 hPa. The 1,000 and 975 hPa levels were not examined, as the average height of their location over the surface may be questionable with respect to their use in the analysis of large parts of the study area. In addition, the mean daily height of isobaric surfaces was checked in order to exclude cases where they were located underground at selected grid points.

The data were 6-hour averages. The use of daily data led to the omission of cases characterized as extremes. This is not an accidental omission, and it did make it possible to realize the principal goal of the study: the climatologic analysis of the vertical differentiation of water vapor content. The use of a 35-year-long daily data series (1981–2015) provides a sample large enough to produce a detailed climatologic analysis focused on extreme conditions. Wypych [23] was able to show differentiation in vertical profiles of specific humidity over Europe on both spatial and seasonal bases. Despite the smoothing of moisture fields in the course of reanalysis, this differentiation shows that moisture variables may be used to perform climatologic analyses while maintaining vital characteristics of mesoscale processes. Although climatologic reanalysis remains a somewhat questionable method due to its lack of full agreement with measured data, it represents a way to study moisture variables in the face of a lack of homogeneous data series. In addition, reanalysis remains insensitive to spatial changes and instrumental changes in the observation network. Numerous comparative studies [4, 17, 18] have shown the usefulness of reanalysis in the study of moisture fields.

The first step in the research study consisted of the preparation of a SHUM profile and the identification of cases of moisture inversion. The depth and strength of moisture inversions were determined using a formula provided by Brunke et al. [18] based on Vihma et al. [14] (1):

$$\text{QIS} = \frac{q(p_{\min}) - q(p_{\max})}{|p_{\min} - p_{\max}|}, \tag{1}$$

where QIS is the inversion strength ($g \cdot kg-1 \cdot (50\,hPa)^{-1}$), q is the specific humidity ($g \cdot kg^{-1}$), pmin is the height of the inversion top (hPa), and p_{max} is the height of the inversion base (hPa)

The denominator of (1) describes the depth of an inversion layer. The strength of the inversion is calculated based on a layer with a depth of 50 hPa. Additionally, an analysis of inversion intensity based on normalized values was performed because of the general exponential decrease in humidity with height resulting in a decrease in inversion strength. Given the methods utilized in this study, only inversions of a certain intensity and those lasting over longer periods of time were examined (the positive gradient q present in averaged daily data). SHUM gradient (Δq) is calculated between consecutive 50 hPa layers. Inversion layers separated by a negative SHUM gradient were treated as separate moisture inversions. The study only considers cases where $\Delta q > 0.009\,g \cdot kg^{-1}$ (10% positive lapse rate). Cases of inversion were classified as surface-based cases (inversion base ≥ 900 hPa) and elevated cases (inversion base < 900 hPa). The same method was used to identify inversions of the air temperature, although cases where $\Delta t < 0.6\,K$ were ignored (10% positive lapse rate). Cases where moisture inversions and temperature inversions occur simultaneously were also noted by calculating the frequency of occurrence and the coefficient of correlation for Δq and Δt. Water vapor transport was characterized via the specific humidity flux index (SHUMF, $g \cdot kg^{-1} m \cdot s^{-1}$) calculated using (2), specific humidity q, and zonal (u) and meridional (v) wind component values for every available vertical level up to 300 hPa:

$$\mathrm{SHUMF} = \sqrt{(q \cdot u)^2 + (q \cdot v)^2}. \qquad (2)$$

The advective flux of specific humidity ($g \cdot kg^{-1}\, m \cdot s^{-1}$) was calculated for the zonal direction ($u \cdot \Delta q|_x$) and for the meridional direction ($u \cdot \Delta q|y$), where u and v are the horizontal wind speeds in the respective directions and $\Delta q|_x$ and $\Delta q|_y$ are the horizontal humidity gradients in the respective direction

The effect of atmospheric circulation on the vertical structure of water vapor content in the air was determined based on SHUM and SHUMF anomalies in advection directions. The study uses a simple division based on advection directions obtained directly from geostrophic wind directions. In addition, advection-free situations were identified with a wind speed not in excess of the 5th percentile, identified separately for each studied pressure level. Thresholds for the studied pressure levels were as follows: $2.0\,m \cdot s^{-1}$ for 950 hPa, $2.2\,m \cdot s^{-1}$ for 850 hPa, $2.7\,m \cdot s^{-1}$ for 700 hPa, $3.7\,m \cdot s^{-1}$ for 500 hPa, and $5.5\,m \cdot s^{-1}$ for the 300 hPa level. SHUM and SHUMF anomalies were calculated relative to mean monthly values and then standardized.

The use of climatologic reanalysis data in the study assures homogeneous spatial information for the study area in the context of water vapor data in the atmosphere. Multiple comparative studies [4, 17, 18] confirm the usefulness of reanalysis data in the examination of water vapor fields. A positive error was detected in temperature analysis results for the near-surface layer, which undoubtedly impacts moisture content values, and this is significant in cases of SHUM inversion occurrence [17, 24, 25]. In addition, research has shown a lack of adequate representativeness in moisture convection in research models [6]. However, despite a lack of agreement with respect to the occurrence of SHUM inversions in reanalysis data and aerologic surveys, Brunke et al. [18] were able to show the usefulness of reanalysis data in climatologic studies or studies with a lower time resolution (daily or monthly averages), representing a larger geographic area. New knowledge on the moisture variable in newer reanalysis data including ERA-Interim has led to much improved parametrization of water circulation patterns. In turn, this results in a reduction in the size of errors in humidity-sensitive variables [4, 17, 18, 26].

RESULTS AND DISCUSSION

Vertical Structure Climatology

Water vapor content in the air varies over time and across geographic space, which is closely linked with changes in temperature conditions. The relationship between air temperature and water vapor pressure (saturated) is described by the Clausius–Clapeyron equation. It assumes a decline of about 7% in water vapor content in the atmosphere due to a 1 K decrease in air temperature. This tendency, best observed over ocean areas, may be affected over land areas by rapid moisture transport, various processes affecting the near-surface zone of the air layer driven by local conditions, and convection.

The stated relationships indicate that the largest amount of water vapor in Europe is found in the area over the Atlantic Ocean at subtropical latitudes and in the south of the continent. Both areas are regions with the smallest fluctuations in water vapor content with high air temperatures throughout the year.

The smallest amount of water vapor is detected in the atmosphere in the northern part of the study area, especially over the ocean, an area characterized by small fluctuations in air temperature on an annual basis. In temperate zones, especially across continental Europe, annual fluctuations in water vapor content in the air are the highest in the study area due to the increasingly continental climate of the interior of Europe.

In the winter, water vapor content over land areas corresponds to that detected at polar latitudes, while in the summer, it is greater than that determined for subtropical regions [21]. Given that the presence of water vapor in the Earth's atmosphere is substantial up to an altitude of about 2 km, determined by a negative temperature gradient, the greatest differentiation in the vertical profile is noted up to the 700 hPa level (Figure 2).

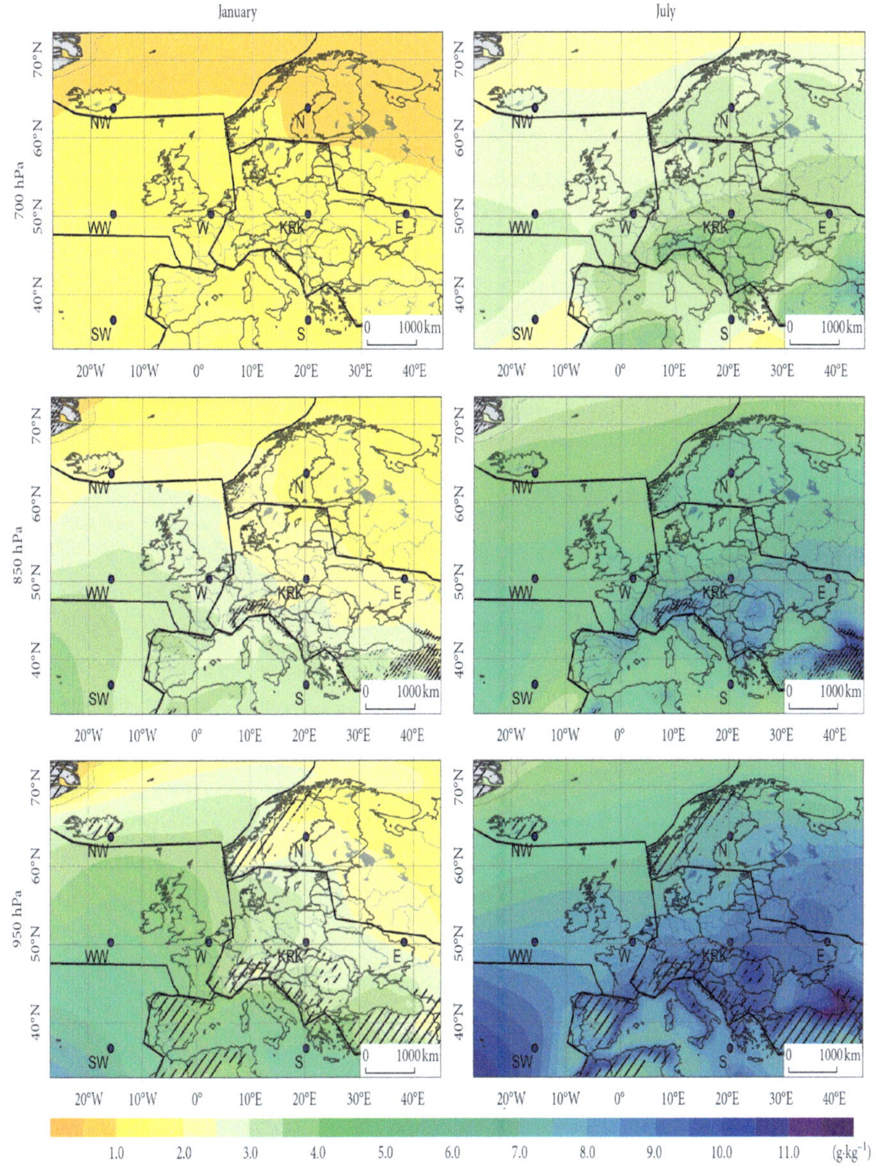

Figure 2: Monthly mean specific humidity at selected pressure levels (1981–2015); slanted lines are used to denote areas situated above the average of selected pressure levels.

Changes in the atmospheric boundary layer yield 950 hPa and 850 hPa levels characterized by the occurrence of significant differences in SHUM values. This applies to differentiation in the vertical profile and also in the spatial sense, as expressed by an areal standard deviation (σ) calculated using all the domain grid points (Figures 2 and 3). In July, its values for mentioned pressure levels were as follows: $\sigma_{areal} = 1.6\,g \cdot kg^{-1}$ and $\sigma_{areal} = 1.2\,g \cdot kg^{-1}$, and in January, its values were $\sigma_{areal} = 1.4\,g \cdot kg^{-1}$ and $\sigma_{areal} = 0.75\,g \cdot kg^{-1}$. In January, the effect of warm North Atlantic and Norwegian currents may be observed along the western and northwestern coasts of the continent up to the 850 hPa level along with related increases in moisture levels across Western Europe (Figure 2). On the contrary, the interior of the European continent is affected by a seasonal high and characterized by much lower SHUM values. In July, the high SHUM content is readily observable in the troposphere over land areas, higher than that over the ocean at all three studied levels (Figure 2).

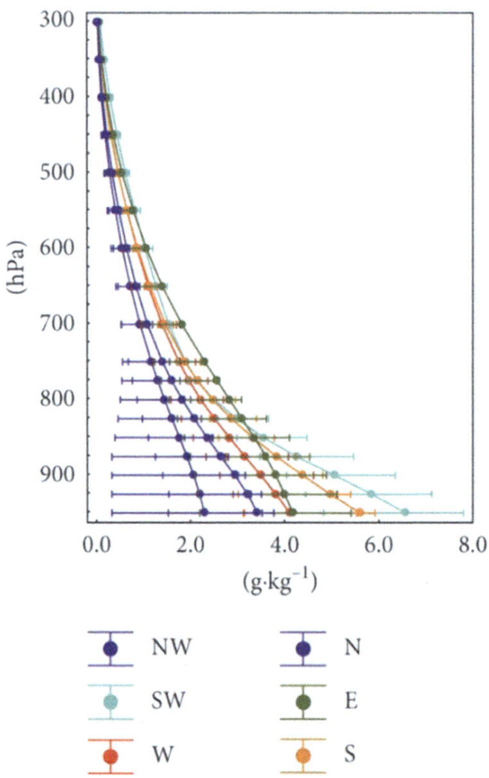

(a)

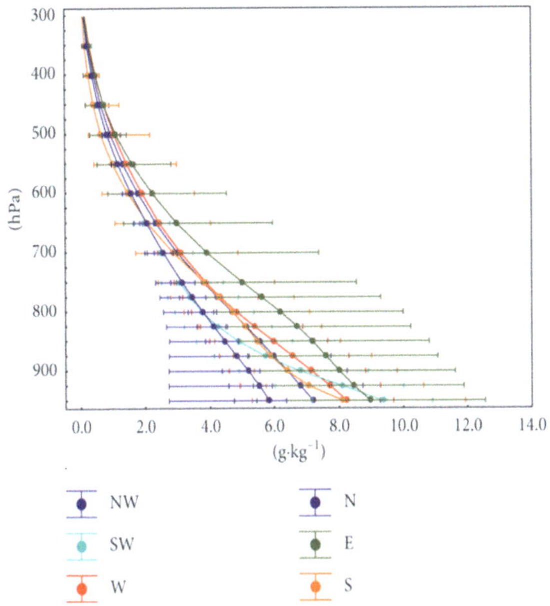

(b)

Figure 3: Monthly mean specific humidity (SHUM) vertical profiles at selected grid points (for details, see Figure 1 and Table 1); whiskers show the range defined by minimum and maximum SHUM values in a moisture region represented by the given grid point: (a) January; (b) July (1981–2015).

The stated characteristics of water vapor content are provided in detail in the vertical SHUM profiles produced for selected grid points in Figure 3.

From a climatologic perspective, the decline in the mean monthly water vapor content in the Earth's atmosphere, based on the Clausius–Clapeyron equation, is exponential in nature. This type of pattern, however, is representative mainly of points located across ocean areas. The vertical SHUM profile is quite similar for January (Figure 3(a)) and July (Figure 3(b)) and shifted in July in the direction of higher values. The Mediterranean Sea is partly affected by the surrounding land masses. The point located on the Mediterranean Sea (S) in July features a higher than oceanic moisture content at levels between 950 and 650 hPa (Figure 3(3)) and follows a similar pattern in January (Figure 3(a)). Northern areas (NW) have low air and water temperatures and are characterized by the smallest differences in SHUM in July. In January, on the contrary, northern areas have a higher water temperature than land temperature and experience evaporation off the ocean's surface, resulting in a moisture supply leading to a vertical SHUM gradient that is larger than that over continental areas (grids N and E). As stated previously, land areas are characterized by the largest differences in water vapor content. The SHUM gradient in layers from 950 hPa to 800 hPa at sites located deep inland equals about $2\,g{\cdot}kg^{-1}{\cdot}(150\,hPa)^{-1}$, while its value in the troposphere over the ocean (grid SW) is about $6\,g{\cdot}kg^{-1}{\cdot}(150\,hPa)^{-1}$ (Figure 3(b)).

There exist special cases in the troposphere characterized by increasing water vapor content in line with increasing altitude. Moisture inversions may assume variable depths and degrees of strength in relation to location and season of the year. In some cases, several inversion layers may be observed within one vertical SHUM profile [11, 15–18].

Given the procedural assumptions made in this study, only inversions lasting longer periods of time are considered (positive SHUM gradient in daily averages) as well as those of appropriate strength (Section 2). The frequency of inversions in the troposphere over Europe varies substantially according to season and across geographic space. Inversions occur much more frequently in the winter (January) over land areas—sometimes in excess of 80% of days in the Scandinavian Peninsula and roughly 50% of days in Eastern and Southeastern Europe (Figures 4 and 5; KRK and E). In addition, inversions occur on 30% to 40% of days across the subtropical areas of the Atlantic (Figure 5; SW). However, in the summer, inversions are almost entirely limited to the subtropical zone across the Atlantic Ocean (40% to 50%) and the Mediterranean Sea (30% to 40%) (Figures 4 and 5; SW and S).

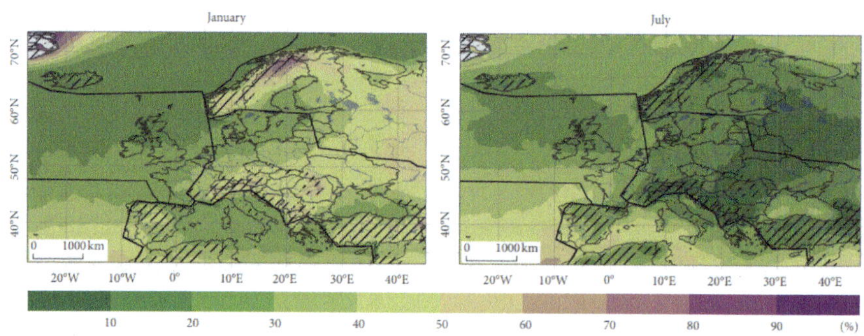

Figure 4: Frequency (%) of humidity inversion occurrence (1981–2015); slanted lines are used to denote areas situated above the average 950 hPa level.

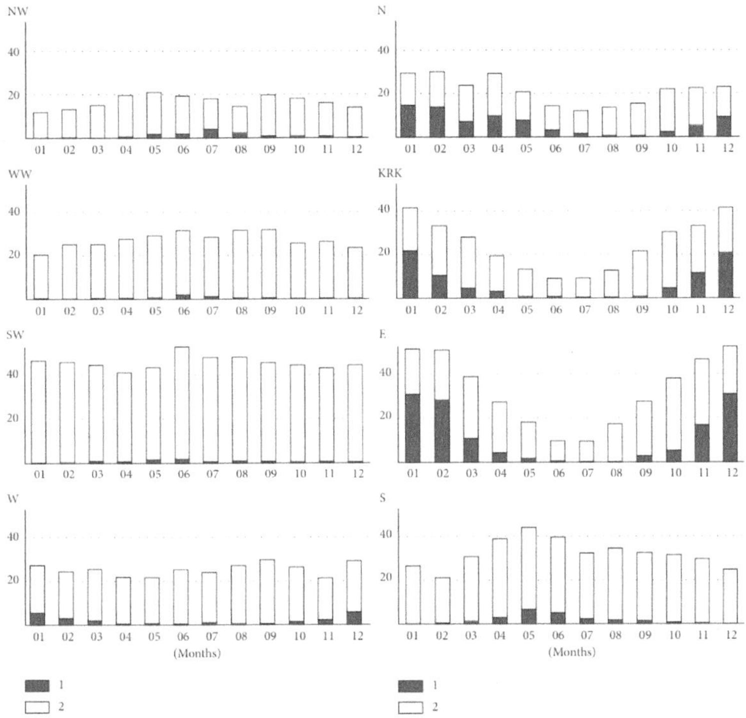

Figure 5: Frequency (%) of humidity inversion occurrence: (1) surface-based and (2) elevated at selected grid points (for details, see Figure 1 and Table 1) (1981–2015).

The temporal differentiation and spatial differentiation of the level of inversion occurrence are readily observable (Figure 6). In January, the mean base of the inversion layer over most land areas does not exceed 850 hPa, while in Western Europe, Southwestern Europe, and western coastal areas, it is 700 hPa. On the contrary, most inversions occurring over ocean areas are characterized by a lower level above 700 hPa (Figure 6). The mean base of the inversion layer in July in Eastern Europe and the Iberian Peninsula exceeds 700 hPa, while over the Caucasus, it is above 550 hPa (Figure 6). A large number of winter inversions take the form of surface-based inversions occurring over continental areas (Figure 5; W, N, KRK, and E). On the contrary, elevated-type inversions are the predominant type over sea and ocean areas regardless of season, while the few surface-based inversions that are detected may be designated marginal in significance (Figure 5; NW, SW, and S).

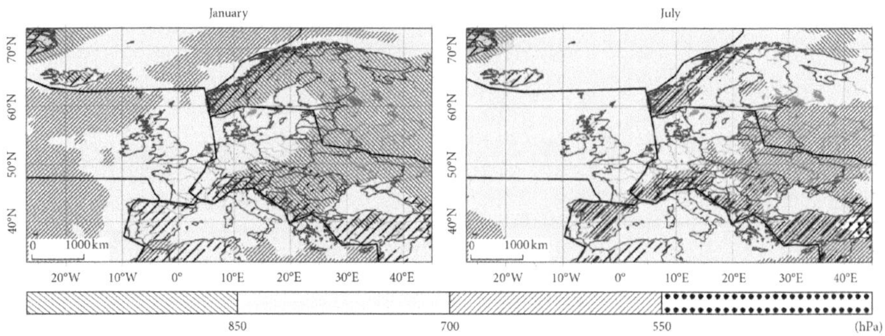

Figure 6: Mean height (hPa) of humidity inversion base (1981–2015); slanted lines are used to denote areas situated above the average 950 hPa level.

The current study has confirmed and expanded on the study by Brunke et al. [18], who used data from 5 different sets of reanalysis data and compared them with weather balloon surveys in order to show the occurrence of inversions in polar areas, mostly in the winter season, and the low base of these inversions. Nygård et al. [15, 16] and Vihma et al. [14] found statistically significant relationships between the occurrence of moisture inversions in relation to temperature inversions in near-polar areas in the northern and southern hemispheres. The similar conditions present in the winter in the atmospheric boundary layer over land areas (i.e., European interior areas) including negligible cloud cover, strong cooling of surfaces, and the associated water vapor condensation all suggest the simultaneous occurrence of moisture and temperature inversions also in polar regions.

Research has shown that surface-based SHUM inversions are accompanied by temperature inversions 70% of the time, which then increase to 90% in the direction of the continent's interior (grid E), while in Western Europe, they do not exceed 60%. This dependency declines to an average of 50% in the layer between 900 and 800 hPa. The correlation coefficient for the moisture gradient and air temperature on days with inversions reaches a statistically significant (=0.05) average value of 0.5. For July, the absence of any type of relationship between the occurrence of moisture and temperature inversions confirms the dynamic origin of this particular linkage resulting from the advection of moisture or convection.

Although SHUM inversions occur more frequently in the winter, their depth is small, especially in privileged areas, and they range from <70 hPa in the Scandinavian Peninsula to about 80 hPa over other land areas (Figure 7). The deepest layer wherein water vapor content grows with altitude is most often detected over the North Atlantic (~100 hPa). Summer inversions are shallow and do not exceed 80 hPa over water and about 60 hPa over land (Figure 7). The strength of inversions varies negligibly over the study area. Somewhat higher values may be detected in July at an average of $0.2\,g\cdot kg^{-1}\cdot(50\,hPa)^{-1}$: from more than $0.3\,g\cdot kg^{-1}\cdot(50\,hPa)^{-1}$ over subtropical ocean areas to $0.1\,g\cdot kg^{-1}\cdot(50\,hPa)^{-1}$ over land. The strength of winter inversions is greatest over the Scandinavian Peninsula ($>0.3\,g\cdot kg^{-1}\cdot(50\,hPa)^{-1}$) and over parts of continental Europe ($0.2\text{-}0.3\,g\cdot kg^{-1}\cdot(50\,hPa)^{-1}$). Other parts of the study area do not exceed $0.15\,g\cdot kg^{-1}\cdot(50\,hPa)^{-1}$.

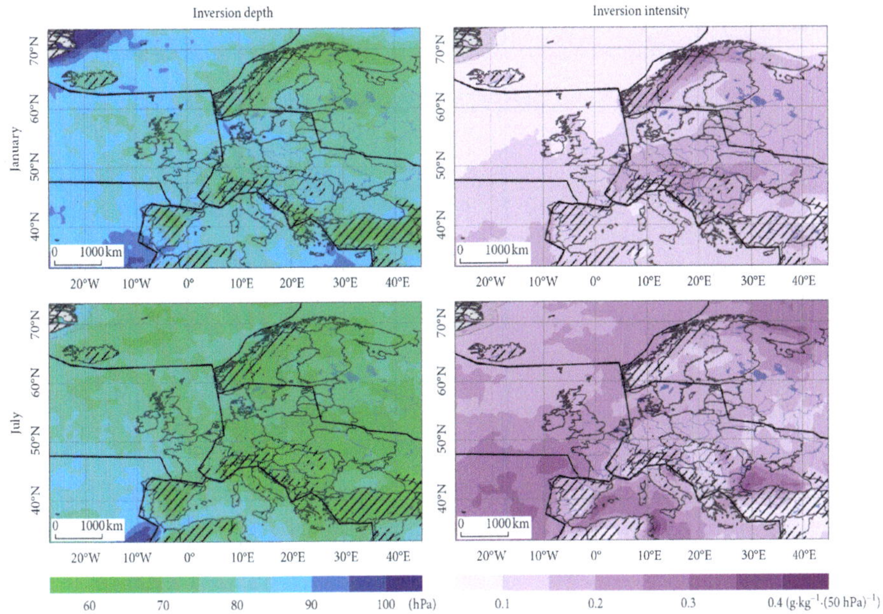

Figure 7: Mean humidity inversion depth (hPa) and intensity ($g\cdot kg^{-1}\cdot(50\,hPa)^{-1}$) (1981–2015); slanted lines are used to denote areas situated above the average 950 hPa level.

In order to reduce bias originating due to the exponential decrease in the amount of water vapor in the atmosphere, the analysis was repeated using standardized values. The procedure confirmed the most intensive inversions occurring in January in the northern (Scandinavian Peninsula) and northeastern parts of Europe (Northern Continental moisture region) but highlighted also the strength of inversions over the Subtropical Atlantic region (Figure 8), which are mostly elevated, with the base between the levels of 700 and 550 hPa (Figure 6). The strength of summer inversions over the Subtropical Atlantic region was noted but was also determined over the continent where it was smoothed over by the moisture decrease bias (Figure 8).

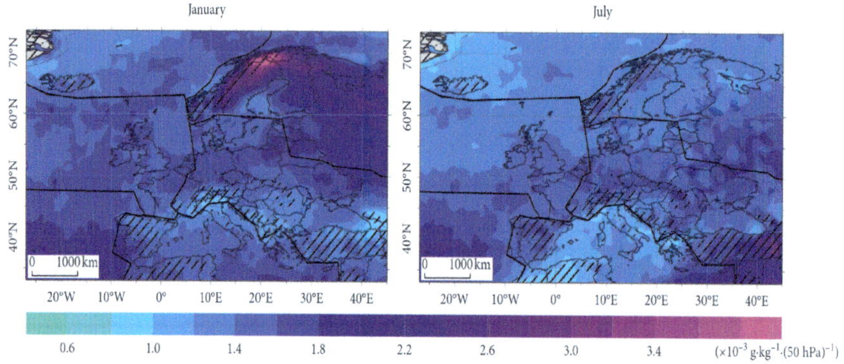

Figure 8: Mean standardized humidity inversion intensity ($g \cdot kg^{-1} \cdot (50\,hPa)^{-1}$) (1981–2015); slanted lines are used to denote areas situated above the average 950 hPa level.

Atmospheric Circulation Impact

Water vapor transport is one of the most important processes determining differences in its distribution across Europe. Existing research strongly suggests that atmospheric circulation, especially the North Atlantic Oscillation (NAO), shapes moisture content in the air in the winter [27–29]. Air saturated with water vapor is transported from over ocean areas in the direction of land, thanks to zonal circulation, and in the northerly direction, also known as meridional flux. Significantly weaker dependencies were noted in the warm half of the year when differences in water vapor content originate in convection, while horizontal moisture flux plays a secondary role.

In the present study, the analysis of water vapor transport concerns the advection of air masses or horizontal moisture flux at principal pressure levels: 950 hPa, 850 hPa, 700 hPa, and 500 hPa. Moisture transport is strongly aided by advection at the 850 hPa level. The limited impact of the atmospheric boundary layer enables the flux of air masses in the direction of the continent, and this includes not only Western Europe, as in the case of altitudes closer to the surface of the Earth (Figure 9), but also Central Europe. Moisture transport strongly varies by season due to the amount of water vapor available per given air temperature and due to circulation in each given season. The substantial role of seasonal pressure highs is observable in January, in particular at the 950 hPa level, limiting the horizontal motion of air masses (advection-free in many cases) and substantially affecting temperature conditions. The role of orographic barriers is also quite significant. Despite a high SHUM in July, SHUM transport is limited. In the south of Europe, this is associated with a shift in the intertropical convergence zone in the northerly direction and a predominance of longitudinal flux with a northerly component in the Mediterranean region. The influx of moisture becomes observable only in the free atmosphere (Figure 9).

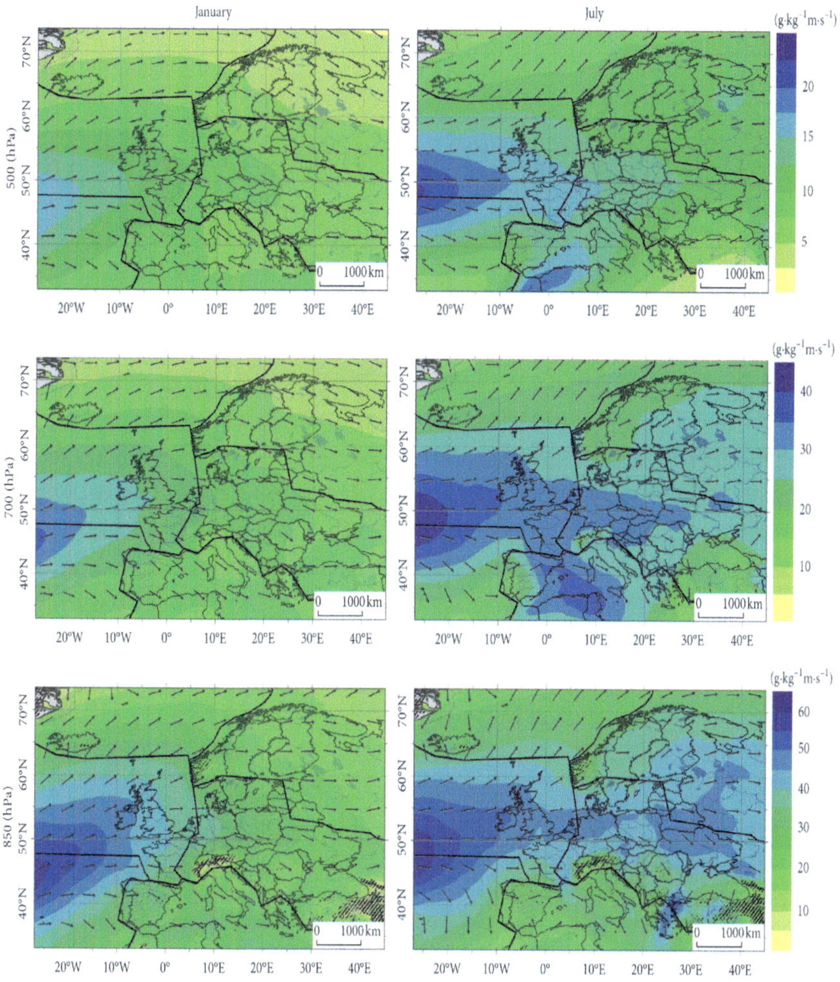

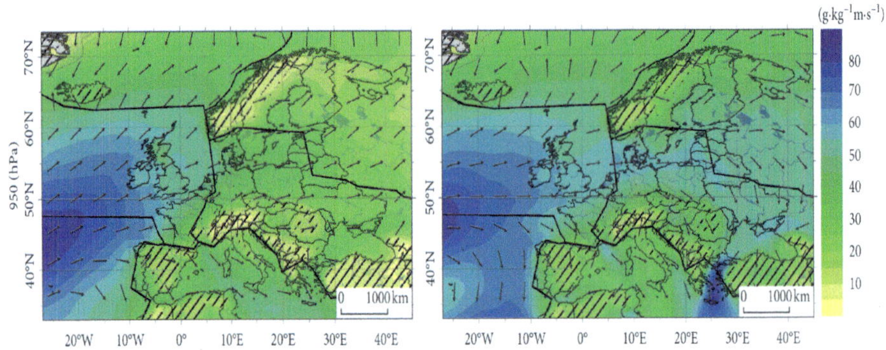

Figure 9: Monthly mean specific humidity flux and wind direction (arrows) at selected pressure levels (1981–2015); slanted lines are used to denote areas situated above the average of selected pressure levels.

Figure 10 uses dashed curves to show the significance of zonal and meridional transport of the moisture content in the air. Vertical profiles of specific humidity advective flux show a virtual lack of advection in January in continental Europe (grids W, N, KRK, and E). Small amounts of SHUM are carried by meridional flux along with relatively small differences in moisture transport over the ocean in the temperate and near-polar zones (grids WW and NW, resp.). On the contrary, large differences do occur in the subtropical zone, in particular over the Mediterranean Sea (grid S). Meridional flux assumes significant negative deviations, while zonal flux assumes equally large positive deviations (Figure 10). July vertical profiles show significant variability in specific humidity advective flux in continental Europe, especially in cases of days with SHUM inversions (solid lines).

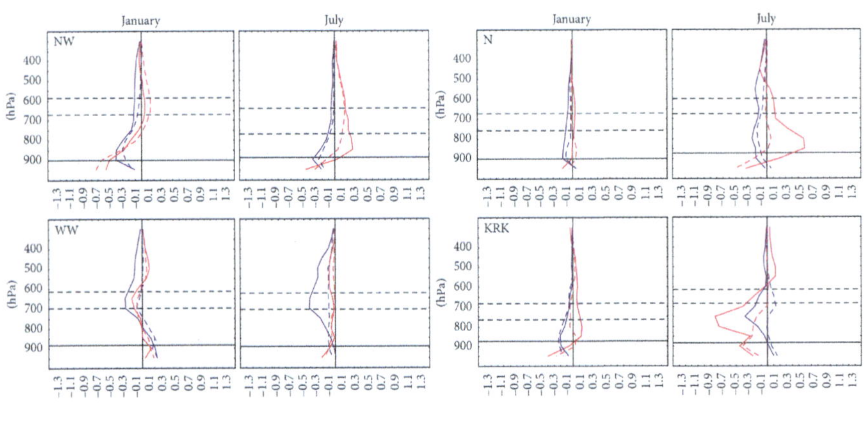

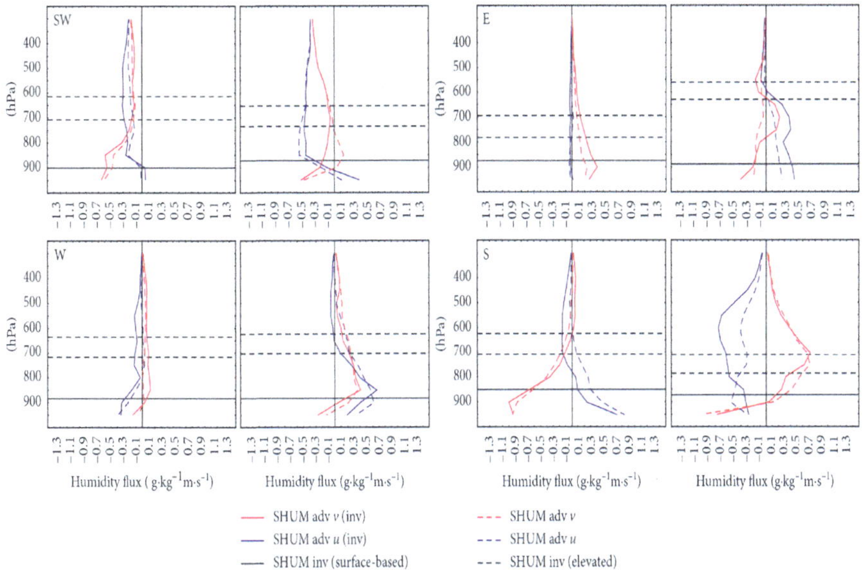

Figure 10: Climatologic mean profiles of specific humidity advective flux: eastward (*u*) and northward () during inversion (solid curve) and noninversion (dashed curve) events (detailed description in the text) at selected grid points (for details, see Figure 1 and Table 1).

Figure 11 shows in detail the role of selected advection directions in moisture transport based on selected grid points. Negative SHUM anomalies in January occur with advection from the north and east at all pressure levels, although in the interior of the continent (grids KRK and E), the deviations are largely closer to the surface of the Earth (levels 950 hPa and 850 hPa).

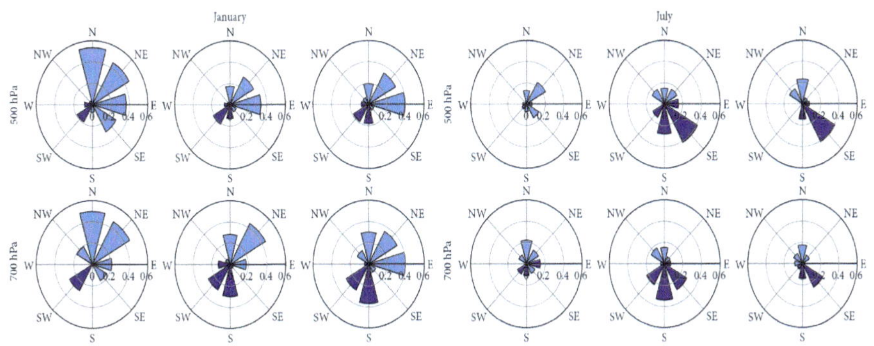

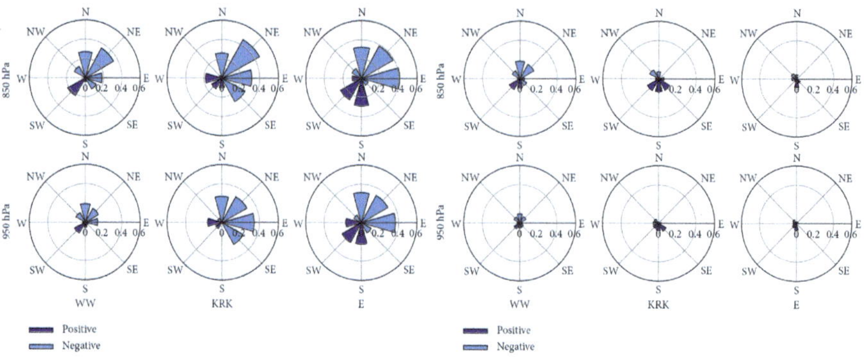

Figure 11: Specific humidity anomalies (g·kg^{-1}) in advection directions (with respect to monthly mean, standardized) at selected pressure levels (1981–2015) and selected grid points (for details, see Figure 1 and Table 1), based on Wypych [23].

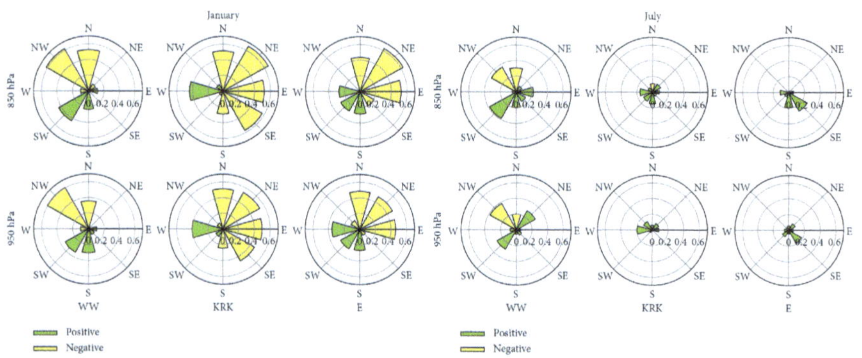

Figure 12: Specific humidity flux anomalies (g·kg^{-1} m·s^{-1}) in advection directions (with respect to monthly mean, standardized) at selected pressure levels (1981–2015) and selected grid points (for details, see Figure 1 and Table 1), based on Wypych [23].

Positive moisture anomalies and moisture transport in the winter are associated with advection from the west—or in the case of points located farthest to the east—and also from the south. The insignificance of air circulation in the summer is confirmed by almost nonexistent SHUM and SHUMF anomalies in July at selected grid points (Figures 11 and 12). One special case of the role of moisture advection is the occurrence of SHUM inversions (Figure 12). The significance of moisture inversions in the formation of total column water vapor (TCWV) has often been noted in the context of near-polar regions [14–16]. Given generally low water vapor content in the air, an inversion column may recharge TCWV up to 40%, whereas given a larger water vapor content, the effect of inversion on the TCWV value declines to about 15%. Therefore, the occurrence of inversions and share in the radiation balance are most substantial in the winter and spring. As noted previously, most cool season moisture inversions are associated with the occurrence of temperature inversions, often accompanying high pressure systems.

The lowering of air temperature near the surface of the ground leads to the condensation of water vapor present in the air or the drying of the air mass resulting in an inversion of air moisture levels. Another factor that favors the occurrence of an inversion is the advection of moist air at higher levels or dry air near the surface of the Earth. Figure 10 shows mean profiles of specific humidity advective flux and mean inversion levels: top of surface-based inversions (solid lines) and bottom and top of elevated inversions (dashed lines). It may be readily observed that, in the case of surface-based inversions in January, the moisture flux effect is not significant. There is no meaningful difference between days with an inversion (solid curve) and days without an inversion (dashed curve).

The grid point *E* is an exception where moisture advection is noted at the top of the surface-based inversion. On the contrary, in July, a month characterized by far fewer moisture inversions and especially by elevated inversions over regions of land, the significant role of meridional moisture flux may be observed on the top of surface-based inversions at points located in the northern part of the study area (grids NW and N) along with the advection of dry air (regardless of moisture flux direction) at the bottom of elevated inversions at the point KRK. Dependencies determined at selected grid points are consistent with results produced by studies in near-polar areas [14–16], as well as on the macroscale [21], which underscore the significance of meridional moisture flux and the convergence and convection of water vapor in the summer season.

CONCLUSIONS

Studies on the vertical structure of SHUM yield better parametrization of both climate models and mesometeorological models. Water vapor condensation in the atmospheric boundary layer is determined first and foremost by its vertical flux, as it is the structure of moisture both in the atmospheric boundary layer and in the free atmosphere that determines the strength and direction of water vapor transport [30]. For example, the presence of moisture inversions plays a decisive role in shifts in convection at subtropical latitudes [5] and also helps support the cloud layer by making evaporation off its upper part impossible [11] and assists in fog formation [12]. The present study constitutes a climatologic analysis based on averaged daily data and shows the existence of differences in the vertical structure of water vapor content in the troposphere over Europe and the North Atlantic, including the presence of moisture inversions beyond 60°N and at temperate and subtropical latitudes. Inversions may be surface-based or elevated, and their occurrence follows a seasonal pattern.

Land areas are characterized by the presence of surface-based inversions in the winter and elevated inversions in the summer, while ocean areas are characterized by lower inversions in the summer versus winter, although these are not true surface-based inversions, which tend to occur sporadically over the ocean. Atmospheric circulation plays an important role in changes in water vapor content, especially in the winter season, and determines both surpluses and shortages of moisture through effects generated by different pressure system types (especially important is the seasonal high) and advection directions. Anomalies in the atmospheric boundary layer occur primarily over land areas. Vertical profiles of water vapor content are characterized by large standard deviations due to the large variety of physical characteristics of incoming air masses. Over ocean areas, the vertical structure of moisture is quite stable at lower levels, with some deviations related to advection direction; however, at higher pressure levels, it becomes differentiated.

In summer months, the vertical structure of moisture remains largely unaffected by changes in circulation conditions, which given a variable SHUM content at subsequent pressure levels indicating the significance of processes occurring in the atmospheric boundary layer in the shaping of moisture content in the troposphere. The results of the present study suggest a need for further research on the subject of variances in the vertical structure of water vapor content over Europe, with a special focus on subdaily data that would make it possible to examine differences in water vapor content in a dynamic sense. The use of reanalysis data bears the burden of near-surface biases that also do impact the structure of moisture; nevertheless, given a large array of variables available, it allows for a comprehensive analysis of processes determining moisture levels and an evaluation of their role in relation to ongoing weather changes.

Conflicts of Interest

The authors declare that they have no conflicts of interest.

ACKNOWLEDGMENTS

The authors would like to thank Mr. Michał Różycki for calculation assistance.

REFERENCES

1. S. Manabe and R. T. Wetherald, "Thermal equilibrium of the atmosphere with a given distribution of relative humidity," *Journal of Atmospheric Sciences*, vol. 24, no. 3, pp. 241–259, 1967.
2. E. K. Schneider, B. P. Kirtman, and R. S. Lindzen, "Tropospheric water vapor and climate sensitivity," *Journal of Atmospheric Sciences*, vol. 56, no. 11, pp. 1649–1658, 1999.
3. S. C. Sherwood, R. Roca, T. M. Weckwerth, and N. G. Andronova, "Tropospheric water vapor, convection, and climate," *Reviews of Geophysics*, vol. 48, no. 2, p. RG2001, 2010.
4. K. E. Trenberth, J. T. Fasullo, and J. Mackaro, "Atmospheric moisture transports from ocean to land and global energy flows in reanalyses," *Journal of Climate*, vol. 24, no. 18, pp. 4907–4924, 2011.
5. K. A. Kloesel and B. A. Albrecht, "Low-level inversions over the tropical Pacific-thermodynamic structure of the boundary layer and the above-inversion moisture structure," *Monthly Weather Review*, vol. 117, no. 1, pp. 87–101, 1989.
6. C. Holloway and J. D. Neelin, "Moisture vertical structure, column water vapor, and tropical deep convection," *Journal of Atmospheric Sciences*, vol. 66, no. 6, pp. 1665–1683, 2009.
7. O. Peters and J. D. Neelin, "Critical phenomena in atmospheric precipitation," *Nature Physics*, vol. 2, no. 6, pp. 393–396, 2006.
8. D. Wagner, E. Ruprecht, and C. Simmer, "A combination of microwave observations from satellites and an EOF analysis to retrieve vertical humidity profiles over the ocean," *Journal of Applied Meteorology*, vol. 29, no. 11, pp. 1142–1157, 1990.
9. M. C. Serreze, R. G. Barry, and J. E. Walsh, "Atmospheric water vapor characteristics at 70°N," *Journal of Climate*, vol. 8, no. 4, pp. 719–731, 1995.
10. M. Gerding, C. Ritter, M. Müller, and R. Neuber, "Tropospheric water vapour soundings by lidar at high Arctic latitudes," *Atmospheric Research*, vol. 71, no. 4, pp. 289–302, 2004.
11. A. Devasthale, J. Sedlar, and M. Tjernström, "Characteristics of water-vapour inversions observed over the Arctic by Atmospheric Infrared Sounder (AIRS) and radiosondes," *Atmospheric Chemistry and Physics*, vol. 11, no. 18, pp. 9813–9823, 2011.

12. A. Solomon, M. D. Shupe, P. O. G. Persson, and H. Morrison, "Moisture and dynamical interactions maintaining decoupled Arctic mixed-phase stratocumulus in the presence of a humidity inversion," *Atmospheric Chemistry and Physics*, vol. 11, no. 19, pp. 10127–10148, 2011.
13. C. Tomasi, B. Petkov, E. Benedetti et al., "Characterization of the atmospheric temperature and moisture conditions above Dome C (Antarctica) during austral summer and fall months," *Journal of Geophysical Research*, vol. 111, Article ID D20305, 2006.
14. T. Vihma, T. Kilpeläinen, M. Manninen et al., "Characteristics of temperature and humidity inversions and low-level jets over Svalbard Fjords in spring," *Advances in Meteorology*, vol. 2011, Article ID 486807, 14 pages, 2011.
15. T. Nygård, T. Valkonen, and T. Vihma, "Antarctic low-tropospheric humidity inversions: 10-yr climatology," *Journal of Climate*, vol. 26, no. 14, pp. 5205–5219, 2013.
16. T. Nygård, T. Valkonen, and T. Vihma, "Characteristics of Arctic low-tropospheric humidity inversions based on radio soundings," *Atmospheric Chemistry and Physics*, vol. 14, no. 4, pp. 1959–1971, 2014.
17. M. C. Serreze, A. P. Barrett, and J. Stroeve, "Recent changes in tropospheric water vapor over the Arctic as assessed from radiosondes and atmospheric reanalyses," *Journal of Geophysical Research*, vol. 117, Article ID D10104, 2012.
18. M. A. Brunke, S. T. Stegall, and X. Zeng, "A climatology of tropospheric humidity inversions in five reanalyses," *Atmospheric Research*, vol. 153, pp. 165–187, 2015.
19. H. Liu, H. Zhang, L. Bian et al., "Characteristics of micrometeorology in the surface layer in the Tibetan Plateau," *Advances in Atmospheric Sciences*, vol. 19, no. 1, pp. 74–87, 2002.
20. D. Liu, J. Yang, S. Niu, and Z. Li, "On the evolution and structure of a radiation fog event in Nanjing," *Advances in Atmospheric Sciences*, vol. 28, no. 1, pp. 223–237, 2010.
21. A. Wypych, B. Bochenek, and M. Różycki, "Atmospheric moisture content over Europe and the Northern Atlantic," *Atmosphere*, vol. 9, no. 1, p. 18, 2018.
22. D. P. Dee, S. M. Uppala, A. J. Simmons et al., "The ERA-Interim reanalysis: configuration and performance of the data assimilation

system," *Quarterly Journal of Royal Meteorological Society*, vol. 137, no. 656, pp. 553–597, 2011.

23. A. Wypych, *Tropospheric Moisture Content over Europe*, Institute of Geography and Spatial Management Jagiellonian University, Krakow, Poland, 2018, in Polish.
24. M. Tjernström and R.G. Graversen, "The vertical structure of the lower Arctic troposphere analysed from observations and the ERA-40 reanalysis," *Quarterly Journal of Royal Meteorological Society*, vol. 135, no. 639, pp. 431–443, 2009.
25. E. Jakobson, T. Vihma, T. Palo, L. Jakobson, H. Keernik, and J. Jaagus, "Validation of atmospheric reanalyses over the central Arctic Ocean," *Geophysical Research Letters*, vol. 39, no. 10, p. L10802, 2012.
26. H. Flentje, A. Dörnbrack, A. Fix, G. Ehret, and E. Hólm, "Evaluation of ECMWF water vapour analyses by airborne differential absorption lidar measurements: a case study between Brazil and Europe," *Atmospheric Chemistry and Physics*, vol. 7, no. 19, pp. 5033–5042, 2007.
27. J. W. Hurrel, "Decadal trends in the North Atlantic Oscillation: regional temperatures and precipitation," *Science*, vol. 269, no. 5224, pp. 676–679, 1995.
28. E. Ruprecht, S.S. Schröder, and S. Ubl, "On the relation between NAO and water vapour transport towards Europe," *Meteorologische Zeitschrift*, vol. 11, no. 6, pp. 395–401, 2002.
29. A. Stohl, C. Forster, and H. Sodemann, "Remote sources of water vapour forming precipitation on the Norwegian west coast at 60°N—a tale of hurricanes and an atmospheric river," *Journal of Geophysical Research*, vol. 113, Article ID D05102, 2008.
30. H. Linné, B. Hennemuth, J. Bösenberg, and K. Ertel, "Water vapour flux profiles in the convective boundary layer," *Theoretical and Applied Climatology*, vol. 87, no. 1–4, pp. 201–211, 2007.

CHAPTER 11

Impacts of Model Spatial Resolution on the Vertical Structure of Convection in the Tropics

Hien Xuan Bui[1], JiaYuh Yu[1], Chia Chou[2,3]

[1] Department of Atmospheric Sciences, National Central University, Taoyuan City, Taiwan

[2] Research Center for Environmental Changes, Academia Sinica, Taipei, Taiwan

[3] Department of Atmospheric Sciences, National Taiwan University, Taipei, Taiwan

ABSTRACT

This study examined the impacts of model horizontal resolution on vertical structures of convection in the tropics by performing sensitivity

Citation: Bui, H.X., Yu, JY. & Chou, C. "Impacts of model spatial resolution on the vertical structure of convection in the tropics". Clim Dyn 52, 15–27 (2019). https://doi.org/10.1007/s00382-018-4125-3.

experiments with the NCAR CESM1. It was found that contributions to the total precipitation between top-heavy and bottom-heavy convection are different among various resolutions. A coarser resolution tends to produce a greater contribution from top-heavy convection and, as a result, stronger precipitation in the western Pacific ITCZ; while there is less contribution from bottom-heavy convection and weaker precipitation in the eastern Pacific ITCZ. In the western Pacific ITCZ, where the convection is dominated by a top-heavy structure, the stronger precipitation in coarser resolution experiments is due to changes in temperature and moisture profiles associated with a warmer environment (i.e., thermodynamical effect). In the eastern Pacific ITCZ, where the convection is dictated by a bottom-heavy structure, the stronger precipitation in finer resolution experiments comes from changes in convection structure (i.e., dynamic effect) which favors a greater contribution of bottom-heavy convection as the model resolution goes higher. The moisture budget analysis further suggested that the very different behavior in precipitation tendencies in response to model resolution changes between the western and eastern Pacific ITCZs are determined mainly by changes in convective structure rather than changes in convective strength. This study pointed out the importance of model spatial resolution in reproducing a reasonable contribution to the total precipitation between top-heavy and bottom-heavy structure of convection in the tropical Pacific ITCZs.

Keywords: Coarse Resolution Experiment, Moisture Budget Analysis, Eastern Pacific, Convective Structures, Community Atmospheric Model Version (CAM5)

INTRODUCTION

The core of tropical dynamics involves strong interaction between cumulus convection and large-scale perturbations. A reasonable representation of the collective effect of subgrid scale cumulus convection process in a large-scale atmospheric model is the key to produce a successful simulation of the time-mean tropical circulation and its associated variability. The cumulus convection can be in general divided into two major categories with different intensity of precipitation, namely, deep convection and shallow convection. In addition, congestus clouds are also an important category whose tops stop at slightly above the freezing level (Johnson et al. 1999).

To illustrate the feedback of convective processes on large-scale flows, Yanai et al. (1973) proposed the diagnosis of apparent heat source (Q_1) and apparent moisture sink (Q_2). Diabatic heating in the atmosphere is a combined consequence of radiation heating from both shortwave and longwave, and latent heating from phases changes associated with convection. Vertical profiles of diabatic heating with global coverages have been estimated from satellite retrievals (Tao et al. 2006). However differences among retrieval algorithms could lead to discrepancies in the vertical structures. Besides, many studies have also employed the reanalysis data to examine the horizontal distributions, amplitude and vertical structure of diabatic heating (Nigam et al. 2000; Chan and Nigam 2009; Hagos et al. 2010; Ling and Zhang 2013). A surprising result in these studies is that the tropical heating profiles among the reanalysis were quite diverse. Many studies, such as Schumacher et al. (2007), Takayabu et al. (2010), Hagos et al. (2010) have discussed the double peak structure in the vertical heating profiles. The upper peak is coming from the stratiform and deep convection while the lower peak is coming from the shallow and congestus clouds, which is also in agreement with recent study of Johnson et al. (2016).

One of the common biases in modeled tropical precipitation is the underestimation of heavy precipitation events due to the coarse spatial resolution of the model (Gregory et al. 1997; Stephens et al. 2010). For example, Deser et al. (2006) compared the mean precipitation simulations for a version of the Community Atmospheric Model (CAM3) using T42 and T85 resolutions, respectively. They found that spatial pattern of precipitation in low-resolution simulation shows significant quantitative errors, even though the major features are consistent with observations. By examining the mean summer precipitation over the tropical ocean, Hack et al. (2006) showed a locally significant meridional redistribution of precipitation as a function of spatial resolution.

Climate models have been extensively used over the past two decades to project future climate changes under the anthropogenic greenhouse warming. Although most of the climate models used in CMIP5 have increased spatial resolution compared to those used in CMIP3, some still keep relatively low resolution, e.g., CMSS-CESM (3.75∘×3.75∘), HadCM3 (3.75∘×2.5∘), FGOALS-g2 and MIROC-ESM (2.8125∘×2.8125∘), GFDL-CM2.1 (2.5∘×2∘) and GISS-E2-H (2∘×2.5∘). It is thus worth knowing how the model's spatial resolution impacts the vertical structure of convection in the Tropics because the efficiency of moist static energy export by cumulus convection depends largely on the structure of convection (Back and Bretherton 2006; Bui et al. 2016).

Differences in physical parameterization and model resolution are two major factors from which most of the biases in the contemporary climate models come. In this study, we examine the impacts of model spatial resolution on vertical structures of convection in climate, focusing on how the contribution between top-heavy and bottom-heavy convection might change in various model resolutions. We run the Community Atmospheric Model (CAM) version 5 with four spatial resolutions: 400, 200, 100 and 50 km to cover the spectra of grid sizes of contemporary climate models. Section 2 presents an overview of the atmospheric model and the design of sensitivity experiments. A brief introduction of satellite and field campaign datasets is also presented in this section. Section 3 describes the diabatic heat source and moisture sink as well as the bin method used to identify top-heavy and bottom-heavy convection cases. Major results are presented and discussed in detail in Sect. 4. Section 5 utilizes the moisture budget analysis to examine the dynamic and thermodynamic feedbacks on precipitation. A summary and discussion is given in Sect. 6.

DESIGN OF SENSITIVITY EXPERIMENTS

The Atmospheric Model

The atmospheric model used in this study is the Community Atmospheric Model version 5 (CAM5) (Neale et al. 2010), which is the atmospheric component of the National Center for Atmospheric Research (NCAR) Community Earth System Model (CESM), with a finite-volume dynamical core (Lin 2004) and an integration time step of 1800 s (30 min). In CAM5, the default number of vertical levels is 30, with four additional layers below 700 hPa compared to the CAM4 due to a new parameterization of the boundary layer (Park and Bretherton 2009). The deep convection scheme (Zhang and McFarlane 1995) has been improved, including the effects of deep convection in the momentum equation (Richter and Rasch 2008) and the interaction between convective air with its environment (Neale et al. 2008). The Park and Bretherton (2009) shallow convection scheme (also known as the UWShCu scheme), in which the fundamental frame work and closure assumption are based on Bretherton et al. (2004), is used to parameterize the collective effect of shallow cumulus convection. The mass flux at the cloud base is determined by the ratio between convective inhibition (CIN) and turbulent kinetic energy (TKE) which is provided by the boundary layer scheme (Bretherton and Park 2009) to ensure the interaction between boundary layer and cumulus convection. In the UWShCu scheme, the entrainment and detrainment processes are determined by buoyancy sorting algorithm from Kain and Fritsch (1990). There is no limitation for the cloud height, however, because the deep convection scheme is performed first in CAM5, which provides a tendency to stabilize the atmosphere, the cloud top height simulated by shallow scheme is thus internally limited (Neale et al. 2010). For the microphysics process, a simple assumption is applied in which the cloud condensation amount larger than 1g/kg is removed from the updraft as precipitation and the remainder is detrained in to the environment. As mentioned in Neale et al. (2010), the current scheme also neglects the radiation effect and evaporation of convective precipitation within the convective updraft.

The Sensitivity Experiments

The CAM5 simulations with a prescribed SST of Year 1850 are tested at four different spatial resolutions from 4°, 2°, 1° to 0.5°, corresponding to a grid spacing of about 400, 200, 100 and 50 km, respectively, in the equatorial region. All the simulations are run for 12 years with daily outputs, but only the outputs from the last 10 years are used for analysis.

Besides the changes in the spatial resolution, there are slight differences in some parameters between the sensitivity experiments as these parameters are made functions of spatial resolution in CAM5 to ensure energy balance at the top of atmosphere (Hack et al. 2006; Boyle and Klein 2010). The most significant one is the value of penetrative updraft entrainment efficiency (r_{pen}) in the shallow convection scheme between 2° spatial resolution run and the others which are 5 and 10, respectively. In addition, there is also a small change in the value of minimum relative humidity for low clouds (0.8975 for 1° and 0.8875 for 2°). To ensure consistency, we carried out some sensitivity tests and found that these differences in practice do not alter the results and conclusions of this study (not shown).

Satellite and Field Campaign Datasets

To validate the model's performance in simulating precipitation climatology, we use the satellite derived precipitation data from the Tropical Rainfall Measuring Mission (TRMM) during the period 1998–2007. The original spatial resolution is 0.25° but we made an interpolation to different resolutions to be consistent with the CESM outputs.

For the vertical profiles of convection, some field campaign datasets are also employed for comparison purpose (Table 1). These include the Tropical Ocean Global Atmosphere Coupled Ocean–Atmosphere Response Experiment (TOGA COARE), the South China Sea Monsoon Experiment (SCSMEX), the Kwajelin Experiment (KWAJEX), the Tropical Warm Pool International Cloud Experiment (TWP ICE), and the ARM MJO Investigation Experiment (AMIE). All the 6-hourly data with a vertical grid at 25 hPa intervals from 1000 to 100 hPa are analyzed.

Table 1: List of the field campaign datasets used in this study

Field campaign	Location	Period	References
TOGA COARE	10°S–10oN, 140°E–180	1 Nov 1992–28 Feb 1993	Webster and Lukas (1992)
SCSMEX	Southern China Sea	5 May–20 Jun 1998	Lau et al. (2000)
KWAJEX	7–10°N, 166–169°E	24 Jul–14 Sep 1999	Sobel et al. (2004)
TWP ICE	Darwin Island	17 Jan–12 Feb 2006	Xie et al. (2010)
AMIE (TRMM)	Gan Island, Indian Ocean	13 Nov–13 Dec 2011	Stokes and Schwartz (1994)

METHOD OF ANALYSIS

Apparent Heat Source (Q_1) and Moisture Sink (Q_2)

Apparent heat source (Q_1) and moisture sink (Q_2) were estimated using gridpoint temperature, moisture and wind (Yanai and Johnson 1993), which can be written as:

$$Q_1 \equiv \frac{\partial \bar{s}}{\partial t} + \bar{\mathbf{v}} \cdot \nabla \bar{s} + \bar{\omega}\frac{\partial \bar{s}}{\partial p} = Q_R + L(\bar{c} - \bar{e}) - \nabla \cdot \overline{s'\mathbf{v}'} - \frac{\partial \overline{s'\omega'}}{\partial p} \qquad (1)$$

and

$$Q_2 \equiv -L\left(\frac{\partial \bar{q}}{\partial t} + \bar{\mathbf{v}} \cdot \nabla \bar{q} + \bar{\omega}\frac{\partial \bar{q}}{\partial p}\right) = L(\bar{c} - \bar{e}) + L\nabla \cdot \overline{q'\mathbf{v}'} + L\frac{\partial \overline{q'\omega'}}{\partial p} \qquad (2)$$

where Q_R is the heating rate due to radiation, c is the rate of condensation per unit mass of air, e is the rate of reevaporation of the suspending cloud water and falling rain drop per unit mass, $s = C_pT + gz$ is the dry static energy, T is temperature, ω is pressure velocity, and q is the mixing ratio of water vapor. The horizontal averages are denoted by $\overline{()}$ and deviations from the horizontal averages are denoted by primes.

Apparent heat source (Q_1) is the heating resulted from the diabatic cumulus convection processes (including net condensation and convergence of eddy heat flux) in the atmosphere. Moisture sink (Q_2) is associated with the moisture loss due to net condensation and divergence of eddy moisture transport (Yanai et al. 1973). The advantage of using these two variables is that they provide the relationship and interaction between the cumulus convection and the large-scale environment.

The Dependence of Convective Vertical Structure on Precipitation Rate

Instead of using the direct precipitation outputs from the model convection (precipitation) schemes, we identify the top-heavy and bottom-heavy convection cases based on different precipitation intensity bins and then examine how model resolution change affects the structure of convection. The motivation behind this design was to highlight the relationship between the vertical structure of convection and precipitation intensity in a climatological (or statistical) sense (Bui et al. 2016). Even though there are many studies focusing on the time-mean rainfall patterns [Neelin (2007) and reference therein], investigation on changes of precipitation characteristics is needed, especially when the intensity and frequency of both heavy and light precipitation events could possibly change under global warming (Chen et al. 2016).

In this study, we practically identify the different convection structures based on regime sorting performed in the large-scale conditions and convection properties (Takayabu et al. 2010). The data are sorted and binned according to precipitation intensity (Hagos et al. 2010) with a total of 100 bins (i.e., each bin contains 1% of the total sampled precipitation events), similar to the percentile ranks. In our analysis, a precipitation event is defined as the precipitation rate greater than 0.1 mm day−1 as in previous studies (Sun et al. 2007; Chou et al. 2012). The first few bins represent light precipitation events, which are presumably dominated by shallow bottom-heavy convection; while the last few bins denote heavy precipitation events, in which both top-heavy and bottom-heavy convection often coexist, in agreement with the results from Hagos et al. (2010). Once obtaining the precipitation bins, we then look at the corresponding vertical profiles.

To extract the convection contribution from the radiative cooling and others effects, we subtract the 1st bin from the very light rain and the 91st bin from the very heavy rain with the assumption that radiation cooling profile does not change too much among precipitation bins. In the heavy-rain regime (rainfall events within 91–100th percentile bins), as the cloud tops are located at various levels, strong cancellations between heating and cooling occur within the convective column, resulting in a relatively uniform radiative profile. Thus, subtracting the 91st bin, to a large extent, may remove the effect associated with radiation. In the light-rain regime (rainfall events within 1–10th percentile bins), the radiative profile is in nature uniform, subtracting the 1st bin not only removes the effect associated with radiation, it also excludes the background climatology of light-rain regime dominated by strong subsidence. This allows us to highlight the convection structure changes associated with shallow cumulus. As in Chen et al. (2016), we also normalized the original profile through dividing it by a normalization factor that represents the magnitude of the profile (i.e., the value x in each level p by dividing by $[\sum_{P_S}^{P_T} x(p)^2]^{1/2}$ with P_T and $_{PS}$ being 100 and 1000 hPa, respectively) to focus only on the vertical structure.

IMPACTS OF SPATIAL RESOLUTION ON SIMULATIONS OF TROPICAL CLIMATE

Contrast between Heavy-rain and Light-rain Regimes

To assess how the model spatial resolution affects the vertical structure of convection, the vertical distributions of Q_1 and Q_2 averaged over the first 10 bins between 20∘S and 20∘N were first examined (Fig. 1a, b). There is a maximum of Q_1 at 900 hPa accompanied by a minimum at 800 hPa for all four experiments (Fig. 1a), indicating that light rain events are generally associated with bottom-heavy convection. We note that the above Q_1 maximum should come from the latent heating of net moisture condensation and the Q_1 minimum is a result of the re-evaporation cooling due to detrainment at the top of shallow convection (see the pressure velocity field in Fig. 1c for comparison). The above feature is consistent with the Q2 profile, i.e., a maximum moisture sink at 900 hPa and a moisture source at 800 hPa. The Q_1 and Q_2 profiles also correspond well with ascending motions below 850 hPa, i.e., a strong low-level convergence (upward motion) near the surface and a divergence (downward motion) maximum at about 850 hPa (Fig. 1c, d) except for the coarsest resolution experiment.

For the temperature field, a warming peak is found near the surface and a maximum cooling at 850 hPa (Fig. 1e). The divergence and moisture fields (Fig. 1d, f) are controlled by convection, which generally show opposite phases with the pressure field (Fig. 1c). These results demonstrate typical dynamic and thermodynamic profiles of bottom-heavy convection, and its role in vertically redistributing the moisture and temperature fields below 800 hPa.

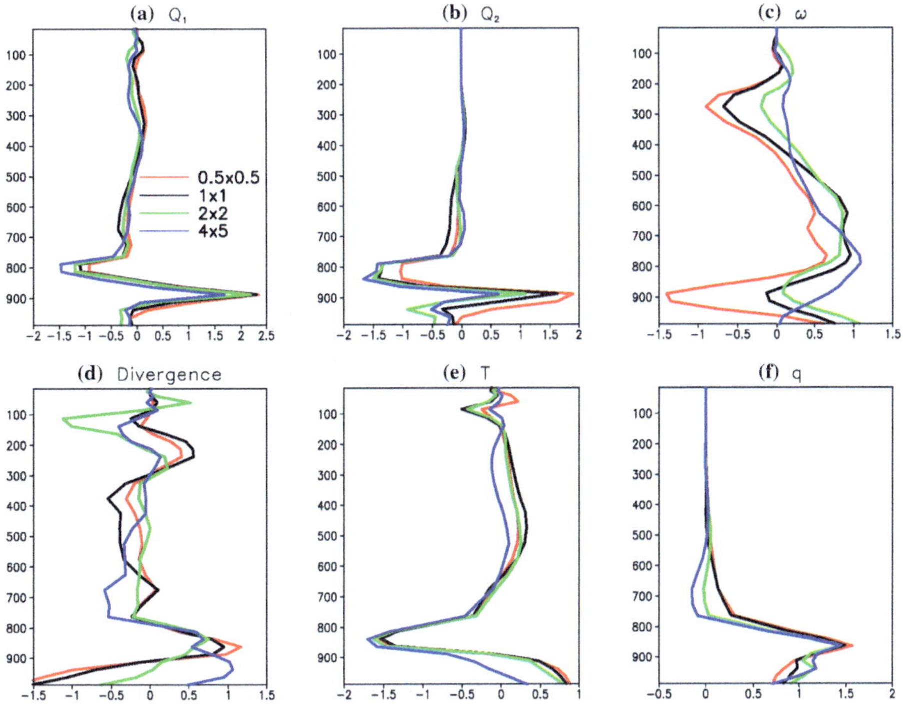

Figure 1: The normalized vertical structures of **a**Q_1, **b**Q_2, **c** ω, **d** divergence, **e** temperature and **f** specific humidity averaged over 1–10th bins.

We next compare results between the experiments with different spatial resolutions. In the vertical structure of kinematic fields, it is clear that the coarsest resolution experiment (blue curves in Fig. 1c, d) behaves differently from the finer resolution ones. It may be due to the fact that parts of shallow convection with smaller scales are not fully represented in the coarsest resolution experiment, so the background subsidence tends to dominate the dynamic fields. In general, the strength of bottom-heavy convection does not have a clear relationship with the spatial resolution without the co-existence of top-heavy convection. We also find that there is a local peak of ascending motion at around 300 hPa for finer resolution runs (see Fig. 1c), even though no corresponding changes of thermodynamic profiles due to the much drier upper-level atmospheric environment for the first 10 precipitation bins. The thermodynamic structures of bottom-heavy convection (e.g., Q_1, Q_2, *T* and *q* profiles) show less dependence on the spatial resolution compared with the kinematic structures (e.g., omega and divergence profiles).

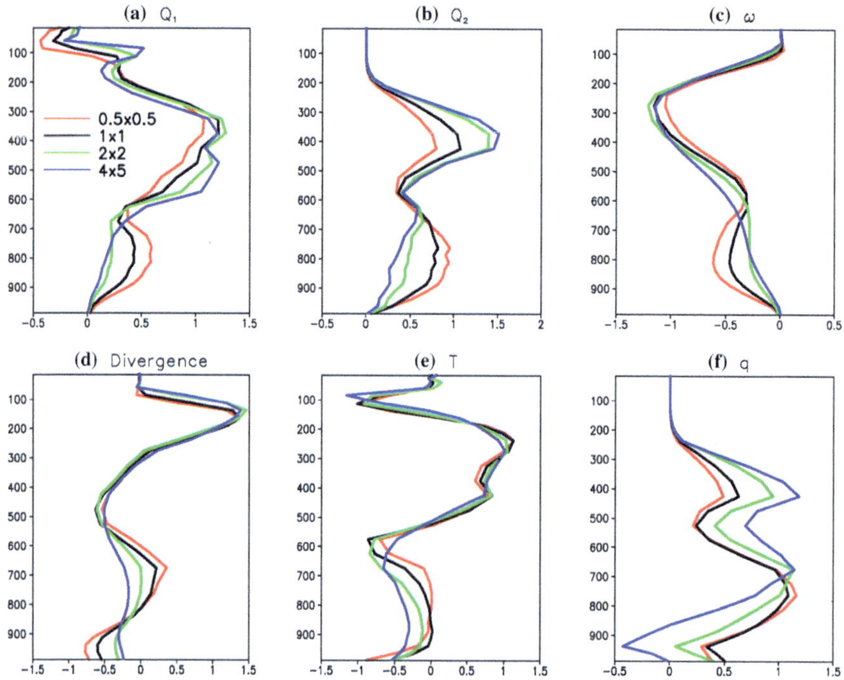

Figure 2: As in Fig. 1 but averaged over 91–100th bins

Likewise, Fig. 2 shows the same dynamic and thermodynamic fields but averaged over the last 10 bins to assess the spatial resolution effect on heavy precipitation events. As shown, all four experiments exhibit 2 peaks, one in the lower troposphere and the other in the upper troposphere, a clear evidence of the coexistence of top-heavy and bottom-heavy convection. Figure 2a shows the structure of Q_1 with double peaks at around 800 and 350 hPa, corresponding to the release of latent heat from top-heavy and bottom-heavy convection, respectively. Q_2 shows a very similar profile to Q_1, with two peaks at around 800 hPa and 400 hPa (Fig. 2b). The ascending motion shows two maxima at around 800 and 300 hPa (Fig. 2c), consistent with the divergence field shown in Fig. 2d. Unlike what happens in light-rain regime (Fig. 1e), the temperature structure in heavy-rain regime shows a weak cooling near the surface and around 600 hPa for all experiments (Fig. 2e), possibly due to re-evaporation cooling of falling raindrops. The significant upper warming above 400 hPa is dominated by the moist adiabatic heating, as discussed in Holloway and Neelin (2007). The moisture field also shows two peaks around 700 and 400 hPa (Fig. 2f).

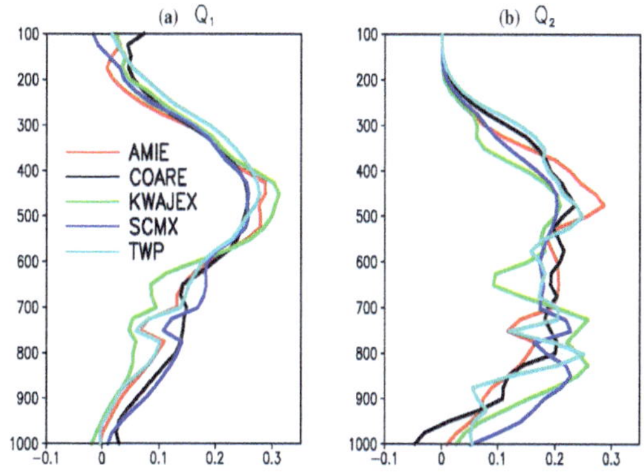

Figure 3: The normalized vertical structures of **a**Q_1, **b**Q_2 from field campaign datasets

As the spatial resolution becomes finer, the lower peak in both dynamics and thermodynamics fields becomes larger, while the corresponding upper peak appears to have a reverse tendency. This implies that a change in the contribution between top-heavy and bottom-heavy convection occurs when the model spatial resolution varies. In short, a greater contribution of bottom-heavy convection and a less contribution of top-heavy convection are found in the finer resolution experiments, which is in agreement with the previous study of Boyle and Klein (2010).

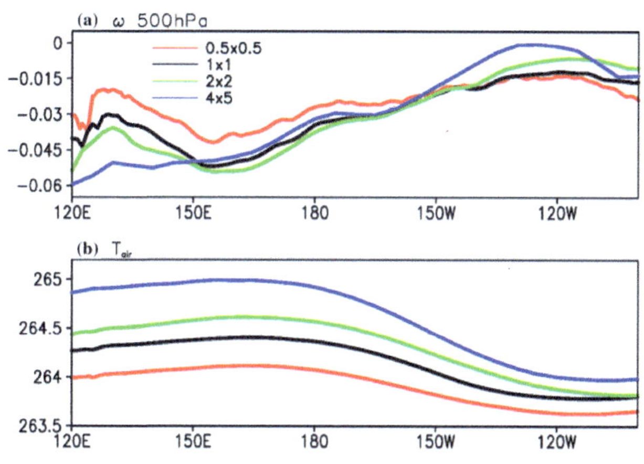

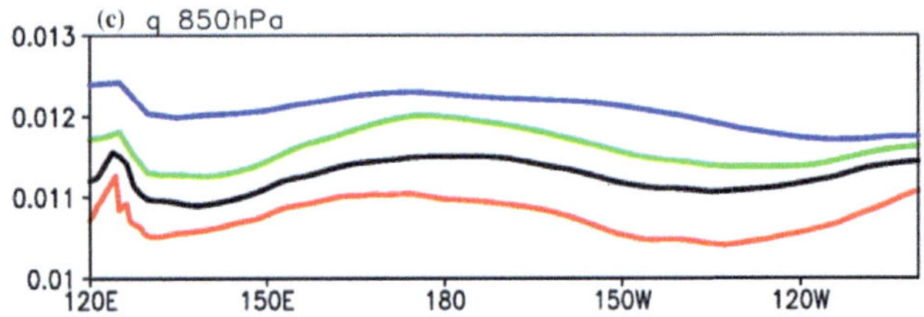

Figure 4: The meridional averages (0–10oN) of **a** omega at 500 hPa (unit is Pa s−1), **b** mean tropospheric temperature from 850–200 hPa (unit is K) and **c** specific humidity at 850 hPa (unit is kg kg−1) for four different spatial resolutions

To see whether the above features exist in observations, we analyze the data from several field observations (Table 1). Note that these data were gathered in a limited area and time period in the Tropics when a particular weather system occurred. Because sampling sizes of these datasets are small, they inclined to exhibit features of heavy precipitation events.

Therefore, we only compare these profiles with Fig. 2. Figure 3 depicts the vertical structures of Q_1 and Q_2 from the field observations. Both Q_1 and Q_2 profiles exhibit a somewhat double peak structure as in the finer resolution experiments of CESM (Fig. 2a, b), except that the maximum levels of heating and moisture sink (at around 450 hPa and 500 hPa, respectively) are slightly lower in observations (Fig. 3a, b).

This is in agreement with the results from Yanai and Johnson (1993), in which the differences in the peaks of Q_1 and Q_2 suggests the presence of eddy vertical transport of moist static energy associated with cumulus convection. These observational results show a clear evidence of the coexistence of top-heavy and bottom-heavy convection in presence of heavy precipitation events, and the finer resolution experiments (1o×1o and 0.5o×0.5o) are more capable than the lower ones (2o×2o and 4o×5o) in reproducing the observed contribution of top-heavy and bottom-heavy convection (see Fig. 2a, b for comparison).

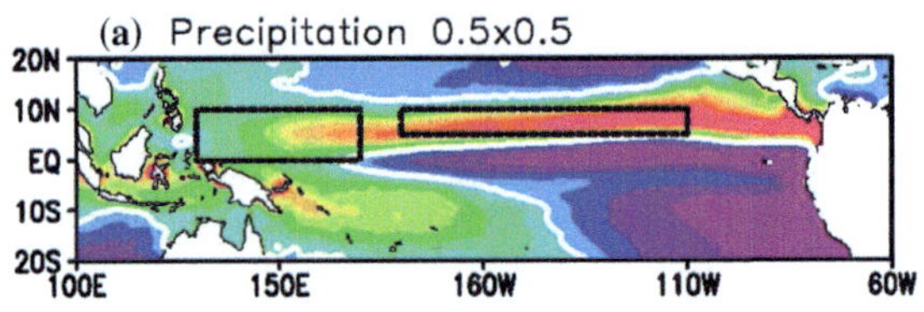

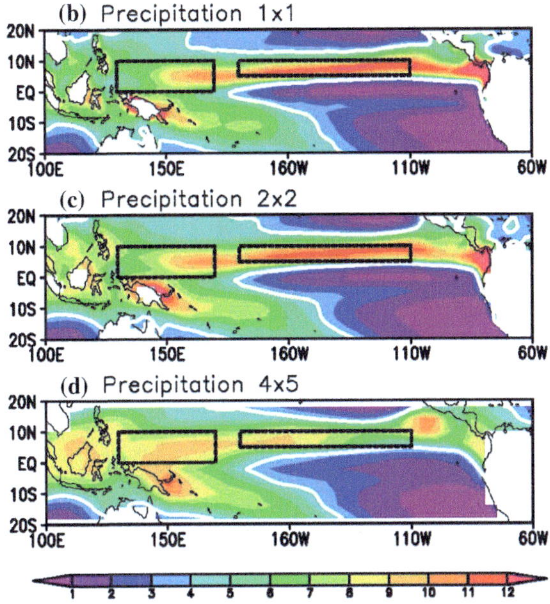

Figure 5: The spatial distribution of precipitation for four different spatial resolutions. The boxes in each panel are western Pacific (0–10oN, 130–170oE) (left) and eastern Pacific (5–10oN, 180–110oW) (right). The unit is mm day−1

Figure 4a shows the zonal distribution of pressure velocity at 500 hPa averaged over 0–10oN for the four experiments. The coarser resolution experiment produces stronger upward motion in the western Pacific, but weaker upward motion in the eastern Pacific ITCZ. For comparison, the climatology of temperature and specific humidity are also shown in Fig. 4b, c, respectively.

In general, a warmer and moister troposphere tends to occur in the coarser resolution experiments. From the entire tropical Pacific point of view, the corresponding model precipitation also shows a consistent result, i.e., a warmer and moister troposphere tends to generate a greater amount of the total tropical precipitation (Fig. 5).

In other words, the thermodynamic effect is a dominant factor in determining the magnitude of tropical precipitation. This is very similar to tropical precipitation changes found in global-warming simulations (e.g., Chou and Neelin 2004; Held and Soden 2006; Chou et al. 2009).

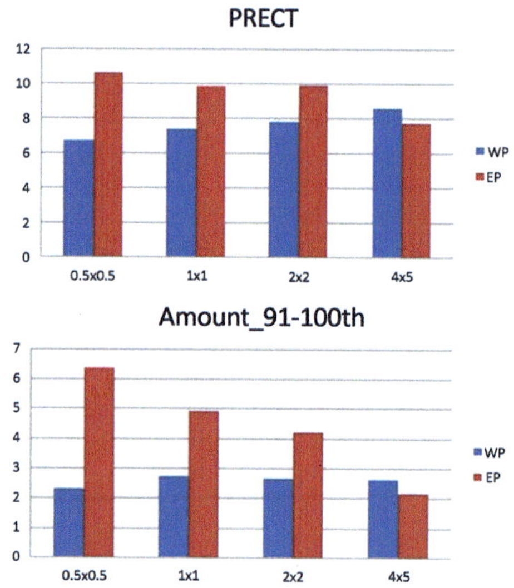

Figure 6: Precipitation averaged over the boxes in Fig. 5 for western Pacific (blue) and eastern Pacific (red) for **a** the total amount and **b** the amount at the last 10 percentiles bins. The unit is mm day−1

From the regional point of view, the dependence of tropical precipitation on model resolution, however, differs significantly between the eastern and western Pacific ITCZs, as shown in Fig. 5. In the western Pacific warm pool, the coarser resolution experiment exhibits a stronger magnitude of precipitation; while in the eastern Pacific ITCZ, the coarser resolution experiment generates a weaker magnitude of precipitation. In the following section, we will provide a detailed analysis to explain why the magnitude of precipitation in response to model resolution changes is so different between the eastern and western Pacific ITCZs.

Contrast between Western and Eastern Pacific ITCZs

The dependence of regional precipitation on model spatial resolution provides an opportunity to identify changes in convective structure and precipitation between the western and eastern Pacific ITCZs. Following Bui et al. (2016), two target domains over the western and the eastern Pacific ITCZs are selected as shown by the rectangular boxes in Fig. 5. These domains are located within heavy convective regions, with precipitation rate all greater than 4 mm day−1. Figure 6a shows the area means of precipitation over these two domains. As the resolution decreases, the precipitation intensity increases over the western Pacific ITCZ but decreases in the eastern Pacific ITCZs, consistent with the results shown in Fig. 5. If we consider only the heavy-rain regime (91–100th bins), the dependence of precipitation on the model spatial resolution exhibits a similar trend (Fig. 6b) except that the increase of precipitation with the decrease (increase) of the spatial resolution in the western (eastern) Pacific ITCZ is slower (faster) than the total precipitation shown in Fig. 6a. The greater sensitivity of precipitation intensity in response to model resolution change in the eastern Pacific ITCZ is an interesting phenomenon that needs to be clarified.

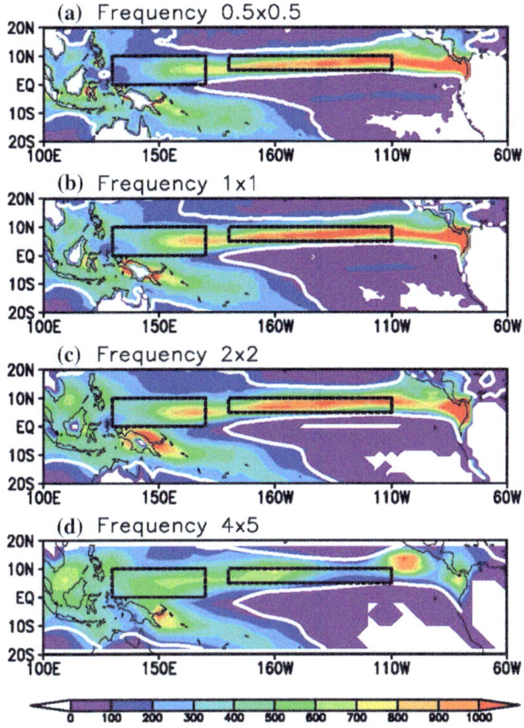

Figure 7: The spatial distribution of the frequency of precipitation over the last 10 percentiles bins. The unit is number of event

In order to understand why there is a difference between these two domains, we further analyzed the characteristics of tropical precipitation, focusing on the intensity and frequency patterns of the heavy-rain regime (91–100th bins). For the occurrence frequency patterns (Fig. 7), the finer resolution experiment tends to have fewer heavy precipitation events than the coarser one over most of the climatologically ascending regions except over the eastern Pacific where more heavy rain events are observed in the finer resolution runs. For the intensity patterns, the finer resolution experiment, however, tends to generate a stronger precipitation intensity almost everywhere in the tropics (Fig. 8), suggesting that finer resolution experiments are more capable of producing small-scale intense cumulus convection. Overall, the change of total precipitation amount associated with the model spatial resolution seems to be dominated by the precipitation frequency. The precipitation intensity only matters in the eastern Pacific ITCZ for higher resolution runs (1° and 0.5°) (see Figs. 5, 7, 8 for comparison).

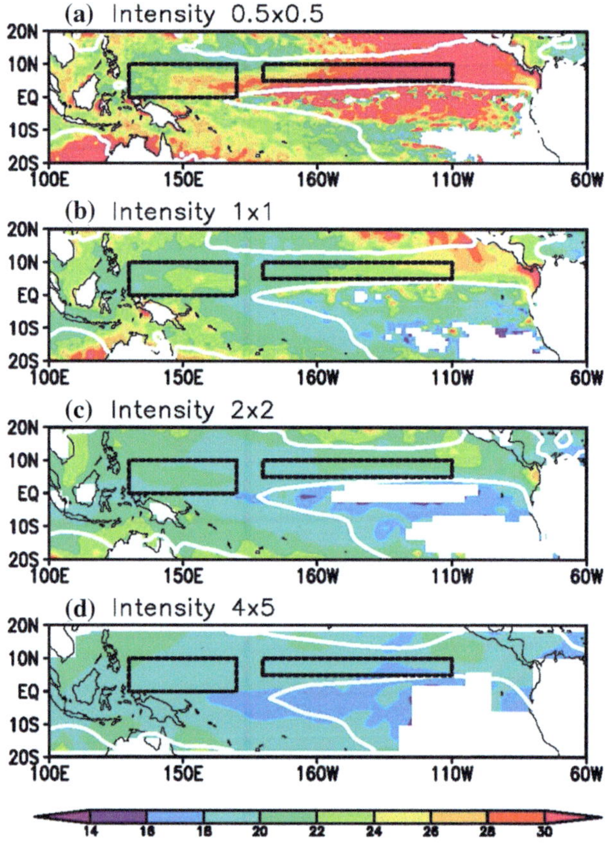

Figure 8: As in Fig. 7, but for precipitation intensity. The unit is mm day−1

Figure 9 shows the domain averages of the precipitation frequency and intensity patterns of the heavy-rain regime (91–100th bins) for different spatial resolution experiments. Over the western Pacific ITCZ, the frequency decreases significantly with the increase of the spatial resolution (Fig. 9a) while the corresponding intensity only increases slightly with the increase of spatial resolution (Fig. 9b), suggesting the dominance of precipitation frequency in determining the precipitation amount (i.e., a weaker magnitude of precipitation over the western Pacific in higher resolution runs shown in Fig. 5). Over the eastern Pacific ITCZ, the frequency of heavy precipitation events in general increases with the increase of spatial model resolution, though the change is marked from 4° to 2° but relatively modest from 2° to 0.5° (Fig. 9a). The precipitation intensity also increases with the increase of model resolution, especially from 1° to 0.5° in determining the precipitation amount. Thus, the higher precipitation amounts in the eastern Pacific in finer resolution experiments appear to come from more frequent occurrences of stronger precipitation events.

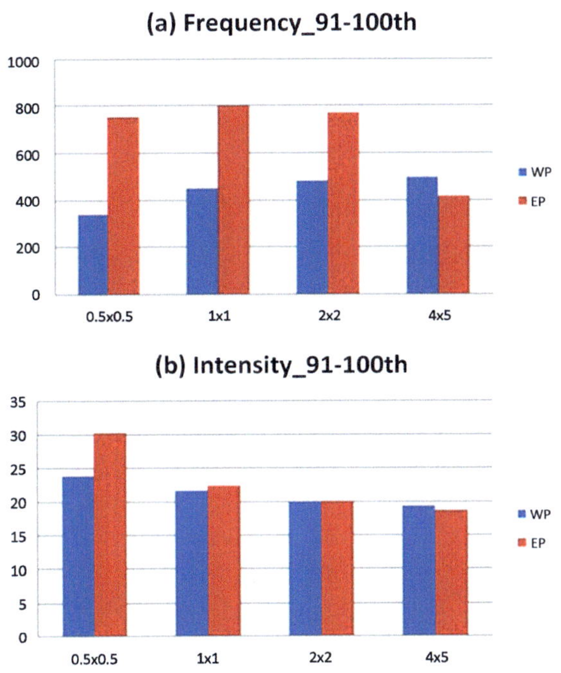

Figure 9: The **a** frequency and **b** intensity of precipitation over the last 10 percentiles bins for western and eastern Pacific

To understand what cause more frequent occurrences of stronger precipitation events over the eastern Pacific ITCZ in finer spatial resolution experiments, we again examine the vertical structure of key parameters, as discussed in Sect. 4.1. Similar to Fig. 2, we compare the Q_1, Q_2 profiles for the last 10 bins between the western and eastern Pacific ITCZs (Fig. 10). For the Q_1 profile, the top-heavy convection structure shows relatively little dependence on model resolution over the western Pacific ITCZ, except for a slightly greater contribution of bottom-heavy convection in the two finest resolution experiments (Fig. 10a). Over the eastern Pacific ITCZ, the contribution of bottom-heavy convection becomes much greater, while the contribution of top-heavy convection becomes much smaller in the two finest spatial resolution experiments (Fig. 10b). Similar patterns are also observed for Q_2, although the contrast between the western and the eastern Pacific ITCZ is not as clear as in Q_1 (Fig. 10c, d). Note that similar changes are also found for the pressure velocity and divergence fields (not shown). Based on the above findings, we may conclude that a greater contribution of bottom-heavy convection is responsible for more frequent occurrences of heavy precipitation events in the eastern Pacific ITCZ in the finer resolution runs.

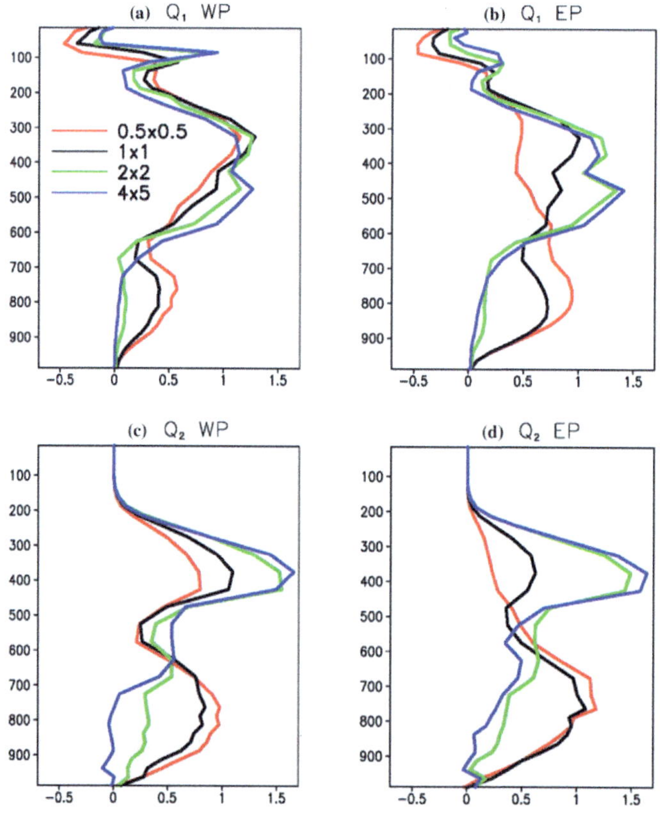

Figure 10: The vertical structure of **a**, **b**Q_1 and **c**, **d**Q_2 for the last 10 percentiles bins of four different spatial resolutions. **a** and **c** are averaged over western Pacific; **b** and **d** are averaged over eastern Pacific

THE MOISTURE BUDGET ANALYSIS

In the previous section, we reported the changes in contribution between top-heavy and bottom-heavy convection and the differences in precipitation frequency and intensity tendencies between the western and eastern Pacific ITCZs when the model spatial resolution varies. Here, we employed the moisture budget analysis to explore the possible causes behind such changes. Following Chou and Neelin (2004) and Chou et al. (2009), the vertically integrated moisture equation can been written as

$$P = -\langle \omega \partial_p q \rangle - \langle \mathbf{v} \cdot \nabla q \rangle + E \quad (3)$$

where P is precipitation, E is evaporation (both in W m−2), **v** is horizontal wind vector, q is the specific humidity (in energy unit, J kg−1, by absorbing the latent heat per unit mass L). The brackets $\langle\rangle$ denotes a mass integration through the troposphere, typically from 1000 hPa to the tropopause. Here, the tropopause level is fixed at 100 hPa as in previous studies (Chen et al. 2016; Bui et al. 2016). In (3), precipitation P is balanced by the vertical moisture advection, $-\langle \omega \partial_p q \rangle$, the horizontal moisture advection, $-\langle \mathbf{v} \cdot \nabla q \rangle$, and the surface evaporation, E. Figure 11 shows the magnitude of each term in (3) averaged over the two target domains for four different resolutions. It is clear that the vertical moisture advection determines the precipitation amount (comparing the first four groups of column), implying a major balance between drying of precipitation and moistening of vertical moisture transport in convective regions. The contributions from surface evaporation and horizontal moisture advection are small (except in the finest resolution experiment), and the negative value of $-\langle \mathbf{v} \cdot \nabla q \rangle$ implies a horizontal advection of dry air into these regions. Of particular interest shown in Fig. 11 is the reverse tendencies in precipitation and vertical moisture transport terms between the eastern and western Pacific ITCZs. Over the western Pacific ITCZ, both the precipitation and vertical moisture transport terms increase as the model resolution becomes coarser. The situation reverses over the eastern Pacific ITCZ where the precipitation and vertical moisture transport terms increase as the model resolution becomes finer. As discussed in previous section, there is a marked difference in vertical structure of convection between the western and eastern Pacific ITCZs that could attribute to the above reverse tendencies. To provide a quantitative evaluation of how the vertical structure of convection affects the vertical moisture transport, we decompose vertical moisture transport anomaly into anomaly due to moisture structure change (thermodynamic effect), anomaly due to convective structure and intensity change (dynamic effect of anomalous circulation), and anomaly due to eddy transport of moisture. The prime term is the departure from the corresponding temporal averaged over the tropical mean:

$$-\langle \omega \partial_p q \rangle' = -\langle \overline{\omega} \partial_p q' \rangle - \langle \omega' \partial_p \overline{q} \rangle - \langle \omega' \partial_p q' \rangle \tag{4}$$

As shown, contributions from eddy and thermodynamic effects are relatively small (the sixth and seventh columns in Fig. 11). In contrast, the term due to convection structure and intensity change (dynamic effect) appears to dictate the vertical moisture advection (see the fifth columns in Fig. 11). We also note that the magnitude of the dynamic effect, $\langle \omega' \partial_p \overline{q} \rangle$, decreases as the spatial resolution becomes coarser over the eastern Pacific ITCZ, consistent with the decrease of total precipitation as shown in Fig. 6. In summary, it is the change in convection structure and intensity (i.e., the $\langle \omega' \partial_p \overline{q} \rangle$ term) that determines change of precipitation among different resolutions.

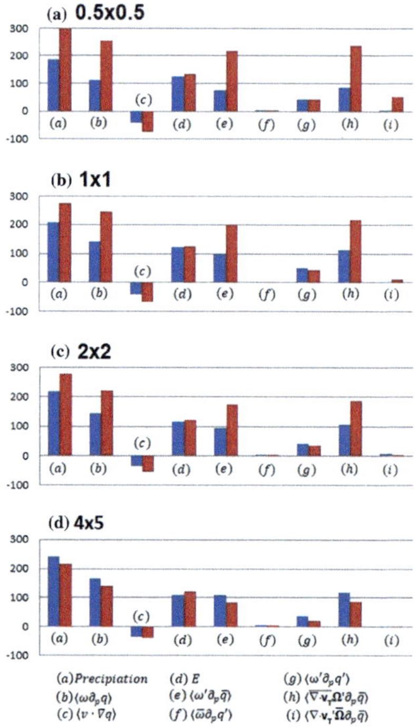

Figure 11: Magnitude of each moisture budget terms and its decomposition: the first four, **a–d**, are from Eq. 3, the next three (**e, f, g**) are from Eq. 4, and the last two (**h, i**) are from Eq. 6, averaged over the western (blue) and eastern (red) Pacific ITCZs for different spatial resolutions. The unit is W m−2

To understand how the convection change affects the vertical moisture transport, we further decompose the pressure velocity ω into a product of its vertical profile and its strength (Yu and Neelin 1997; Chou and Neelin 2004; Chou et al. 2013; Chen et al. 2016):

$$\omega = -\Omega(p)\nabla \cdot \mathbf{v}_T(x, y) \qquad (5)$$

where $\Omega(p)$ describes the vertical structure of vertical velocity (convection), and $\nabla\cdot\mathrm{v}_T$ is the divergence induced by the first baroclinic winds, representing the intensity of vertical velocity (convection) and a positive $\nabla\cdot\mathrm{v}_T$ denotes upward motion. In this study, Ω is estimated from ω by a normalization using the vertical integral of its corresponding intensity at each level, i.e., $\Omega(p) = \frac{\omega}{\sqrt{\sum_{P_S}^{P_T} \omega^2}}$, similar to Chen et al. (2016). Therefore, the second term on the right hand side of (4) can be written as

$$-\langle \omega' \partial_p \bar{q} \rangle = \langle \nabla \cdot \mathbf{v}'_T \bar{\Omega} \partial_p \bar{q} \rangle + \langle \nabla \cdot \bar{\mathbf{v}}_\mathbf{T} \Omega' \partial_p \bar{q} \rangle \qquad (6)$$

Comparing the last two columns in Fig. 11, the contribution of $\langle \nabla \cdot \mathbf{v}'_T \bar{\Omega} \partial_p \bar{q} \rangle$ is relatively small in both regions. It is the change in vertical velocity profile, i.e., $\langle \nabla \cdot \bar{\mathbf{v}}_\mathbf{T} \Omega' \partial_p \bar{q} \rangle$, that accounts for $-\langle \omega' \partial_p \bar{q} \rangle$. One should note that this vertical structure of convection is very sensitive to the spatial resolution used in climate model (as shown in Sect. 4.1). The above results confirm that the differences in precipitation tendencies between the western and eastern Pacific ITCZs among various spatial resolution experiments shown in Sect. 4.2 are determined mainly by changes in total precipitation contribution between the top-heavy and bottom-heavy convection (i.e., changes in the vertical structure of convection).

SUMMARY AND DISCUSSION

This paper examines how the model spatial resolution impacts simulations of tropical climate, with a focus on the vertical structure of convection. We analyze the vertical profiles of apparent heat source (Q_1) and moisture sink (Q_2) at the first and last 10 percentiles of precipitation events, representing respectively the light- and heavy-rain regimes. We also compare the differences in precipitation behaviors between the western and eastern Pacific ITCZs dominated climatologically by top-heavy and bottom-heavy convection, respectively. Some notable findings are summarized as follow:

1. (Contributions to the total precipitation from top-heavy and bottom-heavy convection are sensitive to model's spatial

resolution, especially in the heavy-rain regime (91–100th percentiles). A greater contribution from bottom-heavy convection and less contribution from top-heavy convection are found in the finer resolution experiments (Fig. 2).

2. From the regional point of view, coarser resolution experiments exhibit stronger precipitation in the western Pacific warm pool, but weaker precipitation in the eastern Pacific ITCZ (see Figs. 5, 6). In the former region where the convective structure is dominated by top-heavy convection, precipitation increases in coarser resolution experiments are due to higher occurrence frequencies of heavy rainfall (last 10 bins) events associated with a warmer environment. In the latter region, the heavy precipitation events are dominated by bottom-heavy structure of convection in the higher resolution simulation, and this leads to more frequent occurrence of heavy rainfall events and therefore more precipitation (see Figs. 9, 10).
3. The moisture budget analysis (see Fig. 11) further suggests that, in the heavy-rain regime, the differences in precipitation tendencies between the western and eastern Pacific ITCZs among various spatial resolution experiments (Sect. 4.2) are determined mainly by changes in convective structure (i.e., the shape of convection) rather than changes in convective strength. More contribution from bottom-heavy structure of convection occurs in the finer resolution; while more contribution from top-heavy structure of convection tends to occur in the coarser resolution experiments, which is particularly evident in the eastern Pacific ITCZ (Sect. 4.2).

This study for the first time shows the sensitivity of model's spatial resolution in determining the contribution between top-heavy convection and bottom-heavy convection in tropical climate. A recent study by Bui et al. (2016) indicated that the very different results of the vertical transport of moist static energy (MSE) between the western and eastern Pacific ITCZs (e.g., positive column-integrated vertical MSE transport in the western Pacific and negative column-integrated vertical MSE transport in the eastern Pacific) should come from different shapes of vertical motion in these two regions. A top-heavy (bottom-heavy) structure of vertical motion, such as the case happening in the western (eastern) Pacific ITCZ, favors a positive (negative) value of MSE transport.

Although the stratiform precipitation may have a significant contribution to the total precipitation (Xu 2009; Johnson et al. 2016), however, it seems impossible to separate the convective and stratiform preciptiation in this study as a clear quantitative separation requires high-resolution observations (e.g., meteorological radar) or non-hydrostatic mesoscale numerical simulations [e.g., WRF or CRM (cloud resolving model)]. This is beyond the scope of this study and we prefer to put it in our future work.

Finally, one major concern arising from this study broaches the question of whether there is a minimum spatial resolution for climate models in order to reproduce a reasonable climatology of the structure of tropical convection (or large-scale vertical motions from the sense of model simulations). The results presented here suggest that a spatial resolution finer than 2°×2° is required for climate models to obtain a reasonable precipitation contribution between top-heavy and bottom-heavy convection, especially in the eastern Pacific, along with a fair convective adjustment process associated with the vertical MSE transport in the tropical ITCZs.

SUPPLEMENTARY MATERIAL

Contribution of Precipitation from Various Convection Schemes in CAM5

In this supplementary section, we further examine the contribution of precipitation from various precipitation schemes in CAM5. These include the deep convection scheme (Zhang and McFarlane scheme), the shallow convection scheme, and the microphysics scheme (representing the large-scale condensation processes).

Figure S1 shows the spatial distribution of convective precipitation based on direct outputs from convective schemes (including deep and shallow schemes). The results are similar to the total precipitation shown in Fig. 5 of the paper manuscript, which indicates that convective (deep and shallow) precipitation accounts for most (>90%) of the total precipitation. While large-scale (microphysics) precipitation (Figure S2) seems to contribute only a small fraction (<10%) of the total precipitation, except over the eastern Pacific and SPCZ where the finest resolution experiment (0.5x0.5) produces a comparable amount of precipitation against the convective schemes. We note that the less contribution from large-scale precipitation in the coarser resolution runs is reasonable as the atmospheric environment at a larger grid-size would be more difficult to get saturated compare to that at a smaller grid-size. Unlike the convective and large-scale (microphysics) precipitation, shallow convective precipitation pattern is quite scatter and doesn't match the other two. Most shallow precipitation signals are located in the eastern Pacific ITCZ with higher intensity as the resolution becomes finer (Figure S3).

In summary, the convective (deep and shallow) schemes produce most of the total precipitation, contributing more than 90% of the total precipitation (cf. Figs. S1-3). In this study, we focus on the change of convective structure over the light and heavy precipitation regimes, rather than the partitioning of precipitation between deep convection, shallow convection and microphysics (large-scale condensation) schemes. We argue that examining convective structure changes in terms of various precipitation bins instead of using direct precipitation outputs from various precipitation schemes is a better and more physical approach that can be compared directly with observations because precipitation events in nature exhibit continuous spectra in space and time.

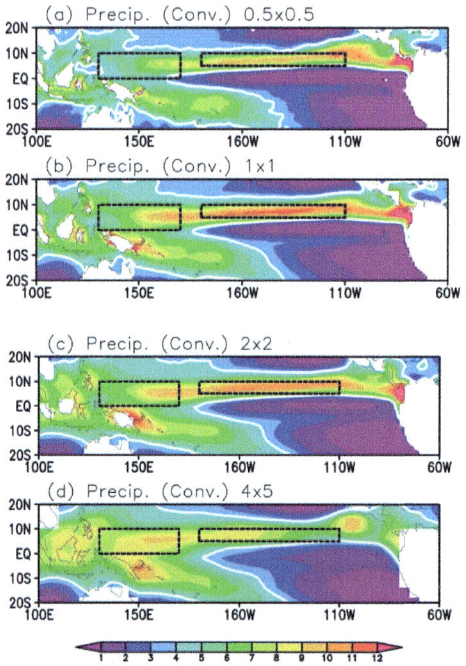

Figure S1: Similar to Fig. 5 in the paper manuscript, but for the precipitation amount from deep and shallow convection schemes. The unit is mm day^{-1} .

Besides, comparing the precipitation tendency among various precipitation schemes on model resolution shows a clear difference between the western and eastern Pacific ITCZ (see Fig. S4). In the western Pacific, precipitation tends to increase as the resolution becomes coarser; while in the eastern Pacific precipitation tends to increase as the resolution becomes finer (Fig. S4d). Most of the total precipitation comes from the deep convective precipitation, while large-scale (microphysics) precipitation and shallow convective precipitation only contribute a small amount. We also note that the resolution dependence appears in all precipitation schemes, in particular, the precipitation from microphysics (large-scale condensation) scheme becomes sizable over the eastern Pacific in the finest resolution experiment (cf. Figs. S4a-d).

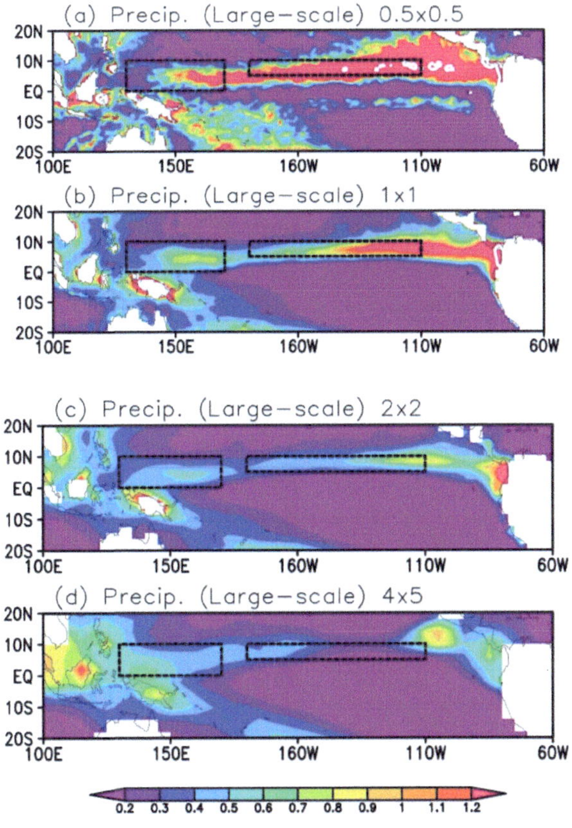

Figure S2: Similar to Fig. 5 in the paper manuscript, but for the precipitation amount from microphysics (large-scale condensation) scheme. The unit is mm day^{-1}.

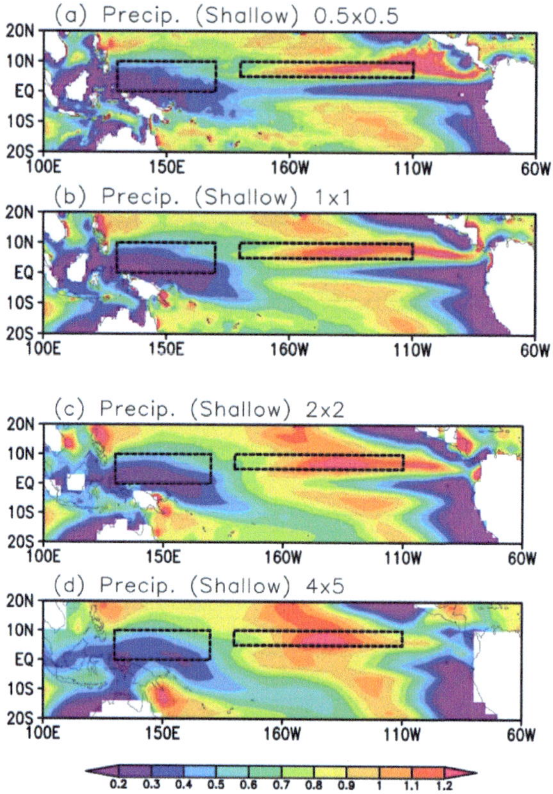

Figure S3: Similar to Fig. 5 in the paper manuscript, but for the precipitation amount from shallow convection scheme alone. The unit is mm day^{-1} .

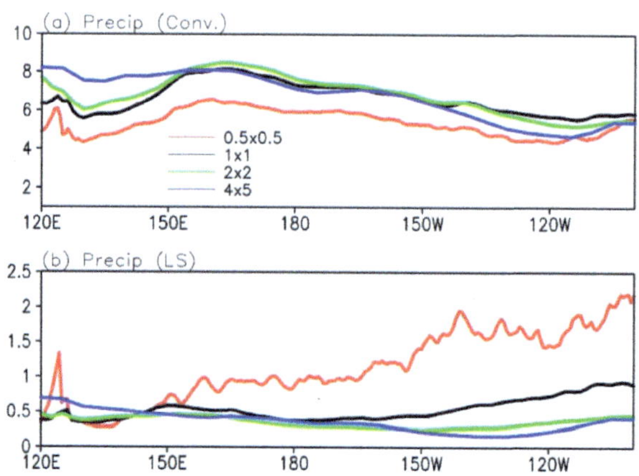

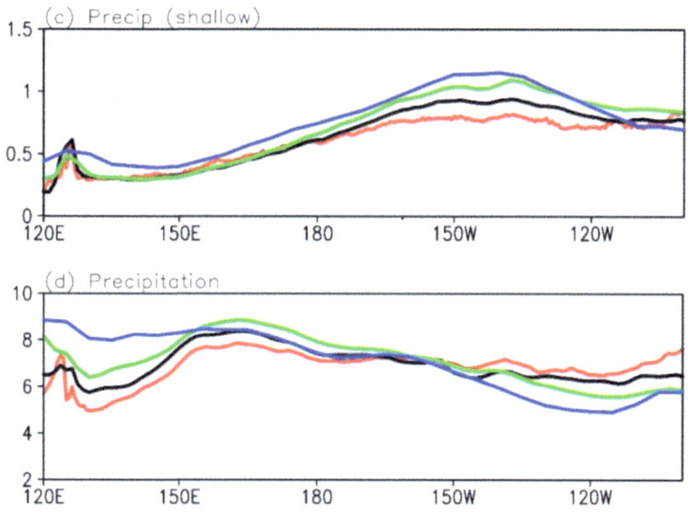

Figure S4: Similar to Fig. 4 in the paper manuscript but for (a) the precipitation amount from deep and shallow convection schemes, (b) the precipitation amount from microphysics (large-scale condensation) precipitation, (c) the precipitation amount from shallow convection scheme and (d) the total precipitation. The unit is mm day^{-1} .

ACKNOWLEDGEMENTS

This research is supported by the Ministry of Science and Technology (MOST) in Taiwan under Grants NSC102-2111-M-001-002-MY3, MOST104-2111-M-008-026-MY2, and MOST105-2111-M-008-025-MY3. The first author (HX Bui) thanks Dr. Wei-Liang Lee for useful discussions. The authors also acknowledge the two anonymous reviewers for their appreciation to our work and many constructive comments to improve the quality of this manuscript.

REFERENCES

1. Back LE, Bretherton CS (2006) Geographic variability in the export of moist static energy and vertical motion profiles in the tropical Pacific. Geophy Res Lett 33(17)
2. Boyle J, Klein SA (2010) Impact of horizontal resolution on climate model forecasts of tropical precipitation and diabatic heating for the TWP-ICE period. J Geophys Res 115(D23)
3. Bretherton CS, McCaa JR, Grenier H (2004) A new parameterization for shallow cumulus convection and its application to marine subtropical cloud-topped boundary layers. Part I: Description and 1D results. Mon Weather Rev 132(4):864–882
4. Bretherton CS, Park S (2009) A new moist turbulence parameterization in the community atmosphere model. J Clim 22(12):3422–3448
5. Bui HX, Yu J-Y, Chou C (2016) Impacts of vertical structure of large-scale vertical motion in tropical climate: Moist static energy framework. J Atmos Sci 73(11):4427–4437
6. Chan SC, Nigam S (2009) Residual diagnosis of diabatic heating from ERA-40 and NCEP reanalyses: intercomparisons with TRMM. J Clim 22(2):414–428
7. Chen C-A, Yu J-Y, Chou C (2016) Impacts of vertical structure of convection in global warming: the role of shallow convection. J Clim 29(12):4665–4684
8. Chou C, Chen C-A, Tan P-H, Chen KT (2012) Mechanisms for global warming impacts on precipitation frequency and intensity. J Clim 25(9):3291–3306
9. Chou C, Neelin JD (2004) Mechanisms of global warming impacts on regional tropical precipitation*. J Clim 17(13):2688–2701
10. Chou C, Neelin JD, Chen C-A, Tu J-Y (2009) Evaluating the "rich-get-richer" mechanism in tropical precipitation change under global warming. J Clim 22(8):1982–2005
11. Chou C, Wu T-C, Tan P-H (2013) Changes in gross moist stability in the tropics under global warming. Clim Dyn 41(9–10):2481–2496
12. Deser C, Capotondi A, Saravanan R, Phillips AS (2006) Tropical Pacific and Atlantic climate variability in CCSM3. J Clim 19(11):2451–2481
13. Gregory D, Kershaw R, Inness PM (1997) Parametrization of momentum transport by convection. II: Tests in single-column and

general circulation models. Q J R Meteorol Soc 123(541):1153–1183
14. Hack JJ, Caron JM, Danabasoglu G, Oleson KW, Bitz C, Truesdale JE (2006) CCSM-CAM3 climate simulation sensitivity to changes in horizontal resolution. J Clim 19(11):2267–2289
15. Hagos S et al (2010) Estimates of tropical diabatic heating profiles: commonalities and uncertainties. J Clim 23(3):542–558
16. Held IM, Soden BJ (2006) Robust responses of the hydrological cycle to global warming. J Clim 19(21):5686–5699
17. Holloway CE, Neelin JD (2007) The convective cold top and quasi-equilibrium*. J Atmos Sci 64(5):1467–1487
18. Houze RA (1997) Stratiform precipitation in regions of convection: a meteorological paradox? Bull Am Meteorol Soc 78:21792196
19. Johnson RH, Rickenbach TM, Rutledge SA, Ciesielski PE, Schubert WH (1999) Trimodal characteristics of tropical convection. J Clim 12(8):2397–2418
20. Johnson RH, Ciesielski PE, Rickenbach TM (2016) A further look at Q1 and Q2 from TOGA COARE. Meteorol Monogr 56:1.1–1.12
21. Kain JS, Fritsch JM (1990) A one-dimensional entraining/detraining plume model and its application in convective parameterization. J Atmos Sci 47(23):2784–2802
22. Lau KM et al (2000) A report of the field operations and early results of the south china sea monsoon experiment (SCSMEX). Bull Am Meteorol Soc 81(6):1261–1270
23. Lin S-J (2004) A vertically lagrangian finite-volume dynamical core for global models. Mon Weather Rev 132(10):2293–2307
24. Ling J, Zhang C (2013) Diabatic heating profiles in recent global reanalyses. J Clim 26(10):3307–3325
25. Neale RB, Richter JH, Jochum M (2008) The impact of convection on ENSO: from a delayed oscillator to a series of events. J Clim 21(22):5904–5924
26. Neale RB et al (2010) Description of the NCAR community atmosphere model (CAM 5.0). NCAR Tech. Note NCAR/TN-486+ STR
27. Neelin JD (2007) Moist dynamics of tropical convection zones in monsoons, teleconnections and global warming, Chap. 10, 400. Princeton University Press

28. Nigam S, Chung C, DeWeaver E (2000) ENSO diabatic heating in ECMWF and NCEP-NCAR reanalyses, and NCAR CCM3 simulation. J Clim 13(17):3152–3171
29. Park S, Bretherton CS (2009) The university of Washington shallow convection and moist turbulence schemes and their impact on climate simulations with the community atmosphere model. J Clim 22(12):3449–3469
30. Richter JH, Rasch PJ (2008) Effects of convective momentum transport on the atmospheric circulation in the community atmosphere model, version 3. J Clim 21(7):1487–1499
31. Schumacher C, Zhang MH, Ciesielski PE (2007) Heating structures of the TRMM field campaigns. J Atmos Sci 64(7):2593–2610
32. Sobel AH, Yuter SE, Bretherton CS, Kiladis GN (2004) Large-scale meteorology and deep convection during TRMM KWAJEX*. Mon Weather Rev 132(2):422–444
33. Stephens GL et al (2010) Dreary state of precipitation in global models. J Geophys Res 115(D24)
34. Stokes GM, Schwartz SE (1994) The atmospheric radiation measurement(ARM) program: programmatic background and design of the cloud and radiation test bed. Bull Am Meteor Soc 75(7):1201–1221
35. Sun Y, Solomon S, Dai A, Portmann RW (2007) How often will it rain? J Clim 20(19):4801–4818
36. Takayabu YN, Shige S, Tao W-K, Hirota N (2010) Shallow and deep latent heating modes over tropical oceans observed with TRMM PR spectral latent heating data. J Clim 23(8):2030–2046
37. Tao W-K et al (2006) Retrieval of latent heating from TRMM measurements. Bull Am Meteorol Soc 87(11):1555–1572
38. Webster PJ, Lukas R (1992) TOGA COARE: the coupled ocean-atmosphere response experiment. Bull Am Meteorol Soc 73(9):1377–1416
39. Xie S, Hume T, Jakob C, Klein SA, McCoy RB, Zhang M (2010) Observed large-scale structures and diabatic heating and drying profiles during TWP-ICE. J Clim 23(1):57–79
40. Xu K (2009) Partitioning mass, heat, and moisture budgets of explicitly simulated cumulus ensembles into convective and stratiform components. J Atmos Sci 52:551573

41. Yanai M, Esbensen S, Chu J-H (1973) Determination of bulk properties of tropical cloud clusters from large-scale heat and moisture budgets. J Atmos Sci 30(4):611–627
42. Yanai M, Johnson R (1993) Impacts of cumulus convection on thermodynamic fields. 39–62
43. Yu J-Y, Neelin JD (1997) Analytic approximations for moist convectively adjusted regions. J Atmos Sci 54(8):1054–1063
44. Zhang G, McFarlane NA (1995) Sensitivity of climate simulations to the parameterization of cumulus convection in the Canadian climate centre general circulation model. Atmo Ocean 33(3):407–446

CHAPTER 12

Winter Thermodynamic Vertical Structure in the Arctic Atmosphere Linked to Large-Scale Circulation

Tiina Nygård[1], Michael Tjernström[2], and Tuomas Naakka[1]

[1]Finnish Meteorological Institute, Helsinki, Finland

[2]Department of Meteorology, Bolin Centre for Climate Research, Stockholm University, Stockholm, Sweden

ABSTRACT

Thermodynamic profiles are affected by both the large-scale dynamics and the local processes, such as radiation, cloud formation and turbulence. Based on ERA5 reanalysis, radiosoundings and cloud cover observations from winters 2009–2018, this study demonstrates manifold impacts of large-scale circulation on temperature and specific humidity profiles in the circumpolar Arctic north of 65° N. Characteristic wintertime circulation types are allocated

Citation: Nygård, T., Tjernström, M., and Naakka, T.: "Winter thermodynamic vertical structure in the Arctic atmosphere linked to large-scale circulation", Weather Clim. Dynam., 2, 1263–1282, https://doi.org/10.5194/wcd-2-1263-2021, 2021.

using self-organizing maps (SOMs). The study shows that influence of different large-scale flows must be viewed as a progressing set of processes: (1) horizontal advection of heat and moisture, driven by circulation, lead to so-called first-order effects on thermodynamic profiles and turbulent surface fluxes, and (2) the advection is followed by transformation of the air through various physical processes, causing second-order effects. An example of second-order effects is the associated cloud formation, which shifts the strongest radiative cooling from the surface to the cloud top. The temperature and specific humidity profiles are most sensitive to large-scale circulation over the Eurasian land west of 90° E and the Arctic Ocean sea ice, whereas impacts over North America and Greenland are more ambiguous. Eurasian land, between 90 and 140° E, occasionally receives warm and moist air from the northern North Atlantic, which, with the support of radiative impacts of clouds, weakens the otherwise strong temperature and specific humidity inversions. Altitudes of maximum temperature and specific humidity in a profile and their variability between the circulation types are good indicators of the depth of the layer impacted by surface–atmosphere processes interacting with the large-scale circulation. Different circulation types typically cause variations of a few hundred metres to this altitude, and the layer impacted is deepest over north-eastern Eurasia and North America.

INTRODUCTION

The spatial distributions of temperature and humidity in the wintertime Arctic atmosphere are to a large extent controlled by the predominantly negative surface net radiation budget and poleward energy transport by the atmosphere; in an average annual sense, the poleward energy transport is necessary to balance the radiation loss to the space at the top of the atmosphere for a stable climate. Horizontal advection of heat and moisture, dictated by large-scale atmospheric circulation (Nygård et al., 2019; Papritz, 2020; Messori et al., 2018), has direct implications on distributions of temperature and humidity but also affects other meteorological conditions such as cloudiness and cloud properties and surface fluxes, in turn modifying temperature and humidity distributions in the atmosphere. The thermodynamic characterization of the lower Arctic atmosphere hence requires understanding of both the local processes, such as radiation, cloud formation and turbulence, and the larger-scale dynamics, responsible for the advection (Morrison et al., 2012).

The warmest anomalies in the wintertime Arctic are connected to horizontal transport of air from regions with higher climatological potential temperature and to diabatic heating (Papritz, 2020). A meridionally oriented circulation pattern, or cyclone track, especially favours intrusions of warm and moist air from the mid-latitudes into the Arctic (Messori et al., 2018; Woods et al., 2013; Fearon et al., 2021; Pithan et al., 2018). Such intrusions are associated with increased cloudiness, cloud water content and downward longwave radiation (Nygård et al., 2019; Liu et al., 2018; Park et al., 2015). The coldest anomalies in the central Arctic, on the other hand, occur after a prolonged period of uninterrupted radiative cooling, when the large-scale circulation pattern shelters the high Arctic from meridional air mass exchange (Papritz, 2020; Messori et al., 2018; Pithan et al., 2018), and cloud-free conditions often prevail.

Due to the radiative cooling at the surface in clear-sky conditions, surface-based temperature inversions are very common in winter over the snow- or sea-ice-covered surfaces (Curry, 1983; Serreze et al., 1992; Devasthale et al., 2010). However, when warm and moist air is advected from the lower latitudes into the Arctic, it experiences radiative cooling, leading to cloud formation (Pithan et al., 2014). In cloudy conditions, the strongest radiative cooling rapidly shifts from the surface to the cloud top, causing buoyancy and mixing from turbulent overturning, while increased longwave radiation from the cloud base warms the surface (Tjernström and Graversen, 2009; LeMone et al., 2019). As a consequence, an elevated temperature inversion forms over a shallow but relatively well-mixed boundary layer. However, continued cooling of the cloud and related phase changes from liquid to ice gradually reduce the emissivity of the cloud (Pithan et al., 2014). Finally, after cloud ice formation has caused the cloud to precipitate out, the clear state with a surface-based temperature inversion again prevails. These shifts between surface-based and elevated temperature inversions are rapid in winter (Tjernström and Graversen, 2009) and are largely governed by the large-scale atmospheric circulation (Stramler et al., 2011).

Specific humidity profiles over snow- or sea-ice-covered surfaces in winter are nearly always characterized by specific humidity inversions (Vihma et al., 2014; Nygård et al., 2014; Naakka et al., 2018; Devasthale et al., 2011). Most of the wintertime specific humidity inversions in the lower troposphere are formed due to the radiative cooling and associated condensation, at the surface or the cloud top, and are thus tied to temperature inversions and saturated conditions (Curry, 1983; Naakka et al., 2018). Above 800 hPa, specific humidity inversions are more commonly independent from temperature inversions, directly related to elevated advection of a moist air (Naakka et al., 2018).

In this study, we associate vertical profiles of temperature and humidity with atmospheric large-scale circulation patterns within the circumpolar Arctic (north of 65° N). The allocation of characteristic circulation types is based on self-organizing maps (SOMs). In the analyses, we have utilized radiosonde, in situ cloud cover and radiation observations as well as ERA5 atmospheric reanalysis fields for the winter months (DJF) of 2009–2018. The focus is on processes and their interactions in the present climate. Specifically, we (i) show how the shape of temperature and specific humidity profiles varies with the large-scale circulation and assess the sensitivity of the profiles to the large-scale circulation in different regions within the Arctic; (ii) identify the vertical extent of the surface–atmosphere interactions in different circulation types; and (iii) assess the roles of horizontal advection, clouds and surface energy budget, all dependent on the large-scale circulation, for the profile shapes. As a side product we also evaluate how well ERA5 reanalysis agrees with radiosonde observations on the response of the profiles to the large-scale circulation, a prerequisite for a meaningful study.

DATA

ERA5 Reanalysis

We used the latest global reanalysis, ERA5 (Hersbach et al., 2020), from the European Centre for Medium-Range Weather Forecasts (ECMWF) for the Copernicus Climate Change Service (C3S, 2017). The spectral model resolution of ERA5 is T639, while the horizontal resolution of the data used in this study is 0.25° × 0.25°. ERA5 has 137 vertical model levels, and the vertical spacing of model levels is denser near the surface. The lowest model level is located at 7–10 m over the surface, with eight levels below ~200 m and 20 below 1 km. We used the lowest 58 levels to cover the whole air column from the surface to approximately 300 hPa. In ERA5, a variety of atmospheric observations were assimilated into a numerical weather prediction model, applying a four-dimensional variational data assimilation method. The assimilated observations include, for example, satellite, surface and radiosonde observations. However, even when a radiosonde observation was assimilated, ERA5 temperature and humidity profiles are not necessarily always similar to observations, as the comparisons in Sect. 4.4 show. This is because impacts of observations on the reanalyses vary regionally due to observation and model uncertainty (Naakka et al., 2019) and availability and quality of other observations.

Large-scale circulation in the Arctic was clustered using SOMs, based on ERA5 mean sea level pressure fields at 6 h intervals for the winter months (DJF) of 2009–2018. We then composite total cloud cover, total column cloud water (TCW), surface fluxes, and vertically integrated heat and moisture transport as well as temperature and specific humidity fields on the model levels on each cluster. The vertical axis of model level data was converted to geopotential heights, defined here as the distance from the surface. Energy fluxes were extracted at hourly resolution from the associated ERA5 forecast fields and averaged over ±3 h around the SOM time steps. The diurnal cycle is ignored since it is essentially absent in winter (LeMone et al., 2019).

Uncertainties related to circulation patterns represented by ERA5 are assumed to be small as atmospheric reanalyses are able to accurately represent atmospheric pressure fields. Uncertainties in reanalysis temperature and specific humidity, and especially in cloud cover, may be considerably larger and are evaluated as a part of this study. The uncertainty for cloud phase and the turbulent surface fluxes, which depend on vertical profiles but are not constrained directly by the data assimilation, is likely also large, but lack of observations preclude an evaluation. It is known from earlier studies that ERA5 performs well for the surface meteorology and water vapour profiles in winter over the Arctic sea ice and ice-free ocean (Graham et al., 2019; Renfrew et al., 2021). However, ERA5, similar to many other atmospheric reanalyses, has a near-surface winter warm bias, strongest at temperatures below −25 °C (Wang et al., 2019; Graham et al., 2019). Graham et al. (2019) concluded that reanalyses have difficulties in resolving vertical temperature gradients in very stable boundary layers. It is known from previous-generation reanalyses that they generally capture spatial distribution and seasonal cycle of specific humidity inversion occurrence (Brunke et al., 2015; Naakka et al., 2018); however, specific humidity inversions are often weaker and less frequent compared to radiosonde observations (Naakka et al., 2018).

Radiosoundings and Cloud Observations

We use temperature and specific humidity profiles between the surface and 500 hPa from 32 Arctic radiosounding stations north of 65° N for the winter months (DJF) of 2009–2018. These were taken from the Integrated Global

Radiosonde Archive version 2 (IGRA V2) (Durre et al., 2006; Durre and Yin, 2008; Ferreira et al., 2019), a global dataset of quality-assured radiosonde data. In IGRA V2, temperature and specific humidity data are provided at the mandatory pressure levels defined by the World Meteorological Organization and at significant levels where a sounding variable deviates a specified amount from linearity.

The level of uncertainty in the data and the number of vertical levels reported vary in time and space as radiosonde types and data analysis procedures vary between countries. At some of the stations, there were also changes in the radiosonde type during the study period. Information on radiosonde types used at individual stations as well uncertainty characterization of radiosonde types is not easily obtainable. Temperature uncertainty in Russian radiosondes is likely a factor of 2 larger than for more modern radiosonde types (Naakka et al., 2019; Ingleby, 2017), and their humidity measurements clearly suffer from cloud contamination above mid-level clouds (Ingleby, 2017). We have not adjusted for inhomogeneities due to changes in instrumentation or observation practices.

Utqiaġvik (formerly known as Barrow) and Sodankylä stations, studied in more detail in this paper, are part of the GCOS Reference Upper-Air Network (GRUAN), established to provide high-quality sounding observations with traceable uncertainty assessment and documented practices. During the study period, Vaisala RS92 radiosondes were used at the stations, replaced with Vaisala RS41 radiosondes in 2017 and 2018 at Sodankylä and Utqiaġvik, respectively. Uncertainty in Vaisala R92 radiosondes is estimated to be approximately 0.5 °C for temperature and 5 % for relative humidity (Dirksen et al., 2014), consistent with manufacturer information. Jensen et al. (2016) compared performance of the Vaisala RS41 radiosonde against the Vaisala RS92 radiosonde using the twin-sounding method, concluding that differences were typically much smaller than these uncertainty estimates.

Cloud cover observations at 27 of the radiosounding stations were taken from the Integrated Surface Database (ISD) at 6 h intervals. For the remaining Arctic stations, the percentage of missing data was ≥30 %, and these stations were not included. Incoming and outgoing longwave radiation, used to define radiatively clear and cloudy cases at Sodankylä and Utqiaġvik, was measured with Kipp&Zonen CG4 and CGR4 pyrgeometers and with the Eppley PIR pyrgeometer, respectively.

METHODS

Large-Scale Circulation Clustering

The identification and clustering of large-scale circulation types were accomplished using the SOM method, a statistical unsupervised learning method determining generalized patterns in data, developed by Kohonen (2001). A SOM is a type of artificial neural network, hence a machine-learning approach, previously applied in climatology of synoptic patterns (Hewitson and Crane, 2002; Cassano et al., 2006), for addressing moisture transport (Mattingly et al., 2016; Nygård et al., 2019; Skific et al., 2009) and analysing temperature and humidity inversions in the Arctic (Yu et al., 2019). The ability of a SOM to provide physically meaningful composites of the circulation patterns has been demonstrated in numerous studies (Hewitson and Crane, 2002; Cassano et al., 2006; Johnson et al., 2008). Here, the SOM method was applied to allocate 12 characteristic atmospheric winter (DJF) circulation types for 10 years (2009–2018) of mean sea level pressure (MSLP) from the ERA5 reanalysis at 6 h intervals. The SOM analysis was made for the region north of 50° N, while the cloud and thermodynamic analyses were made for the region north of 65° N. The SOM area was extended from 65 to 50° N to capture the Icelandic Low and a sufficient part of the Aleutian Low, which are both very essential features in transporting heat and moisture to the Arctic (Nygård et al., 2019). The input MSLP data for the region north of 50° N were first re-gridded to an equal-area grid. The SOM method provides a two-dimensional array of gridded MSLP fields, having the characteristics and probability density of the input MSLP data. The size of the SOM array was subjectively selected to 3×4 to include the main features of regional circulation patterns and yet be sufficiently general to enable conceptualization of the results. The size of the SOM array was subjectively selected to 3×4 to include the main features of regional circulation patterns and yet be sufficiently general to enable conceptualization of the results. In comparisons of 2×3, 3×4 and 4×5 SOM array sizes, it was clear that that the 2×3 SOM does not contain the needed level of details; it especially lacks circulation types (i) where North Atlantic low pressure has a very eastern location over the Barents Sea and (ii) where the high-pressure bridge between Russia and North America across the central Arctic Ocean has broken up. These circulation types, lacking in the 2×3 SOM, are known to be very important for heat and moisture advection (Nygård et al., 2019). Instead, the 4×5 SOM does not display circulation patterns or main characteristics of pressure fields that are totally missing in the 3×4 SOM. We also calculated variances of MSLP within each circulation type. The mean variance within a circulation type is notably reduced in the 3×4 SOM array (123 hPa^2) compared to the 2×3 SOM array (144 hPa^2), but the mean variance does not decrease at the same rate towards the 4×5 SOM array (120 hPa^2).

The principle of the SOM method is described in detail in Kohonen (2001), Hewitson and Crane (2002), Cassano et al. (2006), and Gibson et al. (2017). Briefly, the SOM procedure includes initialization, training, evaluation and visualization. In this study, the random initialization scheme (Hewitson and Crane, 2002) was applied to create an initial SOM array in which each node had an associated reference vector with an equal dimension to the input MSLP data. During the training, each input data vector (i.e. each time step of MSLP data) was compared with the reference vector of each node, and those reference vectors most similar to the input data vector were adjusted towards the input data vector. Training was continued until reference vectors converged. As a result, a SOM array of characteristic MSLP patterns was obtained. The SOM array nodes are organized so that most similar circulation patterns are located close to each other, while the most dissimilar patterns are situated in the corners of the array. As a final result each input MSLP field can be associated with the most similar node of the array. Here, we present composites of the MSLP fields from ERA5, not the reference vectors directly from the SOM (see Figs. 1 and S1).

Analyses

The study covers the whole circumpolar Arctic north of 65°N. Radiosoundings, cloud cover observations and variables from ERA5 were clustered according to their associated circulation types as allocated by the SOM analysis. Anomalies for each circulation type were calculated by subtracting the grid point medians within each circulation type from the corresponding median values for the whole time series. We used medians rather than mean values due to the often skewed non-Gaussian distributions of the analysed variables.

Statistical significance of anomalies within each circulation type was tested using a Monte Carlo approach. First, 3000 random subsets of a given field of ERA5 were constructed, having the same number of samples as each individual SOM circulation type. Then, anomalies in random subsets were compared to the anomalies in the SOM circulation types by applying a two-tailed test at each grid point. An anomaly for a given SOM circulation type was assumed significant at the 99 % level if less than 0.5 % of the anomalies in the random subsets were larger or, if negative, smaller than the anomaly for the SOM circulation type. Similarly, a Monte Carlo approach was also applied to test the statistical significance of the anomalies of radiosonde and cloud cover observations for each circulation type.

In the analyses of cloud cover, clear-sky and overcast conditions were identified from the surface observations (Sect. 2.2) and ERA5 reanalysis (Sect. 2.1). The sky was considered clear when 0– (<0.19 in ERA5) was covered by clouds and overcast when the coverage was 7– (>0.81 in ERA5).

Conditions at two stations, Utqiaġvik and Sodankylä, were studied in more detail. These were selected partly due to location and partly for having a large selection of high-quality observations. Estimates of cloud conditions for these two stations were made from radiation observations, considered more reliable since they directly represent the radiative impacts of clouds. The use of data was maximized by identifying radiatively clear-sky and cloudy (opaque) cases using the methodology of Stramler et al. (2011) based on the bimodal relative frequency distribution (RFD) of surface net longwave radiation flux (LWN). Similar to Sedlar and Tjernström (2019), the thresholds were LWN W m^{-2} for the radiatively clear state and LWN W m^{-2} for the radiatively cloudy state. Our own LWN analysis confirmed these thresholds as appropriate for both Utqiaġvik, and Sodankylä. Temperature and specific humidity gradients at these stations were derived from thermodynamic profiles by linearly interpolating the profiles to a dense equidistant vertical grid and calculating the difference between adjacent levels.

Profile metrics

To characterize thermodynamic profiles, we use the following simplified metrics: (i) maximum temperature and specific humidity in a profile, (ii) the difference between (i) and the corresponding surface values, and (iii) the altitude to (i) above the surface. For simplicity, we will henceforth refer to (ii) as "bulk inversion strength", or just "inversion strength", while recognizing that this metric will not distinguish between elevated inversions and surface inversions and will also miss elevated inversions when the inversion top value is below the near-surface value; however, it works well for surface inversions.

RESULTS

Large-scale circulation has manifold ways to influence profiles of temperature and specific humidity in the Arctic. In addition to direct impacts of horizontal transport (advection) of heat and moisture, largely driven by large-scale circulation, atmospheric circulation also influences cloud conditions and surface fluxes, which in turn strongly modify the temperature and humidity profiles. After introducing the circulation patterns in Sect. 4.1, we first present the impacts of circulation on cloud conditions (Sect. 4.2) and surface fluxes (Sect. 4.3) and then proceed to link circulation, cloud conditions and fluxes to profiles of temperature and specific humidity in Sect. 4.4 and 4.5.

Atmospheric Circulation Patterns

As an outcome of the SOM analysis, 12 characteristic wintertime atmospheric MSLP patterns were identified and organized in a 2-D array according to their similarities (Fig. 1). Corresponding anomalies of MSLP in these 12 circulation types, calculated with respect to the DJF mean, are shown in Fig. S1 in the Supplement.

High pressure over northern Eurasia and low pressures over the northern North Atlantic and Pacific are present in all circulation types, whereas other features only occur in some of the circulation types, with intensity, extent and location varying between the 12 circulation types (Fig. 1). The main large-scale features appearing in the wintertime MSLP fields are (i) an extensive high-pressure area over northern Eurasia; (ii) high pressure over Northern America; (iii) an area of high pressure connecting the Eurasian and Northern American high-pressure areas across the Arctic Ocean, previously referred to as the Arctic Bridge (Nygård et al., 2019); (iv) high pressure over Greenland; (v) Icelandic Low in the northern North Atlantic; and (vi) Aleutian Low in the Pacific sector.

The number of cases in each of the 12 circulation types varies between 147 (4% of the time) and 461 (13% of the time), sufficient for statistical analyses. The most common circulation type, with an extensive high pressure in the Arctic, is type 10, while the most persistent types (i.e. with the longest uninterrupted duration) are 1, 4, 9 and 10 (Fig. 1), in which the Icelandic Low has a relatively western location (types 1, 4 and 9), or there is a strong high pressure over the Arctic Ocean (type 10). The least frequent and least persistent circulation types are situated in the centre of the array (types 5 and 8). The most distinct circulation types are situated in the corners of the

SOM array (Fig. S1); in the following subsections we focus mostly on these patterns. These four circulation types demonstrate the general dependencies between circulation, clouds, fluxes and thermodynamic profiles, which are also commonly found in the other circulation types.

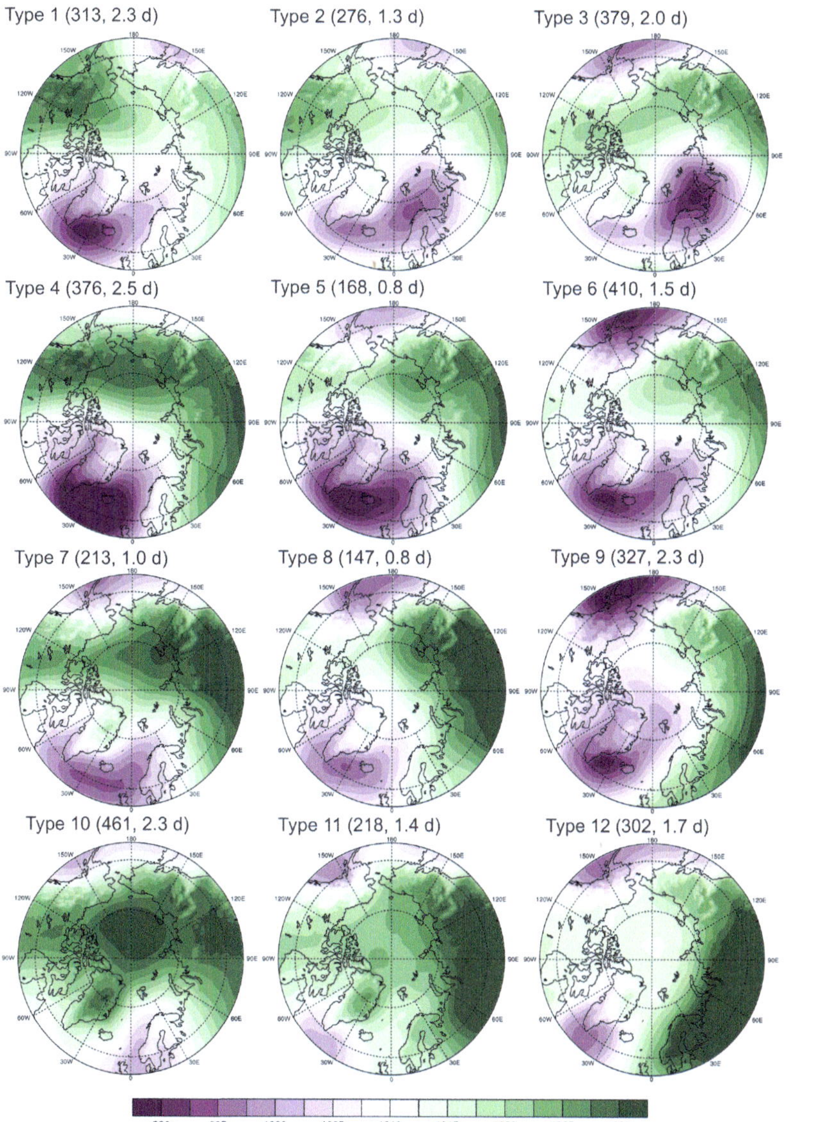

Figure 1: Mean sea level pressure (MSLP) averaged over the cases belonging to each of the circulation types (SOM nodes) in winters (DJF) of 2009–2018. The values in the parentheses indicate the number of cases belonging to the circulation type and the median persistence of the circulation type.

Cloud Conditions Linked to Circulation Patterns

First, we briefly present the median cloud conditions in the circumpolar Arctic in winter to facilitate interpretation of cloud anomalies related to different circulation types. The median cloud cover is high in the winter Arctic, mostly 7– in both ERA5 and cloud cover observations (Fig. 2a); note that cloud cover observations in the polar night may be biased low due to light conditions (Hahn et al., 1995). Because the wintertime cloud fraction distribution in the Arctic is bi-modal, peaking at clear-sky and overcast conditions, we also separately show occurrences of the clear-sky (cloud fraction) and overcast conditions (cloud fraction) (Fig. 2c and d). Overcast conditions occur more than 50% of the time in most of the circumpolar Arctic. Clear-sky conditions are rare, the Canadian Arctic Archipelago and the northern coast of Greenland being exceptions. The median TCW, as analysed from ERA5, displays more spatial variations with median values higher than 0.05 kg m^{-2} only in some parts of the ice-free ocean areas, mainly along the ice edge and over land in Scandinavia and the western part of Russia (Fig. 2b). In these regions, a substantial part (mostly 30%–70%) of the TCW is liquid water (Fig. S2), which is more efficiently suppressing surface radiative cooling than cloud ice. Over Siberia, the central Arctic Ocean, North America and Greenland, the median TCW is low, less than 0.01 kg m^{-2}. In these areas, the clouds mostly consist of ice; the variations are also mostly due to cloud ice (Fig. S2), hence limiting radiative impacts.

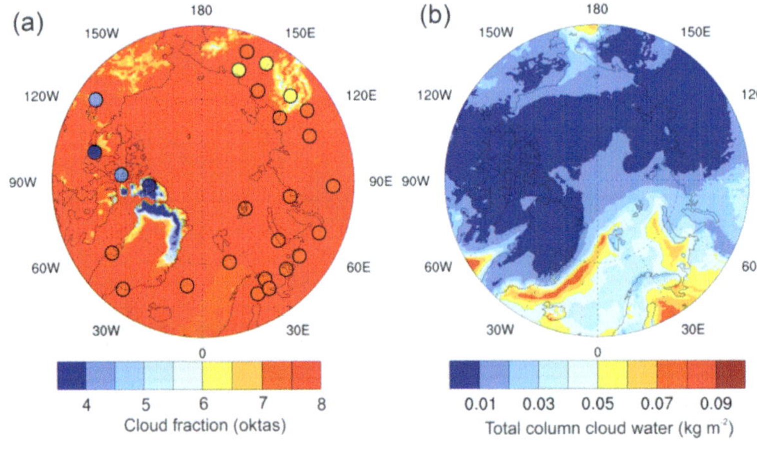

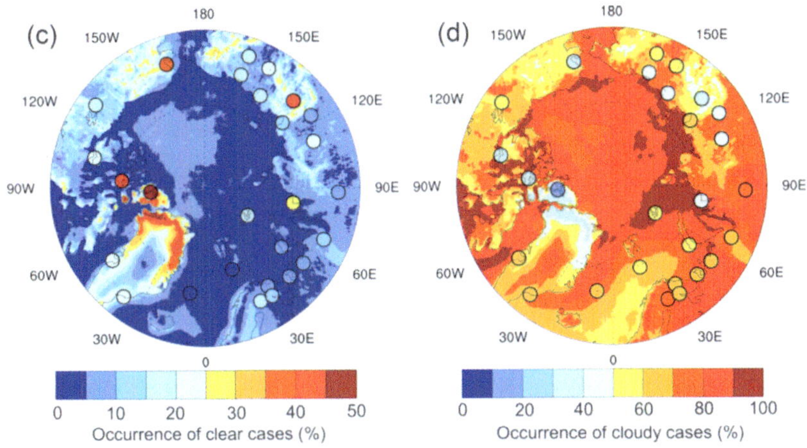

Figure 2(a) Median cloud fraction, **(b)** median total column (liquid and frozen) cloud water, **(c)** occurrence of clear (cloud fraction) cases and **(d)** occurrence of overcast (cloud fraction) cases in December–February 2009–2018. The colours inside the circles in **(a)–(c)** show synoptic cloud observations at the radiosounding stations, whereas the fields in **(a)–(d)** are based on ERA5.

We demonstrate the general relationships between the near-surface circulation and clouds by showing results for the four corner circulation types (Fig. 3). In circulation type 1, characterized by anomalous high pressure over Alaska and lower-than-average MSLP elsewhere in the Arctic (Figs. 3a and S1), the TCW and occurrence of overcast conditions are both anomalously high around the Bering Strait and in the northern North Atlantic (Fig. 3e and m).

These positive anomalies are linked to anomalous northward meridional transport from the northern North Atlantic and North Pacific oceans. Negative anomalies in TCW, especially liquid, and in occurrence of overcast cases (Fig. 3e and m) are found in the north-western parts of Eurasia, linked to a lack of influence from the North Atlantic and weak transports of heat and moisture from open oceans. In circulation type 3, an anomalous low pressure is centred over the Barents Sea, and MSLP is lower than average in the whole of Eurasia (Figs. 3b and S1). This circulation type is associated with substantial positive anomalies in TCW and overcast occurrences in the sector between 30 and 90° E, influenced by a stronger-than-average westerly flow and advection of warm and moist air from the North Atlantic, linked to the low pressure (Fig. 3f and n). Especially between 60 and 90° E, the fraction of liquid cloud water is clearly anomalously high (Fig. S2).

Circulation type 10 has anomalously high MSLP in most of the Arctic, especially over the Arctic Ocean and Greenland (Figs. 3c and S1). This type is associated with a generally lower TCW and increased clear skies (Fig. 3g and k). However, anomalies in cloudiness in the central Arctic cause only weak anomalies in the net longwave radiation (not shown) because TCW is generally low, and the anomalies are mostly due to cloud ice. In circulation type 12, there is an extensive high-pressure area over northern Eurasia, whereas the high pressure connecting Eurasia and Northern America across the Arctic Ocean (Arctic Bridge) is absent (Figs. 3d and S1). In this circulation type, relatively moist and warm air from the North Atlantic flows eastward over the Barents, Kara and Laptev seas and provides moisture, facilitating an anomalously high amount of TCW in these mostly sea-ice-covered regions. Advection of warm air also increases the fraction of liquid water in clouds here (Fig. S2). In contrast, the land area in the north-western parts of Eurasia has anomalously low TCW and more frequent clear skies.

While clear skies are slightly more common in observations than in ERA5, reanalysis and observations mostly agree. Overcast conditions are up to 20 percentage points more common in ERA5 than in observations at many stations. Importantly, ERA5 captures the variability in cloud cover between the circulation types, although the differences at a few stations to the observations are substantial (Fig. 3). For the purpose of this study, where we mainly focus on the clear and cloudy conditions, we conclude that ERA5 adequately represents the cloud anomalies associated with different circulation types in the wintertime Arctic.

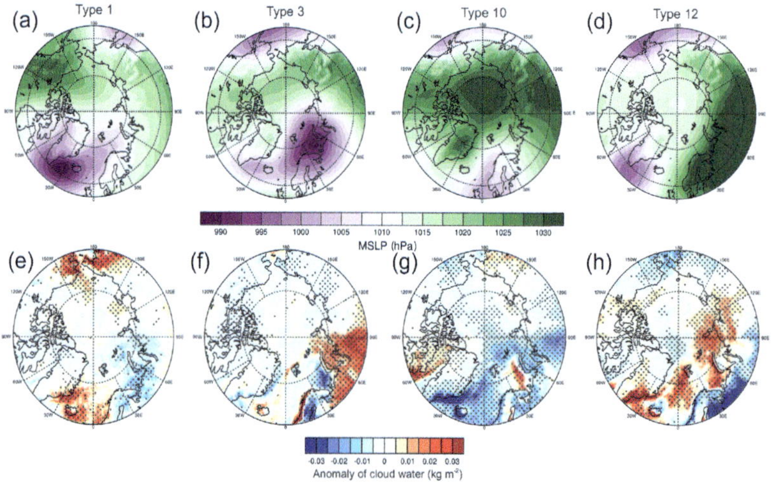

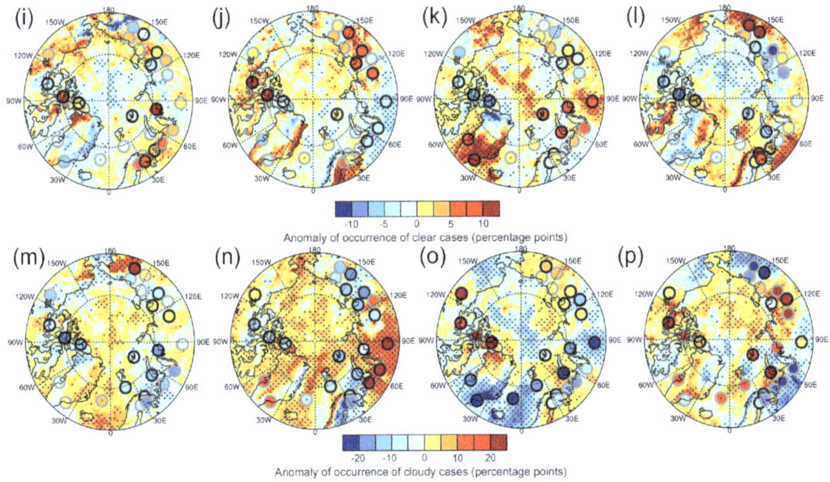

Figure 3: Mean sea level pressure **(a–d)**, anomaly of total column (liquid and frozen) cloud water **(e–h)**, anomaly of occurrence of clear skies (cloud fraction) **(i–l)** and occurrence of overcast conditions (cloud fraction) **(m–p)** in four SOM circulation types (indicated on the top of the column) during December–February 2009–2018. The colours inside the circles in **(i)**–**(p)** show synoptic cloud observations at the radiosounding stations, whereas the fields in **(a)**–**(p)** are based on ERA5. The dotted shadings (ERA5) and black circles (synoptic cloud observations) show the anomalies that are different from zero at the 99 % significance level.

Surface Fluxes Linked to Circulation Patterns

For all surface energy fluxes shown in Fig. 4, we present median values instead of anomalies to clearly display their directions. Since solar radiation is absent, and the surface albedo is very high during a large part of winter, the surface radiation budget is dominated by net longwave radiation; the distribution is bimodal (Stramler et al., 2011), while the median is negative (Fig. 4m–p). Consequently, radiative surface cooling is common over the circumpolar Arctic, especially over the Arctic Ocean. Surface net radiation variations between different circulation types are tightly connected to distributions of TCW; net radiation is less negative when TCW is higher and vice versa (compare Figs. 3e–h and 4m–p).

On snow and ice-covered surfaces, the turbulent heat fluxes compensate for the net radiation at the surface, contributing to shape the temperature and humidity profiles by setting lower-boundary conditions. Therefore, reduced surface radiative cooling immediately leads to surface warming (or melting if the temperature is high enough). The turbulent heat fluxes are small and commonly directed downward; condensation and sublimation of water at the surface are common. Any remaining residual in the energy budget is presumably closed by conduction of heat through the ice. This is not available from atmospheric reanalysis since it requires accurate modelling of ice thickness. Latent heat flux is very small for all circulation types since specific humidity is small to begin with, and the sensible heat flux is mainly responsible for balancing radiative cooling. Hence, over snow and high-concentration sea ice, spatial patterns of sensible heat flux correspond to the patterns of net radiation, in turn closely linked to cloud distributions controlled by large-scale circulation.

Over open water, on the other hand, the large heat capacity of the ocean surface layer prevents rapid changes in the sea surface temperature linked to changes in radiative cooling. Relatively warm open-water surfaces are nearly always a source of atmospheric heat and moisture. Over open ocean, strong upward turbulent heat fluxes lead to efficient vertical mixing and formation of a mixed layer in the lower atmosphere. Sensible and latent heat fluxes over the open ocean have clear variations linked to the large-scale circulation; the largest sensible and latent heat fluxes occur during southward flows in the northern North Atlantic (circulation types 3 and 10) and the smallest during northward flows in the northern North Atlantic (circulation types 1 and 12) (Fig. 4e–l).

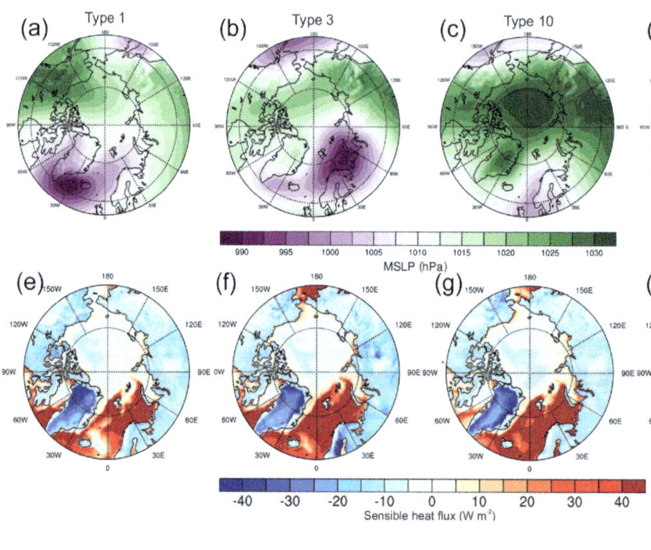

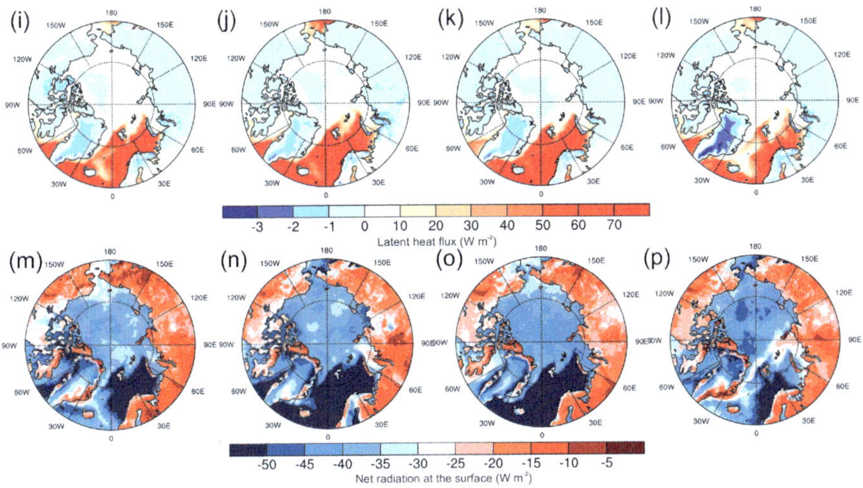

Figure 4: Mean sea level pressure **(a–d)**, sensible heat flux **(e–h)**, latent heat flux **(i–l)** and net radiation at the surface **(m–p)** in four SOM circulation types (indicated on the top of the column) during December–February 2009–2018 based on ERA5. For sensible and latent heat fluxes, positive direction is defined to be upwards and for the net radiation downwards. Note the nonlinear scale for the latent heat flux.

Characteristics of Temperature and Humidity Profiles linked to Circulation Patterns

Within the Arctic the characteristics of the vertical thermodynamic structure is to a first order controlled by the surface type, whereas temporal variations at a certain location are mostly controlled by the synoptic weather conditions. Figures 5–7 show profile metrics (defined in Sect. 3.3) for both ERA5 and observations from selected sounding stations. There is in general a good agreement between the two, not unexpectedly since most of these soundings were assimilated into the reanalysis.

Unsurprisingly, temperature and specific humidity maxima, characterizing the air mass properties, are largest over open ocean (Fig. 5a and d), where they are also typically located at the lowest model level (Fig. 5c and f). This is because the open-water surface is a source of both heat and moisture (see Sect. 4.3). Over land or sea ice, the maxima are typically found at a higher altitude, linked to the cooling of the atmosphere by the surface and resulting inversion formation (see Sect. 4.3). Both temperature and specific humidity maxima are typically found below 1500 m (Fig. 5c and f). In most of the

Arctic, the specific humidity maximum is located slightly higher than that for temperature, and this difference is statistically significant with the 95 % confidence interval; the largest differences between the two are found over the Beaufort Sea and central Siberia. Bulk inversion strengths are largest over the continents for both temperature and specific humidity (Fig. 5b and e); the strengths for temperature exceed 8 °C (with temporal deviations (±1 SD) of 4–7 °C) over most of the North American and Russian Arctic, while the strengths for specific humidity exceed 0.5 g kg^{-1} (with ±1 SD interval 0.2–0.5 g kg^{-1}).

In circulation type 1, anomalously warm and moist air masses are found around the Bering Strait (Figs. 6a and e, 7a and e), associated with increased TCW (Fig. 3e), reduced radiative cooling (Fig. 4m), and weak temperature and specific humidity inversions (Figs. 6i and 7i). The vertical maxima of temperature and specific humidity are at anomalously low levels (Figs. 6m and 7m).

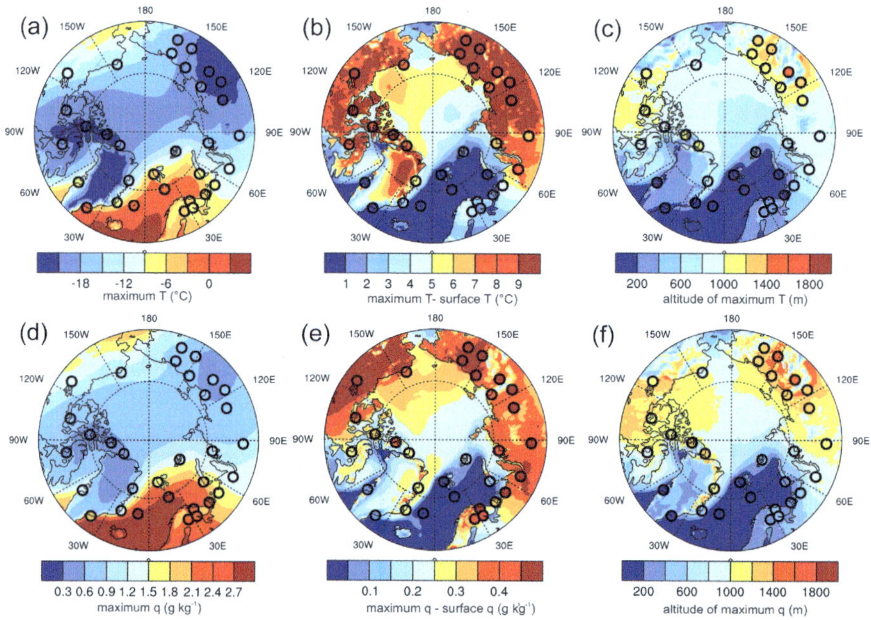

Figure 5 Medians of **(a)** vertical maximum T, **(b)** vertical maximum $T -$ surface T (bulk inversion strength), **(c)** altitude of vertical maximum T, **(d)** vertical maximum q, **(e)** vertical maximum $q -$ surface q (bulk inversion strength) and **(f)** altitude of vertical maximum q in December–February 2009–2018. The colours inside the circles show values based on radiosonde observations, whereas the fields are based on ERA5.

Elsewhere in the Arctic, anomalies are weaker, probably because this circulation type is dominated by the Icelandic Low and a high pressure over North America, and features elsewhere are partly averaged out. At some locations ERA5 and radiosoundings disagree on sign of anomalies: based on ERA5, temperature and specific humidity inversions are anomalously weak, and the vertical maxima of temperature and humidity are located at an anomalously low altitude in most of the Arctic, but radiosonde profiles indicate more commonly positive bulk inversion strength anomalies and higher altitudes of the vertical maxima.

When circulation type 3 dominates, clear linkages between atmospheric circulation and thermodynamic profiles are indicated; here ERA5 and radiosonde profiles agree. Anomalously warm and moist air is found over the Eurasian land due to transport of heat and moisture related to the low pressure (Figs. 6b and f, 7b and f, S3). The positive Eurasian anomaly in TCW and occurrence of overcast cases between 30 and 90° E (Fig. 3f and n) reduce radiative cooling at the surface (Fig. 4n) and contribute to the erosion of temperature and specific humidity inversions in this sector (Figs. 6j and 7j). Consequently, temperature and specific humidity are at their maximum at the lowest model level (Figs. 6n and 7n). However, warm and moist air reaches all the way to 120° E, but its effects on the thermodynamic profiles are rather different in the central Russian Arctic (between 90 and 120° E) compared to farther west. In the central Russian Arctic, the TCW is close to the median, with slightly increased radiative cooling at the surface. As a consequence, inversions are anomalously strong (Figs. 6j and 7j). It is also noteworthy that the anomalously cold and dry air over the northern North Atlantic and Bering Strait, connected to strong upward sensible and latent heat fluxes over the open ocean (Fig. 4f and j), has a very small effect on the thermodynamic profile metrics because our metrics emphasize the inversion layers. However, this cold and dry air is associated with weak negative anomalies in bulk inversion strength of temperature and humidity over the sea ice.

Circulation type 10 leads to isolation of the high Arctic from poleward transport of air from the northern North Atlantic (Fig. S3). During this circulation type, temperature inversions in the central Arctic Ocean are anomalously strong (Fig. 6k), linked to radiative cooling at the surface and probably also to subsidence. The vertical maxima of temperature are found several hundreds of metres higher than the median over the Eurasian Arctic and the Laptev, East Siberian and Chukchi seas (Fig. 6o), presumably due to the uninterrupted surface cooling allowing for vertical growth of inversion layers. Temperature inversions in the central Russian Arctic, where clear-sky conditions are anomalously frequent, are stronger than the median. However, over land westward from 60° E, temperature inversions are anomalously weak, associated with more frequent overcast conditions and hence reduced radiative cooling at the surface, even if the amount of cloud water is low. Hence, the temperature inversions in the cold sector over the Eurasian continent are strongly affected by cloud cover. The cold air limits specific humidity in the sector from 30° W anticlockwise to 120° E (Fig. 6g), and this, possibly together with a drying effect of subsidence, limits the inversion strength of specific humidity (Fig. 7k). However, over the continents here, the vertical maximum of specific humidity is located 100–300 m higher than the median (Fig. 7o), probably due to a long residence time of air over the continent, which has allowed the growth of a persistent inversion layer. Warm and moist air over Greenland and Baffin Bay and over the land in the Pacific sector, from 150° W to 130° E, has ambiguous impacts on temperature and specific humidity profiles. Generally, ERA5 and radiosoundings agree on the main regional features of the profile metrics, but not in all regional details.

For circulation type 12, the largest bulk inversion strength anomalies of temperature and specific humidity are found over Scandinavia and western Russia (20 to 90° E) (Figs. 6l and 7l), where the air is cold and dry, and TCW is low (Figs. 6h and 7h). Likely, the air has spent a relatively long time over the cold continent, facilitating uninterrupted surface cooling over a prolonged time. In this region, temperature inversions are 3–4 °C stronger, while specific humidity inversions are 0.05–0.2 $g\,kg^{-1}$ stronger than their medians.

The high pressure over northern Eurasia steers warm and moist North Atlantic air over the Barents, Kara and Laptev seas, where inversions are partly eroded. This warm and moist air reaches in over the land in central Siberia, where temperature inversions are notably weakened, probably linked to the radiative impacts of clouds. The cold and dry air in the Pacific sector is linked to anomalously strong temperature inversions over land, but specific humidity inversions are anomalously weak in this generally dry air. Over the sea-ice-covered Arctic Ocean with cold and dry air, temperature and specific humidity inversions are weak, linked to the mainly upward-directed sensible heat flux over the sea ice. ERA5 and radiosoundings mostly agree on the main regional features of the profile metrics in this circulation type.

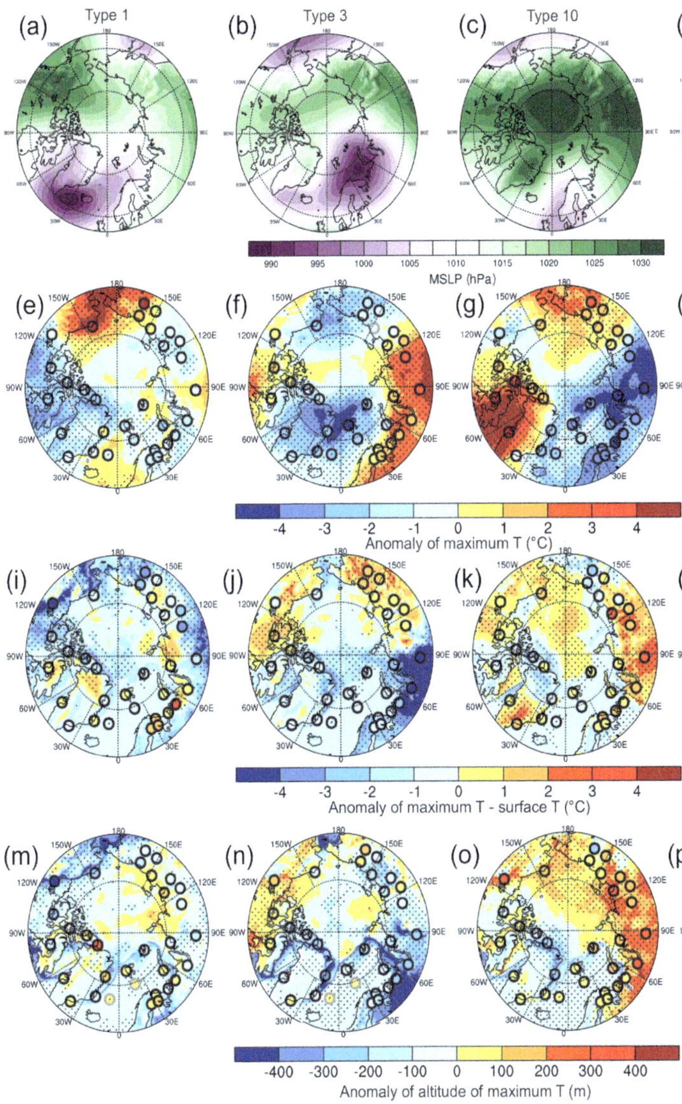

Figure 6: Mean sea level pressure **(a–d)**, anomaly of vertical maximum *T* **(e–h)**, anomaly of vertical maximum *T*−surface *T* (bulk inversion strength) **(i–l)**, anomaly of altitude of vertical maximum *T* **(m–p)** in four SOM circulation types (indicated on the top of the column) in December–February 2009–2018. The colours inside the circles show values based on radiosonde observations, whereas the fields are based on ERA5. The dotted shadings (ERA5) and black circles (radiosonde observations) show the anomalies that are different from zero at the 99% significance level.

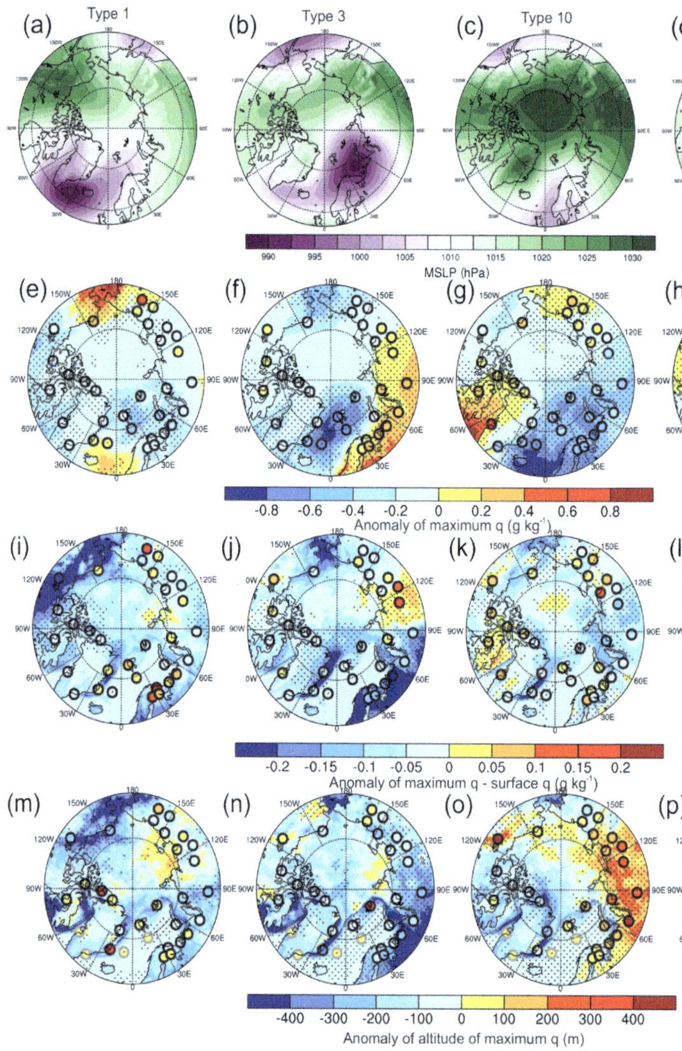

Figure 7Mean sea level pressure **(a–d)**, anomaly of vertical maximum q **(e–h)**, anomaly of vertical maximum $q-$surface q (bulk inversion strength) **(i–l)**, anomaly of altitude of vertical maximum q **(m–p)** in four SOM circulation types (indicated on the top of the column) in December–February 2009–2018. The colours inside the circles show values based on radiosonde observations, whereas the fields are based on ERA5. The dotted shadings (ERA5) and black circles (radiosonde observations) show the anomalies that are different from zero at the 99% significance level.

Vertical Profiles at Utqiaġvik and Sodankylä

Utqiaġvik (formerly Barrow)

While all the flow regimes are distinctly different when examining the pan-Arctic domain, several give rise to similar flow locally when examining a particular geographic location. For this reason, and for brevity, Fig. 8 shows vertical profiles of temperature and specific humidity for Utqiaġvik in Alaska (see location in Fig. 8a–d) for a select set of circulation types: 1, 3, 9 and 10. These are distinct and represent the effects of a variety of large-scale flows at this location well. Gradient profiles of temperature and humidity are shown in Fig. S4. Both radiatively clear and cloudy conditions, based on the radiation observations, are frequent for all of these four circulation types; percentages are 33 %–39 % and 51 %–55 % for cloudy and clear cases, respectively. By investigating profiles separately in radiatively cloudy and clear cases, we are better able to reveal the relative roles of directs impacts of the circulation and local impacts due to cloud conditions.

The radiosounding results for Utqiaġvik (Fig. 8) show that the radiatively cloudy cases are generally warmer and moister in the lowest 3000 m than the radiatively clear cases for all four circulation types. This is presumably partly due to different properties of air masses leading to cloudy and clear conditions (especially circulation type 10) and partly due to radiative effects of clouds, warming the profile below the cloud layer, especially in circulation type 1. Differences in temperature gradients between cloudy and clear conditions are mostly seen in the lowest 400 m, where temperature gradients in cloudy profiles are generally smaller, sometimes even negative (Fig. S4). However, the shape of the temperature profiles varies more between circulation types than between different cloud conditions within a certain circulation type, suggesting a major role of large-scale circulation for shaping the thermodynamic profiles. Furthermore, the cloudy temperature profiles vary more between circulation types than those for radiatively clear situations, suggesting that also properties of clouds depend on large-scale circulation. Even if median near-surface temperature is almost the same across the large-scale circulation types, vertical gradients (Fig. S4) and altitude of temperature maxima vary greatly. A temperature inversion is present in the median profiles for all the four circulation types (Fig. 8). Surface-based temperature inversions are eroded in low-pressure situations (circulation type 9), leading to strong elevated temperature and humidity inversions, with the largest positive temperature and specific humidity gradients between 300 and 800 m (Fig. S4). On the other hand, high pressure (circulation types 1, 3 and 10) is associated with strong vertical gradients of temperature below 500 m, whereas specific humidity gradients are more equally distributed within the lowest 1500 m; circulation type 3 with strong near-surface humidity gradients is an exception. In general, the differences in the profile shape between different large-scale circulation types are mostly seen below 1500 m.

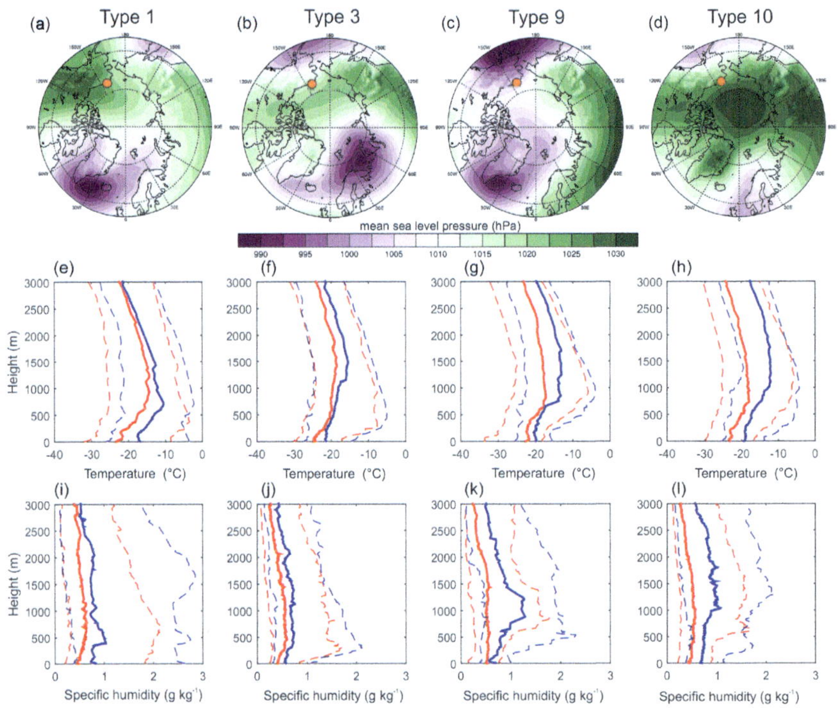

Figure 8: Median (solid lines), 10 % percentile (dashed lines), 90 % percentile (dashed lines) of temperature profiles **(e–h)** and specific humidity profiles **(i–l)** at Utqiaġvik for four atmospheric circulation types **(a–d)** in winter during 2009–2018. The profiles are based on radiosounding observations. Red lines show radiatively clear profiles, blue lines radiatively cloudy profiles. The location of Utqiaġvik is marked with a circle in **(a)–(d)**.

Sodankylä

For Sodankylä in northern Finland (see location in Fig. 9a–d), we explore circulation types 12, 1, 4 and 5 (Fig. 9), which are distinct and well represent a variety of large-scale flow patterns at this location. The different choice of flow patterns compared to those for Utqiaġvik is due to the difference in location relative to the flow. The ordering of circulation types in Fig. 9 is arranged according to the location of the Atlantic low, shifting from west to east. During the four circulation types, percentages of radiatively cloudy and clear cases are 65 %–82 % and 11 %–27 %, respectively.

Variability in the observed temperature and specific humidity profile shapes in Sodankylä is mostly seen below the altitude of 500 m, thus in a 1000 m shallower layer compared to Utqiaġvik; this is evident also when all the 12 circulation types are compared (not shown). Differences in temperature profiles between the cloudy and clear cases are also smaller than in Utqiaġvik, although the absolute differences in specific humidity profiles between the cloudy and clear cases are larger in Sodankylä due to generally higher temperatures.

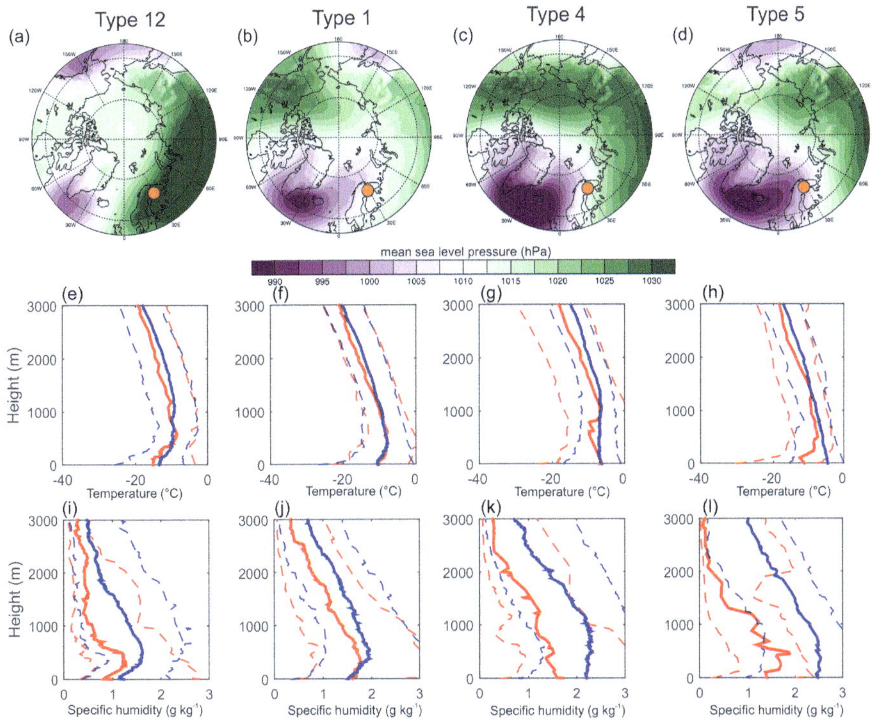

Figure 9 Median (solid lines), 10% percentile (dashed line), 90% percentile (dashed line) of temperature profiles **(e–h)** and specific humidity profiles **(i–l)** at Sodankylä for four atmospheric circulation types **(a–d)** in winter during 2009–2018. The profiles are based on radiosounding observations. Red lines show radiatively clear profiles, blue lines radiatively cloudy profiles. The loca-

tion of Sodankylä is marked with a circle in **(a)–(d)**. Cloudy cases are generally moister than the clear cases, suggesting that cloudy and clear conditions in Sodankylä are associated with different properties of air masses, similar to Utqiaġvik. Strong, positive temperature gradients are only seen in clear conditions, whereas positive specific humidity gradients are found in both clear and cloudy conditions (Fig. S5).

Generally, variability in wintertime temperature profiles is largely related to the circulation, whether Sodankylä is under the influence of Eurasian high-pressure conditions, with weak flow (circulation types 12 and 1), or low-pressure conditions, with an airflow from the northern North Atlantic and southern latitudes (circulation types 4 and 5). High-pressure conditions are linked to colder and drier vertical profiles, in particular in the lowest 500 m, whereas low-pressure conditions erode most of the surface-based temperature and specific humidity inversions.

DISCUSSION

Our results indicate that there are notable regional differences in how atmospheric circulation affects the temperature and specific humidity profiles, especially inversion layers. These regional differences are tightly linked to air mass transformation from warmer and moister mid-latitude air to colder and drier polar air by longwave radiative cooling. Based on the linkages between large-scale circulation and vertical profiles of temperature and humidity, we divide the study area into five regions with specific characteristics:

1. Over the Arctic sea ice, the maximum vertical air temperature varies largely between the circulation types. These variations are connected to different physics: horizontal advection, subsidence, latent heating, or periods with long residence time and uninterrupted radiative cooling, as shown earlier by Messori et al. (2018) and Papritz (2020). Hence the circulation patterns sometimes have a direct effect, like with warm air intrusions (e.g. circulation type 12), and sometimes the effects are due to local physics but enabled by a circulation pattern. Over the central Arctic Ocean, the strongest temperature inversions are associated with the strongest and most persistent high pressures (e.g. circulation type 10), in which TCW is low, and subsidence warms and dries the air, while surface radiation cools the air from below; hence, the warmest air does not reach all the way to the surface. In these

conditions, the air is dry due to the low saturation temperature and also a probable drying effect of the subsidence, mostly keeping the specific humidity inversions weak. In other words, our results indicate that the coldest air masses over the Arctic sea ice are not connected to the strongest temperature inversions in the region. This is probably due to the heat flux through the sea ice, causing the temperature inversions to be weaker over the sea ice compared to land surface (Serreze et al., 1992). We also find that over sea ice, the maximum temperature is typically found between 500 and 1000 m, but the maximum specific humidity is typically found between 600 and 1200 m. For both, the altitude of the maxima varies roughly ±300 m between the circulation types.

2. Over the open ocean, temperature and specific humidity maxima vary largely with large-scale circulation. The relatively warm surface and consequently mostly upward-directed heat fluxes keep the maxima nearly always near the surface; however, warm air advection may lead to stable stratification or even a shallow temperature inversion. Our metrics, focused on bulk inversion strength and vertical maxima, do not reveal the characteristic of the well-mixed layer over open ocean, although this layer is known to be influenced by atmospheric circulation and its effects on the surface fluxes. For example, certain flow patterns linked to pressure dipoles (resembling circulation type 3) lead to cold air outbreaks over open water near the sea ice margin (Kolstad et al., 2009; Pithan et al., 2018). These cold air outbreaks, associated with very large turbulent surface fluxes, lead to a well-mixed boundary layer, initially very shallow but growing with fetch over the open water (Brümmer, 1996). These flows are also associated with some of the largest turbulent surface energy fluxes recorded on Earth, leading to rapid warming and moistening of the air.

3. Over Eurasian land west of 90°E, a large variability in the temperature and humidity profiles is seen, suggesting that the profiles are sensitive to large-scale circulation. This region is occasionally influenced by low-pressure systems facilitating large transport of heat and moisture from the North Atlantic (e.g. circulation type 3). Under those conditions, the temperature and specific humidity inversions are typically eroded, and the vertical maxima of temperature and specific humidity are found at the surface. This erosion of temperature surface inversions has been

pointed out earlier by Serreze et al. (1992). Occasionally, this region is affected by the extensive high-pressure area over Eurasia (e.g. circulation type 12), associated with cold and dry air and low TCW due to a long residence time of the air over the cold continent. Our results indicate that in high-pressure conditions, temperature and specific humidity inversions are much stronger, but still weaker than over most of the Arctic land area; a similar spatial distribution has been reported earlier by Zhang et al. (2011) and Naakka et al. (2018). The maximum temperature in this region is typically found between 200 and 1000 m altitude, and some circulation types cause as large as ±400 m anomalies to this altitude.

4. Over Eurasian land east of 90° E, well-developed polar air with a long residence time over the cold continent dominates. Here, temperature and specific humidity inversions are strong, as shown earlier by Serreze et al. (1992), Zhang et al. (2011) and Naakka et al. (2018). Our results indicate that variability in bulk inversion strengths of temperature is related to the strength of the high pressure and the amount of cloud water, which is mostly ice. Specific humidity inversions are weak when the air is dry, presumably due to the low temperature and any drying effect of subsidence. Interestingly, warm and moist air originating from the northern North Atlantic is occasionally able to reach this region, with an extensive high pressure over northern Eurasia and absence of an Arctic Bridge (high pressure) across the Arctic Ocean (circulation type 12). Hence, the break-up of the Arctic Bridge not only allows for efficient transport of moist and warm northern Atlantic air masses into the central Arctic, as indicated earlier by Nygård et al. (2019), but also enables these air masses to occasionally reach all the way into eastern Eurasia. These warm and moist air masses make the temperature and specific humidity inversions notably weaker but are typically unable to erode them completely. Temperature and specific humidity maxima are mostly found above 1000 m; the different circulation types typically cause variations of a few hundred metres.

5. Over North America and Greenland, impacts of the large-scale circulation on temperature and specific humidity profiles are less systematic than for other regions. This is probably due to mostly modest MSLP variations compared to the rest of the Arctic, expect for in Alaska. Over Alaska, heat advection related

to the Aleutian Low and a high pressure centred east of Alaska as well as anomalous downward longwave radiation due to clouds are known to cause warm extremes (Cassano et al., 2016), while anticyclonic circulation reduces cloudiness and leads to stagnation of the cold air and strong temperature inversions (Sulikowska et al., 2019). Especially the latter is seen also in our results for Utqiaġvik (e.g. circulation type 1). For Greenland, Shahi et al. (2020) reported that surface-based temperature inversions are stronger in the northern and eastern parts of Greenland, suggesting that this spatial distribution is due to the dominance of the Greenland high in the north. In our results, we also see this spatial distribution within Greenland and note that the strongest temperature inversions in northern Greenland are often associated with a strong Icelandic Low south of Greenland (circulation types 4, 5 and 6 in Fig. 1).

CONCLUSIONS

We have explored the impact of large-scale circulation on the vertical structure and clouds in the circumpolar Arctic region using ERA5 in combination with self-organizing maps and observations, especially radiosoundings. We found that radiosonde observations and ERA5 reanalysis mostly agree on the response of the temperature and specific humidity profiles to large-scale circulation. We conclude the following.

- Large-scale circulation determines the origin and pathway of an air mass. The air mass experiences transformation along the pathway, e.g. due to radiative cooling and phase changes in water. Locally, effects of large-scale circulation are manifested as heat and moisture advection, which have first-order effects on temperature and specific humidity profiles and the turbulent surface fluxes. The advection may also lead to cloud formation, which will cause secondary effects on the fluxes and profiles, e.g. by shifting the strongest radiative cooling from the surface to the cloud top and forming an elevated temperature inversion. The secondary effects have a major role especially over snow- and ice-covered surfaces, where temporal variations in spatial patterns of surface sensible heat flux correspond to patterns of net radiation, which are tightly linked to cloud distributions largely controlled by large-scale circulation.

- The most sensitive regions to wintertime large-scale circulation are the Eurasian land area west of 90° E and the Arctic Ocean sea ice region, whereas impacts over the North American Arctic and Greenland are weaker and more ambiguous. Eurasian land between 90 and 140° E is typically not influenced by transport of oceanic air; however, a circulation pattern with an extensive high pressure over northern Eurasia without a continuous high-pressure bridge across the Arctic Ocean enables warm and moist air masses originating from the northern North Atlantic to cross over the Barents, Kara and Laptev seas and reach this Eurasian sector. Along this pathway, temperature and specific humidity inversions are partly eroded over the seas and substantially weakened over the land due to the heat and moisture advection and radiative impacts of clouds.
- The altitudes of maximum temperature and specific humidity and their variability between the circulation types are good indicators of the depth of the layer impacted by surface–atmosphere processes interacting with the large-scale situation, except over the open ocean. This layer is deepest (~1500 m) over Eurasia, east of 90° E, and Northern America and shallowest (~800 m) over western Eurasia, west of 60° E. Different circulation types typically cause variations of a few hundred metres to the altitudes, except over Greenland, where variations are smaller.

Hence, the influence of different large-scale flows in the Arctic must be viewed as a progressing set of processes. The primary reason is the advection – or lack thereof – of air masses into the Arctic, followed by transformation of the air through various physical processes, differently for different circulation types but always facilitated by the circulation. Therefore, an understanding of conditions in a certain region, how it varies and how it may change, must incorporate understanding of both the large-scale circulation and the ensuing physical processes affecting the air mass transformation.

Code Availability

Code for the analysis is available upon request.

Data Availability

ERA5 data are freely available from the Copernicus Climate Data Store (https://cds.climate.copernicus.eu, last access: 15 December 2021; C3S,

2017; DOI of single level data: https://doi.org/10.24381/cds.adbb2d47; Hersbach et al., 2018), IGRA data from the National Climatic Data Center (https://www1.ncdc.noaa.gov/pub/data/igra/) and cloud cover data from the Integrated Surface Database (https://www.ncdc.noaa.gov/isd). The radiation data from Utqiaġvik are provided by the Department of Energy's Atmospheric Radiation Measurement (ARM) programme (https://www.arm.gov/) and from Sodankylä by the Finnish Meteorological Institute (https://en.ilmatieteenlaitos.fi/download-observations).

SUPPLEMENT

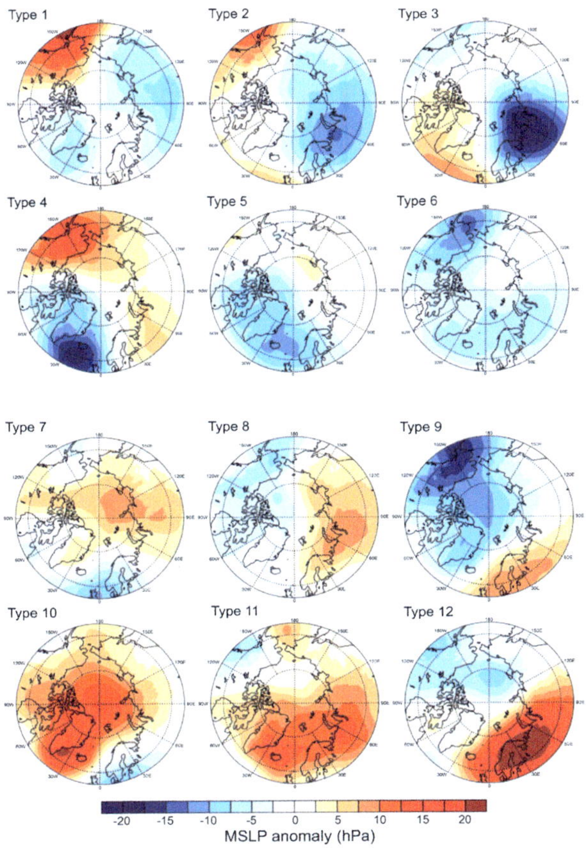

Figure S1: Anomaly of mean sea level pressure (MSLP) averaged over the cases belonging to each of the circulation types (SOM nodes) in winters (Dec-Feb) of 2009 – 2018.

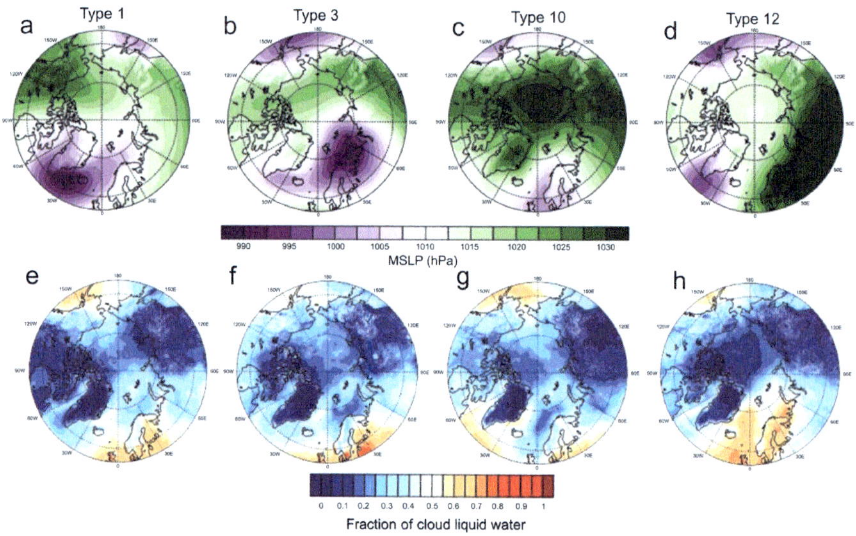

Figure S2: Mean sea level pressure (a–d), and fraction of liquid cloud water of the total cloud water (e–h) in four SOM circulation types (indicated on the top of the column) during December–February 2009–2018 based on ERA5.

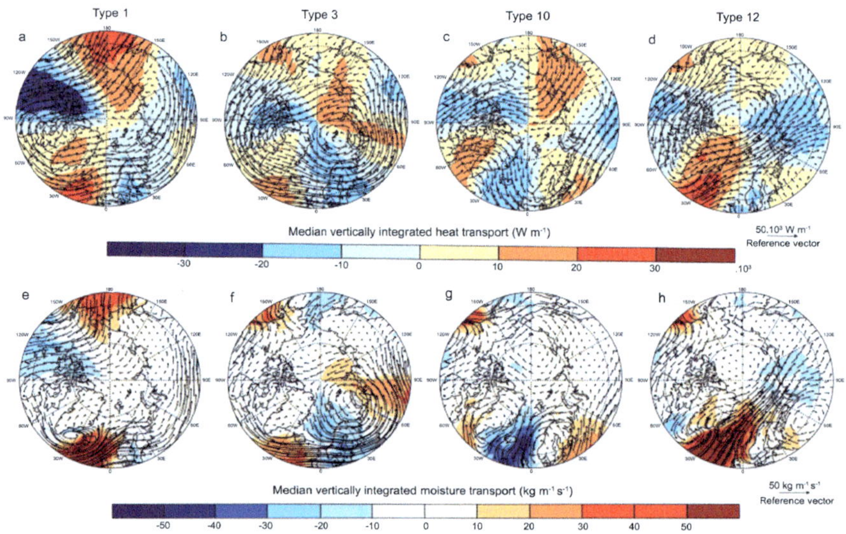

Figure S3: Median vertically integrated heat transport (a–d) and median vertically integrated moisture transport (e–h) in four SOM circulation types (indicated on the top of the column) during December– February 2009–2018 based on ERA5. Vectors show both the zonal and meridional transport whereas the color scale indicates the meridional transport only.

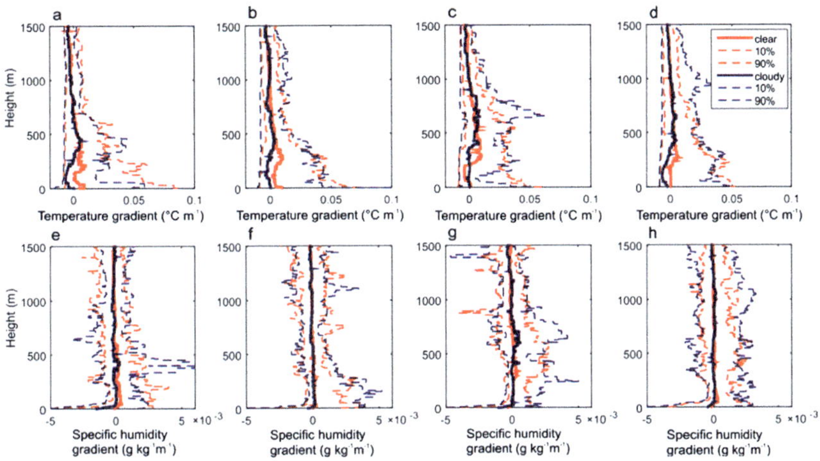

Figure S4: Median (solid lines), 10% percentile (dashed lines), 90% percentile (dashed lines) of vertical temperature gradient profiles (a–d), and vertical specific humidity gradient profiles (e–h) at Utqiaġvik for atmospheric circulation types 1, 3, 9 and 10 (from left to right). The profiles are based on radiosounding observations. Red lines show radiatively clear profiles, blue lines radiatively cloudy profiles. Gradients are defined positive, when the values increase upwards.

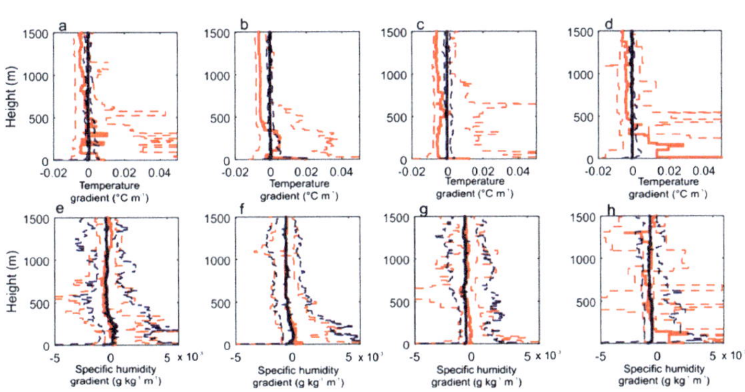

Figure S5: Median (solid lines), 10% percentile (dashed line), 90% percentile (dashed line) of vertical temperature gradient profiles (a–h) and vertical specific humidity gradient profiles (e–h) at Sodankylä for atmospheric circulation types 12, 1, 4 and 5 (from left to right). The profiles are based on radiosounding observations. Red lines show radiatively clear profiles, blue lines radiatively cloudy profiles. Gradients are defined positive, when the values increase upwards.

AUTHOR CONTRIBUTIONS

TiN and MT designed the study. TiN carried out the analysis in collaboration with TuN. All authors contributed to interpretation of the results and writing.

Competing interests

The contact author has declared that neither they nor their co-authors have any competing interests.

Disclaimer

Publisher's note: Copernicus Publications remains neutral with regard to jurisdictional claims in published maps and institutional affiliations.

ACKNOWLEDGEMENTS

Tiina Nygård is grateful for the support of the International Meteorological Institute (IMI) at Stockholm University. The authors thank the data providers, Markku Kangas for assistance with the radiation data from Sodankylä, and Peter B. Gibson and Petteri Uotila for providing the codes for SOM analysis.

Financial Support

This research has been supported by the Academy of Finland (grant no. 308441), the Vetenskapsrådet (grant no. 2016-03807) and the Knut och Alice Wallenbergs Stiftelse (grant no. KAW 2016-0024).

Review Statement

This paper was edited by Tim Woollings and reviewed by two anonymous referees.

REFERENCES

1. Brümmer, B.: Boundary-layer modification in wintertime cold-air outbreaks from the Arctic sea ice, Bound.-Lay. Meteorol., 80, 109–125, https://doi.org/10.1007/BF00119014, 1996.
2. Brunke, M. A., Stegall, S. T., and Zeng, X.: A climatology of tropospheric humidity inversions in five reanalyses, Atmos. Res., 153, 165–187, https://doi.org/10.1016/j.atmosres.2014.08.005, 2015.
3. C3S: ERA5: Fifth generation of ECMWF atmospheric reanalyses of the global climate, Copernicus Climate Change Service Climate Data Store (CDS) [data set], https://cds.climate.copernicus.eu/cdsapp#!/home (last access: 15 December 2021), 2017.
4. Cassano, J. J., Uotila, P., and Lynch, A.: Changes in synoptic weather patterns in the polar regions in the twentieth and twenty-first centuries, part 1: Arctic, Int. J. Climatol., 26, 1027–1049, https://doi.org/10.1002/joc.1306, 2006.
5. Cassano, J. J., Cassano, E. N., Seefeldt, M. W., Gutowski Jr., W. J., and Glisan, J. M.: Synoptic conditions during wintertime temperature extremes in Alaska, J. Geophys. Res.-Atmos., 121, 3241–3262, https://doi.org/10.1002/2015JD024404, 2016.
6. Curry, J. A.: On the formation of continental polar air, J. Atmos. Sci., 40, 2278–2292, 1983.
7. Devasthale, A., Willén, U., Karlsson, K.-G., and Jones, C. G.: Quantifying the clear-sky temperature inversion frequency and strength over the Arctic Ocean during summer and winter seasons from AIRS profiles, Atmos. Chem. Phys., 10, 5565–5572, https://doi.org/10.5194/acp-10-5565-2010, 2010.
8. Devasthale, A., Sedlar, J., and Tjernström, M.: Characteristics of water-vapour inversions observed over the Arctic by Atmospheric Infrared Sounder (AIRS) and radiosondes, Atmos. Chem. Phys., 11, 9813–9823, https://doi.org/10.5194/acp-11-9813-2011, 2011.
9. Dirksen, R. J., Sommer, M., Immler, F. J., Hurst, D. F., Kivi, R., and Vömel, H.: Reference quality upper-air measurements: GRUAN data processing for the Vaisala RS92 radiosonde, Atmos. Meas. Tech., 7, 4463–4490, https://doi.org/10.5194/amt-7-4463-2014, 2014.
10. Durre, I. and Yin, X.: Enhanced Radiosonde Data For Studies of Vertical Structure, B. Am. Meteorol. Soc., 89, 1257–1262, https://doi.org/10.1175/2008bams2603.1, 2008.

11. Durre, I., Russell, S. V., and Wuertz, D. B.: Overview of the Integrated Global Radiosonde Archive, J. Climate, 19, 53–68, 2006.
12. Fearon, M. G., Doyle, J. D., Ryglicki, D. R., Finocchio, P. M., and Sprenger, M.: The Role of Cyclones in Moisture Transport into the Arctic, Geophys. Res. Lett., 48, e2020GL090353, https://doi.org/10.1029/2020GL090353, 2021.
13. Ferreira, A. P., Nieto, R., and Gimeno, L.: Completeness of radiosonde humidity observations based on the Integrated Global Radiosonde Archive, Earth Syst. Sci. Data, 11, 603–627, https://doi.org/10.5194/essd-11-603-2019, 2019.
14. Gibson, P. B., Perkins-Kirkpatrick, S. E., Uotila, P., Pepler, A. S., and Alexander, L. V.: On the use of self-organizing maps for studying climate extremes, J. Geophys. Res.-Atmos., 122, 3891–3903, https://doi.org/10.1002/2016JD026256, 2017.
15. Graham, R. M., Cohen, L., Ritzhaupt, N., Segger, B., Graversen, R. G., Rinke, A., Walden, V. P., Granskog, M. A., and Hudson, S. R.: Evaluation of Six Atmospheric Reanalyses over Arctic Sea Ice from Winter to Early Summer, J. Climate, 32, 4121–4143, https://doi.org/10.1175/jcli-d-18-0643.1, 2019.
16. Hahn, C. J., Warren, S. G., and London, J.: The Effect of Moonlight on Observation of Cloud Cover at Night, and Application to Cloud Climatology, J. Climate, 8, 1429–1446, 1995.
17. Hersbach, H., Bell, B., Berrisford, P., Biavati, G., Horányi, A., Muñoz Sabater, J., Nicolas, J., Peubey, C., Radu, R., Rozum, I., Schepers, D., Simmons, A., Soci, C., Dee, D., and Thépaut, J.-N.: ERA5 hourly data on single levels from 1979 to present, Copernicus Climate Change Service (C3S) Climate Data Store (CDS) [data set], https://doi.org/10.24381/cds.adbb2d47, 2018.
18. Hersbach, H., Bell, B., Berrisford, P., Hirahara, S., Horányi, A., Muñoz-Sabater, J., Nicolas, J., Peubey, C., Radu, R., Schepers, D., Simmons, A., Soci, C., Abdalla, S., Abellan, X., Balsamo, G., Bechtold, P., Biavati, G., Bidlot, J., Bonavita, M., De Chiara, G., Dahlgren, P., Dee, D., Diamantakis, M., Dragani, R., Flemming, J., Forbes, R., Fuentes, M., Geer, A., Haimberger, L., Healy, S., Hogan, R. J., Hólm, E., Janisková, M., Keeley, S., Laloyaux, P., Lopez, P., Lupu, C., Radnoti, G., de Rosnay, P., Rozum, I., Vamborg, F., Villaume, S., and Thépaut, J.-N.: The ERA5 global reanalysis, Q. J. Roy. Meteor. Soc., 146, 1999–2049, https://doi.org/10.1002/qj.3803, 2020.

19. Hewitson, B. C. and Crane, R. G.: Self-organizing maps: applications to synoptic climatology, Clim. Res., 22, 13–26, 2002.
20. Ingleby, B.: An assessment of different radiosonde types 2015/2016, ECMWF Technical Memoranda, Reading, England, 2017.
21. Jensen, M. P., Holdridge, D. J., Survo, P., Lehtinen, R., Baxter, S., Toto, T., and Johnson, K. L.: Comparison of Vaisala radiosondes RS41 and RS92 at the ARM Southern Great Plains site, Atmos. Meas. Tech., 9, 3115–3129, https://doi.org/10.5194/amt-9-3115-2016, 2016.
22. Johnson, N. C., Feldstein, S. B., and Tremblay, B.: The Continuum of Northern Hemisphere Teleconnection Patterns and a Description of the NAO Shift with the Use of Self-Organizing Maps, J. Climate, 21, 6354–6371, https://doi.org/10.1175/2008jcli2380.1, 2008.
23. Kohonen, T.: Self-Organizing Maps, Springer-Verlag, Berlin, Heidelberg, 501 pp., 2001.
24. Kolstad, E. W., Bracegirdle, T. J., and Seierstad, I. A.: Marine cold-air outbreaks in the North Atlantic: temporal distribution and associations with large-scale atmospheric circulation, Clim. Dynam., 33, 187–197, https://doi.org/10.1007/s00382-008-0431-5, 2009.
25. LeMone, M. A., Angevine, W. M., Bretherton, C. S., Chen, F., Dudhia, J., Fedorovich, E., Katsaros, K. B., Lenschow, D. H., Mahrt, L., Patton, E. G., Sun, J., Tjernström, M., and Weil, J.: 100 Years of Progress in Boundary Layer Meteorology, Meteorological Monographs, 59, 9.1–9.85, https://doi.org/10.1175/amsmonographs-d-18-0013.1, 2019.
26. Liu, Y., Key, J. R., Vavrus, S., and Woods, C.: Time Evolution of the Cloud Response to Moisture Intrusions into the Arctic during Winter, J. Climate, 31, 9389–9405, https://doi.org/10.1175/jcli-d-17-0896.1, 2018.
27. Mattingly, K. S., Ramseyer, C. A., Rosen, J. J., Mote, T. L., and Muthyala, R.: Increasing water vapor transport to the Greenland Ice Sheet revealed using self-organizing maps, Geophys. Res. Lett., 43, 9250–9258, https://doi.org/10.1002/2016GL070424, 2016.
28. Messori, G., Woods, C., and Caballero, R.: On the Drivers of Wintertime Temperature Extremes in the High Arctic, J. Climate, 31, 1597–1618, https://doi.org/10.1175/jcli-d-17-0386.1, 2018.
29. Morrison, H., de Boer, G., Feingold, G., Harrington, J., Shupe, M. D., and Sulia, K.: Resilience of persistent Arctic mixed-phase clouds, Nat. Geosci., 5, 11–17, 2012.

30. Naakka, T., Nygård, T., and Vihma, T.: Arctic Humidity Inversions: Climatology and Processes, J. Climate, 31, 3765–3787, https://doi.org/10.1175/jcli-d-17-0497.1, 2018.
31. Naakka, T., Nygård, T., Tjernström, M., Vihma, T., Pirazzini, R., and Brooks, I. M.: The Impact of Radiosounding Observations on Numerical Weather Prediction Analyses in the Arctic, Geophys. Res. Lett., 46, 8527–8535, https://doi.org/10.1029/2019gl083332, 2019.
32. Nygård, T., Valkonen, T., and Vihma, T.: Characteristics of Arctic low-tropospheric humidity inversions based on radio soundings, Atmos. Chem. Phys., 14, 1959–1971, https://doi.org/10.5194/acp-14-1959-2014, 2014.
33. Nygård, T., Graversen, R. G., Uotila, P., Naakka, T., and Vihma, T.: Strong dependence of wintertime Arctic moisture and cloud distributions on atmospheric large-scale circulation J. Climate, 32, 8771–8790, https://doi.org/10.1175/JCLI-D-19-0242.1, 2019.
34. Papritz, L.: Arctic Lower-Tropospheric Warm and Cold Extremes: Horizontal and Vertical Transport, Diabatic Processes, and Linkage to Synoptic Circulation Features, J. Climate, 33, 993–1016, https://doi.org/10.1175/jcli-d-19-0638.1, 2020.
35. Park, H.-S., Lee, S., Son, S.-W., Feldstein, S. B., and Kosaka, Y.: The Impact of Poleward Moisture and Sensible Heat Flux on Arctic Winter Sea Ice Variability, J. Climate, 28, 5030–5040, https://doi.org/10.1175/jcli-d-15-0074.1, 2015.
36. Pithan, F., Medeiros, B., and Mauritsen, T.: Mixed-phase clouds cause climate model biases in Arctic wintertime temperature inversions, Clim. Dynam., 43, 289–303, https://doi.org/10.1007/s00382-013-1964-9, 2014.
37. Pithan, F., Svensson, G., Caballero, R., Chechin, D., Cronin, T. W., Ekman, A. M. L., Neggers, R., Shupe, M. D., Solomon, A., Tjernström, M., and Wendisch, M.: Role of air-mass transformations in exchange between the Arctic and mid-latitudes, Nat. Geosci., 11, 805–812, https://doi.org/10.1038/s41561-018-0234-1, 2018.
38. Renfrew, I. A., Barrell, C., Elvidge, A. D., Brooke, J. K., Duscha, C., King, J. C., Kristiansen, J., Cope, T. L., Moore, G. W. K., Pickart, R. S., Reuder, J., Sandu, I., Sergeev, D., Terpstra, A., Våge, K., and Weiss, A.: An evaluation of surface meteorology and fluxes over the Iceland and Greenland Seas in ERA5 reanalysis: The impact of sea ice distribution, Q. J. Roy. Meteor. Soc., 147, 691–712, https://doi.org/10.1002/qj.3941, 2021.

39. Sedlar, J. and Tjernström, M.: A Process-Based Climatological Evaluation of AIRS Level 3 Tropospheric Thermodynamics over the High-Latitude Arctic, J. Appl. Meteor. Climatol., 58, 1867–1886, https://doi.org/10.1175/jamc-d-18-0306.1, 2019.
40. Serreze, M. C., Kahl, J. D. W., and Schnell, R. C.: Low-level temperature inversions of the Eurasian Arctic and comparisons with Soviet drifting stations, J. Climate, 8, 719–731, 1992.
41. Shahi, S., Abermann, J., Heinrich, G., Prinz, R., and Schöner, W.: Regional Variability and Trends of Temperature Inversions in Greenland, J. Climate, 33, 9391–9407, https://doi.org/10.1175/jcli-d-19-0962.1, 2020.
42. Skific, N., Francis, J. A., and Cassano, J. J.: Attribution of Projected Changes in Atmospheric Moisture Transport in the Arctic: A Self-Organizing Map Perspective, J. Climate, 22, 4135–4153, https://doi.org/10.1175/2009JCLI2645.1, 2009.
43. Stramler, K., Genio, A. D. D., and Rossow, W. B.: Synoptically Driven Arctic Winter States, J. Climate, 24, 1747–1762, https://doi.org/10.1175/2010jcli3817.1, 2011.
44. Sulikowska, A., Walawender, J. P., and Walawender, E.: Temperature extremes in Alaska: temporal variability and circulation background, Theor. Appl. Climatol., 136, 955–970, https://doi.org/10.1007/s00704-018-2528-z, 2019.
45. Tjernström, M. and Graversen, R. G.: The vertical structure of the lower Arctic troposphere analysed from observations and the ERA-40 reanalysis, Q. J. Roy. Meteor. Soc., 135, 431–443, https://doi.org/10.1002/qj.380, 2009.
46. Vihma, T., Pirazzini, R., Fer, I., Renfrew, I. A., Sedlar, J., Tjernström, M., Lüpkes, C., Nygård, T., Notz, D., Weiss, J., Marsan, D., Cheng, B., Birnbaum, G., Gerland, S., Chechin, D., and Gascard, J. C.: Advances in understanding and parameterization of small-scale physical processes in the marine Arctic climate system: a review, Atmos. Chem. Phys., 14, 9403–9450, https://doi.org/10.5194/acp-14-9403-2014, 2014.
47. Wang, C., Graham, R. M., Wang, K., Gerland, S., and Granskog, M. A.: Comparison of ERA5 and ERA-Interim near-surface air temperature, snowfall and precipitation over Arctic sea ice: effects on sea ice thermodynamics and evolution, The Cryosphere, 13, 1661–1679, https://doi.org/10.5194/tc-13-1661-2019, 2019.

48. Woods, C., Caballero, R., and Svensson, G.: Large-scale circulation associated with moisture intrusions into the Arctic during winter, Geophys. Res. Lett., 40, 4717–4721, https://doi.org/10.1002/grl.50912, 2013.
49. Yu, L., Yang, Q., Zhou, M., Zeng, X., Lenschow, D. H., Wang, X., and Han, B.: The Intraseasonal and Interannual Variability of Arctic Temperature and Specific Humidity Inversions, Atmosphere, 10, 214, https://doi.org/10.3390/atmos10040214, 2019.
50. Zhang, Y., Seidel, D. J., Golaz, J. C., Deser, C., and Tomas, R. A.: Climatological characteristics of Arctic and Antarctic Surface-Based Inversions, J. Climate, 24, 5167–5186, https://doi.org/10.1175/2011JCLI4004.1, 2011.

CHAPTER 13

Vertical Structure of the Lower Troposphere Derived from MU Radar, Unmanned Aerial Vehicle, and Balloon Measurements During ShUREX 2015

Hubert Luce[1] , Lakshmi Kantha[2] , Hiroyuki Hashiguchi[3], Dale Lawrence[2] , Tyler Mixa[2] , Masanori Yabuki[3] and Toshitaka Tsuda[3]

[1]Meditterranean Insitute of Oceanography, CNRS/INSU, UMR7294, IRD, Université de Toulon, La Garde, France.

[2] Research Institute for Sustainable Humanosphere, Kyoto University, Kyoto, Japan.

[3] Department of Aerospace Engineering Sciences, University of Colorado Boulder, Boulder, CO, USA.

Citation: Luce, H., Kantha, L., Hashiguchi, H. et al. "Vertical structure of the lower troposphere derived from MU radar, unmanned aerial vehicle, and balloon measurements during ShUREX 2015". Prog Earth Planet Sci 5, 29 (2018). https://doi.org/10.1186/s40645-018-0187-4.

ABSTRACT

The ShUREX (Shigaraki UAV Radar Experiment) 2015 campaign carried out at the Shigaraki Middle and Upper atmosphere (MU) observatory (Japan) in June 2015 provided a unique opportunity to compare vertical profiles of atmospheric parameters estimated from unmanned aerial vehicle (UAV), balloon, and radar data in the lower troposphere. The present work is intended primarily as a demonstration of the potential offered by combination of these three instruments for studying the small-scale structure and dynamics in the lower troposphere. Here, we focus on data collected almost simultaneously by two instrumented UAVs and two meteorological balloons, near the MU radar operated continuously during the campaign. The UAVs flew along helical ascending and descending paths at a nearly constant horizontal distance from the radar (~1.0 km), while the balloons launched from the MU radar site drifted up to ~3–5 km in the altitude range of comparisons (~0.5 to 4.0 km) due to wind advection. Vertical profiles of squared Brünt-Väisälä frequency N^2 and squared vertical gradient of generalized potential refractive index M^2 were estimated at a vertical resolution of 20 m from pressure, temperature, and humidity data collected by UAVs and radiosondes. Profiles of M^2 were also estimated from MU radar echo power at vertical incidence at a vertical sampling of 20 m and various time resolutions (1–4 min). The balloons and the MU radar provided vertical profiles of wind and wind shear S so that two independent estimates of the gradient Richardson number ($Ri=N^2/S^2$) could be obtained at a range resolution of 150 m. The two estimates of Ri profiles also showed remarkable agreement at all altitudes. We show that all three instruments detected the same prominent temperature and humidity gradients, down to decameter scales in stratified conditions. These gradients extended horizontally over a few kilometers at least and persisted for hours without significant changes, indicating that the turbulent diffusion was weak. Large discrepancies between N^2 and M^2 profiles derived from the balloon, UAV, and radar data were found in a turbulent layer generated by a Kelvin-Helmholtz (KH) shear flow instability in the height range from 1.80 to 2.15 km. The cause of these discrepancies appears to depend on the stage of the KH billows.

Keywords: MU radar, Unmanned aerial vehicle (UAV), Lower troposphere and boundary layer, Refractive index gradient, Kelvin-Helmholtz instability, Temperature and moisture fine structures, Sheets and layers, Gradient Richardson number

INTRODUCTION

Nearly co-located and simultaneous measurements from instrumented Unmanned Aerial Vehicles (UAV) (Lawrence and Balsley 2013), balloon-borne radiosondes, and the 46.5 MHz middle and upper atmosphere (MU) radar were made for the first time at the Shigaraki MU observatory during the ShUREX (Shigaraki UAV Radar Experiment) campaign in June 2015. Kantha et al. (2017) describe the experimental set up and present preliminary results for illustrating the potential of the collected datasets. Luce et al. (2018) took advantage of the UAV echoes in the MU radar Doppler spectra for an ultimate validation and evaluation of a frequency domain interferometry technique used to improve the range resolution of the radar. Luce et al. (2017) compared profiles of the squared generalized potential refractive index gradient M^2 (Ottersten 1969) derived from pressure, temperature, and humidity (PTU) data collected by the UAVs with those derived from MU radar echo power by using a model based on partial reflection mechanism. They concluded that the MU radar provides faithful high-resolution profiles of the vertical temperature and humidity gradients down to decameter scale, at a time resolution of only a few tens of seconds, in at least stratified and clear air conditions. Their study confirmed the qualitative results obtained by Luce et al. (2010) from comparisons with simultaneous and co-located lidar measurements of humidity profiles.

In the present work, additional comparisons of atmospheric parameters are presented by including balloon data not presented in Luce et al. (2017). The balloons were launched at the MU radar observatory site almost simultaneously to the flights of two UAVs, at a time interval of about 2 h on 07 June 2015. The concurrent use of all three instruments enabled us to make comparisons between independent estimates of various atmospheric parameters: squared Brünt-Väisälä frequency N^2, squared generalized potential refractive index gradient M^2, horizontal wind (V), and wind shear ($S = \sqrt{(du/dz)^2 + (dv/dz)^2}$), where u and v are the zonal and meridional wind components, respectively, and the Richardson number ($Ri = N^2/S^2$). The definitions of N^2 and M^2 are given in Appendix 'Definition and estimation of N^2 and M^2' section. For clarity and simplicity, Table 1 assembles all possible combinations described in the present study. In addition to these combinations, comparisons between profiles estimated from UAV data during ascents and descents are also presented. In principle, it is possible to estimate horizontal winds from UAV data, but this possibility has yet not been realized.

Table 1: List of atmospheric parameters measured or calculated from UAV, balloon, and radar data (resolution of the profiles are given in meters)

	UAV	Balloon	MU radar
PTU	(~2)	(~2–3)	
N^2	20, 150 [a]	20, 150	
M^2	20	20	20
V		20, 150	150
S		20, 150	150
$Ri - N^2/S^2$		20, 150	150 [a]

[a]The Richardson number was obtained from the wind shear measured by the MU radar and N^2 estimated from UAV data at a vertical resolution of 150 m

It will be shown that UAV and balloon measurements provide virtually the same profiles of the atmospheric parameters in stably stratified conditions above the altitude of 2.15 km. These results are important for two reasons. On the one hand, they confirm that UAV measurements made from helical paths are as reliable as balloon measurements for profiling these atmospheric parameters down to decameter scale, at least, for the atmospheric conditions met during the observations. On the other hand, they suggest that this kind of comparisons can also provide additional insights into the characteristics of the thin layers of temperature and humidity gradients in the lower atmosphere. The existence of these gradients is well known, especially in the stable boundary layer and in the lower stratosphere. They can be even much thinner than those described here (e.g., Misme et al. 1958; Saxton et al. 1964; Du Castel 1966; Gossard et al. 1985; Dalaudier et al. 1994; Luce et al. 1995; Muschinski and Wode 1998). However, little is known about their characteristics, time evolution, and formation mechanisms because observations are still sparse and challenging (e.g., Fritts and Rastogi 1985; Luce et al. 2001a). A better knowledge of these characteristics is desirable because they can potentially affect all wave propagation (light, sound, radio waves) and can have an impact on the vertical transport of properties in stratified conditions. Owing to their flexibility of use, UAVs, in association with additional observations from radars, can contribute to this goal.

It will also be shown that the largest discrepancies between the profiles occurred below the height of 2.15 km, where turbulence associated with convective clouds and Kelvin-Helmholtz shear instabilities was detected by the radar. The discrepancies in the KH turbulent layer could be partly due to the horizontal scale of KH billows, found to be of the order of, or even smaller than, the horizontal distance between the measurements made by the three instruments. The present work describes in detail, the characteristics of the various profiles in such conditions and complements the discussion presented by Luce et al. (2017) based on M^2 comparisons only.

The layout of the paper is as follows. In the section "Experimental/ Methods," instruments and datasets are first briefly described [see also Kantha et al. 2017] and a short outline of the methods used for deriving the atmospheric parameters from the different datasets is given among with references for more details. In the section "Results," time-height cross-sections of parameters measured by the MU radar are first analyzed in order to document the conditions met during the UAV/balloon flights. Then, the results of comparisons between profiles of atmospheric parameters are described. Discussions and concluding remarks are given in the section "Conclusions."

EXPERIMENTAL/METHODS

Instruments and Data

MU Radar

The MU radar is a 46.5 MHz beam-steering, Doppler pulsed radar located at Shigaraki MU Observatory (34.85°N, 136.10°E) in Japan (Fukao et al. 1990). The radar parameters used during the campaign are listed in Table 2. The range resolution defined by the transmitted pulse length was 150 m, but a frequency domain interferometry technique, called range imaging, was applied in order to get an effective range resolution of the echo power profiles of a few tens of meters at vertical incidence (e.g., Luce et al. 2001b). The radar parameters were set up so that one high-resolution profile of echo power at vertical incidence was acquired from the altitude of 1.270 up to 20.465 km above the mean sea level (ASL) every ~4 s at an effective time resolution of ~8 s (see Table 2). In addition, five oblique beams steered at 10° off zenith were included for estimating additional parameters at the same time resolution and at a range resolution of 150 m. The radar measurements were interrupted every ~35 min for a few minutes for technical reasons.

Table 2: MU radar parameters used during ShUREX 2015 campaign

Parameter	
Beam directions	(0°,0°),(0°,0°),(0°,0°),(0°,0°),(0°,0°),(0°,10°),(45°,10°),(90°,10°),(135°,10°),(180°,10°)
Radar frequencies (MHz)	46.00,46.25,46.50,46.75,47.00
Interpulse period (**μs**)	400
Subpulse duration (**μs**)	1
Pulse coding	1 16-bit optimal code [b]
Range resolution (m)	150
Height sampling (FII) (m)	5
Number of gates	128
Coherent integration number	32
Incoherent integration number	- (time series)
Number of FFT points	128
Acquisition time (s)	16.384[a]
Time sampling (s)	4.096
Nyquist frequency (Hz)	3.9063
Velocity aliasing (ms^{-1})	12.6

[a]Acquisition time of a time series of 128 points for each direction but the effective time resolution is about half (~8 s) due to Hanning windowing of the time series. *FII*, frequency domain radar interferometric imaging. *FFT*, fast Fourier transform

[b]Spano and Ghebrebhran (1996)

UAV

Small (less than 1 m wingspan and about 700 g mass) UAVs equipped with autopilots were pre-programmed to execute a preplanned trajectory near the MU radar. They use GPS for navigation so that they could be operated precisely at given distances from the radar. Lawrence and Balsley (2013) and Balsley et al. (2013) describe the design of the UAVs, the characteristics of ground support components, and some data collected from these systems.

Kantha et al. (2017) describe the specific characteristics of the UAVs during the ShUREX 2015 campaign. The UAVs were equipped with various sensors for measurements of various atmospheric parameters.

In particular, the sensors on the IMET radiosonde mounted on the UAV provided measurements of pressure, temperature, and relative humidity (PTU) at a sampling rate of 1 Hz. The trajectories of the two flights described in the present work (hereafter, called UAV4 and UAV5 because they corresponded to the fourth and fifth flights of the campaign, respectively) were helical during ascent and descent over a spot located at a horizontal distance of about 1.0 km southwest of MU radar.

The ascent and descent of UAV4 (UAV5) will be hereafter denoted UAV4A (UAV5A) and UAV4D (UAV5D), respectively. The diameter of helical patterns was ~100 to 150 m, and the ascent (descent) rate of the UAVs was about 2 ms^{-1}.

Balloons

Two meteorological balloons (hereafter called B1 and B2) equipped with Vaisala RS92G radiosondes were launched from the MU radar site. They mainly provided PTU, zonal, and meridional winds at a sampling rate of 1 Hz. The balloons were underinflated so that they ascended virtually at the same velocity as the UAVs (i.e., ~2 ms^{-1}).

They climbed to about the altitude of 4.0 km within an hour. B1 was launched almost at the same time as UAV4 and B2 was launched about 1 h after the take-off of UAV5. Table 3 shows the main characteristics of the UAV and balloon flights, and Fig. 1a, b shows their heights vs time (red curves, UAV; blue curves, balloons).

Table 3: Brief description of the selected UAV flights (07 June 2015) and Vaisala radiosondes

UAV flight	Time (LT)	Maximum height (km)
UAV4	05: 44: 00	3.9
UAV5	07: 14: 11	4.3
Balloon flight	Time (LT)	Maximum horizontal drift (km)
B1	05: 53: 43	~3.3
B2	08: 12: 48	~5.2

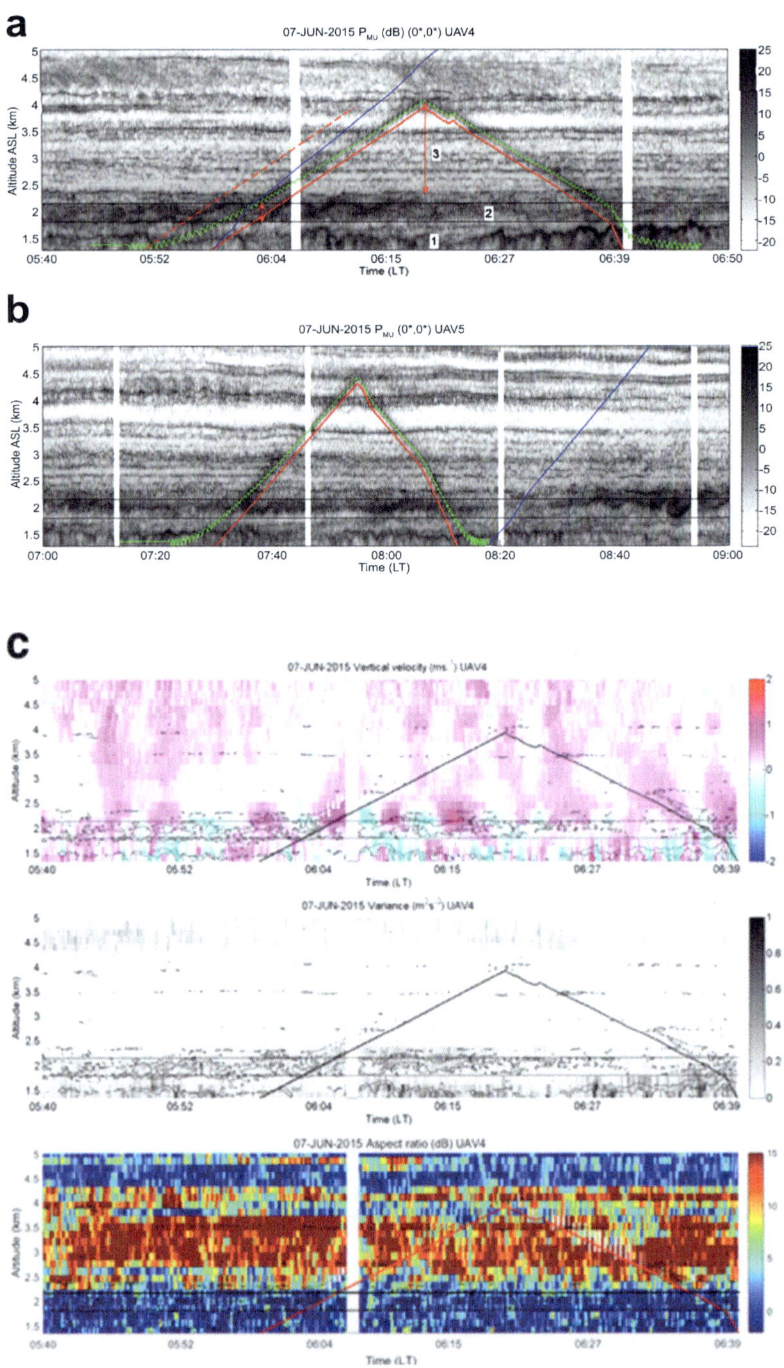
a
07-JUN-2015 P_MU (dB) (0°,0°) UAV4
Altitude ASL (km)
Time (LT)
b
07-JUN-2015 P_MU (0°,0°) UAV5
Altitude ASL (km)
Time (LT)
c
07-JUN-2015 Vertical velocity (ms⁻¹) UAV4
Altitude (km)
Time (LT)
07-JUN-2015 Variance (m²s⁻²) UAV4
Altitude (km)
Time (LT)
07-JUN-2015 Aspect ratio (dB) UAV4
Altitude (km)
Time (LT)

d

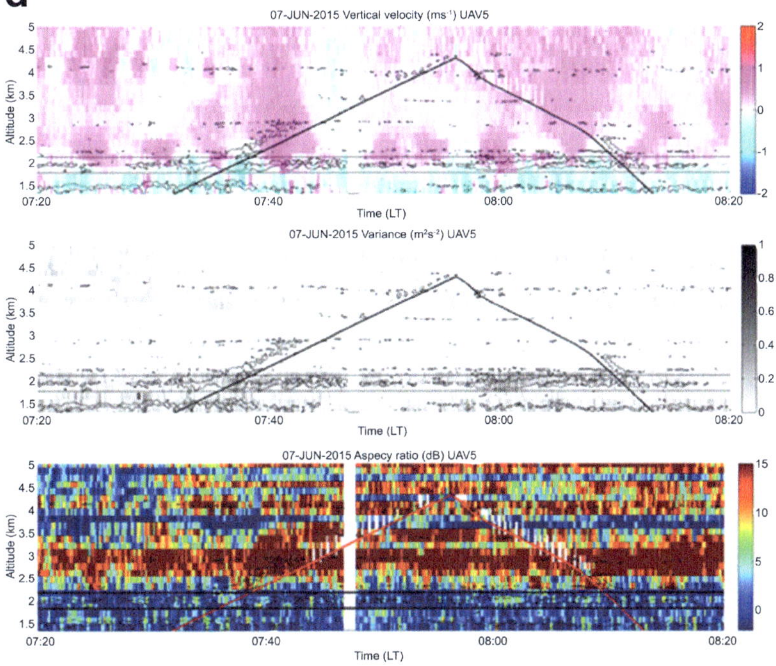

Figure 1: **a** (Top) Time-height plot of MU radar echo power at vertical incidence and high-range resolution from 05:40 to 06:50 LT on 7 June 2015 corresponding to UAV4 and B1. The red (green) line shows the altitude (distance) of UAV4 vs. time. The blue line shows the altitude of B1 vs. time. The label 1 refers to cloud region, 2 to the KH region, and 3 to the stratified region, respectively. These 3 regions are separated by horizontal black lines at 1.80 and 2.15 km, for clarity. See the text for more details. **b** As in Fig. 1,a but for UAV5 and B2. **c** (Top) Time-height plot of the vertical velocity measured by MU radar from 05:40 to 06:40 LT on 7 June 2015 during the flight of UAV4. (Center) The corresponding time-height plot of Doppler variance corrected for beam broadening measured at vertical incidence. (Bottom) Aspect ratio defined as the difference (in decibel) between the vertical echo power and the average of the five oblique echo powers. Superimposed are isolines of echo power (Fig. 1a) for P_{MU} = 10 dB showing the edges of the KH braids and the top of the cloudy cells. The horizontal solid lines delineate the KH region (2). The black or red curves show the altitude of UAV4 vs time, as in Fig. 1.a. **d** As Fig. 1c, but for the radar observations during UAV5

Data Processing Methods

MU Radar Data

For each sampled altitude from 1.345 km above the sea level (ASL), echo power, Doppler velocity, spectral width, and signal to noise ratio (SNR) were estimated from Doppler spectra at vertical and oblique (10° off zenith) incidences at a range resolution of 150 m using the moment method (e.g., Yamamoto et al. 1988). The vertical air velocity W (ms^{-1}) was directly estimated from the Doppler shift at vertical incidence. By convention, a positive (negative) value of W indicates upward (downward) motions. The zonal and meridional wind components u and v were obtained by combining the radial velocities in the different beam-pointing directions. The vertical shear of horizontal wind S and the shear of horizontal wind speed (or "speed shear") $S_V = |dV/dz|$ were calculated from time-averaged (about 1 h) profiles of u, v, and V, using a 3-point central scheme for the vertical derivatives.

The spectral width or variance of the Doppler spectra is partly due to clear air turbulence. The extraction of the variance σ^2 expected to be due to turbulence only was performed using the method described by Wilson et al. (2014). The details of the data processing are not given here, since σ^2 is simply used as an index of turbulence for the present purpose.

Aspect ratio AR of echo power is defined as the ratio between echo powers at vertical and oblique incidences, usually expressed in decibel. AR is a statistical parameter giving information about the anisotropy of the scatterers at the scale to which the radar is sensitive (i.e., half the radar wavelength) (e.g., Gage 1990). When close to 0 dB, backscattering from isotropic turbulence should be dominant. When AR is large (say >10 dB), echoes are believed to arise from stable gradient layers that are horizontally quasi-coherent. Here, AR was calculated as the difference (in decibel) between the vertical echo power and the average power measured by the five oblique beams.

Because the estimations of Doppler velocity and, to a greater extent, width of Doppler spectra are very sensitive to contaminations (e.g., ground clutter, airplane echoes, electromagnetic interferences), especially with the moment method, all the Doppler spectra obtained from the six beam directions were edited and carefully corrected manually by replacing contaminated portions of the spectra by an average noise level. The procedure was tedious but necessary, especially for removing the UAV echo contaminations. It was only possible when the UAV echo peak did not overlap the Doppler peak of atmospheric echoes. Overlapping could be avoided most of the time, owing the configuration of flight used but not always. In such cases, the data were discarded.

Vertical profiles of echo power P_{MU} at a range resolution of a few tens of meters were also obtained at vertical incidence owing to the application of a multiple frequency domain interferometry technique, called range imaging, with the adaptive filter-bank Capon processing method. Luce et al. (2001b) described the principle of the technique that became operational on the MU radar in 2004 after upgrades (Luce et al. 2006). Profiles of M^2 were estimated from P_{MU} at vertical incidence at a vertical sampling of 20 m using the theoretical derivations and practical methods already described by Luce et al. (2017). Basically, the M^2 profiles were estimated from the average of P_{MU} over 1 and 4 min, along the time-height position of the UAVs (shown by the red lines in Fig. 1a, b).

Balloon and UAV Data

PTU data gathered by the UAVs and balloons were used for estimating vertical profiles of N^2 and M^2 (Kantha et al. 2017; Luce et al. 2017). The PTU profiles were first resampled at a constant altitude step and smoothed at a vertical resolution of 20 m after applying a low-pass filter (cut-off=40 m). The same procedure was applied on the wind profiles collected by the radiosondes for estimating S, S_V, and *Ri* from balloon data at vertical resolution of 20 m. When comparing with the radar-derived profiles of S and S_V at range resolution of 150 m, the zonal and meridional wind components measured by radiosondes were first low-pass filtered (cut-off =300 m) and resampled before calculating S and S_V at the same resolution of 150 m.

RESULTS AND DISCUSSIONS

Radar Images

Radar Echo Power At High Resolution

Figure 1a, b shows time-height cross-sections of P_{MU} (decibel, arbitrary levels) at vertical incidence and in the height range 1.27–5.00 km during the flight period of UAV4/B1 and UAV5/B2, respectively. As shown by Luce et al. (2018), the radar continuously detected the echoes backscattered from the UAVs. The green curves showing distances of the UAVs to the radar antenna array estimated from the onboard GPS measurements perfectly overlap the UAV echo peaks, indicating a suitable calibration in altitude of the radar. The twisted appearance of the curves results from the helical motions of the UAVs. B1 was almost synchronized with UAV4A. Both instruments probed the same altitudes with a time lag of a few minutes only but with a horizontal separation of a few kilometers due to the balloon drift. Figure 2 shows the horizontal trajectories of B1 and B2 with respect to MU radar site, the UAV flight paths (nearly identical for UAV4 and UAV5), and additional information described later on.

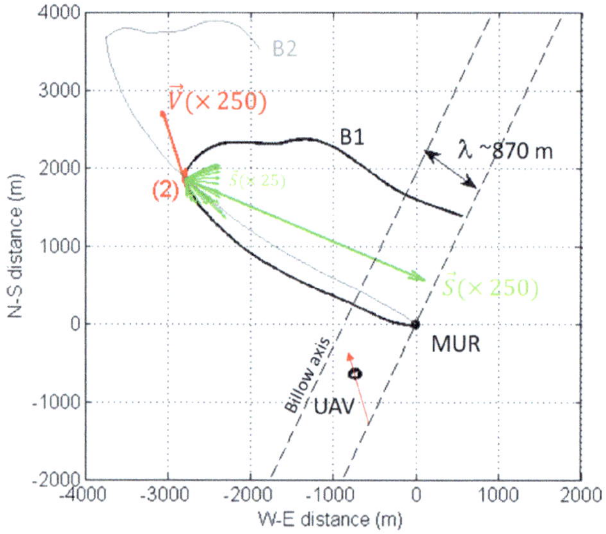

Figure 2: Horizontal trajectories of B1 (black line), B2 (gray line), and the UAVs (black circles at the bottom of the figure). The red (green) arrow shows the mean wind (shear) vector in the KH region (2) according to B1 data and scaled by 250. The variations of the wind shear scaled by 25 in the KH region (2) are given by the short green arrows. The dashed lines show the expected billow crest axes assumed to be perpendicular to the wind shear direction.

Labels and horizontal black lines in Fig. 1a refer to altitude ranges (hereafter, called "regions") in which radar echoes exhibit distinct patterns. These regions were associated with different atmospheric states and dynamics according to UAV and balloon data analyses shown by Figs. 3, 4, and 5. These figures are now utilized as complement to the radar observations for their interpretations. The results of comparisons between the profiles will be described in the next section.

- Label (1): A region in which an enhanced and irregular echo layer was embedded, corresponding to the top edges of convective cloud cells. The relative humidity data collected from UAV and balloon measurements confirm the presence of saturated air in that region (Fig. 3). The bumpy structures recall those observed by FMCW radars in convective boundary layers (e.g., Gossard 1990) and might be a signature of cumulus clouds. Entrainment of subsaturated air by the convective motions likely produced the enhanced echoes at the troughs of the cells. The altitude of the top edge was very irregular with time, from ~1.3 km around 06:20 LT up to ~1.8 km at 06:45 LT. Similar but less prominent structures were observed during UAV5 (Fig. 1b).
- Label (2): An intense echo layer showing evidence of braided cat's-eye pattern believed to result from Kelvin-Helmholtz (KH) instability. The enhanced wind shear and the minimum of *Ri* in this region (Figs. 4 and 5, respectively) are consistent with this interpretation. The KH instability topped the humid region (1) and occurred at the bottom of the temperature inversion that was probably affected itself by the KH instability (Fig. 3). During UAV4A, the crest-to-trough amplitude of the KH braids, indicated by the horizontal black lines, did not exceed 350 m between 1.80 and 2.15 km. The KH braids were capped by a persistent and thin echo layer at the altitude of ~2.45 km and by more or less well-defined enhanced echo layers between them (Fig. 1a, b). The KH braids faded away after 06:15 LT and were replaced by relatively indistinct echo power patterns. During UAV4D, KH breaking wave of smaller dimensions may have occurred just above 1.8 km between 06:35 and 06:40 LT. The enhanced echo layers around 2.0 km after B2 flight (~08:30 LT) were also likely due to the KH instabilities but the billows appeared clearly only after 08:50 LT. B2 ascent was when radar echoes were significant weaker.

The depth of the region (2) was defined according to the morphology of the large amplitude KH braids before and during UAV4A. They do not necessarily fit their vertical extent at a later time, but the same ranges were delineated in Fig. 1b for an easier description.

- Label (3): A region of persistent multi-layer fine structure characteristic of stratified conditions. Figure 3 shows that this region corresponded to much drier conditions compared to regions (1) and (2). It has been limited at the top by the maximum altitude of the UAVs. The thin echo layers were quite regularly spaced vertically without evidence of merging. They were quite stable in height during UAV4, but they showed vertical wave-like oscillations during UAV5.

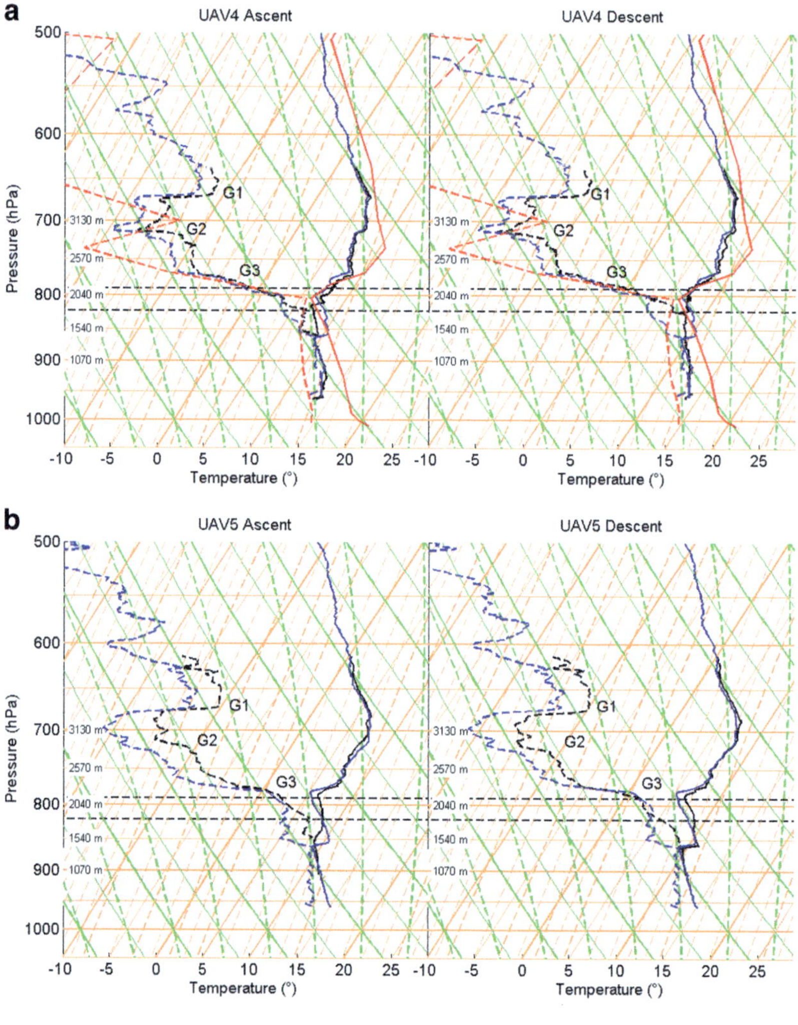

Figure 3: **a** Thermodynamic diagrams showing temperature (solid lines) and dew-point temperature (dashed lines) profiles for UAV4 (black), B1 (blue), and for a balloon (red) launched on 7 June 2015 09:00 LT from Shionomisaki station, about 160 km south of the MU radar site. The left (right) panel shows UAV4 ascent (descent) profile. The B1 profiles are the same for both panels. The solid (dashed) green isolines show potential temperature and moist satu-

rated potential temperature, respectively. G1, G2, and G3 refer to moisture gradients discussed in the text. The two horizontal dashed lines delineate the KH region (2) between the altitudes of 1.80 and 2.15 km. **b** As Fig. 3a, but for B2 and UAV5

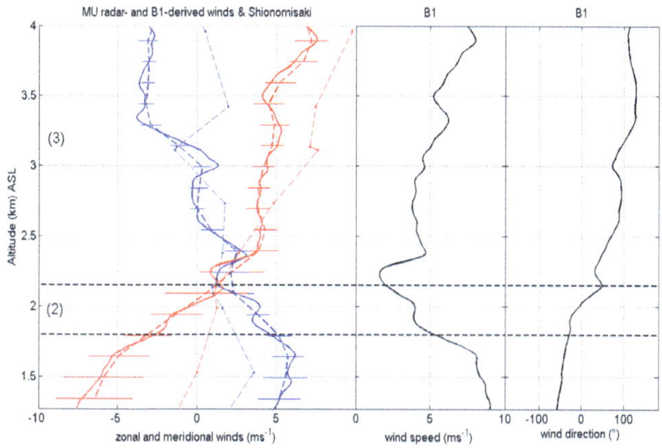

Figure 4: (Left) Vertical profiles of zonal (red) and meridional (blue) winds derived from B1 (solid), MU radar (thick dashed), and a balloon launched at Shionomisaki station (light dashed). The MU radar-derived profiles have been averaged over 1 h (05:40–06:40 LT). The horizontal bars show the standard deviations of the radar estimates during B1 flight time. (Center) Wind speed profile from B1. (Right) Wind direction profile from B1. The horizontal dashed lines delineate the KH region (2)

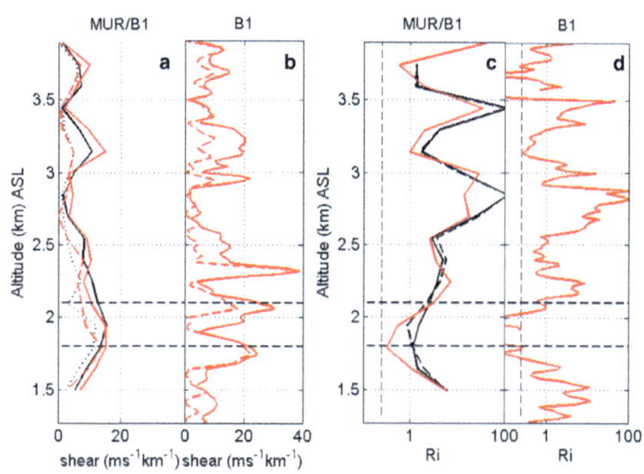

Figure 5: **a** Vertical profiles of vertical shears of horizontal wind (solid, wind shear; dashed, speed shear) at a vertical resolution of 150 m from MU radar data (05:40–06:40 LT) (black) and B1 data (red). **b** Same as (a) at a vertical resolution of 20 m from B1 data. **c** *Ri* profiles estimated at a vertical resolution of 150 m

from B1 data (red) and from MU radar data and N^2 estimated from UAV4A data (black solid) and UAV4D data (black dashed). **d** *Ri* profile estimated from B1 data at the vertical resolution of 20 m. The horizontal thick dashed lines delineate the KH region (2). The vertical thin dashed lines show the critical value of *Ri* (=0.25)

Additional Radar Observations

Figure 1c, d shows time-height cross-sections of W (ms^{-1}) (top panel), σ^2 (m^2s^{-2}) (middle panel), and AR (dB) (bottom panel) for the two periods shown in Fig. 1a, b, respectively.

In the cloudy region (1), the radar images show the following:

1. W fluctuations (up to +/−0.5 ms^{-1}) with downward motions in the trough of the convective cells in Fig. 1c. Downward motions ($W < 0$) dominate in Fig. 1d.
2. Sporadic enhancements of σ^2 (up to 1.0 m^2s^{-2} after 06:27 LT).
3. AR close to 0 dB.

In the KH region (2), the radar images show the following:

1. Nearly periodic W disturbances up to +/− 1.0 ms^{-1} around 06:00 LT, with a possible phase shift of these disturbances around 2.0 km from 06:05 to 06:20 LT, suggesting a critical level in accordance with the morphology of the KH braids. Such W disturbances are usually met on both sides of KH layers (e.g., Luce et al. 2012), and they can be seen in Fig. 1c up to 2.5 km at least.
2. A layer of enhanced σ^2 of up to ~0.5 m^2s^{-2} (more persistent than in region (1)) with some local maxima coinciding with echo power maxima.
3. AR close to 0 dB.

These features are additional clues suggesting that the MU radar detected convective and dynamical turbulence in the regions (1) and (2), respectively.

In the stratified region (3), AR was high and ^{2} was small, with longer period vertical wind oscillations during UAV5, consistent with a calm, weakly turbulent region.

In summary, these characteristics confirm the intuitive interpretation of the radar echo power images of Fig. 1a, b, i.e., turbulence in convective (saturated) cells in region (1), turbulence associated with KH instability in region (2) and stable stratification in region (3). Incidentally, the faint blurred echoes in the range 4.2–4.8 km above region (3) in Fig. 1a and before 07:40 LT in Fig. 1b were also related to turbulence (likely another KH instability) as confirmed by enhanced σ^2 and low AR in Fig. 1c, d in the same range.

Characteristics of the Possible KH Billows

In Fig. 2, the wind shear vectors (green vectors) derived from B1 for the KH region (2) were superimposed on the balloon trajectory. The red and green vectors show the averaged wind and shear vectors, $\vec{V}$ and $\vec{S}$, respectively. The values were scaled for easy reading. Due to the wind veering, $\vec{V}$ and $\vec{S}$ were not collinear (their baselines formed an angle of ~49°). The dashed lines show the possible (idealized) billow axis, expected to be perpendicular to $\vec{S}$. For the mean shear direction shown in Fig. 2, taking the wind speed V equal to 4 ms^{-1} and T, the time difference between two consecutive KH braids observed by the MU radar equal to 330 s (Fig. 1a), the horizontal wavelength λ was estimated to be $\sim 1320 \times \cos(49^{\circ}) \approx 870$ m, where 1320 m was the apparent wavelength ($= V \times T$). Since the amplitude A, defined as the maximum depth of the KH braids, was 350 m, we have $A/\lambda = 0.4$. This ratio is consistent with $0.1 < A/\lambda < 0.5$ reported in the literature (e.g., Fukao et al. 2011).

Results of Comparisons between Profiles of Atmospheric Parameters

Vertical Profiles of Temperature and Dew Point Temperature from UAV and Balloon Data

The left and right panels of Fig. 3a, b shows the temperature profiles (solid black lines) and dew point temperature profiles (dashed black lines) obtained from UAV4A and D, and UAV5A and D, respectively, using a skew-T log-p emagram. Superimposed are the corresponding profiles obtained from B1 and B2 (blue lines).

On the one hand, the height variations of the two temperature profiles measured by the UAVs and balloons were very similar down to the smallest visible scales in the stably stratified region (3). The nearly constant difference (bias) between the UAV- and balloon-derived dew-point temperatures should be ignored, because it was very likely due to a difference in calibration and sensitivity of the IMET and Vaisala humidity sensors when humidity was weak. Therefore, the two instruments virtually measured the same temperature profiles in region (3) despite the distance of a few km separating these instruments (see Fig. 2). In particular, there were noticeable humidity (and temperature) gradients, hereafter-noted G1, G2, and G3 crossed by B1, B2, UAV4, and UAV5 in region (3). "G1" and "G2" refer to the positive and negative humidity gradients at ~3.5 and ~3.0 km, respectively. "G3" refers to the top of the humidity gradient corresponding to the capping inversion around 2.4 km. A quantitative description will be made from comparisons of N^2 and M^2 in the section "Vertical profiles of N^2 and M^2 derived from UAV and balloon data."

On the other hand, there were significant differences between the UAV- and balloon-derived temperature profiles near the top of region (1) and in the KH region (2). There were nearly dry-adiabatic gradients surrounded by more or less steep temperature inversions around the KH region (2) in the four panels of Fig. 3, but the depth of these nearly neutral layers and the steepness of the gradients at the edges differed. These differences were by and large due to the higher horizontal and temporal variability of the fine scale structures generated by the KH instability and turbulence, as suggested by the radar measurements. The UAV and balloons may have crossed the KH billows of various depths or their remnant structures at different locations so that the vertical profiles of temperature were necessarily different.

The red curves in Fig. 3a show the temperature and dew-point temperature profiles measured by a meteorological balloon, launched on 07 June 00:00 UT (09:00 LT) from Shionomisaki station (33.45°N, 135.76°E) about 160 km south of the MU radar site. These profiles exhibit gross features that are very similar to those revealed by the UAVs and balloons, i.e., a nearly moist (saturated) adiabatic gradient capped by a steep (negative) humidity gradient, and a temperature inversion above 800 hPa (~2.2–2.4 km) and a drier layer above. Therefore, the temperature and humidity structures of the lower troposphere were likely similar over tens of kilometers at least.

Vertical Profiles of Horizontal Winds from MU Radar and Balloon Data

Figure 4 shows the MU radar-derived zonal and meridional wind profiles (dashed lines) averaged over 1 h (05:40–06:40 LT) around the time of B1 flight at a resolution of 150 m. The horizontal bars indicate the variability of the wind data during the averaging time. The solid lines show the corresponding profiles measured by B1 at a vertical resolution of 20 m. The radar-derived winds compared well with those measured by B1. Almost all measured values from B1 were within the interval of variability of the radar estimates, despite the better resolution of the balloon measurements. This observation indicates that the dynamical conditions met by B1 were very similar to those measured by the radar in spite of the balloon drift (see Fig. 2). The wind profiles measured by B1 show a decreasing wind speed, with wave-like disturbances below 2.15 km in the KH region (2). The wind direction was nearly constant at the bottom of the region (2), but strong veering occurred above 1.95 up to 2.15 km inside the region (2).

The corresponding profiles measured from Shionomisaki are also superimposed (light dashed lines). The horizontal wind profiles were also similar, despite a significantly weaker lower level jet around 1.5 km. Therefore, the background dynamics of the lower troposphere were also similar over tens of kilometers at least, and were likely features of synoptic scale conditions.

Vertical Profiles of Wind Shear and Ri from MU Radar and Balloon Data

Results of comparisons between horizontal winds measured by radars and balloons have been reported many times in the literature and for a larger range in altitudes [see, e.g., Table 1 of Luce et al. 2001c]. However, to our knowledge, comparisons between wind shear and *Ri* profiles are not common, and this work constitutes the first attempt to compare two independent estimations of these parameters from the combination of UAV, balloon, and radar data.

The corresponding profiles of S (red, solid) and speed shear S_V (red, dashed) derived from B1 are shown in Fig. 5b. When $S \approx S_V$, the turning shear is negligible and the wind vector and wind shear vector directions coincide.

The profiles of S and S_V measured by B1 at a vertical resolution of 20 m are shown in Fig. 5b. S was maximum at the bottom and top of the KH region (2). The wind shear was mainly a speed shear ($S \approx S_V$) at the bottom and a combination of speed and turning shear at the top. B1 measured a strong speed shear (~40 ms^{-1} km^{-1}) at the top of the capping inversion around 2.3 km. In the regions (1) and (3), S did not locally exceed ~20 ms^{-1}km^{-1} (~10 ms^{-1}km^{-1} typically) and was mainly turning shear.

The low-resolution (150 m) radar- and balloon-derived profiles of shears S and S_V shown in Fig. 5a agreed well and provided consistent information on the nature of the shears. For example, both balloon- and radar-derived profiles revealed that a turning shear was dominant in the height range 3.0–3.5 km, and a speed shear was dominant in the height range 3.5–4.0 km. This result indicates that the dynamical conditions at vertical scale of 150 m were homogeneous up to the altitude of 4.0 km, in a horizontal domain of at least ~3 km (maximum distance of the balloon from MU radar site) and stationary (for more than 1 h).

It is important to point out the substantial differences between the shear profiles at the two resolutions of 20 and 150 m. The differences between the radar-derived shear profiles at the resolution of 150 m and the balloon-derived shear profiles at the resolution of 20 m, especially below 2.3 km, were not caused by the horizontal inhomogeneity and/or intermittency of the wind field. It was rather due to the difference of height resolution, because both profiles are similar at the resolution of 150 m. The shear was found to be maximum at the center of the region (2) at the resolution of 150 m, whereas it was found to be minimum, with two maxima at its edges at the resolution of 20 m. In addition, a thin and strong speed shear layer at the altitude of 2.3 km and at the resolution of 20 m did not appear anymore at the resolution of 150 m in both radar- and balloon-derived profiles. At this altitude, the zonal and meridional wind profiles measured by B1 (left panel of Fig. 4) showed a clear nearly monochromatic wavy pattern, removed after applying the low-pass filtering used to get the vertical resolution of 150 m. The contribution of these small wave disturbances was thus eliminated in the shear profiles at the resolution of 150 m. The scale dependence of the shear profiles can have an important impact on the analysis of the dynamical background conditions. Balsley et al. (2008) pointed out the importance of the scale dependence of *Ri*.

The *Ri* profile estimated from B1 data at vertical resolution of 20 m is shown in Fig. 5d. The profile of N^2 used for the calculations is shown by the black lines of Fig. 6a [in Fig. 6, and the subsequent figures, the label CT refers to the location of the cloud top, identified by sharp negative humidity gradients]. *Ri* was smaller than the critical value of 0.25 (and even locally negative) almost everywhere in the region (2), consistent with our expectations of a KH instability. *Ri* was much larger than 0.25 (up to ~100) almost everywhere in the stratified region (3) in accordance with the "calm" conditions of this region except above 3.5 km, where shear-generated turbulence was also suspected. In the cloud region (1), the large values of *Ri* are not meaningful, because N^2 for subsaturated air was used and is not representative of saturated conditions.

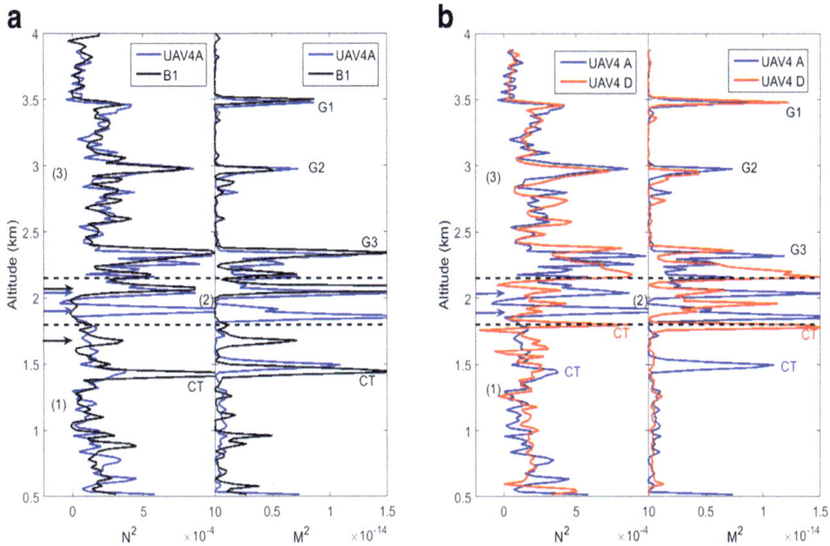

Figure 6: **a** (Left) (dry) N^2 profiles at a vertical resolution of 20 m estimated from UAV4A data (blue) and B1 data (black). (Right) The corresponding M^2 profiles. The CT labels refer to the N^2 and M^2 peaks in the UAV4A- and B1-derived profiles at cloud tops. The two blue (black) arrows indicate the N^2 and M^2 peaks that may coincide with the top and bottom of KH billows crossed by UAV4A (B1). The horizontal dashed lines delineate the KH region (2). **b** (Left) N^2 profiles estimated from UAV4A data (blue) and UAV4D data (red). (Right) The corresponding M^2 profiles. The CT labels refer to the N^2 and M^2 peaks at cloud tops that differed by about 300 m between ascent and descent. The two blue arrows indicate the N^2 and M^2 peaks that may coincide with the top and bottom of KH billows crossed by UAV4A. The horizontal dashed lines delineate the KH region (2)

The corresponding *Ri* profiles at low resolution (150 m) are shown in Fig. 5c. The profile estimated from B1 data is shown in red. The black curves show *Ri* profiles estimated independently from radar-derived shear *S* and UAV-derived N^2 (solid, ascent; dashed, descent). The two independent *Ri* profiles agreed well confirming that the three instruments provided very reliable data at a range resolution of at least 150 m, and the vertical structure and dynamics of the lower troposphere encountered by B1, UAV4, and MU radar were very similar at a vertical scale of 150 m. The low-resolution profiles exhibited a minimum (~1) in the region (2) suggesting the possibility of KH instability, even if the values did not reach the critical level (0.25).

Vertical Profiles of N^2 and M^2 Derived from UAV and Balloon Data

Figures 6a and 7a show the vertical profiles of N^2 and M^2 at a vertical resolution of 20 m estimated from B1 and UAV4A data, and B2 and UAV5D data, respectively. For these comparisons, we selected UAV4A, because it almost coincided with B1 in time; the time difference between UAV4A and B1 did not exceed a few minutes (Fig. 1a). We selected UAV5D, closest to B2 in time. The time difference between UAV5D and B2 measurements was ~40 min at the top and ~3 min at the bottom (Fig. 1b).

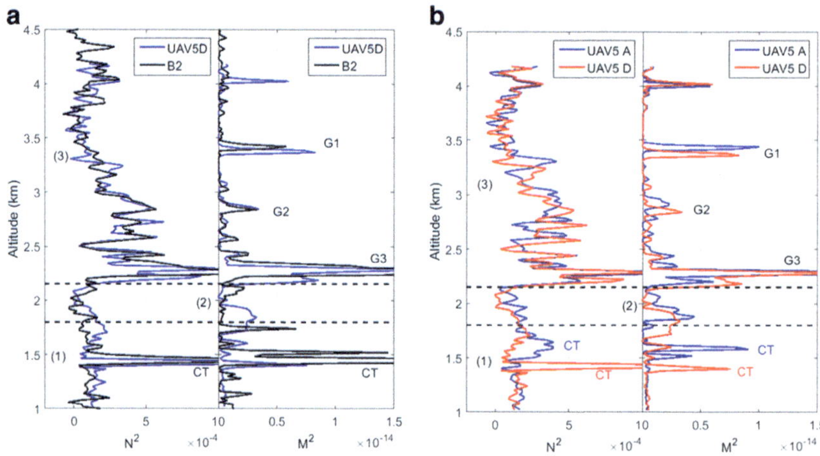

Figure 7: **a** As in Fig. 6a, but for UAV5D and B2. UAV5D data are missing below 1.0 km. **b** As in Fig. 6,b but for UAV5. UAV5D data are missing below 1.0 km

Overall, the balloon- and UAV-derived profiles of N^2 and M^2 almost overlap everywhere. In the stratified region (3), the two instruments virtually detected the same temperature and humidity gradients with surprisingly similar intensities. In particular, the two instruments crossed practically the same gradients G1, G2, and G3 at very similar altitudes (~3.5, ~3.0, and ~2.3 km, respectively). These characteristics are consistent with the persistency and steadiness of the thin echo layers monitored by the radar during the UAV flights in the stratified region (3) (Fig. 1a). Such similarities between the profiles indicate that UAVs probing the temperature and humidity fields along helical paths of ~100–150 m in diameter provide virtually the same information as balloon-borne radiosondes at vertical scales down to 20 m at least. These comparisons also suggest that the most prominent temperature and humidity gradients detectable at a vertical resolution of 20 m can persist over several hours without significant changes due to erosion processes. They can extend horizontally over a few kilometers or even a few tens of kilometers if frozenly advected, so that their aspect ratio (horizontal extent/depth) can exceed 100–1000.

There is a noticeable exception in the KH region (2), where two strong peaks of N^2 and M^2 were detected during UAV4A. Two blue arrows in Fig. 6a indicate these two peaks. Considering that UAV4A crossed KH billows according to the radar observations, they might have been the signature of the top and bottom edges of a KH billow, where temperature and humidity gradients should be enhanced (e.g., Woods 1968). B1 crossed the same top gradient at ~2.05 km but not the bottom one (at ~1.9 km). On the one hand, because the horizontal wavelength of the KH billows was less than 1 km and the horizontal drift of the balloon exceeded 3 km in the KH region (2), B1 and UAV4A could not have crossed the same KH braid at the same time. The identical peaks of N^2 detected at the altitude of 2.05 km by B1 and UAV4A cannot be due to the same billow; the agreement is necessarily fortuitous but could reveal a remarkable regularity between certain characteristics of the billows. On the other hand, B1 may have crossed the bottom edge of a KH billow at a lower altitude (~1.65 km), where peaks of N^2 and M^2 were detected (black arrow in Fig. 6a). This is plausible because these peaks coincided with the peak of maximum speed shear measured also by B1 (Fig. 5b). In that case, the KH billows detected by B1 would have been at least 150 m deeper than those monitored by the radar, depending on the location of B1 with respect to the core of the KH billows.

In the cloud region (1), the strong peaks of N^2 and M^2 below the altitude of 1.5 km in Figs. 6a and 7a were the signatures of cloud tops. They also showed a remarkable agreement in altitude.

Vertical Profiles of N^2 **and** M^2 *Derived from UAV Data during Ascents and Descents*

Figures 6b and 7b show comparisons between the profiles of N^2 and M^2 estimated from UAV4A and D, and UAV5A and D, respectively. A close agreement was also found between the profiles. Indeed, in the stratified region (3), the profiles were virtually identical during UAV4A and D, and during UAV5A and D. However, there were slight height mismatches, in particular, between profiles collected during UAV5A and D in the height range 2.5–3.5 km (Fig. 7b). For example, the heights of G1 were 3.43 and 3.37 km during UAV5A and D at 07:49 and 08:01 LT, respectively, corresponding to a height offset of −60 m. The heights of G2 were 2.92 and 2.85 km during UAV5A and D at 07:45 and 08:05 LT, respectively, corresponding to a height offset of −70 m. These mismatches may reveal a true difference in altitude between ascent and descent, and are consistent with wave-like oscillations visible in the radar echo power image above the altitude of 2.5 km during UAV5 (Fig. 1b). The peaks of echo power corresponding to G1 (G2) were located at the altitudes of 3.43 km (2.93 km) and 3.49 km (2.87 km) during UAV5A and D, respectively, corresponding to a height offset of −40 m (−60 m). It is very close to the values found for the peaks of M^2 (−60 and −70 m). The height agreement in the range 2.5–3.5 km was better between UAV5D and B2 (Fig. 7a). The height offsets for G1 and G2 were +40 and 0 m, respectively, with a time difference of 33 and 25 min. The corresponding height variations of the peak of radar echo power were also weak (less than 20 m) at the times of UAV5D and B2, indicating that the agreement in height between UAV5D and B2 may not have been fortuitous.

In the KH region (2), the N^2 and M^2 profiles significantly differed between UAV4A and D, but were very similar and showed minimum values around 2.0 km during UAV5A and D. An interpretation of the N^2 and M^2 profiles in the KH region (2) during UAV4D is difficult, because no clear and prominent structure emerged and peaks of N^2 and M^2 associated with the cloud top were located just at the bottom side of that region at ~1.8 km (Fig. 6b). However, a careful analysis of the radar echo power image (Fig. 1a) around 06:40 LT, when UAV5D crossed the region (2), shows that the cloud top was indeed close to 1.8 km and that possible smaller scale and less intense KH braids could be present just above (as suggested earlier). We speculate that the two M^2 peaks at the heights of ~1.95 and ~2.1 km in the UAV4D profile (Fig. 6b) may be the signature of these smaller scale KH billows.

Vertical Profiles of M^2 Derived from UAV, Balloon, and Radar Data

Figures 8a, b shows superimposed profiles of radarderived M^2 (M^2_{radar}) (black lines), UAV-derived M^2 (M^2 UAV) (blue lines), and balloon-derived M^2 (M^2_B) (red lines) during UAV4A and B_1 and during UAV5D and B2, respectively.

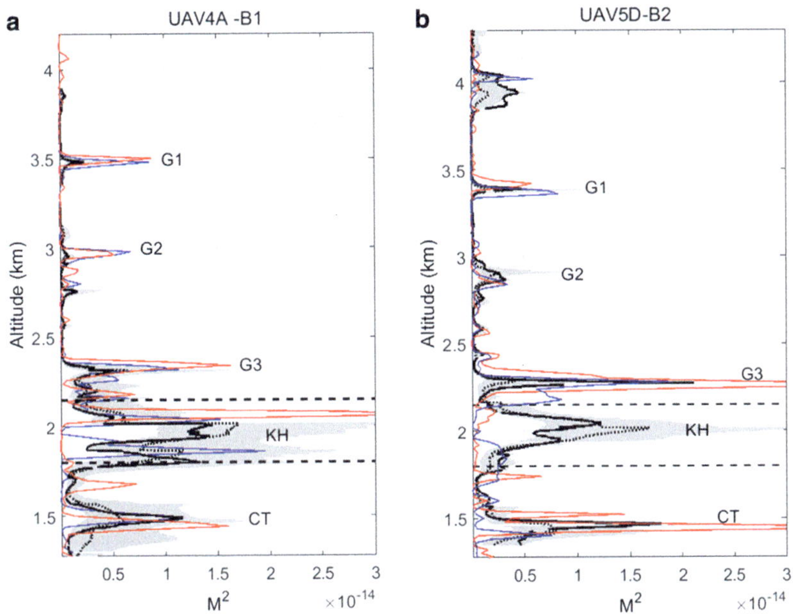

Figure 8: M^2 $_{UAV4A}$ (blue), M^2 $_{B1}$ (red), and M^2 radar (solid black, 1 min averaged; dotted black, 4 min averaged) profiles. Gray color shows the standard deviation of M^2 radar over 4 min. The horizontal dashed lines delineate the KH region (2). b M^2 $_{UAV5D}$ (blue), M^2 $_{B2}$ (red), and M^2 radar (solid black, 1 min averaged; dotted black, 4 min averaged) profiles. Gray color shows the standard deviation of M^2 radar over 4 min. The horizontal dashed lines delineate the KH region (2)

The strongest time variability of M^2 $_{radar}$ (shown by the gray areas in Fig. 8a, b) was found in regions (1) and (2) in accordance with the more variable aspect of the echo patterns. The M^2 $_{UAV4A}$ and M^2 $_{B1}$ peaks associated with the humidity gradient at the cloud top around the height of 1.5 km (see Fig. 3a) coincided well with a M2 radar peak in altitude and amplitude (Fig. 8a). There was also a good agreement between the peaks at the cloud top in Fig. 8b, even though M^2 $_{UAV5D}$ peak appeared to be weaker. This is consistent with the relatively constant altitude of the cloud top (~ 1.5 km) suggested by the radar echoes in Fig. 1b. In the stratified region (3), the three main gradients G1, G2, and G3 detected by UAV4, UAV5, B1, and B2 coincided well with the strong peaks of M^2 $_{radar}$. The agreement was even better between the heights of 2.3 and 2.8 km, where more marginal peaks were observed by all three instruments. These noticeable coincidences, already described by Luce et al. (2017) from the sole comparisons between UAV and radar measurements, confirm the ability of the MU radar to monitor and quantify decameter scale refractive index gradients in stratified conditions. The good agreements reported by Luce et al. (2017) were definitely not fortuitous, since the additional balloon data provided similar M^2 profiles as radar and UAV data, and confirmed the large horizontal extent of the temperature and humidity gradient layers at decameter scale (Fig. 3).

In the KH region (2), the M^2_{radar} profile during UAV4A or B1 showed a double peak (a thin peak at the height of 1.80 km and a broader one around the height of 2.0 km) consistent with the profile obtained from UAV4A, but not exactly at the same altitude (Fig. 8)a. Therefore, the three instruments seem to have captured KH billows of various depths at different locations with respect to their core. In Fig. 8b, the strong enhancement of M^2_{radar} in the core of KH region (2) was associated with a minimum of M^2_{UAV5D} and M^2_{B2}. This noticeable difference had already been noted by Luce et al. (2017) from the comparison of M^2_{radar} and M^2_{UAV5D2}, and is now confirmed from an additional and independent estimation from B2 data. Because two independent instruments (UAV and balloon) revealed M^2 minima in the same range, it is very unlikely that the differences can be explained solely by the horizontal inhomogeneity. Because the enhanced radar echo power producing the enhancements of M^2 in Fig. 8b was due to isotropic turbulence (as suggested by Fig. 1d), it is more likely that the model of M^2 comparisons (basically based on the hypothesis of specular reflection) is not appropriate for these conditions.

CONCLUSIONS

The present work was intended primarily as a demonstration of the potential offered by the combination of radar, UAV, and balloon measurements for studying the structure and dynamics of the lower troposphere down to decameter scales. We used data collected almost simultaneously by two UAVs and two balloon-borne radiosondes near the MU radar operated continuously during the ShUREX 2015 campaign. Vertical profiles of N^2 estimated from balloon and UAV data, and vertical profiles of M^2 estimated from balloon, UAV, and radar data were compared at a vertical resolution of 20 m. In stratified conditions up to ~4.0 km, the profiles derived from the three instruments were nearly identical, with similar peaks in amplitude, position, and depth. This study constitutes a unique cross-validation of the three instruments for estimating vertical gradients of temperature, humidity, and refractive index, down to decameter scales. Even though this analysis covers only a small amount of the data collected during the ShUREX campaign, a first partial lesson is that these gradients can have horizontal extents exceeding a few kilometers at least (likely much more) with very little change in depth and strength, indicating that their generation mechanism may be related to much larger-scale processes. According to radar observations, they seemed to persist for hours with little change in echo power intensity, suggesting that turbulent diffusion was weak, in accordance with the dynamically stable conditions observed during the experiment. These properties are consistent with partial conclusions by Dalaudier et al. (1994) in the stratosphere, and Saxton et al. (1964) in the stable boundary layer. They have long been well recognized, and literature is replete with attempts at modeling their effects on radio wave propagation, for example (e.g., Gage 1990 and references therein). However, the properties of the small-scale stratification of the lower atmosphere are still poorly documented, and the use of UAV observations in combination with balloon and radar data offers new insights.

The balloon and MU radar data also provided wind and wind shear profiles so that two independent estimates of *Ri* could be obtained from additional N^2 estimated from UAV measurements at a vertical resolution of 150 m. The two independent *Ri* profiles also showed remarkable agreement at all altitudes indicating that these *Ri* profiles were truly representative of the dynamic conditions at a vertical scale of 150 m in the region surrounding the MU radar, for several kilometers at least.

Large discrepancies between $N^2$and M^2 profiles derived from balloon, UAV, and radar data were found in a turbulent layer generated by a Kelvin-Helmholtz shear flow instability (region 2). The causes of these discrepancies differ according to the stage of turbulence generated by KH instability. When coherent KH billows occurred (UAV4A), the differences between the various profiles of M^2 derived from the three instruments can be explained by the horizontal inhomogeneity of the refractive index field due to the limited extent of the billows (the horizontal spacing between the billows was less than 1 km and therefore, smaller than the horizontal separation between the instruments). When a later stage of turbulence was observed (UAV5), balloon- and UAV-derived M^2 values were minimum, while radar-derived M^2 values were strongly enhanced. This observation, already reported by Luce et al. (2017) from UAV data only, was thus confirmed by additional balloon data. It indicates that the model used for retrieving M^2 from radar data is not suitable for the turbulent conditions met during the UAV5 flight.

APPENDIX

Definition and estimation of N^2 and M^2

The vertical gradient M of the generalized potential refractive index is defined for non-saturated air as (Ottersten 1969):

$$M = -77.6 \times 10^{-6}\left(\frac{p}{T}\right) \times \left\{\frac{N^2}{g} + \frac{15500q}{T}\left[\frac{N^2}{g} - \left(\frac{1}{2q}\right)\frac{dq}{dz}\right]\right\} \quad \text{(A1)}$$

where

$$N^2 = \frac{g}{T}\left(\frac{dT}{dz} + \Gamma\right), \quad \text{(A2)}$$

Γ, the dry adiabatic lapse rate (km^{-1}); T, the temperature (K); p, the pressure (hPa); q, the specific humidity (gg^{-1}); and g, the gravitational acceleration. In a moist lower troposphere, M^2 peaks are mainly due to moisture gradients (e.g., Tsuda et al. 1988). N^2 and M^2 profiles thus provide information on the vertical structure of the temperature and humidity fields, respectively. In the present work, the N^2 and M^2 profiles were also estimated at vertical resolution of 20 m, because it is expected to be the very best range resolution that can be achieved by the MU radar for high signal to noise ratios in range imaging mode (e.g., Luce et al. 2018).

It also minimizes the noise contribution and sensor time response effects on N^2, while keeping a good vertical resolution. Luce et al. (2017) describe the results of comparisons between UAV- and radar-derived M^2 for 7 successive UAV flights, including UAV4 and UAV5.

ACKNOWLEDGEMENTS

The authors thank JSPS for providing partial funding for the ShUREX 2015 campaign.

Funding

This work was supported by JSPS KAKENHI Grant Number JP15K13568. The MU radar belongs to and is operated by the Research Institute for Sustainable Humanosphere (RISH), Kyoto University.

Availability of Data and Materials

Made available on request from principal author.

CONTRIBUTIONS

HL analyzed the MU radar, balloon, and UAV data with assistance from LK, HH, MY, and TT. LK conceived and led the campaign, and participated in analysis and synthesis of study results. DL and TM flew the UAV and collected data. All authors read and approved the final manuscript.

ETHICS DECLARATIONS

Competing Interests

The authors declare that they have no competing interests.

Publisher's Note

Springer Nature remains neutral with regard to jurisdictional claims in published maps and institutional affiliations.

REFERENCES

1. Balsley BB, Lawrence DA, Woodman RF, Fritts DC (2013) Fine-scale characteristics of temperature, wind, and turbulence in the lower atmosphere (0-1,1300 m) over the south Peruvian coast. Boundary-Layer Meteorol 147:165–178. https://doi.org/10.1007/s10546-012-9774-x
2. Balsley BB, Svenson G, Tjernström M (2008) On the scale-dependence of the gradient Richardson number in the residual layer. Boundary-Layer Meteorol 127:52–72. https://doi.org/10.1007/s10546-007-9251-0
3. Dalaudier F, Sidi C, Crochet M, Vernin J (1994) Direct evidence of sheets in the atmospheric temperature field. J Atmos Sci 51:237–248
4. Du Castel F (1966) Tropospheric radio wave propagation beyond the horizon. Pergamon, tarry-town, N.-Y
5. Fritts DC, Rastogi PK (1985) Convective and dynamical instabilities due to gravity wave motions in the lower and middle atmosphere: theory and observations. Radio Sci 20:1247–1277
6. Fukao S, Luce H, Mega T, Yamamoto MK (2011) Extensive studies of large-amplitude Kelvin-Helmholtz billows in the lower atmosphere with the VHF middle and upper atmosphere radar (MUR). Q J R Meteorol Soc 137:1019–1041
7. Fukao S, Sato T, Tsuda T, Yamamoto M, Yamanaka MD (1990) MU radar–new capabilities and system calibrations. Radio Sci 25:477–485
8. Gage KS (1990) In: Atlas D (ed) Radar observations of the free atmosphere: structure and dynamics, radar in meteorology, chap. 28a, pp 534–565
9. Gossard EE (1990) In: Atlas D (ed) Radar research on the atmospheric boundary layer, radar in meteorology, chap. 27a, pp 477–527
10. Gossard EE, Gaynor JE, Zamora RJ, Neff WD (1985) Fine structure of elevated stable layers observed by sounder and in situ tower sensors. J Atmos Sci 42:2156–2169
11. Kantha L, Lawrence D, Luce H, Hashiguchi H, Tsuda T, Wilson R, Mixa T, Yabuki M (2017) Shigaraki UAV-radar experiment (ShUREX 2015): an overview of the campaign with some preliminary results. Prog Earth Planet Sci 4:19. https://doi.org/10.1186/s40645-017-0133-x

12. Lawrence DA, Balsley BB (2013) High-resolution atmospheric sensing of multiple atmospheric variables using the DataHawk small airborne measurement system. J Atmos Ocean Tech 30:2352–2366
13. Luce H, Crochet M, Dalaudier F (2001a) Temperature sheets and aspect sensitive radar echoes. Ann Geophys 19:899–920
14. Luce H, Crochet M, Dalaudier F, Sidi C (1995) Interpretation of VHF ST radar vertical echoes from in situ temperature sheet observations. Radio Sci 30:1002–1025
15. Luce H, Fukao S, Yamamoto M, Sidi C, Dalaudier F (2001c) Validation of winds measured by MU radar with GPS radiosondes during the MUTSI campaign. J Atmos Ocean Technol 18:817–829
16. Luce H, Hashiguchi H, Kantha L, Lawrence DA, Tsuda T, Mixa T, Yabuki M (2018) On the performance of the range imaging technique using UAVs during the ShUREX 2015 campaign. IEEE Trans Geosci Remote Sens 56:2033–2042. https://doi.org/10.1109/TGRS.2017.2772351
17. Luce H, Hassenpflug G, Yamamoto M, Fukao S (2006) High-resolution vertical imaging of the troposphere and lower stratosphere using the new MU radar system. Ann Geophys 24:791–805
18. Luce H, Kantha L, Hashiguchi H, Lawrence DA, Yabuki M, Tsuda T, Mixa T (2017) Comparisons between high-resolution profiles of squared refractive index gradient M^2 measured by the MU radar and UAVs during the ShUREX 2015 campaign. Ann Geophys 35:423–441. https://doi.org/10.5194/angeo-35-423-2017
19. Luce H, Nishi N, Caccia JL, Fukao S, Yamamoto MK, Mega T, Hashiguchi H, Tajiri T, Nakazato M (2012) Kelvin-Helmholtz billows generated at a cirrus cloud base within a tropopause fold/upper-level frontal system. Geophys Res Lett 39:L04807. https://doi.org/10.1029/2011GL050120
20. Luce H, Takai T, Nakamura N, Yamamoto M, Fukao S (2010) Simultaneous observations of thin humidity gradients in the lower troposphere with a Raman lidar and the VHF MU radar. J Atmos Ocean Tech 27:950–956. https://doi.org/10.1175/2010JTECHA1372.1
21. Luce H, Yamamoto M, Fukao S, Hélal D, Crochet M (2001b) A frequency radar interferometric imaging applied with high resolution methods. J Atmos Sol Terr Phys 63:221–234

22. Misme P, du Castel F, Voge J (1958) Réflexions partielles dans l'atmosphère et propagation à grande distance. Première partie: Mesures météorologigues. Ann Télécom 13:209–214
23. Muschinski A, Wode C (1998) First in situ evidence for coexisting submeter temperature and humidity sheets in the lower free troposphere. J Atmos Sci 55:2893–2906
24. Ottersten H (1969) Atmospheric structure and radar backscattering in clear air. Radio Sci 4:1179–1193
25. Saxton JA, Lane JA, Meadows RW, Matthews RA (1964) Layer structure of the troposphere. Proc Inst Electr Eng 111:275–283
26. Spano E, Ghebrebhran O (1996) Complementary sequences with high sidelobe suppression factors for ST/MST radar applications. IEEE Trans Geos Rem Sens 34:317–329. https://doi.org/10.1109/36.485110
27. Tsuda T, May PT, Sato T, Kato S, Fukao S (1988) Simultaneous observations of reflection echoes and refractive index gradient in the troposphere and lower stratosphere. Radio Sci 23:655–665
28. Wilson R, Luce H, Hashiguchi H, Nishi N, Yabuki M (2014) Energetics of persistent turbulent layers underneath mid-level clouds estimated from concurrent radar and radiosonde data. J Atmos Sol Terr Phys 118:78–89
29. Woods JD (1968) Wave-induced shear instability in the summer thermocline. J Fluid Mech 32:791–800
30. Yamamoto M, Sato T, May PT, Tsuda T, Fukao S, Kato S (1988) Estimation error of spectral parameters of mesosphere-stratosphere-troposphere radars obtained by least squares fitting method and its lower bound. Radio Sci 23:1013–1021

SECTION 4
CLOUD DROPS AND MIXTURES

CHAPTER 14

Cloud Processing of Aerosol Particles in Marine Stratocumulus Clouds

Andrea I. Flossmann and Wolfram Wobrock

Université Clermont Auvergne, CNRS, LaMP, F-63000 Clermont-Ferrand, France

ABSTRACT

Cloud processing of aerosol particles is an important process and is, for example, thought to be responsible for the so-called "Hoppel-minimum" in the marine aerosol particle distribution or contribute to the cell organization of marine boundary layer clouds. A numerical study of the temporal and spatial scales of the processing of aerosol particles by typical marine stratocumulus clouds is presented. The dynamical framework is inspired by observations during the VOCALS (Variability of the American Monsoon System Ocean-Cloud-Atmosphere-Land Study) Regional Experiment in the

Citation: Flossmann, Andrea I., and Wolfram Wobrock. 2019. "Cloud Processing of Aerosol Particles in Marine Stratocumulus Clouds" Atmosphere 10, no. 9: 520. https://doi.org/10.3390/atmos10090520.

Southeast Pacific. The 3-D mesoscale model version of DESCAM (Detailed Scavenging Model) follows cloud microphysics of the stratocumulus deck in a bin-resolved manner and has been extended to keep track of cloud-processed particles in addition to non-processed aerosol particles in the air and inside the cloud drops. The simulation follows the evolution of the processing of aerosol particles by the cloud. It is found that within one hour almost all boundary layer aerosol particles have passed through at least one cloud cycle. However, as the in-cloud residence times of the particles in the considered case are only on the order of minutes, the aerosol particles remain essentially unchanged. Our findings suggest that in order to produce noticeable microphysical and dynamical effects in the marine boundary layer clouds, cloud processing needs to continue for extended periods of time, exceeding largely the time period considered in the present study. A second model study is dedicated to the interaction of ship track particles with marine boundary layer clouds. The model simulates quite satisfactorily the incorporation of the ship plume particles into the cloud. The observed time and spatial scales and a possible Twomey effect were reproduced.

Keywords: aerosol particles, cloud processing, marine boundary layer

INTRODUCTION

Aerosol particles are an essential part of the Earth's atmosphere. They cover sizes from a few nanometers up to tens of micrometers in diameter and are suspended in and transported by air. They are emitted directly as particles at the ground (primary particles) or formed in the atmosphere by gas-to-particle conversion [**1**]. Aerosol particles are removed from the atmosphere either by gravitational settling (dry deposition) or their incorporation into drops and ice particles and formation of precipitation (wet deposition). Their sizes and chemical composition are largely determined by their formation mechanism [**2**], comprising salts, organic or biological compounds as well as silicates. Often, their chemical composition is used as a fingerprint to identify the source region of observed aerosol particles [**3**].

It is commonly accepted that aerosol particles change in size and composition [**4**,**5**] due to condensation of gases, coagulation as well as cloud processes [**6**]. In particular, cloud processing is believed to be an important process in aerosol particle modification. Part of the particulate population can serve as cloud condensation nuclei (CCN) or ice nucleation particles (INP) and are incorporated into the hydrometeors upon their formation [7]. Subsequent uptake of gases and liquid phase chemical reactions, as well as collisions among hydrometers release altered aerosol particles in the air upon cloud hydrometeor evaporation. This evaporation happens at cloud dissipation but also during cloud evolution at the cloud edges and its frequency of occurrence and spatial extension is rather poorly understood and will certainly vary with cloud type and environment.

The long persistence, for example, of marine stratocumulus decks seems to suggest that their drops, and the (particulate or dissolved) pollution mass therein have rather long lifetimes. This assumption, however, is quite incorrect, as even in long-lived clouds renewal and exchange processes can be quite rapid.

Wood [**8**] in his review summarizes our understanding of marine stratocumulus clouds. He indicates the influence of aerosol particles on cloud droplet distribution, cloud albedo and drizzle formation and states that aerosols in turn are strongly modified by the cloud processes. As [**9**] suggests, the aerosol particles and the cloud induced changes are the principal driver for the organization of cells in the marine boundary layer. Large-eddy simulations [**10**,**11**] have demonstrated that drizzle, initiated by low aerosol concentrations in model simulations, can trigger the formation of open cellular structures. The process is supposed to be accelerated by the steady depletion of the aerosol concentration by coalescence scavenging, that further increases drizzle until a super-clean state is reached [**12**].

In [**13**,**14**] the bi-modal form of marine aerosol particle spectra is studied, and it is attributed to cloud processing. In [**13**] it is proposed that on the average, aerosol particles participate in 10–20 nonprecipitating cloud cycles over a 3–10 day period, growing to a size of about 0.1 μm before being removed by precipitation scavenging in a rain cloud.

Thus, given the potential importance of cloud processing, the current study aims to shed some light on the temporal and spatial scales of the cycling of aerosol particles in those marine stratocumulus clouds.

In contrast to previous model studies in a similar context (e.g., [**11**]) our study focuses on the processes at a short time and a small spatial scale, complementing studies, e.g., of [**15**]. The dynamical framework was largely inspired by observations taken in stratocumuli in the tropical southeast Pacific, such as observed during the VOCALS Regional Experiment [**16**], in order to provide a realistic context. In particular, we used the thermodynamics and aerosol parameters observed during the C-130 Flight RF06 [**17**] for a region of open cells with an observed cloud cover of around 60%.

MODEL DESCRIPTION AND INITIALIZATION

The meso-scale model used for this study is a 3D non-hydrostatic model [**18**] coupled with the Detailed Scavenging Model DESCAM [7] using a bin-resolved microphysics approach [**19**]. In contrast to all other bin-resolved studies of a similar dynamical context (e.g., [**15**]), our model follows explicitly the scavenged aerosol particle mass in the drops and releases them to replenish the ambient aerosol spectrum in case of a complete droplet evaporation.

For the detailed microphysics module, we used the extension of the model that allows to distinguish two different types of aerosol particles [**20**]. Two aerosol particle size density distribution functions $fA^{Pa,type}(r)$, where r is the moist radius of the aerosol particle of *itype* = 1 or 2, were considered in the air and their time evolution is calculated from:

$$\frac{\partial f_{APa,ityp}(r)}{\partial t} = \left.\frac{\partial f_{APa,ityp}(r)}{\partial t}\right|_{dyn} + \left.\frac{\partial f_{APa,ityp}(r)}{\partial t}\right|_{act/deact} + \left.\frac{\partial f_{APa,ityp}(r)}{\partial t}\right|_{con/eva} + \left.\frac{\partial f_{APa,ityp}(r)}{\partial t}\right|_{AP,coll,d}$$

as well as two mass density distribution functions for aerosol particles of *itype* = 1 and 2 inside the drops $g_{APd,ityp}(a)$, where a is the drop radius, and their time evolution is calculated from:

$$\frac{\partial g_{APd,ityp}(a)}{\partial t} = \left.\frac{\partial g_{APd,ityp}(a)}{\partial t}\right|_{dyn} + \left.\frac{\partial g_{APd,ityp}(a)}{\partial t}\right|_{act/deact} + \left.\frac{\partial g_{APd,ityp}(a)}{\partial t}\right|_{con/eva} + \left.\frac{\partial g_{APd,ityp}(a)}{\partial t}\right|_{AP,coll,d} + \left.\frac{\partial g_{APd,ityp}(a)}{\partial t}\right|_{d,coal} + \left.\frac{\partial g_{APd,ityp}(a)}{\partial t}\right|_{d,break}$$

As indicated by the subscripts in both equations, the different terms treat dynamical changes: $()|_{dyn;}$ activation of drops and their deactivation back to aerosol particles: $()|_{act/deact}$(; evolution involving phase changes due to water vapor diffusion: $()|_{con/eva}$, as well as collision process for: $()|_{AP,coll,d}$ aerosol/drop collision and $()|_{d,coal}$ for drop/drop collision; and finally breakup of drops: $()|_{d,break}$. The mathematical treatment of the different microphysical terms can be found in [7].

The *itype = 1* aerosol particles were the marine boundary layer particles of the initial background conditions (later also called "pristine" particles, indicated in red in **Figure 1**). They served as CCN for the forming clouds. The *itype = 2* aerosol particles were all those that evaporated after having been inside a cloud drop (later also called "processed" particles, indicated in green in **Figure 1**), regrouping all scavenged aerosol mass into the released particle. They are now treated separately and can later serve also again as CCN, for example. This method allows to follow in detail the fate of the boundary layer aerosol and its cycling through the stratocumulus clouds, in contrast to all previous studies in the literature. As the cloud region is relatively low and warm, no ice phase processes occur.

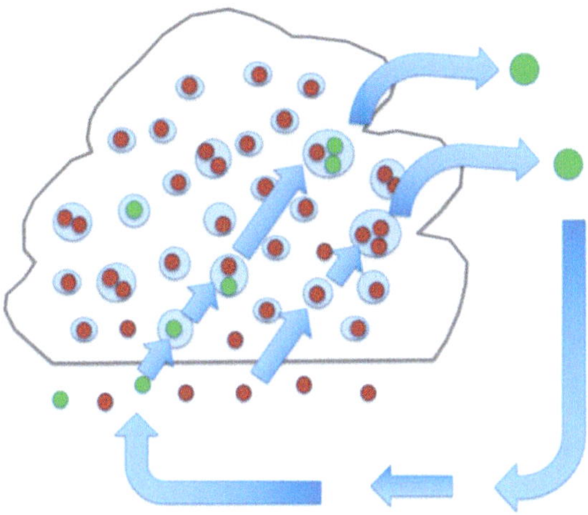

Figure 1. Schematic display of the treatment of cloud processing of aerosol particles in the model (initial "pristine" particles in red and "cloud-processed" particles in green).

In a second study, we studied another typical mixing process occurring in marine stratocumulus clouds, associated to ship tracks. We applied the identical setup as before to study the mixing and processing of aerosol particles that were released by an emission line of ship track aerosol (*itype = 2*), as was observed, for example, during the MAST (Monterey Area Ship Track) experience [**21**]. The explicit modelling allows to study in detail the mixing of a pollution line into a cloudy marine boundary layer and verify the mixing capacity of the model through comparison with observations, for example, regarding the signature of the plume width.

Due to the rather sophisticated and expensive treatment of the microphysics, the modelling simulation was restricted to only a few hours. Thus, no large-scale changes were considered, even though the model holds a simplified radiation code [**22**].

In the configuration for simulating a typical shallow marine stratocumulus deck, the model domain covers a volume of 6.4 × 6.4 × 3 km^3, with a horizontal grid resolution of $\Delta x = \Delta y = 50$ m and a vertical of $\Delta z = 20$ m. The time step Δt is selected to be 1–1.5 s, with a time splitting for the rapid condensation processes. The initial conditions were adapted from

[23] (their **Figure 1**) using a uniform vertical thermodynamic profile with a constant adiabatic temperature of θ = 288 K and a constant water vapour mixing ratio of 7.8 g/kg. Only between 1 km and the inversion at 1500 m potential temperature was slightly increased and humidity decreased, in order to keep the initial atmosphere just below saturation. At the ground a surface latent heat flux of 115 W m^{-2} and a surface sensible heat flux = 15 W m^{-2} were imposed, within the range of the observed values [**17**], and random perturbations of 0.1 K and 0.2 g/kg for potential temperature and water vapour mixing ratio were added. No horizontal wind was considered, which corresponds to a model domain drifting along in a large-scale homogeneous environment, moving at a uniform horizontal windspeed.

The initial background aerosol particle size distribution (*itype = 1*) was fitted to the observations of [**17**] at 1584 m (their Figure 15a) using a superposition of two lognormal distributions for the number of dry aerosol particles N_{AP} of radius r:

$$\frac{dN_{AP}}{d\ln r} = \sum_{i=1}^{2} \frac{n_i}{\sqrt{2\pi}\log\sigma_i \ln 10} \exp\left(-\frac{[\log r/R_i]^2}{2(\log\sigma_I)^2}\right)$$

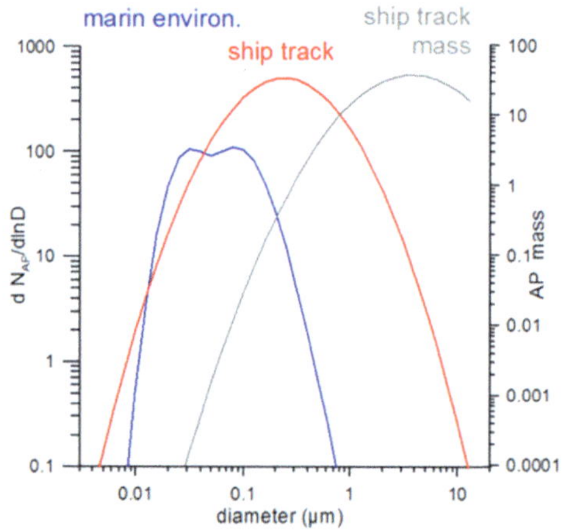

Figure 2. Initial dry aerosol particle number concentration (cm^{-3}) for the marine environment (in blue) and the ship track particles (in red); in gray: the mass of the ship track particles in µg m^{-3} (right axis).

With the values: n_1 = 80 cm^{-3}, n_2 = 137 cm^{-3}, R_1 = 0.014 µm, and R_2 = 0.04 µm, log σ_1 = 0.146 and log σ_2 = 0.216, the total number concentration is 220 cm^{-3} (**Figure 2**, blue curve). This particle distribution was observed above the boundary layer, even though observations were equally made inside the boundary layer. We refrained, however, from using those, as they are probably already affected by cloud presence, as will become obvious later. The aerosol particles were initialized constant over the entire domain.

For the ship-track calculations, an additional plume aerosol (*itype* = *2*) with a total of 1200 cm^{-3} (n_p = 1200 cm^{-3}, R_p = 0.12 μm, log σ_p = 0.416) was added in the second vertical grid box above ground (between 20 and 40 m) in the center grid line (X = 3 km; Δx = 50 m; Y = 0.8–5.6 km) of the domain (**Figure 2**, red and gray curve) during 15 s. These aerosol particles are less numerous and slightly bigger than observed by [**21**], as they are assumed already homogeneous in a grid box of 50 × 50 × 20 m^3, and thus somewhat dispersed with respect to the chimney exhausts that were observed.

RESULTS AND DISCUSSION

After 120 min of simulation time, a stable situation is reached where cloud cover varies around 65% in agreement with the observations of [**17**]. A spin-up influence of 2–3 h is also reported by [**11**]. At 200 min the stratocumulus field just below the inversion is displayed in Figure 3. The liquid water content (Figure 3a) is generally below 0.5 g m^{-3} in agreement with the values reported by [17], with maxima as high as 0.7 g m^{-3} in isolated locations. These regions of high liquid water content correlate to the regions of maximum positive vertical velocity (Figure 3b). Several cells exist, driven by the imposed surface fluxes.

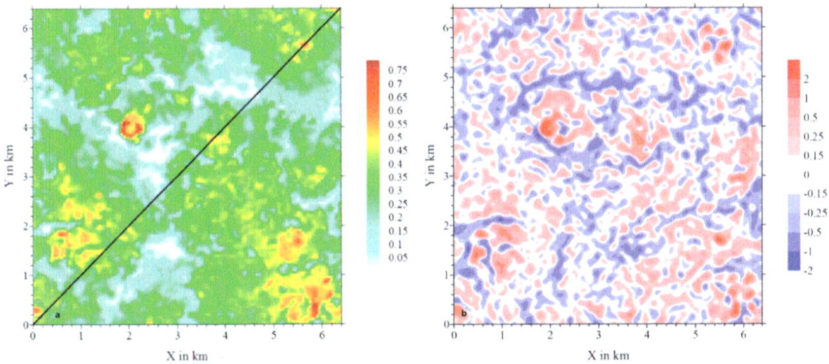

Figure 3. Horizontal cross section of (**a**) the liquid water content in g m^{-3} and (**b**) the vertical wind speed in m s^{-1} in the domain at 1520 m altitude after 200 min of simulation; the black line indicates the location of the vertical cross sections of Figure 4, Figure 5 and Figure 6.

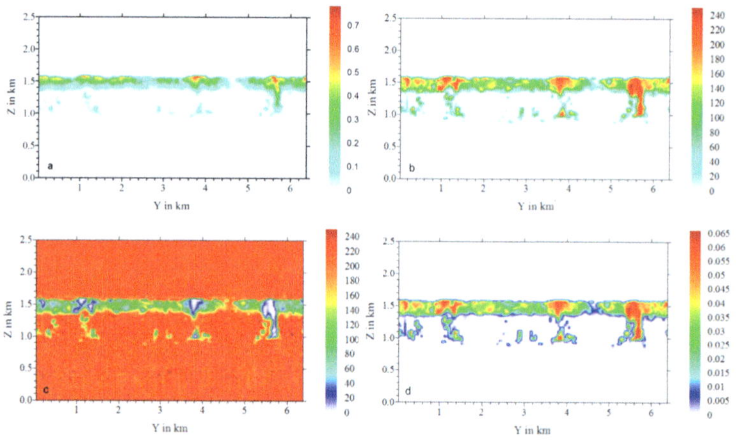

Figure 4. Vertical cross section of (a) the liquid water content (g cm^{-3}), (b) the number of drops (cm^{-3}), (c) the number of aerosol particles (cm^{-3}) and (d) the aerosol particle mass inside the drops (µg m^{-3}) after 200 min along the cross section indicated in Figure 3a and projected onto the Y-axis.

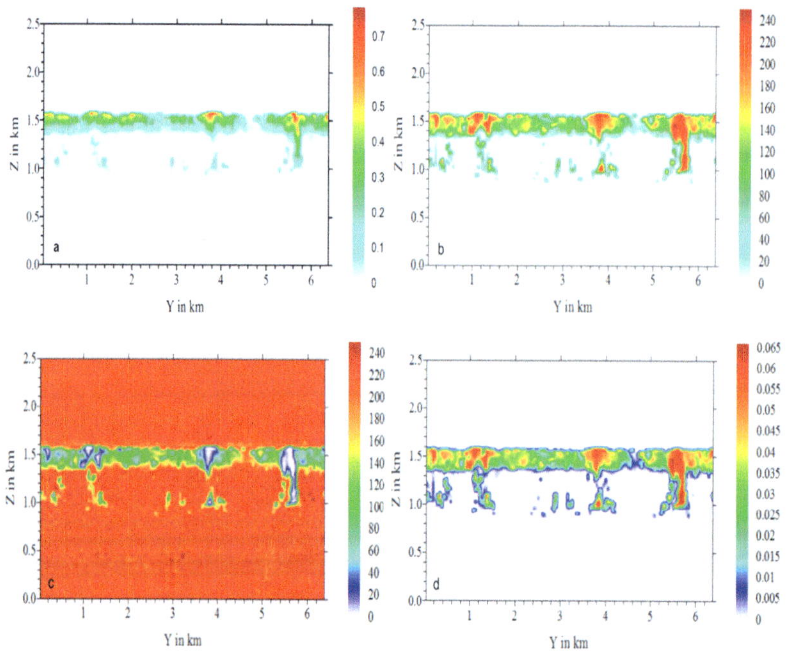

Figure 5. Vertical cross section of (a) the liquid water content (g cm^{-3}), (b) the number of aerosol particles (cm^{-3}), (c) the number of "pristine" aerosol particles

(cm^{-3}), (d) the number of "processed" aerosol particles (cm^{-3}), (e) the "pristine" aerosol particle mass in drops (μg m^{-3}) and (f) the "processed" aerosol particle mass in drops (μg m^{-3}) after 205 min along the cross section indicated in Figure 3a and projected onto the Y-axis.

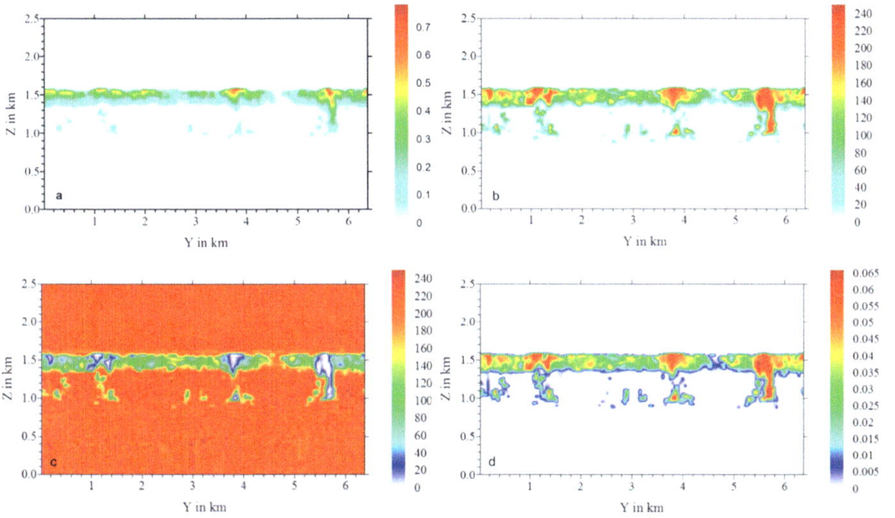

Figure 6. Vertical cross section of (a) the number of "pristine" aerosol particles (cm^{-3}), (b) the number of "processed" aerosol particles (cm^{-3}), (c) the "pristine" aerosol particle mass in drops (μg m^{-3}) and (d) the "processed" aerosol particle mass in drops (μg m^{-3}) after 240 min along the cross section indicated in Figure 3 and projected onto the Y-axis.

Figure 4 displays some microphysical parameters along the vertical cross section indicated in Figure 3a (black line), but it is projected onto the Y-direction for easier reference. Figure 4a for the liquid water content supports an average value below 0.5 g m^{-3} with a cloud top limited by the height of the inversion. Cloud base can extend down to 1 km while most cloud water can be found in the 300 m just below cloud top, in agreement with [17]. The number of drops (Figure 4b) in the cloud varies with liquid water content and vertical velocity (Figure 3). The highest number concentrations exceeding 200 cm^{-3} are found in the regions of maximum Liquid water content (LWC), while in general drop concentrations are around 100 cm^{-3}, exceeding the observed values during this particular VOCALS flight (<50 cm^{-3}), while in other similar flights values up to 100 cm^{-3} were also documented.

Figure 4c displays the total number of aerosol particles in the ambient air. The red color represents the initial background particle concentration of 220

cm^{-3} while inside the cloud layer the number concentration has decreased to values below 100 cm^{-3} due to nucleation of droplets. In the regions of high LWC, less than 20 cm^{-3} stay outside the cloud drops in the air, in agreement with the observations of [17]. Figure 4d displays the uptaken aerosol particle mass which can now be found inside the cloud drops.

The cloud at 200 min (Figure 3 and Figure 4) represents the starting point for the following studies. From here on, in the simulation a second aerosol particle type (itype = 2) was opened which accumulates all aerosol particles that are from now on released back into the air after drop evaporation. In the following, these particles are called "processed" particles and they allow to study the evolution of the number and mass of particles that were cycled through a cloud.

The situation after 5 min is given in Figure 5. Figure 5a,b correspond to Figure 4a,c and confirm a globally stable cloud situation. However, Figure 5b for the total number of aerosol particles in the air is now composed of the sum of "pristine" (itype = 1) particles (Figure 5c) and "processed" (itype = 2) particles (Figure 5d) which have already transited in these 5 min of simulation time through one or more cloud stages. The maxima in Figure 5d reach values of 140 cm^{-3}, indicating that locally already half of the aerosol number population has been activated into a cloud drop. These particles can be found above 800 m, thus close to the lowest cloud base. Figure 5e and f show the scavenged aerosol mass in the drops of "pristine" (Figure 5e) and already "processed" particles (Figure 5f). The mass contribution of the "processed" particles to the overall uptaken particulate mass on the average is around 10%.

Figure 6a–d shows the same information 35 min later. We note that the ambient aerosol particle concentration of "pristine" and "processed" particles is now affected almost down to the sea surface, even though the geometry of the clouds themselves, as can be perceived from Figure 6c–d, has not changed much and cloud base stays above 900 m. Comparing Figure 6a,b shows that locally almost all particles have already been inside the cloud drops (>200 cm^{-3}). In addition, the "processed" aerosol particle mass inside the drops indicates that most of the particle mass has gone through previous cloud events.

A study of the drop and aerosol particle spectra in the model during the simulation, however, has revealed almost no evolution. Figure 7 shows the mean drop size distribution, averaged over all grid points with a liquid water content exceeding 0.5 mg m^{-3}. One curve (dashed line) shows the mean spectrum at the beginning of the processing study (200 min) while the solid one shows the situation 40 min later. The very similar curves show that almost no collision-coalescence broadening of the drop spectum could be observed during this time. Thus, even though Figure 6 demonstrates that a rapid cycling of the aerosol particles through the cloud is happening, the individual particles are in fact affected only very little. Their residence time in the liquid phase is so short (on the order of minutes assuming an average updraft of 1m s^{-1} (Figure 3b) and a cloud depth of 300m (Figure 4a)) that the drops stay small and almost no collision and coalescence is happening. Thus, the cloud condensation nuclei are released after drop evaporation almost at the same size as they entered the cloud. Very few larger drops have been formed, as the little maximum in the drop size distribution between 40 and 60 μm diameter in Figure 7 shows. However, over the studied time intervals (up to 4 h of simulation), no significant drizzle formation could be observed.

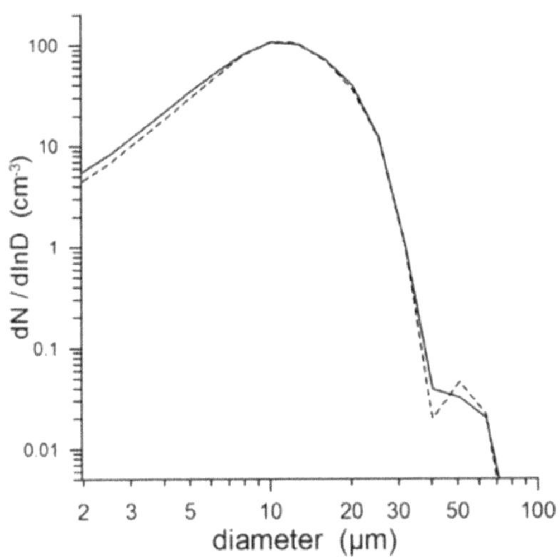

Figure 7. Drop size number density distribution function in cm^{-3} at 1370 m altitude, averaged over all cloudy grid points (LWC > 0.03 g m^{-3}) of the domain at 200 min (dashed curve) and at 240 min (solid curve).

In a second set of simulations, we studied the capacities of the marine boundary layer to mix and process aerosol particles of different origin. Here, the second considered aerosol type (itype = 2) are plume particles as released by ship exhausts [**20**].

The additional aerosol particles follow a log-normal distribution, as given in Figure 2 by the red curve with a total of 1200 particles cm^{-3}. Starting from the same initial state as presented in Figure 3 and Figure 4, these particles are introduced along a line of grid boxes in the middle of the domain during 15 s (see Section 2). Figure 8a gives a horizontal cross section at the height of the release grid box after 5 min. The release line is still very well defined and overall air motion has only somewhat mixed the initial concentrations.

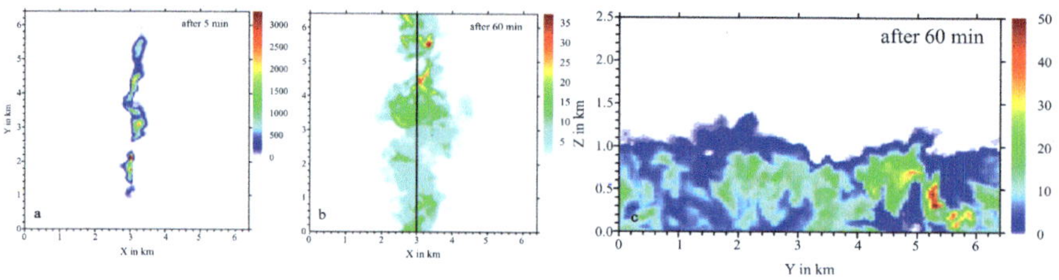

Figure 8. Horizontal cross section of the number of plume aerosol particles (cm^{-3}) (**a**) 5 min and (**b**) 60 min after their release at 30 m altitude; (**c**) vertical cross section of the number of plume particles (cm^{-3}) in the air 60 min after their release along the black line in (b).

After 1 h of simulation time (Figure 8b,c), the plume has widened horizontally from 50 m to over 2 km, in agreement with observation of [24]. Dilution has decreased the number concentrations to values below 35 cm^{-3}. The mixing in an almost neutral boundary layer, however, has transported the particles also vertically. There seems to be a net limit to the transport as the plume number concentrations (Fig.8c) above roughly 1 km are mostly zero. However, as discussed above, the transport continues above 1 km. Here, the large plume particles (see Figure 2 red and gray curve) have been activated into cloud drops.

Figure 9a,b present a horizontal cross section of the aerosol particle mass inside the cloud drops at 1520 m. Here, the signature of the plume aerosols is clearly evident.

Mass concentrations are 10 times higher than the background concentrations and clearly show the increase of mass due to the plume particles in a region of 3 km width in the X-direction, along the release line. Figure 10 shows the vertical cross section along the black line in Figure 9b where we note an increase of the number concentration of drops by around 50 particles cm^{-3} (Figure 10b) with respect to the background cloud (Figure 10a), in agreement with observations of [20], and an increase in the pollution mass in the drops (Figure 10c,d).

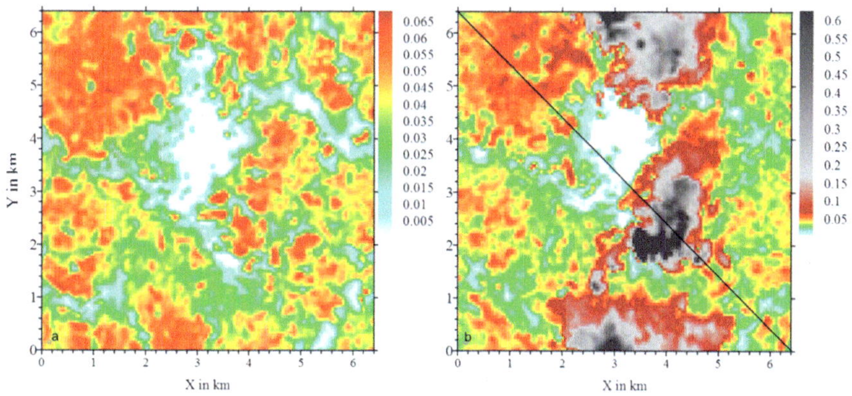

Figure 9. Horizontal cross section of the aerosol mass in drops in µg m^{-3} near cloud top at 1520 m altitude after 260 min of simulation; (**a**) without the presence of the plume; (**b**) with the plume; the black line indicates the location of the vertical cross sections of Figure 10.

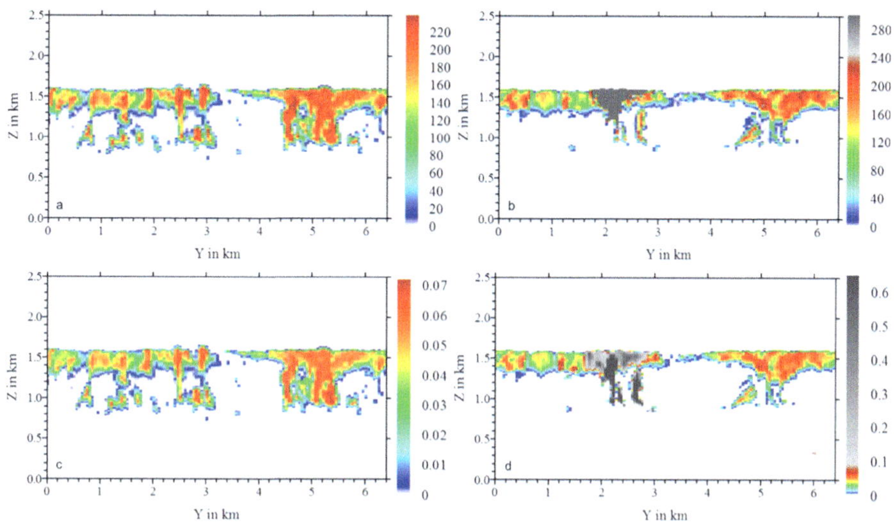

Figure 10. Vertical cross section of the drop number concentration (cm^{-3}): (**a**) without the plume; (**b**) with the plume; vertical cross section of the aerosol particle mass in drops (µg m^{-3}): (**c**) without the plume; (**d**) with the plume; after 260 min along the cross section indicated in Figure 9 and projected onto the Y-axis.

CONCLUSIONS

A small-scale numerical study of the processing of aerosol particles by marine stratocumulus clouds is presented. The dynamical framework is based on the observations during the VOCALS Regional Experiment [**16**] in the Southeast Pacific to guarantee realistic meteorological conditions. The 3-D mesoscale model version of DESCAM follows cloud microphysics of the stratocumulus deck in a bin-resolved manner and has been extended to keeping track of two types of aerosol populations in the air and inside the cloud drops.

The aerosol particle spectrum in the boundary layer was initialized with the spectrum observed in the free troposphere with 220 cm^{-3}. The formed cloud deck shows an average liquid water content, a vertical extension and a cloud cover within the observed values.

The number of drops formed exceeds somewhat the observed values, probably due to the fact that the initial aerosol particle spectrum (observed in the free troposphere) held more particles than observed during this particular VOCALS flight in the boundary layer (100–150 cm^{-3}).

The model was able to simulate a rapid cycling of the aerosol particles through the cloud affecting almost all particles in the boundary layer within less than an hour. However, with residence times in the liquid phase on the order of a few minutes, the individual particles were much less affected than expected and almost no coalescence effect was observed. During the simulated hour, the aerosol particle spectrum did not broaden, and no "Hoppel"-minimum [**13**] was yet formed. Furthermore, no formation of drizzle-sized drops was simulated and drop sizes stayed well below 100 μm diameter. Sensitivity studies varying the boundary layer particle spectrum as model input did not change these conclusions.

In the current configuration and with simulation times on the order of only some hours, the stratocumulus cloud, even though very thoroughly cycling the aerosol particle population, did not reduce the number concentration or increase the size of the particles.

Considering this small effect of cloud cycling on the aerosol particle population, boundary layer cloud residence times extending towards the longer time estimates of [**13**] (10 days) seem more realistic to achieve a modification of the aerosol population, while their suggestion of 10 to 20 cloud cycles seems to be largely underestimated. The simulated time scales were not able to change the organization of the boundary layer in cells as has been suggested by [**9**] for much larger domain sizes and simulation times.

Thus, we can conclude that the cycling of aerosol particles through marine stratocumulus clouds for several hours is a lot more rapid than expected, affecting the entire boundary layer, but also a lot less efficient with respect to the broadening of the size distribution of the aerosol particles.

The second study of the mixing of a ship track plume in the presence of the marine stratocumulus deck demonstrated that overall the model performs correctly. The dispersion of the plume in space and time agrees with the available observations [**20**] and an increase in the number of droplets in the plume track was modelled, suggesting the observed "Twomey effect" [**25**] as modelled by [**26**].

Obviously, the cycling times of aerosol particles through clouds depend significantly on the dynamics and thermodynamics of the studied cloud. Here, only a certain type of marine boundary layer clouds was modelled, and the reported behavior is limited to the conditions that were studied. In order to generalize certain findings, different types of clouds in different geographical locations need to be investigated. More findings can be expected when comparing the microphysical response of different clouds with the transition times of individual air parcels through the cloud. In the future, the technique of following different types of particles thorough the cloud phase and the associated mixing will also be used to study other cloud effects due to distinguishable aerosol particle population, such as used for planned and inadvertent pollution events.

AUTHOR CONTRIBUTIONS

Conceptualization, Methodology and Writing—original draft, A.F.; Formal analysis, Software and Visualization, W.W.

Funding

This research was partly funded by Projet AC-AHC2 grant number ANR-15-CE01-0015 of the French National Agency of Research ANR.

ACKNOWLEDGMENTS

The model calculations have been done on French computer facilities of the "Institut du Développement des ressources en Informatique Scientifique" (IDRIS) CNRS at Orsay, the "Centre Informatique National de l'Enseignement Supérieur" (CINES) at Montpellier under the project

A0060105056; as well as on the Mésocentre of the Université Clermont Auvergne. The authors acknowledge with gratitude the hours of computer time and the support provided.

Conflicts of Interest

The authors declare no conflict of interest

REFERENCES

1. Seinfeld, J.H.; Pandis, S.N. *Atmospheric Chemistry and Physics*; John Wiley and Sons: Hoboken, NJ, USA, 1998; p. 1326.
2. Warneck, P. *Chemistry of the Natural Atmosphere*; Academic Press: Cambridge, MA, USA, 2000; p. 923.
3. Deguillaume, L.; Charbouillot, T.; Joly, M.; Vaïtilingom, M.; Parazols, M.; Marinoni, A.; Amato, P.; Delort, A.-M.; Vinatier, V.; Flossmann, A.; et al. Classification of clouds sampled at the puy de Dôme (France) based on 10 yr of monitoring of their physicochemical properties. *Atmos. Chem. Phys.* **2014**, *14*, 1485–1506.
4. Levin, Z.; Cotton, W.R. *Aerosol Pollution Impact on Precipitation*; A Scientific Review; Springer: Berlin, Germany, 2008; ISBN 978-1-4020-8690-8.
5. Hobbs, P.V. *Aerosol-Cloud-Climate Interactions*; Academic Press: Cambridge, MA, USA, 1993; p. 233.
6. Hatzianastassiou, N.; Wobrock, W.; Flossmann, A.I. The effect of cloud-processing of aerosol particles on clouds and radiation. *Tellus* **1998**, *50B*, 478–490.
7. Flossmann, A.I.; Wobrock, W. A review of our understanding of the aerosol-cloud interaction from the perspective of a bin resolved cloud scale modelling. *Atmos. Res.* **2010**, *97*, 478–497.
8. Wood, R. Stratocumulus clouds. *Mon. Weather Rev.* **2012**, *140*, 2373–2423.
9. Rosenfeld, D.; Kaufman, Y.J.; Koren, I. Switching cloud cover and dynamical regimes from open to closed Benard cells in response to the suppression of precipitation by aerosols. *Atmos. Chem. Phys.* **2006**, *6*, 2503–2511.
10. Xue, H.; Feingold, G.; Stevens, B. Aerosol effects on clouds, precipitation, and the organization of shallow cumulus convection. *J. Atmos. Sci.* **2008**, *65*, 392–406.
11. Wang, H.; Feingold, G. Modeling Mesoscale Cellular Structures and Drizzle in Marine Stratocumulus. Part I: Impact of Drizzle on the Formation and Evolution of Open Cells. *J. Atmos. Sci.* **2009**, *66*, 3237–3256.
12. Sharon, T.M.; Albrecht, B.A.; Jonsson, H.H.; Minnis, P.; Khaiyer, M.M.; van Reken, T.M.; Seinfeld, J.; Flagan, R. Aerosol and cloud microphysical characteristics of rifts and gradients in maritime stratocumulus clouds. *J. Atmos. Sci.* **2006**, *63*, 983–997.

13. Hoppel, W.A.; Fitzgerald, J.W.; Frick, G.M.; Larson, R.E. Aerosol Size Distributions and Optical Properties Found in the Marine Boundary Layer Over the Atlantic Ocean. *J. Geophys. Res.* **1990**, *95*, 3659–3686.
14. Hudson, J.G.; Noble, S.; Tabor, S. Cloud supersaturations from CCN spectra Hoppel minima. *J. Geophys. Res. Atmos.* **2015**, *120*, 3436–3452.
15. Chen, Y.C.; Xue, L.; Lebo, Z.J.; Wang, H.; Rasmussen, R.; Seinfeld, J.H. A comprehensive numerical study of aerosol-cloud-precipitation interactions in marine stratocumulus. *Atmos. Chem. Phys.* **2011**, *11*, 9749–9769.
16. Wood, R.; Mechoso, C.R.; Bretherton, C.S.; Weller, R.A.; Huebert, B.; Straneo, F.; Albrecht, B.A.; Coe, H.; Allen, G.; Vaughan, G.; et al. The VA-MOS Ocean-Cloud-Atmosphere-Land Study Regional Experiment (VOCALS-REx): Goals, platforms, and field operations. *Atmos. Chem. Phys.* **2011**, *11*, 627–654.
17. Wood, R.; Bretherton, C.S.; Leon, D.; Clarke, A.D.; Zuidema, P.; Allen, G.; Coe, H. An aircraft case study of the spatial transition from closed to open mesoscale cellular convection over the Southeast Pacific. *Atmos. Chem. Phys.* **2011**, *11*, 2341–2370.
18. Clark, T.L.; Hall, W.D. Multi-domain simulations of the time dependent Navier–Stokes equations: Benchmark error analysis of some nesting procedure. *J. Comput. Phys.* **1991**, *92*, 456–481.
19. Planche, C.; Wobrock, W.; Flossmann, A.I.; Tridon, F.; Van Baelen, J.; Pointin, Y.; Hagen, M. The influence of aerosol particle number and hygroscopicity on the evolution of convective cloud systems and their precipitation: A numerical study based on the COPS observations on 12 August 2007. *Atmos. Res.* **2010**, *98*, 40–56.
20. Hiron, T.; Flossmann, A.I. A study of the role of the parameterization of heterogeneous ice nucleation for the modeling of microphysics and precipitation of a convective cloud. *J. Atmos. Sci.* **2015**, *72*, 3322–3339.
21. Frick, G.M.; Hoppel, W.A. Airship Measurements of Ship's Exhaust Plumes and Their Effect on Marine Boundary Layer Clouds. *J. Atmos. Sci.* **2000**, *57*, 2625–2648.
22. Stevens, B.; Moeng, C.; Ackerman, A.S.; Bretherton, C.S.; Chlond, A.; Roode, S.R.; Edwards, J.; Golaz, G.J.; Jiang, H.; Khairoutdinov, M.F.; et al. Evaluation of Large-Eddy Simulations via Observations of Nocturnal Marine. *Mon. Weather Rev.* **2005**, 1443–1462.

23. Wang, H.; Feingold, G.; Wood, R.; Kazil, J. Modelling microphysical and meteorological controls on precipitation and cloud cellular structures in Southeast Pacific stratocumulus. *Atmos. Chem. Phys.* **2010**, *10*, 6347–6362.
24. Durkee, P.A.; Chartier, R.E.; Brown, A.; Trehubenko, E.J.; Rogerson, S.D.; Skupniewicz, C.; Nielsen, K.E.; Platnick, S.; King, M.D. Composite Ship Track Characteristics. *J. Atmos. Sci.* **2000**, *57*, 2542–2553.
25. Twomey, S. The influence of pollution on the shortwave albedo of clouds. *J. Atmos. Sci.* **1977**, *34*, 1149–1154.
26. Possner, A.; Zubler, E.; Lohmann, U.; Schär, C. Real-case simulations of aerosol–cloud interactions in ship tracks over the Bay of Biscay. *Atmos. Chem. Phys.* **2015**, *15*, 2185–2201.

CHAPTER 15

Cloud Drop Number Concentrations Over the Western North Atlantic Ocean: Seasonal Cycle, Aerosol Interrelationships, and other Influential Factors

Hossein Dadashazar[1], David Painemal[2,3], Majid Alipanah[4], Michael Brunke[5], Seethala Chellappan[6], Andrea F. Corral[1], Ewan Crosbie[2,3], Simon Kirschler[7], Hongyu Liu[8], Richard H. Moore[2], Claire Robinson[2,3], Amy Jo Scarino[2,3], Michael Shook[2], Kenneth Sinclair[9,10], K. Lee Thornhill[2], Christiane Voigt[7], Hailong Wang[11], Edward Winstead[2,3], Xubin Zeng[5], Luke Ziemba[2], Paquita Zuidema[6], Armin Sorooshian[1,5]

[1]Department of Chemical and Environmental Engineering, University of Arizona, Tucson, AZ, USA

[2]NASA Langley Research Center, Hampton, VA, USA

Citation: Dadashazar, H., Painemal, D., Alipanah, M., Brunke, M., Chellappan, S., Corral, A. F., Crosbie, E., Kirschler, S., Liu, H., Moore, R. H., Robinson, C., Scarino, A. J., Shook, M., Sinclair, K., Thornhill, K. L., Voigt, C., Wang, H., Winstead, E., Zeng, X., Ziemba, L., Zuidema, P., and Sorooshian, A.: "Cloud drop number concentrations over the western North Atlantic Ocean: seasonal cycle, aerosol interrelationships, and other influential factors", Atmos. Chem. Phys., 21, 10499–10526, https://doi.org/10.5194/acp-21-10499-2021, 2021.

[3]Science Systems and Applications, Inc., Hampton, VA, USA
[4]Department of Systems and Industrial Engineering, University of Arizona, Tucson, AZ, USA

[5]Department of Hydrology and Atmospheric Sciences, University of Arizona, Tucson, AZ, USA

[6]Rosenstiel School of Marine and Atmospheric Science, University of Miami, Miami, FL, USA

[7] Institute of Atmospheric Physics, German Aerospace Center, Oberpfaffenhofen, Germany

[8]National Institute of Aerospace, Hampton, VA, USA

[9]NASA Goddard Institute for Space Studies, New York, NY, USA

[10]Universities Space Research Association, Columbia, MD, USA

[11]Atmospheric Sciences and Global Change Division, Pacific Northwest National Laboratory, Richland, WA, USA

ABSTRACT

Cloud drop number concentrations (N_d) over the western North Atlantic Ocean (WNAO) are generally highest during the winter (DJF) and lowest in summer (JJA), in contrast to aerosol proxy variables (aerosol optical depth, aerosol index, surface aerosol mass concentrations, surface cloud condensation nuclei (CCN) concentrations) that generally peak in spring (MAM) and JJA with minima in DJF. Using aircraft, satellite remote sensing, ground-based in situ measurement data, and reanalysis data, we characterize factors explaining the divergent seasonal cycles and furthermore probe into factors influencing N_d on seasonal timescales. The results can be summarized well by features most pronounced in DJF, including features associated with cold-air outbreak (CAO) conditions such as enhanced values of CAO index, planetary boundary layer height (PBLH), low-level liquid cloud fraction, and cloud-top height, in addition to winds aligned with continental outflow. Data sorted into high- and low-N_d days in each season, especially in DJF, revealed that all of these conditions were enhanced on the high-N_d days, including reduced sea level pressure and stronger wind speeds. Although aerosols may be more abundant in MAM and JJA, the conditions needed to activate those particles into cloud droplets are weaker than in colder months, which is demonstrated by calculations of the strongest (weakest) aerosol indirect effects in DJF (JJA) based on comparing N_d to perturbations in four different aerosol proxy variables (total and sulfate aerosol optical depth, aerosol index, surface mass concentration of sulfate). We used three machine learning models and up to 14 input variables to infer about most influential factors related to N_d for DJF and JJA, with the best performance obtained with gradient-boosted regression tree (GBRT) analysis. The model results indicated that cloud fraction was the most important input variable, followed by some combination (depending on season) of CAO index and surface mass concentrations of sulfate and organic carbon. Future work is recommended to further understand aspects uncovered here such as impacts of free tropospheric aerosol entrainment on clouds, degree of boundary layer coupling, wet scavenging, and giant CCN effects on aerosol–N_d relationships, updraft velocity, and vertical structure of cloud properties such as adiabaticity that impact the satellite estimation of N_d.

INTRODUCTION

Aerosol indirect effects remain the dominant source of uncertainty in estimates of total anthropogenic radiative forcing (Boucher et al., 2013; Myhre et al., 2013). Central to these effects is knowledge about cloud drop number concentration (N_d), as it is the connection between the subset of particles that activate into drops (cloud condensation nuclei, CCN) and cloud properties. It is widely accepted that warm clouds influenced by higher number concentrations of aerosol particles have elevated N_d and smaller drops (all else held fixed), resulting in enhanced cloud albedo at fixed liquid water path (Twomey, 1977) and potentially suppressed precipitation (Albrecht, 1989) and increased vulnerability to overlying air resulting from enhanced cloud-top entrainment (Ackerman et al., 2004).

Reducing uncertainty in how aerosols and clouds interact within a given meteorological context requires accurate estimates of N_d and aerosol concentrations and properties. Since intensive field studies struggle to obtain broad spatial and temporal coverage of such data, satellite remote sensing and reanalysis datasets are relied on for studies examining intra- and interannual features over large spatial areas. Limitations of satellite retrievals are important to recognize. N_d is not directly retrieved but derived using other parameters (e.g., cloud optical depth, cloud drop effective radius, cloud-top temperature) and with assumptions about cloud adiabatic growth and N_d being vertically constant (Grosvenor et al., 2018). Aerosol number concentrations are usually represented by a columnar parameter such as aerosol optical depth (AOD) and thus not directly below clouds, which is the aerosol layer most likely to interact with the clouds. Furthermore, aerosol data are difficult to retrieve in cloudy columns. Reanalysis datasets circumvent issues for the aerosol parameters as they provide vertically resolved data (e.g., surface layer and thus below clouds) and are available for cloudy columns.

Of special interest in this work is the western North Atlantic Ocean (WNAO) where decades of extensive research have been conducted for topics largely unrelated to aerosol–cloud interactions (Sorooshian et al., 2020), thereby providing an opportunity for closing knowledge gaps for this area in a region with a wide range of aerosol and meteorological conditions (Corral et al., 2021; Painemal et al., 2021). Past work showed different seasonal cycles of AOD and N_d in this region (Grosvenor et al., 2018; Sorooshian et al., 2019), which partly motivates this study to unravel why N_d behaves differently on seasonal timescales. A previous study investigating seasonal cycles of N_d in the North Atlantic region found that cloud microphysical properties were primarily dependent on CCN concentrations while cloud macrophysical properties were more dependent on meteorological conditions (e.g., Sinclair et al., 2020). However, due to the complexity of interactions involved and the co-variability between individual components, the magnitude and sign of these feedbacks remain uncertain.

This study uses a multitude of datasets to characterize the N_d seasonal cycle and factors related to N_d variability. The structure of the results and discussion is as follows: (i) case study flight highlighting the wide range of N_d in wintertime and factors potentially affecting that variability; (ii) seasonal cycle of N_d and aerosol concentrations based on different proxy variables; (iii) seasonal cycles of factors potentially influential for N_d such as aerosol size distribution, vertical distribution of aerosol, humidity effects, and aerosol–cloud interactions; (iv) composite analysis of influential factors on high- and low-N_d days in each season; (v) modeling analysis to probe more deeply into N_d relationships with other parameters for winter and summer seasons; and (vi) discussion of other factors relevant to N_d unexplored in this work.

METHODS

Study Region

We focus on the WNAO, defined here as being bounded by 25–50° N and 60–85° W. A subset of the results focuses on six individual sub-domains representative of different parts of the WNAO (shown later), with five just off the East Coast extending from south to north (south: S; central-south: C-S; central: C; central-north: C-N; north: N) and one over Bermuda.

Datasets

Satellite Observations (CERES-MODIS/CALIPSO)

Relevant cloud parameters were obtained from the Clouds and the Earth's Radiant Energy System (CERES) edition 4 products (Minnis et al., 2011, 2020), which are based on the application of CERES's retrieval algorithms on the radiances measured by the MODerate resolution Imaging Spectroradiometer (MODIS) instrument aboard the Aqua satellite. Aqua observations used to estimate N_d were from the daytime overpasses of the satellite around 13:30 local time (LT). Level 3 daily cloud properties at 1° × 1° spatial resolution (listed in Table 1) were used for the period between January 2013 and December 2017 from CERES-MODIS edition 4 Single Scanning Footprint (SSF) products (Loeb et al., 2016). The CERES-MODIS SSF Level 3 product includes 1° × 1° averaged data according to the cloud-top pressure of individual pixels: low (heights below 700 hPa), mid-low (heights within 700–500 hPa), mid-high (heights within 500–300 hPa), and high (heights above 300 hPa) level clouds. For this study, we only use low-cloud averages.

Table 1. Summary of various data products used in this study.

Parameter	Data source	Spatial resolution	Vertical level	Date range	Spatial area	Temporal resolution
Cloud optical thickness	CERES-MODIS	1° × 1°	NA	1 January 2013–31 December 2017	25–50° N, 60–85° W	Daily
Cloud effective radius	CERES-MODIS	1° × 1°	NA	1 January 2013–31 December 2017	25–50° N, 60–85° W	Daily
Cloud fraction	CERES-MODIS	1° × 1°	NA	1 January 2013–31 December 2017	25–50° N, 60–85° W	Daily

Cloud-top temperature	CERES-MODIS	1° × 1°	NA	1 January 2013–31 December 2017	25–50° N, 60–85° W	Daily
Cloud effective height	CERES-MODIS	1° × 1°	NA	1 January 2013–31 December 2017	25–50° N, 60–85° W	Daily
Cloud-top pressure	CERES-MODIS	1° × 1°	NA	1 January 2013–31 December 2017	25–50° N, 60–85° W	Daily
Precipitation	PERSIANN-CDR	1° × 1°	NA	1 January 2013–31 December 2017	25–50° N, 60–85° W	Daily
Aerosol extinction (532 nm)	CALIPSO/CALIOP	5 km	NA	1 January 2013–31 December 2017	25–50° N, 60–85° W	Daily
Total aerosol extinction AOT (550 nm)	MERRA-2	1° × 1°	NA	1 January 2013–31 December 2017	25–50° N, 60–85° W	Daily
Total aerosol Ångström parameter (470–870 nm)	MERRA-2	1° × 1°	NA	1 January 2013–31 December 2017	25–50° N, 60–85° W	Daily
Sulfate extinction AOT (550 nm)	MERRA-2	1° × 1°	NA	1 January 2013–31 December 2017	25–50° N, 60–85° W	Daily
Sea salt extinction AOT (550 nm)	MERRA-2	1° × 1°	NA	1 January 2013–31 December 2017	25–50° N, 60–85° W	Daily
Dust extinction AOT (550 nm)	MERRA-2	1° × 1°	NA	1 January 2013–31 December 2017	25–50° N, 60–85° W	Daily

Organic carbon extinction AOT (550 nm)	MERRA-2	1° × 1°	NA	1 January 2013–31 December 2017	25–50° N, 60–85° W	Daily
Black carbon extinction AOT (550 nm)	MERRA-2	1° × 1°	NA	1 January 2013–31 December 2017	25–50° N, 60–85° W	Daily
Sulfate surface mass concentration	MERRA-2	1° × 1°	Surface	1 January 2013–31 December 2017	25–50° N, 60–85° W	Daily
Sea salt surface mass concentration	MERRA-2	1° × 1°	Surface	1 January 2013–31 December 2017	25–50° N, 60–85° W	Daily
Dust surface mass concentration	MERRA-2	1° × 1°	Surface	1 January 2013–31 December 2017	25–50° N, 60–85° W	Daily
Organic carbon surface mass concentration	MERRA-2	1° × 1°	Surface	1 January 2013–31 December 2017	25–50° N, 60–85° W	Daily
Black carbon surface mass concentration	MERRA-2	1° × 1°	Surface	1 January 2013–31 December 2017	25–50° N, 60–85° W	Daily
Sea level pressure	MERRA-2	1° × 1°	Surface	1 January 2013–31 December 2017	25–50° N, 60–85° W	Daily
Geopotential height	MERRA-2	1° × 1°	850 hPa	1 January 2013–31 December 2017	25–50° N, 60–85° W	Daily
Sea surface temperature	MERRA-2	1° × 1°	Sea surface	1 January 2013–31 December 2017	25–50° N, 60–85° W	Daily

Air temperature	MERRA-2	1° × 1°	Surface, 850, 700 hPa	1 January 2013–31 December 2017	25–50° N, 60–85° W	Daily
Relative humidity	MERRA-2	1° × 1°	1000–500 hPa	1 January 2013–31 December 2017	25–50° N, 60–85° W	Daily
Wind speed	MERRA-2	1° × 1°	2 m, 950 hPa	1 January 2013–31 December 2017	25–50° N, 60–85° W	Daily
Planetary boundary layer height	MERRA-2	1° × 1°	NA	1 January 2013–31 December 2017	25–50° N, 60–85° W	Daily
Vertical pressure velocity	MERRA-2	1° × 1°	800 hPa	1 January 2013–31 December 2017	25–50° N, 60–85° W	Daily
Aerosol/ cloud	Airborne: ACTIVATE	–	NA	22 February 2020	34.08–37.16° N, 72.31–76.64° W	1 s
CCN-Cape Cod	Ground-based measurement	Point measurement	Surface	16 July 2012–4 May 2013	41.67° N, 70.30° W	1 s

N_d is estimated based on an adiabatic cloud model (Grosvenor et al., 2018):

$$N_{\mathrm{d}} = \frac{\sqrt{5}}{2\pi k}\left(\frac{f_{\mathrm{ad}}C_{\mathrm{w}}\tau}{Q_{\mathrm{ext}}\rho_{\mathrm{w}}r_{\mathrm{e}}^{5}}\right)^{1/2}, \quad (1)$$

where is cloud optical depth and r_e is cloud drop effective radius, both of which are obtained from CERES-MODIS for low-level (i.e., surface to 700 hPa) liquid clouds. Q_{ext} is the unitless extinction efficiency factor, assumed to be 2 for liquid cloud droplets, and ρ_w is the density of water (1 g cm^{-3}). Methods described in Painemal (2018) were used to estimate parameters in Eq. (1) as follows. (i) Adiabatic water lapse rate (C_w) was determined using cloud-top pressure and temperature provided by CERES-MODIS. (ii) The N_d estimation is often corrected for the sub-adiabatic profile by applying the adiabatic value (f_{ad}), but in this work, a value of $f_{ad} = 1$ was assumed due to both lack of consensus on its value and its relatively minor impact on N_d estimation (Grosvenor et al., 2018). (iii) k is the parameter representing the width of the droplet spectrum and was assumed to be 0.8 over the ocean. Statistics of N_d are often estimated after screening daily observations based on cloud fractions (Wood, 2012; Grosvenor et al., 2018). The purpose of such filters is to reduce the uncertainties associated with the estimation of N_d (Eq. 1) driven by the errors in the retrieval of r_e and from MODIS's observed reflectance in a highly heterogeneous cloud field. However, this may inadvertently mask the effects of cloud regime on aerosol–cloud interactions by only including certain low-level cloud types in the analyses (e.g., closed-cell stratocumulus). Therefore, we use all N_d data regardless of cloud fraction with exceptions being Sects. 3.5 and 4.2 where a filter of low-level liquid cloud fraction (i.e., $CF_{low\text{-}liq.} \geq 0.1$) was applied.

The Cloud-Aerosol Lidar with Orthogonal Polarization (CALIOP) instrument aboard the Cloud-Aerosol Lidar and Infrared Pathfinder Satellite Observations (CALIPSO) provides data on the vertical distribution of aerosols (Winker et al., 2009). Nighttime extinction profiles were acquired from Level 2 version 4.20 products (i.e., 5 km aerosol profile data), between January 2013 and December 2017. We averaged the Level 2 daily extinctions in different 4° × 5° sub-domains (shown later) to obtain the seasonal profiles after applying the screening scheme outlined in Tackett et al. (2018).

MERRA-2

Aerosol data were obtained from the Modern-Era Retrospective Analysis for Research and Applications-Version 2 (MERRA-2) (Gelaro et al., 2017). MERRA-2 is a multidecadal reanalysis where meteorological and aerosol observations are jointly assimilated into the Goddard Earth Observation System version 5 (GEOS-5) data assimilation system (Buchard et al., 2017; Randles et al., 2017). Aerosols in MERRA-2 are simulated with a radiatively coupled version of the Goddard Chemistry, Aerosol, Radiation, and Transport model (GOCART; Chin et al., 2002; Colarco et al., 2010). GOCART treats the sources, sinks, and chemistry of 15 externally mixed aerosol mass mixing ratio tracers, which include sulfate, hydrophobic and hydrophilic black and organic carbon, dust (five size bins), and sea salt (five size bins). MERRA-2 includes assimilation of bias-corrected Collection 5 MODIS AOD, bias-corrected AOD from the Advanced Very High Resolution Radiometer (AVHRR) instruments, AOD retrievals from the Multiangle Imaging Spectro-Radiometer (MISR) over bright surfaces, and ground-based Aerosol Robotic Network (AERONET) direct measurements of AOD (Gelaro et al., 2017). In this study we used total and speciated (i.e., sea salt, dust, black carbon, organic carbon, and sulfate) AOD at 550 nm between January 2013 and December 2017 at times relevant to Aqua's overpass time (13:30 LT). Aerosol index was calculated as the product of AOD and the Ångström parameter. MERRA-2 also provides surface mass concentrations of aerosol species including sea salt, dust, black carbon, organic carbon, and sulfate, which were used as a measure of aerosol levels in the planetary boundary layer (PBL).

MERRA-2 data were also used for environmental variables including both thermodynamic (e.g., temperature and relative humidity) and dynamic parameters (e.g., sea level pressure (SLP) and geopotential heights) (Gelaro et al., 2017) listed in Table 1. Bilinear interpolation was applied to transfer all MERRA-2 variables (Table 1) from their original 0.5° × 0.625° spatial resolution to the equivalent 1° × 1° grid in CERES-MODIS Level 3 data.

Precipitation Data

Daily precipitation data were obtained from the Precipitation Estimation from Remotely Sensed Information using Artificial Neural Networks–Climate Data Record (PERSIANN-CDR) data product (Ashouri et al., 2015; Nguyen et al., 2018). Bilinear interpolation was applied to convert the PERSIANN-CDR data from their native spatial resolution (i.e., 0.25° × 0.25°) to equivalent 1° × 1° grids in CERES-MODIS Level 3 data. It is important to note that we use daily averaged PERSIANN-CDR precipitation and, therefore, there is some temporal mismatch with the daily N_d value from MODIS-Aqua that comes at one time of the day. This can contribute to some level of uncertainty for the discussions based on analyses involving relationships between precipitation and N_d.

Surface-based CCN Data

Cloud condensation nuclei (CCN) data were obtained from the U.S. Department of Energy's Two-Column Aerosol Project (TCAP) (Berg et al., 2016) to examine the seasonal variations in CCN number concentration at a representative site by Cape Cod, Massachusetts (41.67° N, 70.30° W), over the US East Coast. TCAP was a campaign conducted between June 2012 and June 2013 to investigate aerosol optical and physicochemical properties and interactions between aerosols and clouds (Berg et al., 2016; Liu and Li, 2019). CCN data were available between July 2012 and May 2013 at multiple supersaturations with some gaps in the data collection (i.e., November–December); for simplicity, we focused on CCN data measured at a single supersaturation of 1% owing to relatively better data coverage compared to lower supersaturations. We note that this higher supersaturation is not necessarily representative of that relevant to the clouds of interest but is still insightful for understanding the seasonal cycle of CCN concentration. The qualitative seasonal cycle of CCN concentration at 1% matches those at lower supersaturations (e.g., 0.15%–0.8%).

Airborne in situ Data

We used airborne in situ data collected during the fifth research flight (RF05) of the Aerosol Cloud meTeorology Interactions oVer the western ATlantic Experiment (ACTIVATE) campaign. One flight is used both for simplicity and because it embodied conditions relevant to the discussion of other results. The mission concept involves joint flights between the NASA Langley UC-12 King Air and HU-25 Falcon such that the former flies around 8–10 km, and the latter flies in the boundary layer to simultaneously collect data on aerosol, cloud, gas, and meteorological parameters in the same column (Sorooshian et al., 2019). The Falcon flew in a systematic way to collect data at different vertical regions relative to cloud, including the following of relevance to this study: BCB – below cloud base; ACB – above cloud base; BCT – below cloud top; and Min. Alt – minimum altitude the plane flies at (~ 150 m).

This study makes use of the HU-25 Falcon data from the following instruments: fast cloud droplet probe (FCDP; D_p ~ 3–50 μm) (SPEC Inc.) aerosol and cloud droplet size distributions for quantification of cloud liquid water content (LWC), N_d, and aerosol number concentrations with D_p exceeding 3 μm in cloud-free air (termed FCDP-aerosol); two-dimensional stereo (2DS; D_p ~ 28.5–1464.9 μm) (SPEC Inc.) probe for estimation of rain water content (RWC) by integrating raindrop ($D_p \geq 39.9$ μm) size distributions; cloud condensation nuclei (CCN; DMT) counter for CCN number concentrations; laser aerosol spectrometer (LAS; TSI model 3340) and condensation particle counter (CPC; TSI model 3772) for aerosol number concentrations with D_p between 0.1–1 μm and above 10 nm, respectively; highresolution time-of-flight aerosol mass spectrometer

(AMS; Aerodyne) for submicrometer non-refractory aerosol composition (DeCarlo et al., 2008), operated in 1 Hz Fast-MS mode and averaged to 25 s time resolution; and turbulent air-motion measurement system (TAMMS) for winds and temperature (Thornhill et al., 2003).

CCN, LAS, CPC, and AMS data were collected downstream of an isokinetic double diffuser inlet (BMI, Inc.), whereas the AMS and LAS also sampled downstream of a counterflow virtual impactor (CVI) inlet (BMI, Inc.) when in cloud (Shingler et al., 2012). However, a filter was applied to remove LAS data when the CVI inlet was used. Measurements from the CCN counter, LAS, CPC, and FCDP-aerosol are only shown in cloud-free and rain-free conditions, distinguished by LWC <0.05 gm^{-3} and RWC <0.05 gm^{-3}, respectively, and also excluding data collected 20 s before and after evidence of rain or cloud. Estimation of supermicrometer particles from FCDP measurements was performed after conducting the following additional screening steps to minimize cloud droplet artifacts: (i) only samples with RH <98% were included; (ii) data collected during ACB and BCT legs were excluded. CCN, LAS, CPC, and AMS measurements are reported at standard temperature and pressure (i.e., 273K and 101.325 kPa) while FCDP and 2DS measurements correspond to ambient conditions.

Regression Analyses

Regression modeling was conducted to investigate relationships between environmental variables and N_d. The gradient-boosted regression trees (GBRT) model, classified as a machine learning (ML) model, is used, consisting of several weak learners (i.e., regression trees with a fixed size) that are designed and subsequently trained to improve prediction accuracy by fitting the model's trees on residuals rather than response values (Hastie et al., 2009). Desirable characteristics of the GBRT model include both its capacity to capture non-linear relationships and being less vulnerable to overfitting (Persson et al., 2017; Fuchs et al., 2018; Dadashazar et al., 2020). Two separate GBRT models were trained using daily CERES-MODIS N_d data (1° × 1°) in winter (DJF) and summer (JJA) to reveal potential variables impacting N_d. Winter and summer are chosen as they exhibit the highest and lowest N_d concentrations, respectively, among all seasons over the WNAO.

Many variables were picked as input parameters (Table 2) for the GBRT model, categorized as being aerosol, dynamic/thermodynamic, or cloud variables. Aerosol parameters included MERRA-2 surface mass concentrations for sulfate, sea salt, dust, and organic carbon. Black carbon concentration was removed from input parameters because of its high correlation (R^2 = 0.6) with organic carbon. The following is the list of thermodynamic/dynamic input parameters derived from MERRA-2: vertical pressure velocity at 800 hPa (ω_{800}), planetary boundary layer height (PBLH), cold-air outbreak (CAO) index, wind speed and wind direction at 2 m (wind$_{2\ m}$ and wind-dir$_{2\ m}$), and relative humidity (RH) in the PBL and free troposphere represented by RH$_{950}$ and RH$_{800}$, respectively. CAO index is defined as the difference between skin potential temperature ($_{skt}$) and air potential temperature at 850 hPa ($_{850}$) (Papritz et al., 2015). Updraft velocity plays a crucial role in the activation of aerosol into cloud droplets in warm clouds (Feingold, 2003; Reutter et al., 2009). Since the direct representation of updraft speed is not available from reanalysis data, near-surface wind speed (i.e., wind$_{2\ m}$) is used as a representative proxy parameter as an input parameter to the regression models. CERES-MODIS cloud parameters include liquid cloud fraction and cloud-top height for low-level clouds. In addition, PERSIANN-CDR daily precipitation (Rain) was included as a relevant cloud parameter.

Table 2. List of input parameters used as predictor variables in the GBRT and linear models. Variables are grouped into three general categories.

	Parameter
Aerosol	Sulfate surface mass concentration (Sulfate$_{sf\text{-}mass}$) Sea salt surface mass concentration (Sea-salt$_{sf\text{-}mass}$) Dust surface mass concentration (Dust$_{sf\text{-}mass}$) Organic carbon surface mass concentration (OC$_{sf\text{-}mass}$)
Cloud	Low-level liquid cloud fraction (CF$_{low\text{-}liq.}$) Low-level liquid cloud-top effective height (Cloud-top$_{low\text{-}liq.}$) Precipitation rate (Rain)
Dynamic/thermodynamic	Cold-air outbreak index (CAO$_{index}$): θ*skt-θ850 Relative humidity at 950 hPa (RH$_{950}$) Relative humidity at 800 hPa (RH$_{800}$) Vertical pressure velocity at 800 hPa (ω_{800}) Wind speed at 2 m (Wind$_{2\ m}$) Wind direction at 2 m (Wind-dir$_{2\ m}$) Planetary boundary layer height (PBLH)

*Skin potential temperature.

Data were split into two sets: training/validation (70 %) and testing (30 %). Five-fold cross-validation was implemented to train the GBRT model using the training/validation data. Furthermore, both performance and generalizability of the trained models were tested via the aid of the test set, which was not used in the training process. Hyperparameters of the GBRT models were optimized through a combination of both random and grid search methods. Table S1 in the Supplement shows the list of important hyperparameters of the GBRT model and associated ranges tested via random and grid search methods. The optimized model hyperparameters can also be found in Table S1. The GBRT models were performed using the scikit-learn module designed in Python (Pedregosa et al., 2011).

The regression analyses were not performed solely to construct and provide a highly accurate model useful for prediction, but rather to disclose and examine the possible effects of the relevant input variables on N_d considering all the shortcomings of such analyses. For instance, there is some level of interdependency between input variables. To reduce unwanted consequences of correlated features, the interpretation of the results was done with the aid of accumulated local effect (ALE) plots, which are specifically designed to be unbiased to the correlated input variables (Apley and Zhu, 2020). ALE plots illustrate the influence of input variables on the response parameter in ML models. The ALE value for a particular variable s at a specific value of x_s (i.e., $f_{s,\mathrm{ALE}}(x_s)$) can be calculated as follows:

$$f_{s,\mathrm{ALE}}(x_s) = \int_{z_{0,1}}^{x_s} \int_{x_c} f^s(z_s, x_c)\, P(x_c \mid z_s) \mathrm{d}x_c \mathrm{d}z_s - \text{constant}, \tag{2}$$

where $f^s(z_s x_c)$ is the gradient of model's response with respect to variable s (i.e., local effect) and $P(x_c|z_s)$ is the conditional distribution of x_c, where c denotes the other input variables rather than s, and x_c is the associated point in the variable space of c. $z_{0,1}$ is chosen arbitrarily below the smallest observation of feature s (Apley and Zhu, 2020). The steps in Eq. (2) can be summarized as follows (Molnar, 2019; Apley and Zhu, 2020): (i) the average change in the model's prediction is calculated using the conditional distribution of features; (ii) the average change will then be accumulated by integrating it over feature s; and (iii) a constant will be subtracted to vertically center (i.e., the average of ALE becomes zero) the ALE plot. The aforementioned steps, although seemingly complex, assure the avoidance of undesired extrapolation (especially an issue for correlated variables) occurring in alternative approaches such as partial dependence

(PD) plots. The value of $f_{s,\text{ALE}}(x_s)$ can be viewed as the difference between the model's response at x_s and the average prediction. We used the source code available in https://github.com/blent-ai/ALEPython (Jumelle et al., 2021) for the calculation of ALE plots.

RESULTS AND DISCUSSION

Aircraft Case Study of N_d Gradient

ACTIVATE Research Flight 5 (RF05) on 22 February 2020 demonstrates the wide range in N_d offshore in the PBL (1.6 km) over the WNAO (Fig. 1). On this day, the ACTIVATE study region was dominated by a surface high-pressure system centered over the southeastern US, with a significant ridge axis extending from the main high to the eastnortheast off the Virginia–North Carolina coast and into the WNAO.

Aloft, the flight region was located in northwesterly flow behind a trough offshore. This setup led to subsidence in the region and generally clear skies, except where scattered to broken marine boundary layer clouds formed along and east of the Gulf Stream.

The 2 d NOAA HYSPLIT (Stein et al., 2015; Rolph et al., 2017) back trajectories using the "model vertical velocity" method and "REANALYSIS" meteorology data indicate air in the flight region (between 0–3 km) had wrapped around the surface high from the north and left the New England coast 12–24 h before-hand (with a descending profile).

Along the flight segment shown, winds were approximately 6 m s^{-1}, out of the northnorthwest during the initial descent, Min. Alt. 1, and BCB1 legs and primarily from the northeast for the other sections of the flight. Sea surface temperatures were 6–9 °C near the coast during the descent and Min. Alt. 1 leg (readers are referred to Fig. 1's caption for the definition of different legs); 21–25 °C over the Gulf Stream during the BCB1, ACB1, and BCB2 legs; and 17–20 °C for the remainder of the flight segment shown. The majority of the segment was in or below the boundary layer clouds, with cloud base around 900–1100 m and cloud top around 1750 m. Note that the initial BCB1 leg was much lower at around 460 m, likely reflecting a shallower marine boundary layer and cloud base near the much colder waters close to the coast. Static air temperature ranged between 0–10 °C, except for the BCT1 leg where temperatures were around −2.3 °C.

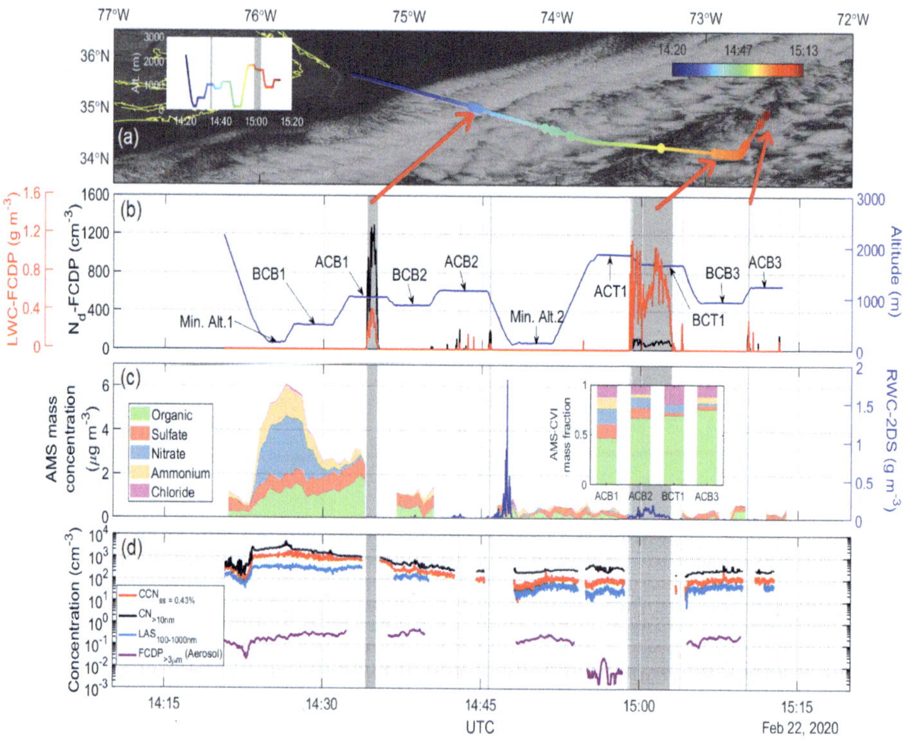

Figure 1. Time series of selected parameters measured by the HU-25 Falcon aircraft during a selected segment of RF05 on 22 February 2020. **(a)** Overlayed flight track on GOES 16 visible imagery obtained at 14:55:04 UTC. **(b)** Altitude, cloud liquid water content (LWC), and N_d, with the latter two obtained from the FCDP. **(c)** Rain water content (RWC) measured by 2DS probe, AMS speciated mass concentration in cloud/rain-free air, and AMS mass fractions for droplet residual particles in cloud as measured downstream of a CVI inlet. **(d)** Number concentrations for CCN at 0.43% supersaturation and particles for three diameter ranges: above 10 nm (CPC), 100–1000 nm (LAS), and above 3 μm (FCDP). Shaded gray areas in panels **(b)–(d)** highlight cloudy periods identified as having LWC ≥ 0.05 g m^{-3}. Locations of the cloudy regions are pointed to with red arrows in the satellite imagery. Level legs are defined as follows: BCB: below cloud base; ACB: above cloud base; Min. Alt.: minimum altitude the plane flies at (~ 150 m); ACT: above cloud top; BCT: below cloud top.

N_d values from the FCDP ranged from a maximum value of 1298 cm^{-3} closer to the coast during the ACB1 leg (35.00° N, 74.55° W) to a minimum of 19 cm^{-3} farther away in the BCT1 leg (34.32° N, 72.73° W). The minimum N_d value in the ACB3 leg was 85 cm^{-3} (34.11° N, 72.80° W), which is a fairer comparison to the ACB1 leg compared to the BCT1 leg in terms of being closer to cloud base. The mean N_d values (cm^{-3}) in the cloudy portions of the ACB1, BCT1, and ACB3 legs were as follows: 849, 77, and 143.

Based on the nearest BCB legs adjacent to the maximum and minimum N_d values (BCB1 = 35.31° N, 74.95° W; BCB3 = 34.41° N, 72.70° W), there was a significant offshore gradient in LAS submicrometer particle number concentration and AMS non-refractory aerosol mass, ranging from as high as 424 cm^{-3} and 5.60 $\mu g\ m^{-3}$ (during BCB1) to as low as 21 cm^{-3} and 0.27 $\mu g\ m^{-3}$ (during BCB3). The mean values of submicrometer particle number concentration and AMS non-refractory aerosol for the two BCB legs were as follows: 277 cm^{-3}/3.64 $\mu g\ m^{-3}$ (BCB1) and 48 cm^{-3}/0.42 $\mu g\ m^{-3}$ (BCB3). The higher N_d value (1298 cm^{-3}) relative to LAS aerosol concentration (424 cm^{-3}) at the near-shore point is suggestive of aerosol smaller than 0.1 μm activating into drops.

This is supported by the fact that both CCN (supersaturation = 0.43 %) and CPC number concentrations with D_p > 10 nm exhibited mean values of 980 and 1723 cm^{-3} in the BCB1 leg, respectively, dropping to 98 and 260 cm^{-3} in the BCB3 leg. For the duration of the flight portion shown in Fig. 1, supermicrometer concentrations varied over 2 orders of magnitude (0.002–0.51 cm^{-3}) and expectedly did not exhibit a pronounced offshore gradient as it is naturally emitted from the ocean.

Closer to shore during the Min. Alt. 1 leg, nitrate was the dominant aerosol species (~ 70% mass fraction). Farther offshore during both the BCB1 leg and cloud-free portion of the ACB1 leg, organics were the dominant constituent (~ 46% mass fraction), whereas farther during the BCB3 leg, the mean mass fraction of sulfate was the highest (75 %). Droplet residual particle data show a greater contribution of organics farther offshore, increasing from 46% to 75% between the ACB1 and ACB3 legs, respectively.

These composition results, albeit limited to the non-refractory portion of submicrometer aerosol particles, reveal significant changes with distance offshore indicative of varying chemical properties of particles activating into droplets.

The cloudy portions of ACB1 are characterized as having little or no rain with a maximum RWC value of 0.02 $g\ m^{-3}$ and mean value of 0.003 $g\ m^{-3}$. There is a notable RWC peak at the beginning of the Min. Alt. 2 leg, reaching as high as 1.81 $g\ m^{-3}$ associated with clouds aloft. The precipitation occurrence was also evident in a subsequent BCT1 leg where RWC reached as high as 0.18 g m^{-3}. GOES satellite imagery of the study region (Fig. 1) also reflects the effect of precipitation on cloud morphology where clouds farther offshore resemble open-cell structures. Associated scavenging of particles through the washout process is presumed to contribute to the decline in aerosol concentrations with distance offshore.

Figure 1 shows changes in aerosol characteristics coincident with the large gradient in N_d. While ACTIVATE airborne data collection is ongoing to build flight statistics over multiple years, the wide changes in microphysical properties in RF05 motivate looking at other datasets with broader spatiotemporal coverage to learn about potential seasonally dependent drivers of N_d, including meteorological parameters that vary throughout the year. Furthermore, other datasets can provide insight into the source(s) of seasonal discrepancy between columnar aerosol remote sensing parameters and N_d.

Seasonal Cycles of N_d and AOD

Figure 2 illustrates the seasonal differences in MERRA-2 AOD and CERES-MODIS N_d over the WNAO that partly motivate this study. Seasonal mean values (±standard deviation) of AOD/N_d (cm^{-3}) were as follows for the entire WNAO: DJF = 0.11 ± 0.03/64.1 ± 18.0; MAM = 0.16 ± 0.03/60.4 ± 13.1; JJA = 0.15 ± 0.03/49.1 ± 10.1; and SON = 0.11 ± 0.03/50.3 ± 13.9. In contrast to AOD, N_d values and low-cloud fraction (Fig. 2c) were highest in DJF and lowest in JJA.

DJF showed notably high N_d near the coast, qualitatively consistent with the airborne data. The seasons with the greatest AOD values, accompanied by the most pronounced spatial gradient offshore, were JJA and MAM. The offshore gradient owes to continental pollution outflow (Corral et al., 2021, and references therein).

In contrast, DJF and SON exhibited lower AOD values with a distinct area of higher AOD values offshore between ~ 35–40° N accounted for by sea salt. MERRA-2 speciated AOD data (Fig. 3) indicate that sea salt and sulfate dominate total AOD regardless of season and that sulfate, organic carbon, and black carbon most closely follow the offshore gradient pattern owing to continental sources. Dust and sea salt have different spatial distributions with the former derived from sources such as North Africa leading to enhanced AODs < 30° N in particular in JJA and sea salt being enhanced offshore in particular in JJA.

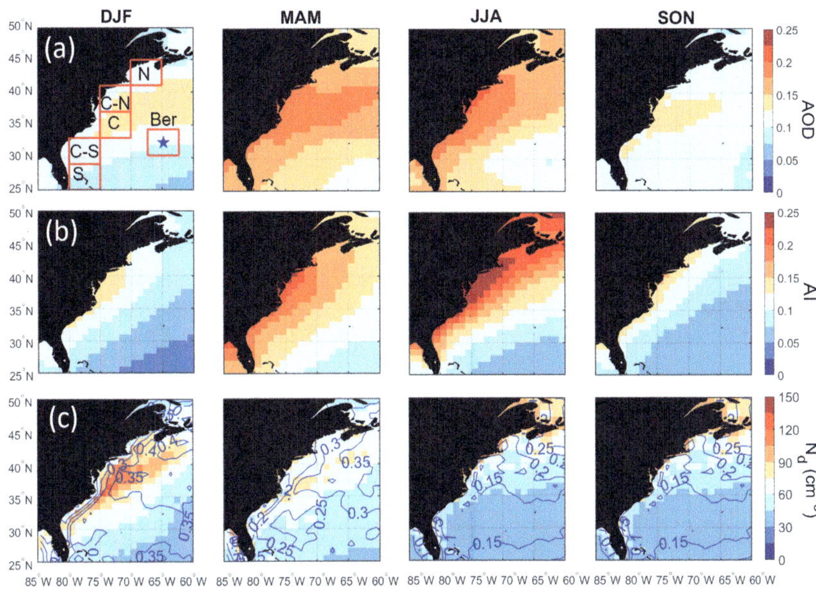

Figure 2. Seasonal spatial maps for **(a)** MERRA-2 aerosol optical depth (AOD), **(b)** MERRA-2 aerosol index (AI), and **(c)** cloud drop number concentration (N_d) over the western North Atlantic Ocean (WNAO). Contours in panel **(c)** represent low-level (cloud-top pressure > 700 hPa) liquid cloud fraction ($CF_{low\text{-}liq.}$). Cloud data are based on daily Level 3 data from CERES-MODIS. The maps are based on data between January 2013 and December 2017. The boxes in the top left panel represent sub-domains examined in more detail throughout the study, with the blue star denoting Bermuda.

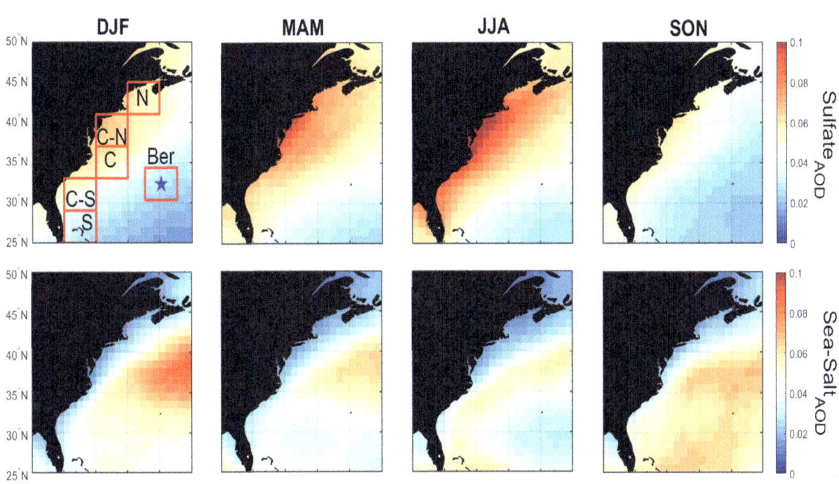

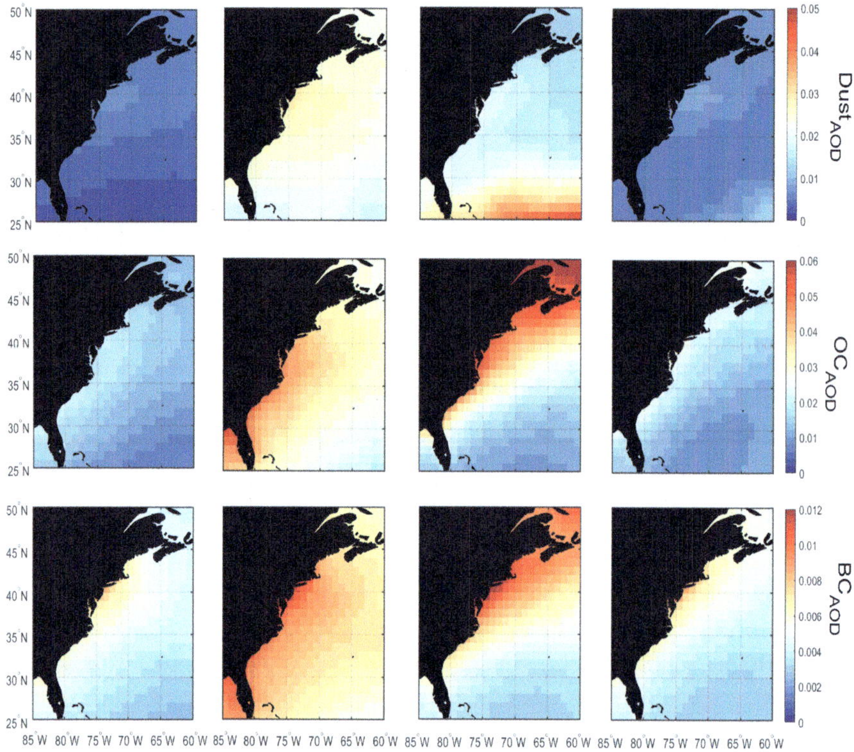

Figure 3. Seasonal maps of MERRA-2 speciated AOD based on data between January 2013 and December 2017. The boxes in the top left panel represent sub-domains examined in more detail throughout the study, with the blue star denoting Bermuda.

Table 3 probes deeper into individual WNAO sub-domains to compare seasonal AOD and N_d values. For the six sub-domains in Fig. 2, MERRA-2 AOD peaks in MAM and JJA, while N_d peaks in DJF. The Bermuda sub-domain was unique in that mean N_d was slightly higher in MAM (53 cm^{-3}) compared to DJF (48 cm^{-3}). We attribute the slightly different seasonal cycle over Bermuda to its remote nature, leading to differences in meteorology and aerosol sources between seasons.

One factor that could bias AOD towards higher values with disproportionately less impact on N_d is aerosol hygroscopic growth in humid conditions. Table 3 summarizes mean MERRA-2 RH values in the PBL and free troposphere (FT). Results show that while RH is highest in JJA (except for FT of DJF in sub-domain N), differences between seasons were not very large. The maximum difference among the four seasons when considering mean RH in the PBL and FT for all sub-domains ranged between 3 %–9% and 7 %–25 %, respectively. Consequently, humidity effects on remotely sensed aerosol parameters are less likely to be the sole explanation of the dissimilar seasonal cycle of N_d and AOD, but can plausibly contribute to some extent.

Table 3. Average drop number concentration (N_d), MERRA-2 AOD, and vertically resolved AOD characteristics from CALIOP for each season over the sub-domains shown in Fig. 2. Total CALIOP AOD is shown outside parentheses, and numbers inside are the percent AOD fraction in the planetary boundary layer followed by in the free troposphere. Also shown are PBLHs (shown in Fig. 4) and the relative humidity in the PBL and FT.

	S	**C-S**	**C**	**C-N**	**N**	**Bermuda**
$AOD_{MERRA-2}/N_d$ (cm^{-3})						
DJF	0.10/56	0.11/74	0.13/91	0.12/97	0.11/78	0.10/48
MAM	0.14/55	0.17/62	0.18/72	0.19/75	0.16/70	0.14/53
JJA	0.14/41	0.16/43	0.17/47	0.19/68	0.17/73	0.11/37
SON	0.11/42	0.12/53	0.13/62	0.13/74	0.11/73	0.11/36
AOD_{CALIOP} (%PBL, %FT)						
DJF	0.11 (64,36)	0.11 (67,33)	0.15 (68,32)	0.09 (61,39)	0.13 (59,41)	0.14 (72,28)
MAM	0.11 (54,46)	0.10 (53,47)	0.12 (58,42)	0.10 (30,70)	0.07 (30,70)	0.12 (58,42)
JJA	0.11 (53,47)	0.11 (44,56)	0.10 (46,54)	0.11 (20,80)	0.08 (11,89)	0.08 (49,51)
SON	0.09 (63,37)	0.10 (57,43)	0.10 (65,35)	0.08 (47,53)	0.07 (35,65)	0.10 (69,31)
PBLH (m)/RH_{PBL} (%)/RH_{FT} (%)						
DJF	1018/ 78/37	1156/ 76/43	1364/ 79/46	1013/ 76/52	926/76/58	1198/80/43
MAM	903/ 77/41	955/ 72/43	1043/ 75/48	722/ 72/53	568/79/55	966/79/50
JJA	775/ 81/62	725/ 81/60	697/ 81/59	481/ 78/53	351/85/55	713/82/58
SON	1018/ 80/50	1094/ 76/45	1181/ 76/42	825/ 71/43	593/77/51	1095/81/48

One factor that could drive the seasonal variation in N_d is the unwanted effects of retrieval errors in the estimation of N_d at low-cloud-coverage conditions. Uncertainty associated with the estimation of N_d from MODIS observation increases as cloud fraction decreases (Grosvenor et al., 2018).

This is mainly because of the overestimation of droplet effective radius (r_e) in the retrieval algorithm due to the interference of cloud-free pixels and also high spatial inhomogeneity in low-cloud-coverage conditions that violates horizontal homogeneity assumptions in the retrieval of r_e and τ from radiative transfer modeling (Zhang et al., 2012, 2016). To test whether retrieval errors in N_d are the main driver of seasonal trends, Fig. S1 shows the seasonal cycle of N_d at various low-level liquid cloud fractions. The results show that as cloud fraction increases the average N_d increases, regardless of season. Perhaps the more important result is that the seasonal trend in spatial maps of N_d remains similar regardless of cloud fraction. This finding is important as it confirms that the seasonal cycle in N_d cannot be solely explained by the uncertainties associated with the retrieval of N_d at low cloud fraction.

Contrasting AOD and Aerosol Index

While previous studies have pointed to the limitations of AOD as an aerosol proxy (e.g., Stier, 2016; Gryspeerdt et al., 2017; Painemal et al., 2020), the N_d–AOD anticorrelation at seasonal scale over the WNAO is at odds with findings for other regions supporting the relationship between these two parameters (Nakajima et al., 2001; Sekiguchi et al., 2003; Quaas et al., 2006, 2008; Grandey and Stier, 2010; Penner et al., 2011; Gryspeerdt et al., 2016) and also that between sulfate and N_d (Boucher and Lohmann, 1995; Lowenthal et al., 2004; Storelvmo et al., 2009; McCoy et al., 2017, 2018; MacDonald et al., 2020). Values of N_d are influenced by the number concentration of available CCN, which is determined by aerosol properties (size distribution and composition) and supersaturation level. AOD is an imperfect CCN proxy variable because it does not provide information about composition and size distribution and is sensitive to relative humidity.

Aerosol index (AI) is more closely related to CCN as it partially accounts for the size distribution of aerosols (Deuze et al., 2001; Nakajima et al., 2001; Breon et al., 2002; Hasekamp et al., 2019). The sensitivity of AI to size is evident in spatial maps for each season showing more of an offshore gradient (like sulfate AOD in Fig. 3) in each season and lacking both the offshore peak in sea salt between ~ 35–40° N and the maximum AOD for dust south of 30° N in JJA. However, when comparing absolute values between the four seasons in Fig. 2b, AI exhibits a similar seasonal cycle to AOD, thereby indicating that size distribution alone cannot explain diverging seasonal cycles for N_d and AOD. We next compare N_d to aerosol data in the PBL where CCN more relevant to droplet activation are confined. Size distribution effects in the PBL can instead be more of a factor, especially as sea salt is abundant.

Aerosol Size Distribution and Vertical Aerosol Distribution

Vertical profiles of aerosol extinction coefficient estimated from CALIOP nighttime observations are shown in Fig. 4 for the six sub-domains. Also shown are the seasonally representative planetary boundary layer heights (PBLHs) from MERRA-2, with numerical values of both PBLH and fractional AOD contributions to the PBL and FT in Table 3.

Although here we used nighttime observations from CALIOP because of having higher signal-to-noise ratio than daytime observations, we expect the general seasonal trends discussed here to remain the same regardless of the observation time. The CALIOP results indicate that aerosol extinction more closely follows the N_d seasonal cycle with the highest (lowest) values in the PBL during DJF (JJA).

However, aerosol extinction coefficient is sensitive to aerosol size distribution, and a plausible scenario is that DJF extinction in the PBL is primarily contributed by coarse sea salt particles, which are especially hygroscopic but do not contribute significantly to number concentration as demonstrated clearly by airborne observations (i.e., $FCDP_{>3\mu m}$ time series shown in Fig. 1d). This is supported in part by how DJF is marked by the highest fractional AOD contribution from the PBL (59 %–72 %) where sea salt is concentrated. In contrast, JJA has the lowest fractional AOD contribution from the PBL (11.3 %–52.6 %).

It is also possible that the higher fractional AOD contribution from the PBL in winter is partly owed to aerosol particles being more strongly confined to the PBL compared to the summer. Sub-domains C-N and N exhibit the greatest changes in AOD fraction in the PBL between seasons with a maximum in DJF (59 %–61 %) and a minimum in JJA (11%–19%), suggesting they are relatively more sensitive to the aerosol vertical distribution in leading to contrasting AOD and N_d seasonal cycles. Bermuda stands out as having the highest AOD fractional contributions in the PBL in DJF (72 %) and SON (69 %) and among the highest seasonal total AODs in those two seasons (0.14 in DJF and 0.10 in SON) assisted in large part by sea salt (Fig. 3) (Aldhaif et al., 2021), coincident with high seasonal wind speeds (Corral et al., 2021).

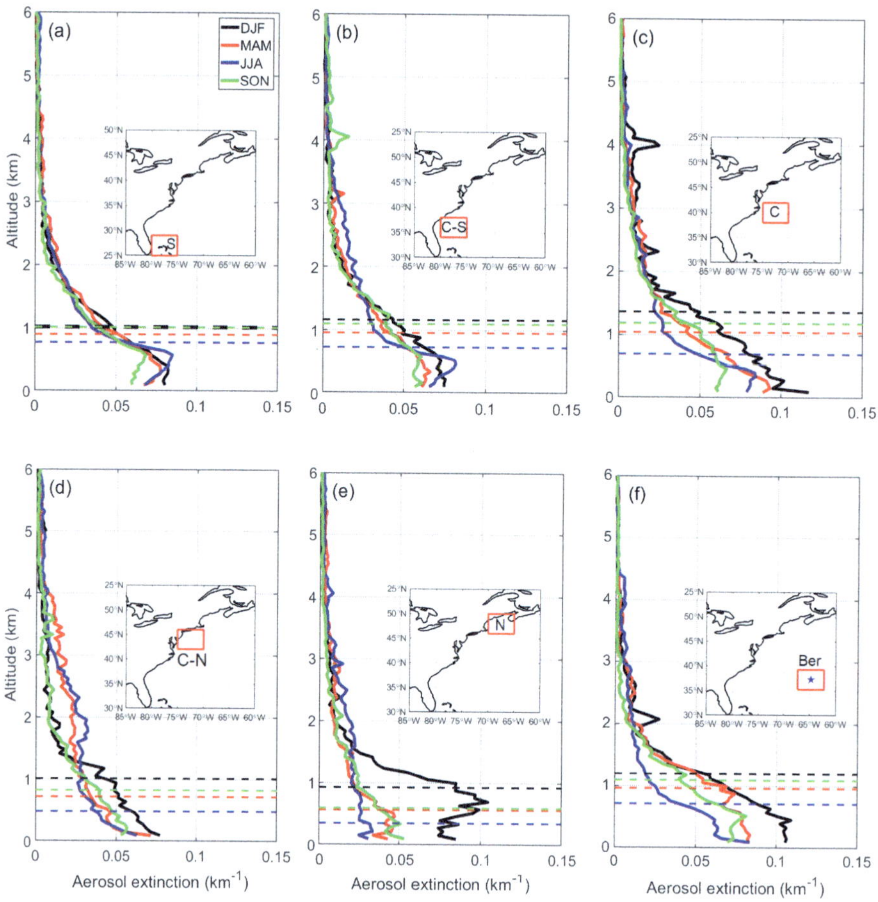

Figure 4. Vertical profiles of CALIPSO aerosol extinction for different seasons in **(a)–(f)** six different sub-domains of the WNAO. Average seasonal planetary boundary layer heights (PBLH) from MERRA-2 are denoted with dashed lines.

To explore aerosol number concentration characteristics in the PBL in different seasons, we next discuss results from an opportune dataset over the US East Coast (Cape Cod, MA) providing an annual profile of CCN concentration at 1% supersaturation (Fig. 5). Cape Cod is a coastal location representative of the outflow, providing an important fraction of the CCN impacting offshore low-level clouds. As the supersaturation examined is relatively high (1 %), the measured CCN include smaller particles representing high number concentrations that would not appreciably contribute to the high aerosol extinctions from CALIOP in the PBL in direct contrast to sea salt (i.e., high extinction due to fewer but larger particles).

Seasonal mean CCN values do not follow the seasonal cycle of N_d nor CALIOP extinction in the PBL, with values being as follows: DJF = 1436 cm^{-3}; MAM = 1533 cm^{-3}; JJA = 1895 cm^{-3}; and SON = 1326 cm^{-3}. These results suggest the following: (i) size distribution effects are significant in the PBL when comparing extinction to number concentration, and (ii) aerosol vertical distribution behavior cannot alone explain the divergent seasonal cycles of N_d and aerosol parameters (e.g., AOD, AI, surface number concentrations).

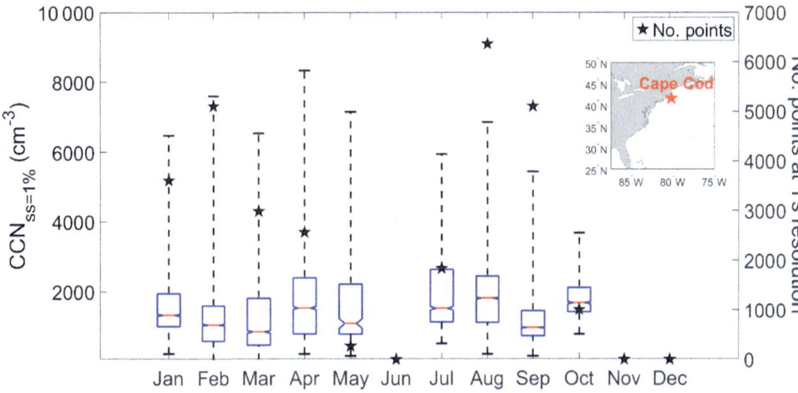

Figure 5. Monthly statistics of CCN concentration (1% supersaturation) measured at Cape Cod between July 2012 and May 2013. Red lines represent the median, whiskers are the monthly range, and the top and bottom of the boxes represent the 75th and 25th percentiles, respectively. The notches in the box plots demonstrate whether medians are different from each other with 95% confidence. Boxes with notches that do not overlap with each other have different medians with 95% confidence.

We next compare MERRA-2 speciated aerosol concentrations at the surface (Fig. 6) to those of speciated AOD (Fig. 3). Surface mass concentrations have the limitation of being biased by larger particles (similar to extinction). The seasonal cycle of mean values for speciated AOD and surface concentration for individual sub-domains generally agree with the exception that there was disagreement for sulfate in each sub-domain (see seasonal mean values in Table S2). Sulfate exhibited higher AODs in JJA but with surface concentrations usually being highest in DJF or MAM; although differences in seasonal mean mass concentrations were relatively small (< 1 μg m^{-3}). A plausible explanation is enhanced secondary production of sulfate via oxidation of SO_2 or DMS convectively lifted to the free troposphere in JJA. An important result confirmed by the surface mass concentrations is that sea salt is an order of magnitude higher than the other species, supporting the previous speculation that sea salt dominates the aerosol extinction in the PBL from CALIOP.

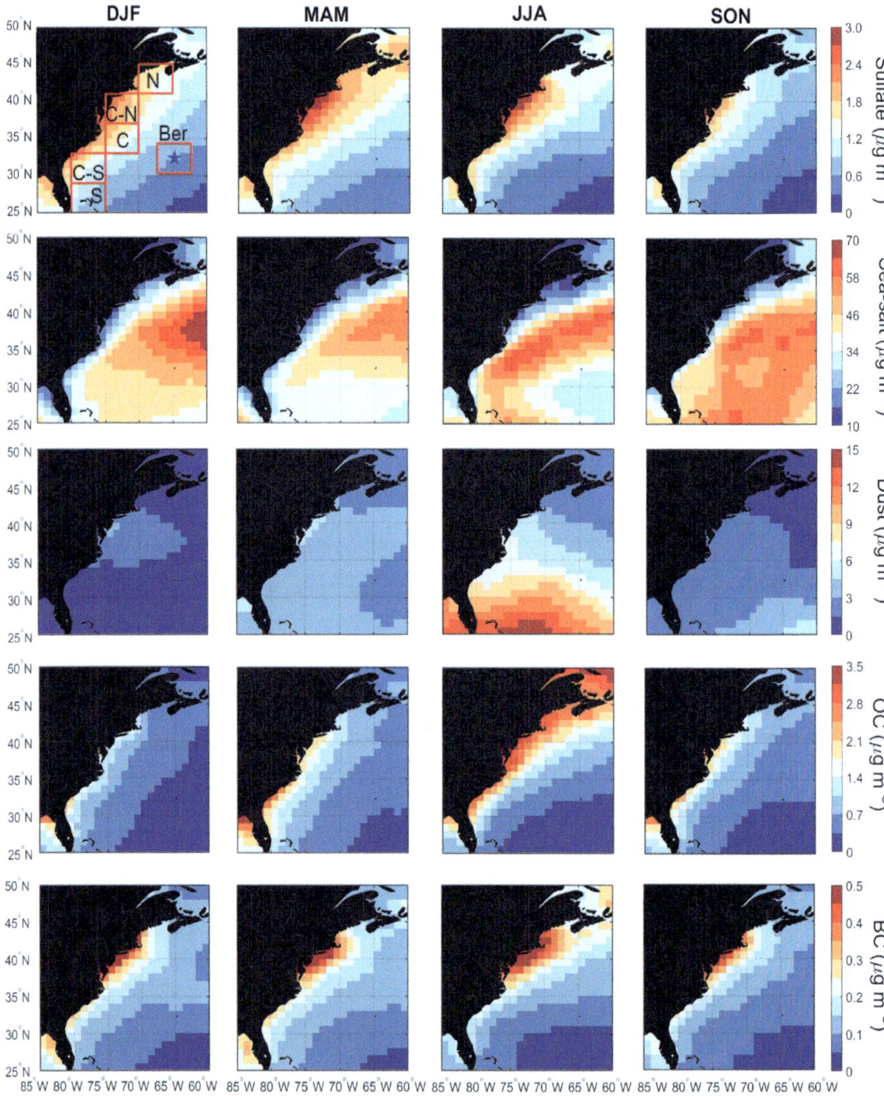

Figure 6. Seasonal maps of MERRA-2 speciated aerosol concentrations at the surface based on data between January 2013 and December 2017. The boxes in the top left panel represent sub-domains examined in more detail throughout the study, with the blue star denoting Bermuda.

Aerosol–cloud Interactions

Studies of China's east coast have shown that the aerosol indirect effect is especially strong in wintertime, whereby pollution outflow leads to high N_d and suppressed precipitation (Berg et al., 2008; Bennartz et al., 2011). It is hypothesized that a similar effect is taking place off of North America's east coast, which could in part explain enhanced N_d without a significant jump in aerosol parameter (e.g., AOD, AI) values necessarily. Grosvenor et al. (2018) suggested that high cloud fractions in wintertime off these east coasts relative to other seasons are coincident with strong temperature inversions usually associated with cold-air outbreaks that serve to concentrate and confine surface layer aerosols. We examine the relative seasonal strength of the aerosol indirect effect via spatial maps of the following metric commonly used in aerosol–cloud interaction (ACI) studies:

$$ACI = d\ln(N_d)/d\ln(\alpha), \qquad (3)$$

where α represents an aerosol proxy parameter that is represented here as AI, AOD, the speciated sulfate AOD ($Sulfate_{AOD}$), and sulfate surface mass concentration ($Sulfate_{sf\text{-}mass}$). The expected range by common convention is 0–1, with higher values suggestive of greater enhancement in N_d for the same increase in the aerosol proxy parameter.

Table 4 shows that DJF always exhibits the highest ACI values regardless of the aerosol proxy used, consistent with a stronger aerosol indirect effect in DJF over East Asia. The mean ACI values in DJF using AI, AOD, $Sulfate_{AOD}$, and $Sulfate_{sf\text{-}mass}$ ranged from 0.25 to 0.55, 0.28–0.59, 0.25–0.53, and 0.22–0.47, respectively, depending on the sub-domain. Spatial maps of ACI (Fig. 7) do not point to significant geographic features. Coefficients of determination (R^2) for the linear regression between $\ln(N_d)$ and ln( ) when computing seasonal ACI values were generally low (≤ 0.30), with spatial maps of R^2 and data point numbers in Fig. S2.

Poor correlations are suggestive of the non-linear nature of aerosol–cloud interactions (e.g., Gryspeerdt et al., 2017) and the influence of other likely factors such as dynamical processes and turbulence, data spatial resolution and dataset size, cloud adiabaticity, wet scavenging effects, and aerosol size distribution (McComiskey et al., 2009). The results of this section suggest though that aerosol indirect effects could be strongest in DJF, meaning that N_d values increase more for the same increase in aerosol. Factors that can contribute to higher ACI values in winter than summer include seasonal differences in the following: (i) dynamical processes and turbulent structures of the marine boundary layer; (ii) aerosol size distributions and consequently varying particle number concentrations for a fixed mass concentration; and (iii) hygroscopicity of particles, especially as a result of changes in the composition of the carbonaceous aerosol fraction. Regarding dynamical processes and the effects of turbulence, Fig. 2 in Painemal et al. (2021) shows that heat fluxes (i.e., latent and sensible fluxes) are strongest (lowest) in the winter (summer) over the WNAO. The greater heat fluxes in DJF can contribute to more turbulent and coupled marine boundary layer conditions in winter than summer, presumably resulting in more efficient transport and activation of aerosol in the marine boundary layer, leading to higher ACI values. Forthcoming work will probe this issue in greater detail.

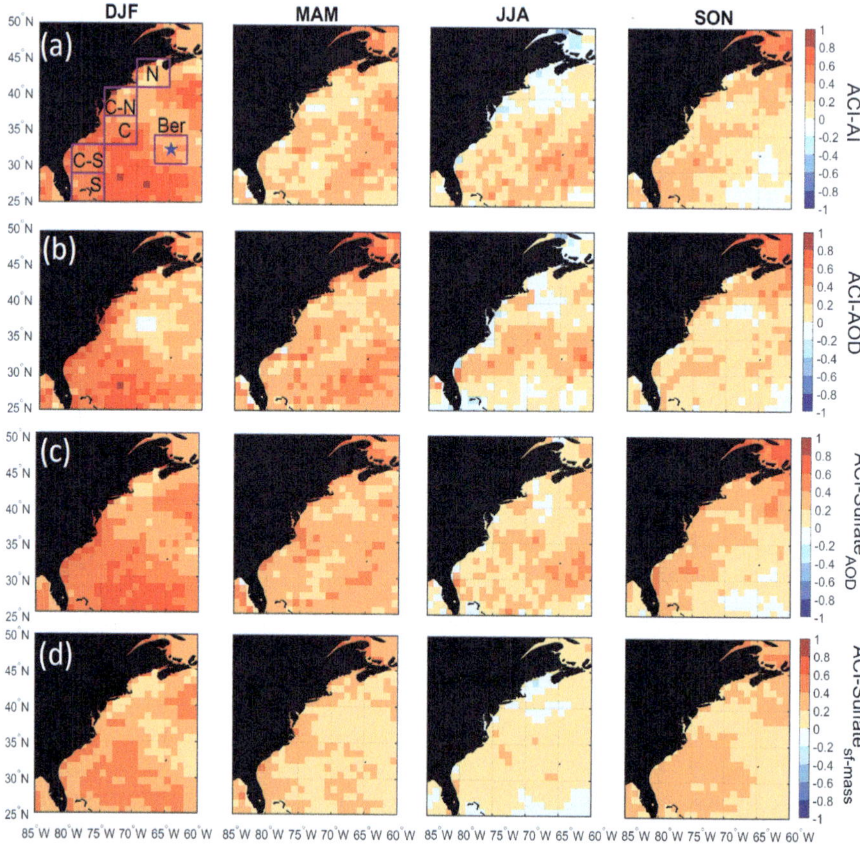

Figure 7. Seasonal maps of the aerosol–cloud interaction (ACI) parameters over the WNAO using daily N_d and four different aerosol proxy parameters (AI, AOD, $Sulfate_{AOD}$, $Sulfate_{sf\text{-}mass}$) from CERES-MODIS and MERRA-2, respectively. ACI statistics associated with the six sub-domains shown are summarized in Table 4.

Table 4. Estimated values of ACI calculated four ways (d log(N_d)/d log(AOD); d log(N_d)/d log(AI); d log(N_d)/d log($Sulfate_{AOD}$); d log(N_d)/d log($Sulfate_{sf\text{-}mass}$)) for the sub-domains shown in Fig. 2. The ACI values were obtained from log–log regression on average daily values of N_d and each of the aerosol proxy variables including only the pixels with $CF_{low\text{-}liq.}$ greater than 0.1. Numbers in parentheses, in order, are R^2 and the number of points used for linear regression. Statistically insignificant ACI values with p values greater than 0.05 are marked by bold font.

	S	C-S	C	C-N	N	Bermuda
ACI-AI						
DJF	0.55 (0.24,440)	0.53 (0.17,421)	0.53 (0.14,403)	0.33 (0.05,418)	0.25 (0.04,403)	0.42 (0.09,422)
MAM	0.21 (0.03,451)	0.13 (0.01,439)	0.30 (0.06,422)	0.17 (0.02,426)	0.31 (0.05,428)	0.28 (0.04,437)
JJA	0.25 (0.02,437)	0.20 (0.03,437)	0.28 (0.07,424)	0.11 (0.01,430)	−0.12 (0.01,408)	0.38 (0.09,443)
SON	0.23 (0.03,435)	0.20 (0.03,428)	0.26 (0.05,431)	0.19 (0.04,412)	0.24 (0.06,394)	**0.00 (0.00,428)**
all	0.27 (0.05,1763)	0.16 (0.02,1725)	0.22 (0.04,1680)	0.12 (0.01,1686)	0.12 (0.01,1633)	0.23 (0.04,1730)
ACI-AOD						
DJF	0.59 (0.13,440)	0.53 (0.12,421)	0.47 (0.10,403)	0.39 (0.06,418)	0.28 (0.04,403)	0.37 (0.08,422)
MAM	0.26 (0.02,451)	0.22 (0.01,439)	0.43 (0.07,422)	0.30 (0.04,426)	0.40 (0.06,428)	0.32 (0.03,437)
JJA	**0.02 (0.00,437)**	0.24 (0.02,437)	0.36 (0.07,424)	0.15 (0.01,430)	**−0.06 (0.00,408)**	0.30 (0.04,443)
SON	0.14 (0.01,435)	0.18 (0.02,428)	0.17 (0.02,431)	0.16 (0.02,412)	0.27 (0.05,394)	0.18 (0.02,428)
all	0.13 (0.01,1763)	0.12 (0.01,1725)	0.22 (0.03,1680)	0.15 (0.01,1686)	0.16 (0.02,1633)	0.31 (0.05,1730)
ACI-$Sulfate_{AOD}$						
DJF	0.53 (0.25,440)	0.53 (0.21,421)	0.53 (0.19,403)	0.37 (0.08,418)	0.25 (0.05,403)	0.43 (0.13,422)
MAM	0.29 (0.05,451)	0.27 (0.04,439)	0.42 (0.14,422)	0.32 (0.07,426)	0.41 (0.11,428)	0.34 (0.07,437)
JJA	0.21 (0.02,437)	0.19 (0.03,437)	0.33 (0.09,424)	0.20 (0.04,430)	**0.04 (0.00,408)**	0.39 (0.09,443)
SON	0.16 (0.02,435)	0.23 (0.04,428)	0.29 (0.07,431)	0.28 (0.09,412)	0.35 (0.13,394)	**0.07 (0.00,428)**
all	0.23 (0.04,1763)	0.19 (0.03,1725)	0.30 (0.07,1680)	0.23 (0.05,1686)	0.22 (0.05,1633)	0.25 (0.05,1730)

ACI-Sulfate$_{sf\text{-}mass}$						
DJF	0.44 (0.29,440)	0.41 (0.22,421)	0.47 (0.22,403)	0.22 (0.04,418)	0.23 (0.06,403)	0.32 (0.14,422)
MAM	0.24 (0.07,451)	0.25 (0.08,439)	0.29 (0.12,422)	0.24 (0.05,426)	0.36 (0.09,428)	0.16 (0.04,437)
JJA	0.11 (0.01,437)	0.12 (0.03,437)	0.23 (0.11,424)	0.19 (0.06,430)	−0.12 (0.01,408)	0.20 (0.07,443)
SON	0.32 (0.16,435)	0.36 (0.18,428)	0.34 (0.19,431)	0.19 (0.06,412)	0.21 (0.05,394)	0.17 (0.07,428)
all	0.32 (0.13,1763)	0.30 (0.12,1725)	0.36 (0.17,1680)	0.19 (0.04,1686)	0.15 (0.02,1633)	0.25 (0.11,1730)

DISCUSSION OF POTENTIAL INFLUENTIAL FACTORS

We probe deeper into factors related to the N_d seasonal cycle by using (Sect. 4.1) composite analyses based on high- and low-N_d days and (Sect. 4.2) advanced regression techniques tackling non-linear relationships. We focus the analyses on one sub-domain (C-N) for both simplicity and intriguing characteristics: (i) among the highest anthropogenic AOD values over the WNAO; (ii) significant seasonal changes in fractional AOD contribution to the PBL; (iii) close to the Cape Cod site where CCN data were shown; and (iv) the aerosol indirect effect (Table 4) strongest (weakest) in DJF (JJA).

Composite Analysis

Discussion first addresses the behavior of different environmental parameters on days with the highest and lowest N_d values. Seasonal histograms of averaged daily N_d were generated for sub-domain C-N. The histograms are based on the natural logarithm of N_d to better resemble a normal distribution. We assign values as being low in each season if they are less than 1 standard deviation below the seasonal value; conversely, high values are those exceeding 1 standard deviation above the seasonal mean. Cut-off N_d values (cm^{-3}) are as follows (low/high): 33/153 (DJF), 29/118 (MAM), 38/100 (JJA), and 31/115 (SON). Next, composite maps for these groups were created (Figs. 8-12) for sea level pressure, near-surface wind, low-level cloud fraction, cold-air outbreak index, and AOD. The figures contrast the low- and high-N_d maps with those showing mean seasonal values to investigate potential factors that contribute to seasonal N_d variability. Interested readers are referred to Figs. S3-S20 where similar composite map results are shown for N_d itself and other parameters including those in Table 2.

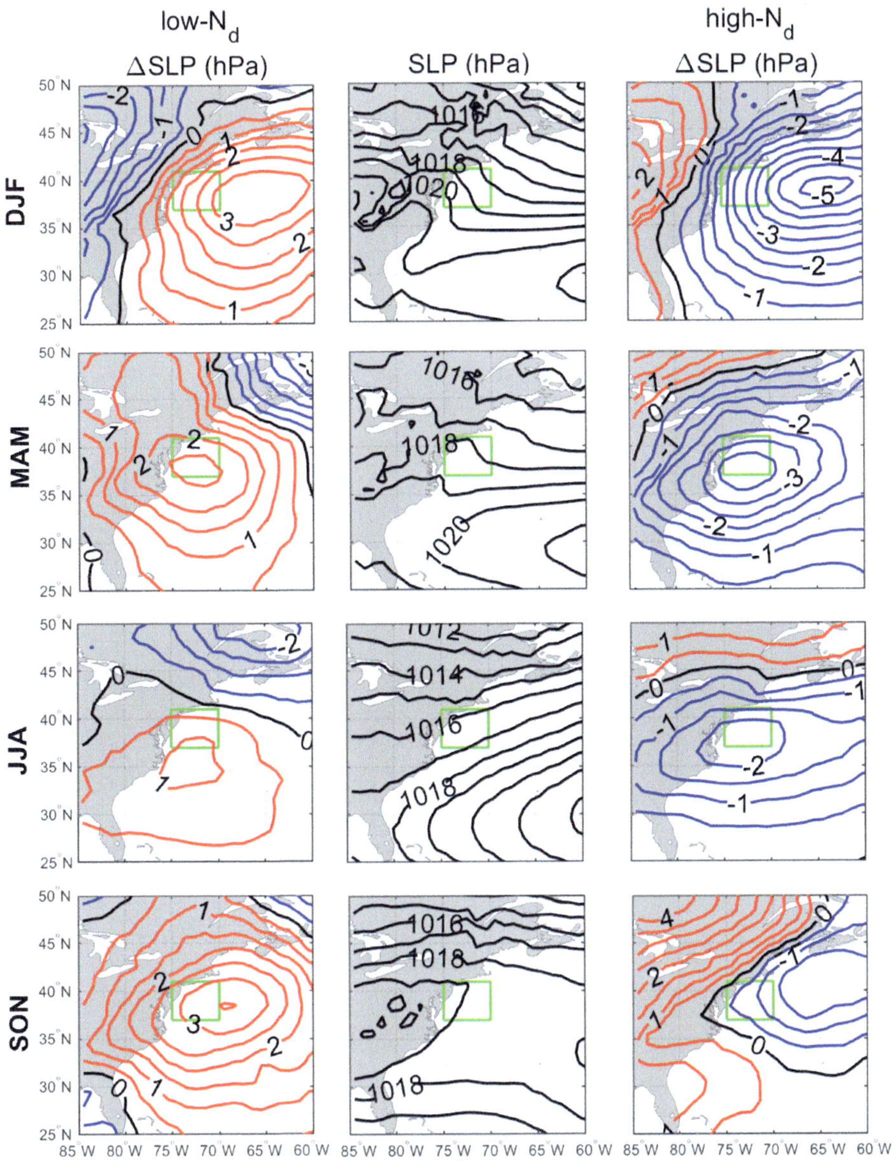

Figure 8. Seasonal climatology of sea level pressure (SLP) (middle column) and anomalies from seasonal averages for low-N_d days (left column) and high-N_d days (right column). In the left and right columns, red and blue contours are associated with positive and negative anomalies from the climatology, respectively. The green box represents sub-domain C-N for which the analysis was conducted.

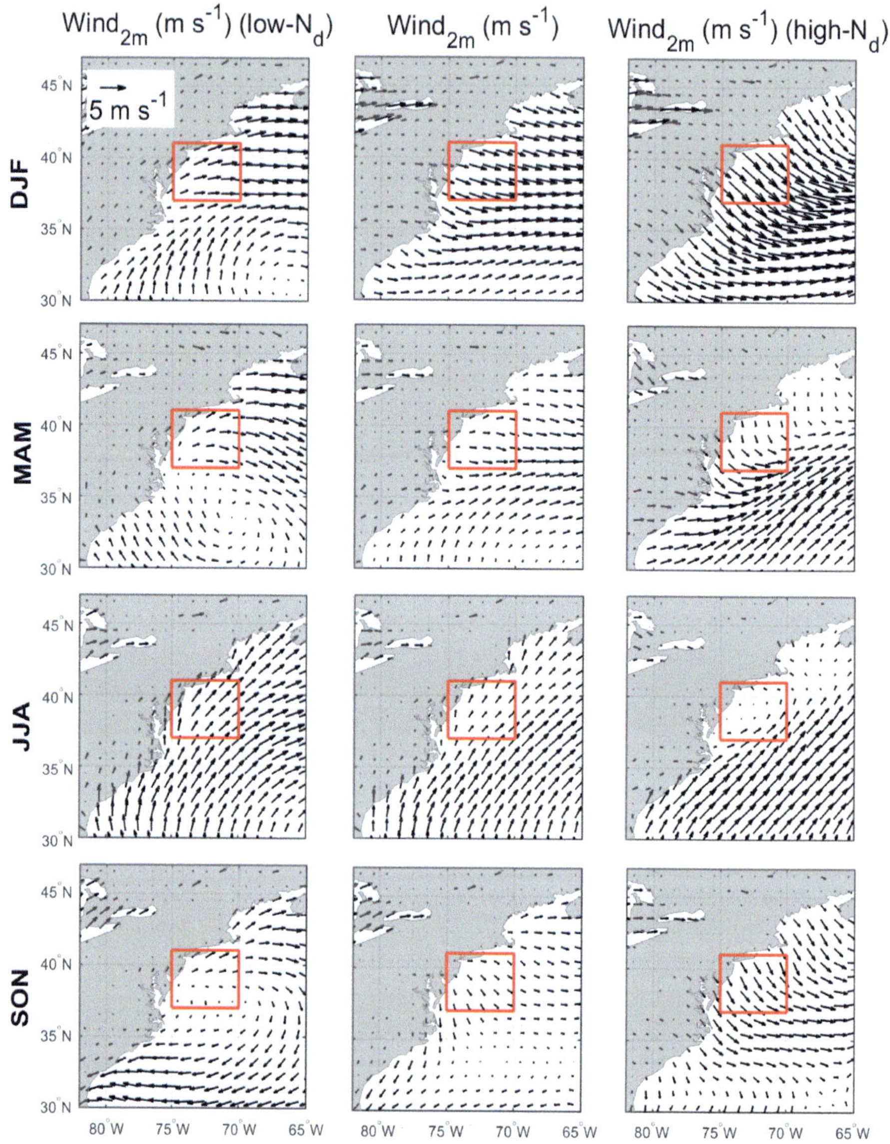

Figure 9. Seasonal climatology of near-surface (2 m above ground) wind speed (middle column) and mean values for low-N_d days (left column) and high-N_d days (right column). The reference wind vector is shown in the top left panel. The red box represents sub-domain C-N for which the analysis was conducted.

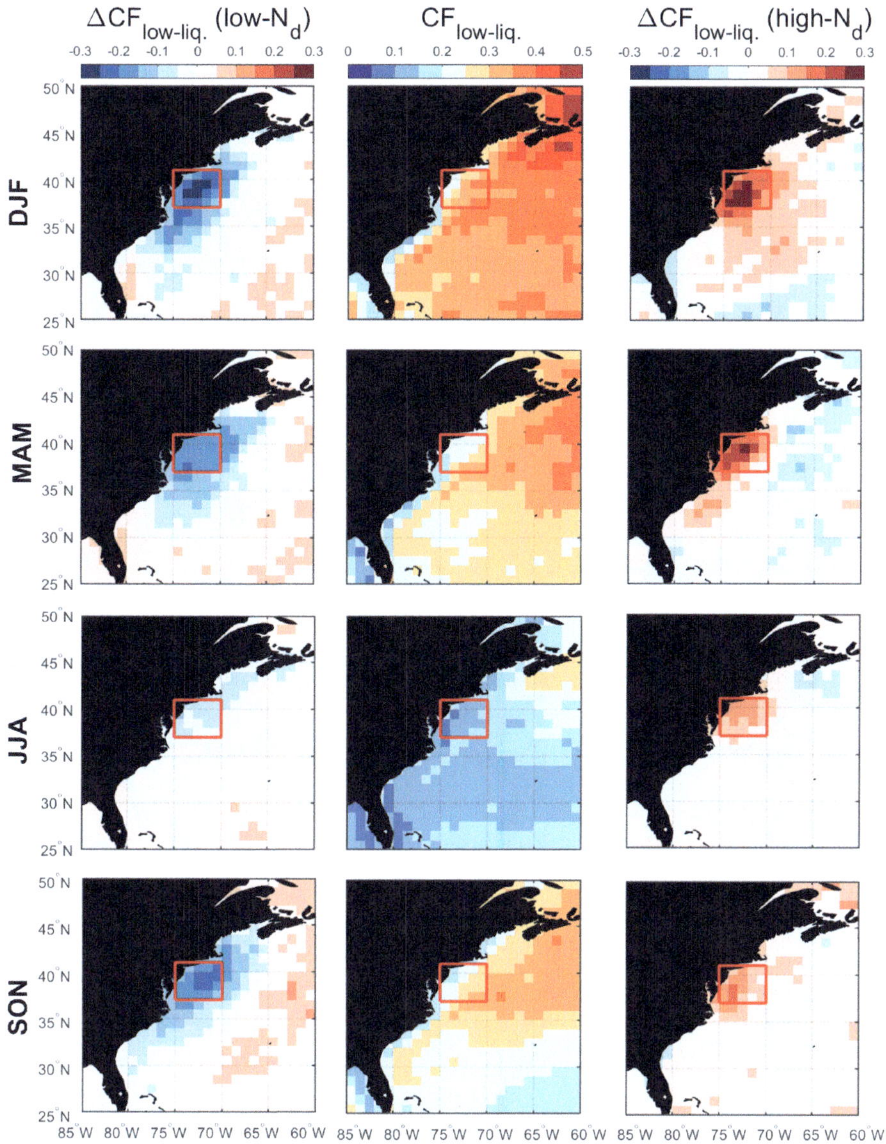

Figure 10. Seasonal averages of low-level liquid cloud fraction (middle column) and associated anomalies on low-N_d days (left column) and high-N_d days (right column). The red box represents sub-domain C-N for which the analysis was conducted.

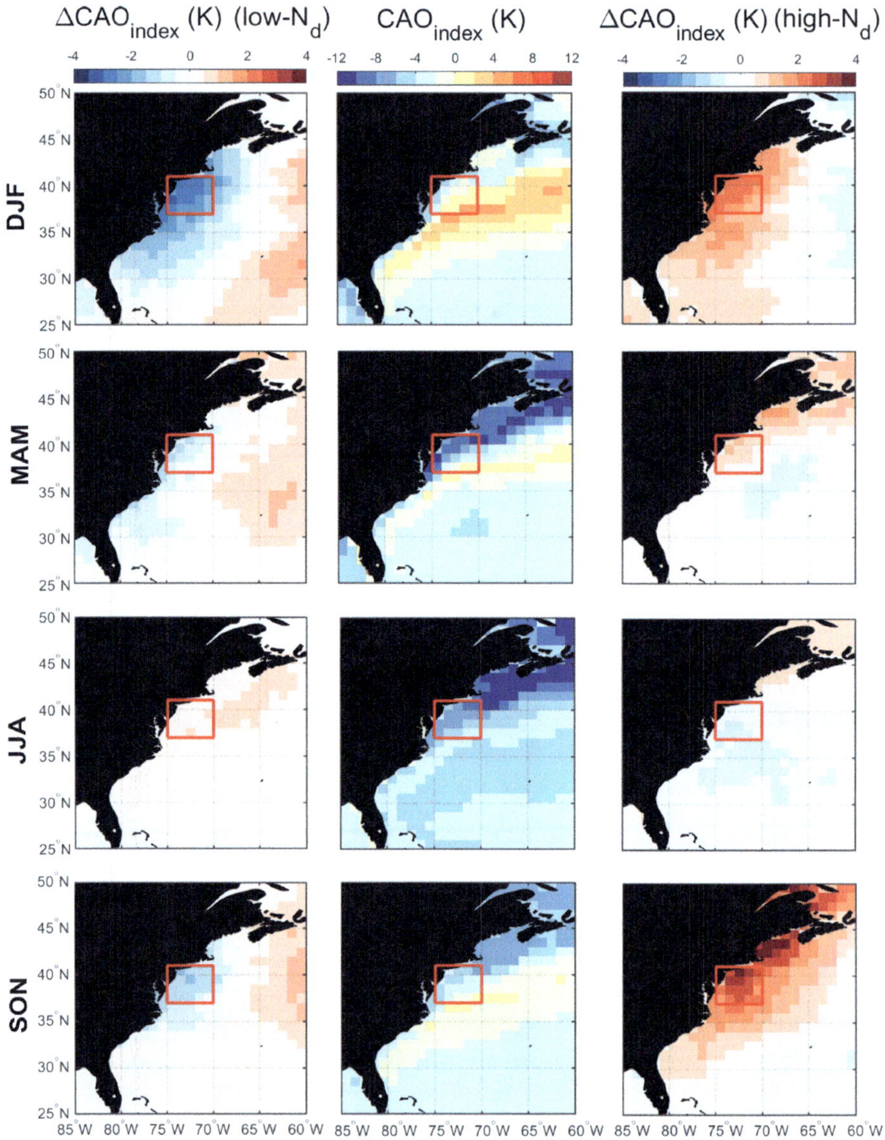

Figure 11. Seasonal averages of cold-air outbreak (CAO) index (middle column) and associated anomalies on low-N_d days (left column) and high-N_d days (right column). The red box represents sub-domain C-N for which the analysis was conducted.

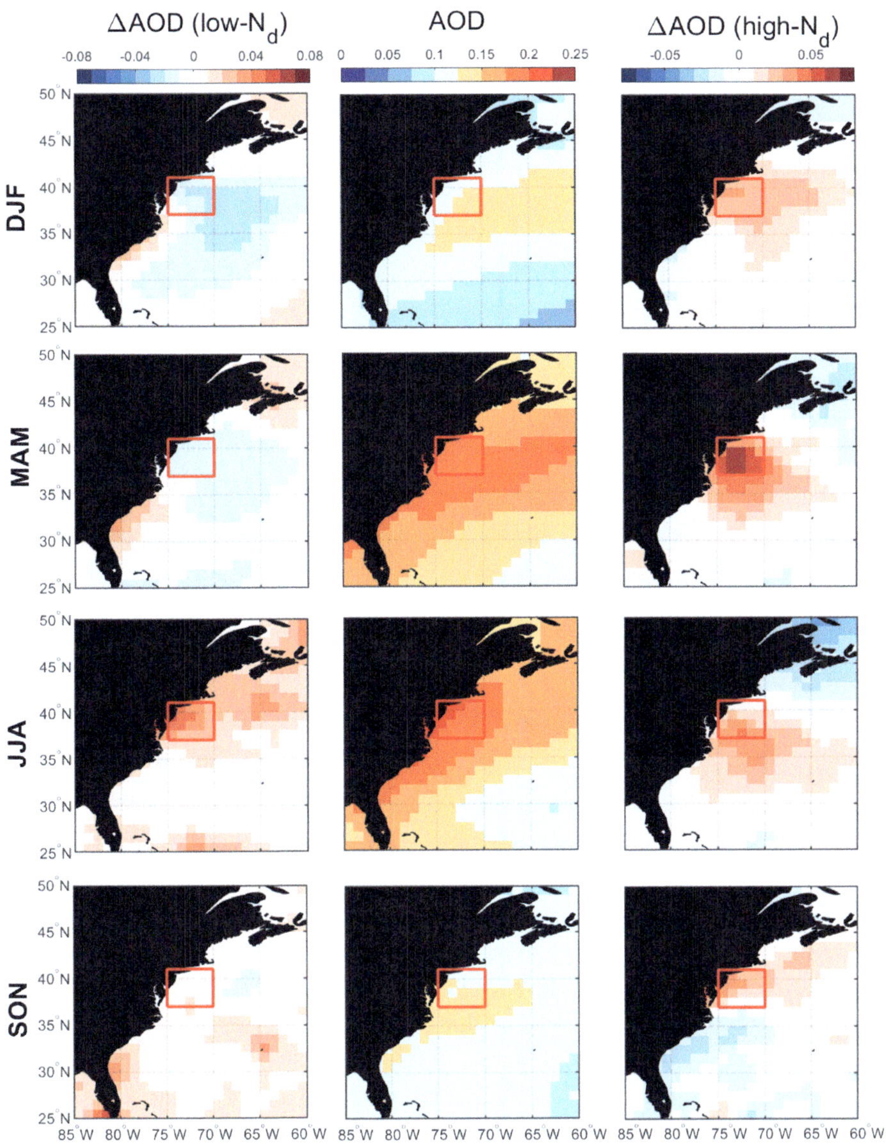

Figure 12. Seasonal averages of MERRA-2 AOD (middle column) and associated anomalies on low-N_d days (left column) and high-N_d days (right column). The red box represents sub-domain C-N for which the analysis was conducted.

The resulting composite maps indicate high-N_d days are characterized by (i) reduced SLP; (ii) more northerlynorthwesterly flow for all seasons (except JJA) and especially stronger winds in DJF and SON; (iii) higher low-level liquid cloud fraction, especially in DJF; (iv) higher CAO index in the seasons when CAO events occur more frequently (DJF, SON, MAM); and (v) enhanced AOD. Low-N_d days generally exhibited opposite conditions when compared to seasonal mean values: (i) enhanced SLP; (ii) wind ranging from southerly to westerly without any significant wind speed enhancement; (iii) reduced low-level liquid cloud fraction, especially in DJF; (iv) lower CAO index in DJF, SON, and MAM; and (v) reduced AOD in DJF and MAM, enhanced AOD in JJA, and limited change in SON. Noteworthy results from Figs. S3-S20 included the enhancement/reduction of PBLH on high-/low-N_d days (least pronounced in JJA), higher/lower RH at 950 and 800 hPa on high-/low-N_d days, and higher/lower sulfate AOD and surface concentrations on high-/low-N_d days for DJF and MAM. Furthermore, there was a general reduction in rain on low-N_d days for most seasons except SON, with rain enhancement on high-N_d days except for DJF (Fig. S6); this was unexpected as wet removal was hypothesized to be a reason for reduced N_d for at least the low-N_d days. This may be attributed to the rain product being for surface precipitation (and thus not capturing all drizzle) and for all cloud types, including more heavily precipitating clouds deeper and higher than the low-level clouds examined for N_d. Another factor potentially contributing to the observed counterintuitive trends is the temporal offset between N_d estimations from MODIS-Aqua and precipitation data from PERSIANN-CDR.

The mean seasonal climatological values and anomalies suggest that high-N_d cases are marked by continental outflow, high cloud fractions, high PBLH, and low SLP, all of which occur most commonly in DJF and are associated with cold-air outbreaks. These events are marked by cold air over the warm ocean leading to strong surface heat fluxes, boundary layer deepening, weakened inversion strength, and high and deep clouds (Brummer, 1996; Kolstad et al., 2009; Fletcher et al., 2016; Abel et al., 2017; Naud et al., 2018). Coincident with these features is the Icelandic Low, which is a significant climatological feature of the North Atlantic whereby subpolar low pressure builds in extratropic areas beginning in the fall with westerly winds in the boundary layer that shift more to northerly in the winter (Sorooshian et al., 2020; Painemal et al., 2021). This low-pressure system seems to be stronger on high-N_d days, resulting in more continental outflow and high number concentrations of CCN; the greater CAO index values near the coast promote high cloud coverage, affording more opportunity for cloud processing of particles to ultimately enhance droplet activation. While there can be considerable enhancement in N_d as cold-air outbreak air masses evolve over warmer waters, precipitation scavenging farther downwind will be an efficient method of boundary layer aerosol (and N_d) removal (Abel et al., 2017; Lloyd et al., 2018), which contributes at least in part to the sharp N_d gradients offshore demonstrated in Fig. 1.

Multivariate Regression Analysis

Modeling analysis focuses on the two seasons (DJF and JJA) with the extremes in terms of seasonal mean values for N_d and aerosol parameters. Added motivation for examining those two seasons stems from spatial maps of R^2 based on ACI analysis (Fig. S2). Using the surface sulfate concentration as the aerosol proxy generally yielded higher R^2 values in three seasons (DJF = 0.13, MAM = 0.05, SON = 0.08) except JJA (0.02) for which the choice did not matter owing to low R^2 (≤ 0.03) values for all four aerosol proxy variables tested. Although the R^2 values are all generally low, DJF and JJA are the seasons when surface sulfate levels are the most and least capable of explaining N_d, with R^2 among the four proxy variables exhibiting the widest (DJF values: 0.07–0.13) and narrowest range (JJA: 0.01–0.03) of values. We address here how much improvement is gained in modeling N_d by advancing from linear regressions based on one input variable to (i) adding more input variables and (ii) moving to a more sophisticated model (GBRT) that captures non-linear relationships.

We show in Table 5 the performance of two linear models based on a single linear regression (with sulfate mass concentration) and a multi-regression that uses 14 input variables listed in Table 2. In addition, Table 5 also lists the performance of the GBRT model that ingests 14 input variables, similar to the linear multi-regression model. The average R^2 scores of the test set for predicting N_d based on a linear regression using only sulfate surface mass concentration were 0.17 and 0.09 in DJF and JJA, respectively. In contrast, R^2 between the multi-regression linear model and the test dataset increased to 0.28 and 0.25 for DJF and JJA, respectively.

This increase in predictive capability was helpful to reduce the gap between seasons by presumably accounting for factors more important in JJA aside from surface concentration of sulfate. The R^2 scores increased even more to 0.47 and 0.43 for DJF and JJA, respectively, for the GBRT model. Therefore, accounting for non-linear relationships improved predictive capability in both seasons. It is important to note that the GBRT model was robust in terms of overfitting and especially generalizability as R^2 values of the test and validation sets were similar for both seasons.

Table 5. Performance of different models in predicting N_d assessed based on average R^2 scores on both validation and test sets. The models were fitted separately for DJF and JJA seasons. Table 2 has the complete list of variables used in the GBRT model.

Model	Model type	Number of predictor variables	R^2 score (DJF/JJA)	
			Validation set	Test set
$N_d \sim f(\text{Sulfate}_{sf\text{-}mass})$	Linear	1	0.17/0.09	0.17/0.09
$N_d \sim f(\text{Sulfate}_{sf\text{-}mass}, \text{CF}_{low\text{-}liq}, \ldots)$	Linear	14	0.27/0.24	0.28/0.25
$N_d \sim f(\text{Sulfate}_{sf\text{-}mass}, \text{CF}_{low\text{-}liq}, \ldots)$	GBRT	14	0.48/0.43	0.47/0.43

We next discuss the importance ranking of different parameters from Table 2 in terms of influencing N_d for DJF and JJA (Fig. 13). Low-level liquid cloud fraction was the most important parameter in both seasons with some commonality in the next three parameters for both seasons. In DJF, sulfate surface mass concentrations were the second most important factor, followed by organic carbon surface concentrations and low-level liquid cloud-top effective height. As sulfate is secondarily produced via gas-to-particle conversion processes, this result is consistent with those from Fig. 1 showing the presumed strong impact of particles smaller than 100 nm in impacting N_d values close to shore. In JJA, the CAO index was the second most important, followed by organic carbon and sulfate surface concentrations. Also, our results throughout the study and supported by modeling are in agreement with Quinn et al. (2017) that sulfate particles contribute more to the CCN budget than sea salt particles. In DJF and JJA, the fifth most important factor was CAO index (second most important in JJA) and PBLH (11th most important in DJF), respectively.

Figures 14 and 15 show accumulated local effect (ALE) plots for the various parameters ranked in Fig. 13. In both seasons, but especially DJF, enhanced surface concentrations of sulfate and organic carbon coincide with higher N_d, whereas there was not any obvious positive association between N_d and either sea salt or dust (Fig. 14). Dust in JJA and sea salt in DJF, seasons of which each respective aerosol type is most predominant, exhibited negative relationships with N_d.

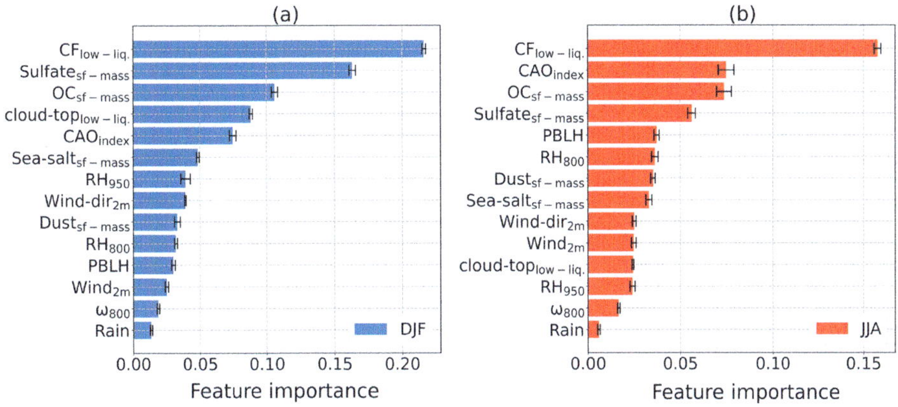

Figure 13. Average permutation feature importance of input parameters for **(a)** DJF and **(b)** JJA based on GBRT models trained in each season. Feature importance values were calculated based on using the test set. Error bars exhibit the range of feature importance values stemming from the variability of the obtained models from the cross-validation resampling procedure.

Such a negative relationship is plausibly related to differences between ACI when calculated using AOD versus AI (Painemal et al., 2021); for instance, coarse sea salt can expedite collision–coalescence and thus reduce N_d, which has the effect of reducing ACI (Eq. 3) and even possibly yielding negative values (Table 4). Negative values of other ACI constructs coincident with poor R^2 values have previously been attributed to potential effects of giant CCN (Terai et al., 2015; Dadashazar et al., 2017), but further research needs to examine this in more detail.

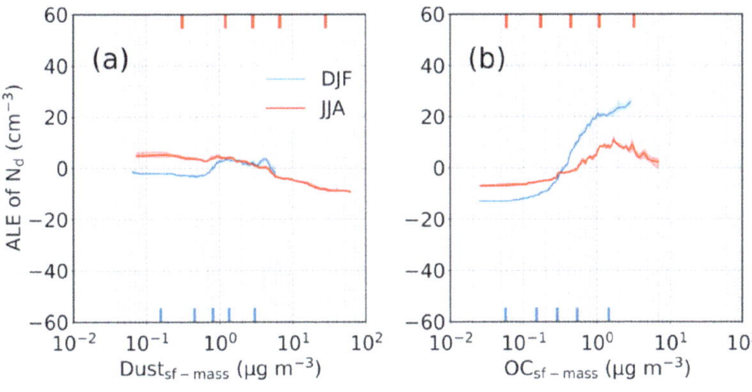

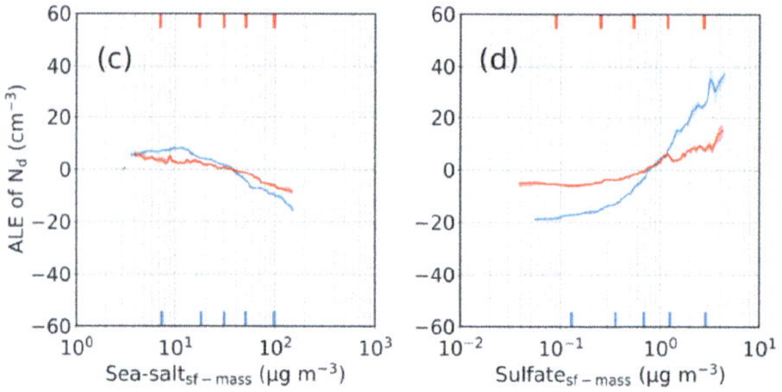

Figure 14. Average local accumulated effect (ALE) profiles based on GBRT modeling for surface mass concentrations of the following parameters: **(a)** dust, **(b)** organic carbon, **(c)** sea salt, and **(d)** sulfate. Blue and red profiles represent ALEs of DJF and JJA, respectively. Shaded areas show the ALE ranges stemming from the variability of the obtained models from the cross-validation resampling procedure. Markers on the bottom and top *x* axes denote the values of the 5th, 25th, 50th, 75th, and 95th percentiles for each input variable.

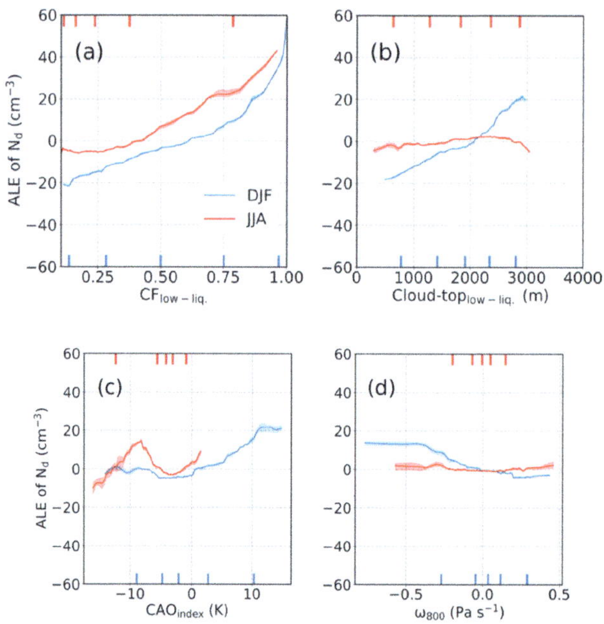

Figure 15. Same as Fig. 14 but for the following input parameters: **(a)** low-level liquid cloud fraction ($CF_{low-liq.}$), **(b)** cloud-top effective height of low-level liquid cloud (cloud-top$_{low-liq.}$), **(c)** cold-air outbreak (CAO) index, and **(d)** vertical pressure velocity at 800 hPa (ω_{800}).

Figure 15 shows the similarity in the positive relationship between cloud fraction and N_d in both seasons. Only in DJF did cloud-top effective height exhibit a clear relationship with N_d (positive), likely linked to the common phenomenon of CAOs noted in Sect. 4.1 based on heightened CAO index values, deepening of the boundary layer, and weakened inversion strength. This is supported by enhanced N_d values coincident with negative values for ω_{800} (i.e., rising motion) and CAO index values above 0 in DJF without such relationships in JJA (Fig. 15). The six parameters in Fig. S21 (PBLH, RH_{950}, RH_{800}, Rain, $Wind_{2\ m}$, $Wind\text{-}dir_{2\ m}$) did not reveal very pronounced trends with N_d in either season consistent with how they did not rank highly in importance (Fig. 13). Of particular interest is $Wind_{2\ m}$, which is used here as a proxy variable for updraft speed in the marine boundary layer, which is expected to have a high impact on N_d via its effect on in-cloud supersaturation. Although the ALE plot of $Wind_{2\ m}$ suggested a small increase of about ~ 10 cm^{-3} in N_d as the wind speed increased, $Wind_{2\ m}$ did not come out as a very important parameter in either season. This may be due to the fact that environmental conditions representing updraft speed were already included in parameters such as cloud fraction and CAO index. Another explanation can be the shortcomings and high uncertainties associated with the use of $Wind_{2\ m}$ as a proxy for updraft speed.

The results of regression analysis highlight the high sensitivity of N_d to cloud fraction regardless of season. As discussed earlier, this can be attributed largely to two factors: (i) the relationship between cloud type (e.g., stratocumulus, shallow cumulus) and cloud fraction, which can, in turn, influence cloud microphysical properties like N_d, and (ii) uncertainties associated with N_d estimates from satellite observations that can result in negative biases in N_d for low-cloud-coverage conditions. To further test the relative influence of various variables at different cloud fractions, two sensitivity tests with GBRT modeling were conducted using subsets of data with varying cloud fraction ($0.2 \leq CF_{low\text{-}liq.} \leq 0.4$ and $CF_{low\text{-}liq.} \geq 0.7$).

Beginning with results for $CF_{low\text{-}liq.} \geq 0.7$ (Figs. S22-S25), the average R^2 scores for validation and test sets for these runs were 0.47/0.39 (DJF/JJA) and 0.49/0.38 (DJF/JJA), respectively. A feature that stands out is that for both DJF and JJA, surface mass concentrations of sulfate became the most important factor. ALE plots presented in Fig. S23 suggest that N_d has a very similar sensitivity to sulfate concentration in high-cloud-coverage conditions regardless of season in contrast to the results of the original run where N_d was more sensitive to the changes in sulfate level in DJF than JJA. These results are in agreement with previous studies where N_d values for marine boundary layer clouds were highly sensitive to sulfate concentrations at the level close to cloud base (Boucher and Lohmann, 1995; Lowenthal et al., 2004; Storelvmo et al., 2009; McCoy et al., 2017, 2018; MacDonald et al., 2020). The second most important factor for DJF was the surface mass concentrations of organic carbon followed by $CF_{low\text{-}liq.}$ and sea salt surface mass concentrations. On the other hand, the second most important factor in JJA was CAO index followed by $CF_{low\text{-}liq.}$ and wind direction. ALE plots presented in Figs. S23-S25 showed similar relationships between N_d and input parameters as observed for the original runs where full datasets were used as the input. Figure S26 shows the results of the GBRT model using input data with a cloud fraction between 0.2 and 0.4, the condition relatively more prevalent in JJA than DJF. The average R^2 scores for validation and test sets for these runs were 0.30/0.30 (DJF/JJA) and 0.33/0.31 (DJF/JJA), respectively. It is interesting to see that for both seasons, an aerosol parameter emerged as the most important factor. Mass concentrations of OC appeared as the most important factor in JJA (the fourth most important factor in DJF) while in DJF, sulfate concentration came out as the most important factor (the fourth most important factor in JJA) consistent with the results of previously discussed models for DJF. It should be noted that ALE plots also suggested less sensitivity of N_d to sulfate in JJA than DJF, similar to the results observed in the original model run including all the data points. The second most important factor in DJF turned out to be the cloud-top effective height of low-level liquid clouds followed by CAO index. On the other hand, CAO index was the second most important factor in JJA followed by PBLH. ALE plots presented in Figs. S27-S29 also showed similar qualitative trends observed in original and high-cloud-coverage runs.

Unexplored Factors

Additional factors impacting the relationship between aerosol and N_d seasonal cycles are discussed here that warrant additional research with more detailed data at finer scales such as with aircraft. We are cognizant that this list is not fully exhaustive. As low-level cloud fraction impacted model results of Sect. 4.2 so substantially, the dynamics of the studied clouds require further characterization. As cloud fraction and CAO index are well related, especially in DJF, aerosol–cloud interactions are likely stronger than other seasons (as implied by Sect. 3.5) due in part to enhanced surface fluxes and turbulence and thus more droplet activation with higher cloud supersaturations (Painemal et al., 2021); in contrast, the smaller shallow cumulus clouds in summertime may be less favorable for droplet activation due to factors such as reduced turbulence and more lateral entrainment.

Entrainment of free tropospheric aerosol can impact N_d values, with potentially varying degrees of influence between seasons. It is presumed that with summertime convection, the more broken cumulus scenes are less adiabatic through the cloudy column and more affected by entrainment and mixing; hence, N_d values derived using data that remote sensors retrieve near cloud top could be considerably lower than values lower by cloud base. Satellite remote sensing studies of aerosol–cloud interactions will presumably be more challenging in winter periods versus the summer with regard to the spatial and temporal mismatch between cloud and aerosol retrievals. More specifically, it is easier to get nearly coincidental sampling in summertime due to lower cloud fractions, while in winter the frontal regions with high cloud fractions make it challenging to get aerosol retrievals. There is complexity in understanding how aerosols relate to N_d due to how giant CCN can reduce N_d and also since wet scavenging can remove aerosols efficiently. As aircraft data are limited and difficult to use for assessing seasonal cycles, new techniques of retrieving CCN and N_d from space will greatly assist such types of studies in the future.

CONCLUSIONS

This work investigates the seasonal cycle of N_d over the WNAO region in terms of concentration statistics and with discussion of potential influential factors. The results of this work have implications for increased understanding of aerosol–cloud interactions and meteorological factors influencing concentration of cloud droplets in the marine boundary layer. The results and interpretations can be summarized as follows in the order of how they were presented.

- An ACTIVATE case flight during the DJF season shows a sharp offshore N_d gradient ranging from > 1000 to < 50 cm^{-3} explained in part by particles smaller than 100 nm activating into drops during a cold-air outbreak with post-frontal clouds. There were significant changes in aerosol composition in cloud-free air and also in droplet residual particles as a function of offshore distance. These changes included a sharp decrease in aerosol number concentration, a decrease in mass fraction of sulfate in droplet residual particles, and an increase in mass fraction of organic and chloride of droplet residual particles moving offshore.

- N_d is generally highest (lowest) in DJF (JJA) over the WNAO, but aerosol parameters such as AOD, AI, surface-based aerosol mass concentrations for most species, and CCN concentrations (1% supersaturation) are generally highest in JJA and MAM and are at (or near) their lowest values in DJF. While aerosol extinction in the PBL is highest in DJF, it is driven largely by sea salt (large but few in number) and thus cannot explain the N_d peak in wintertime.
- While relative humidity was generally highest in JJA across the WNAO, the differences between seasons in the PBL and FT were not sufficiently large to explain the divergent seasonal cycles of AOD and N_d.
- The susceptibility of N_d to aerosols (Eq. 3) was strongest in DJF using four different proxy variables for aerosols, suggestive of at least one reason why N_d can be highest when aerosol proxy variables for concentration are typically near or at their lowest values.
- Composite maps of high- versus low-N_d days across the WNAO reveal that conditions associated with the highest-N_d days, regardless of season (but especially DJF), are reduced sea level pressure, stronger winds aligned with continental outflow, high low-level liquid cloud fraction, higher CAO index and PBLH, and enhanced AOD. Cold-air outbreaks are coincident with all of these conditions, especially in the colder months of DJF in sharp contrast to JJA when N_d is lowest.
- Gradient-boosted regression analysis shows that the most important predictors of N_d in DJF and JJA vary to some extent, but with cloud fraction being the most important parameter, followed by either (for DJF) surface mass concentrations of sulfate and organic carbon and CAO index or (for JJA) CAO index, surface mass concentrations of organic carbon, and sulfate concentrations. Accumulated local effect plots confirm that sulfate and organics help drive the high N_d values via continental outflow, which is assisted in large part by conditions associated with CAOs such as high cloud fraction and high CAO index.

Therefore, the combination of continental pollution outflow and turbulence changes contributed by surface fluxes (manifested in the strongest CAO index values in DJF and weakest in JJA) markedly influence the N_d cycle, leading to differing annual cycles in cloud microphysics and aerosols. More detailed data such as from aircraft and modeling can help extend this line of research to confirm these findings and speculations such as how (i) the aerosol indirect effect is strongest in DJF due to boundary layer dynamics such as with more turbulence and mixing than other seasons (Painemal et al., 2021); (ii) enhanced giant CCN in forms such as sea salt and dust can reduce N_d via expediting the collision–coalescence process; and (iii) substantial aerosol removal can occur far offshore as postfrontal clouds associated with CAOs build and then begin to precipitate. The latter hypothesis may help explain why Bermuda (> 1000 km offshore the US East Coast) was the only selected sub-domain in this study to not have a seasonal N_d peak in DJF.

SUPPLEMENTARY MATERIAL

Table S1. Range of model hyperparameters tested during training/validation of the GBRT models through a combination of grid and random searches. Final model values are also listed in the last column.

Model parameter	Range of values tested	Final model values (DJF/JJA)
Learning rate	0.001-0.1	0.05/0.05
Number of estimators	100-5000	400/400
Maximum depth of a tree	2-35	9/11
Minimum number of samples to split an internal node	20-100	66/45
Minimum number of samples at a leaf node	20-60	31/66

Table S2: Mean seasonal values of speciated AOD and surface mass concentration for black carbon, dust, organic carbon, sulfate, and sea-salt for the six sub-domains in Figure S1.

	Speciated AOD/Surface mass concentration ($\mu g\ m^{-3}$)					
	S	C-S	C	C-N	N	Bermuda
			Sulfate			
DJF	0.03/0.91	0.05/1.37	0.05/1.49	0.06/1.99	0.06/1.50	0.03/0.59
MAM	0.05/0.99	0.06/1.37	0.07/1.54	0.08/2.38	0.07/1.84	0.04/0.72
JJA	0.05/0.75	0.07/1.10	0.07/1.25	0.09/2.26	0.08/1.45	0.04/0.50
SON	0.04/0.73	0.04/1.00	0.04/1.08	0.05/1.56	0.05/1.09	0.02/0.41
			Sea-salt			
DJF	0.04/38.10	0.04/39.12	0.05/45.89	0.03/29.55	0.03/27.45	0.05/49.51
MAM	0.04/36.72	0.04/38.04	0.04/44.28	0.03/28.52	0.02/23.22	0.04/41.76
JJA	0.05/46.12	0.05/48.83	0.05/56.56	0.03/30.91	0.02/17.99	0.03/35.95
SON	0.06/48.44	0.05/43.76	0.06/54.24	0.04/36.20	0.03/27.24	0.06/54.60

Dust						
DJF	<0.01/0.74	<0.01/1.06	0.01/1.64	0.01/2.64	0.01/1.24	<0.01/0.98
MAM	0.02/3.90	0.02/3.86	0.02/4.03	0.03/4.47	0.02/2.90	0.02/2.81
JJA	0.03/12.05	0.02/8.22	0.02/6.08	0.02/5.02	0.02/2.41	0.02/6.58
SON	0.01/2.56	0.01/1.79	0.01/2.01	0.01/2.51	0.01/1.20	0.01/2.01
Organic carbon						
DJF	0.01/0.69	0.02/0.92	0.02/0.67	0.02/1.07	0.02/0.58	0.01/0.25
MAM	0.03/1.00	0.04/1.37	0.04/1.02	0.04/1.53	0.03/0.92	0.03/0.42
JJA	0.01/0.55	0.03/1.16	0.03/1.25	0.05/2.49	0.05/2.43	0.02/0.35
SON	0.01/0.67	0.02/1.03	0.02/0.87	0.02/1.43	0.02/0.98	0.01/0.34
Black carbon						
DJF	<0.01/0.10	<0.01/0.15	0.01/0.15	0.01/0.37	0.01/0.20	<0.01/0.08
MAM	0.01/0.12	0.01/0.16	0.01/0.17	0.01/0.38	0.01/0.21	0.01/0.09
JJA	<0.01/0.08	0.01/0.12	0.01/0.15	0.01/0.41	0.01/0.31	<0.01/0.06
SON	<0.01/0.09	<0.01/0.12	<0.01/0.14	0.01/0.32	0.01/0.20	<0.01/0.06

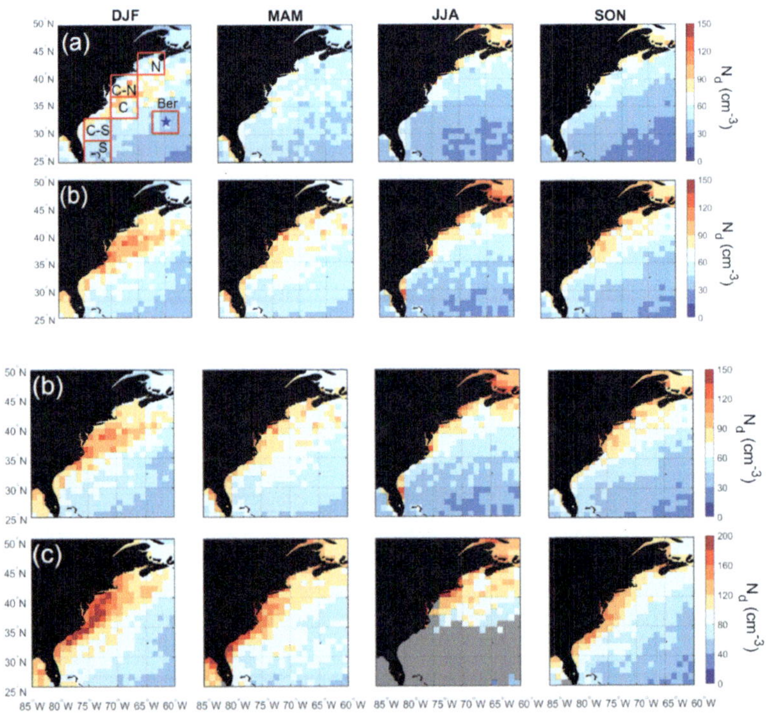

Figure S1: Seasonal maps of cloud drop number concentration for different ranges of low level liquid cloud fraction (CFlow-liq.) as follows: (a) $0.1 \leq CF_{low\text{-}liq.} < 0.3$, (b) $0.3 \leq CF_{low\text{-}liq.} < 0.6$, and (c) $CF_{low\text{-}liq.} \geq 0.7$. Gray pixels represent regions without sufficent sample points (less than 10 points) for calculating averages.

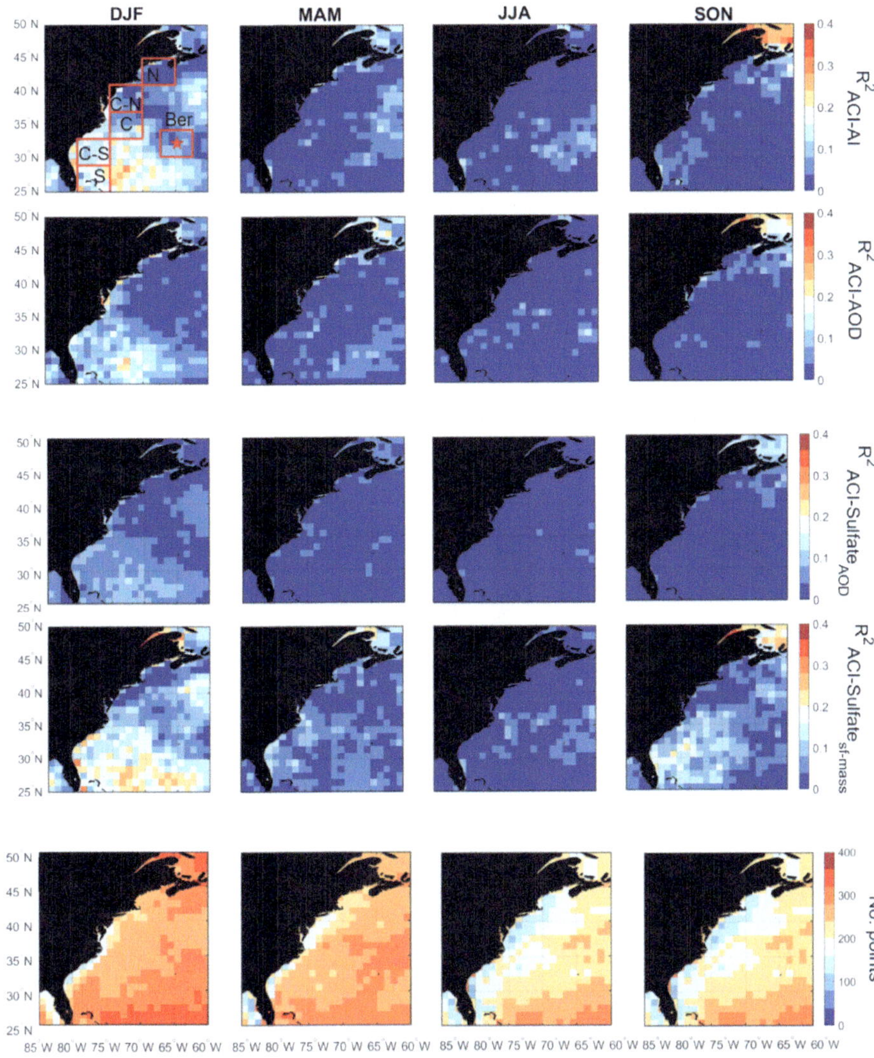

Figure S2: Seasonal maps of the coefficient of determination (R^2) and number of points for the analysis of the aerosol-cloud interaction (ACI) parameters over the WNAO using daily Nd and four different aerosol proxy parameter values (AI, AOD, SulfateAOD, Sulfatesf-mass) from CERES-MODIS and MERRA-2, respectively. ACI statistics associated with the six sub-domains shown are summarized in Table 4.

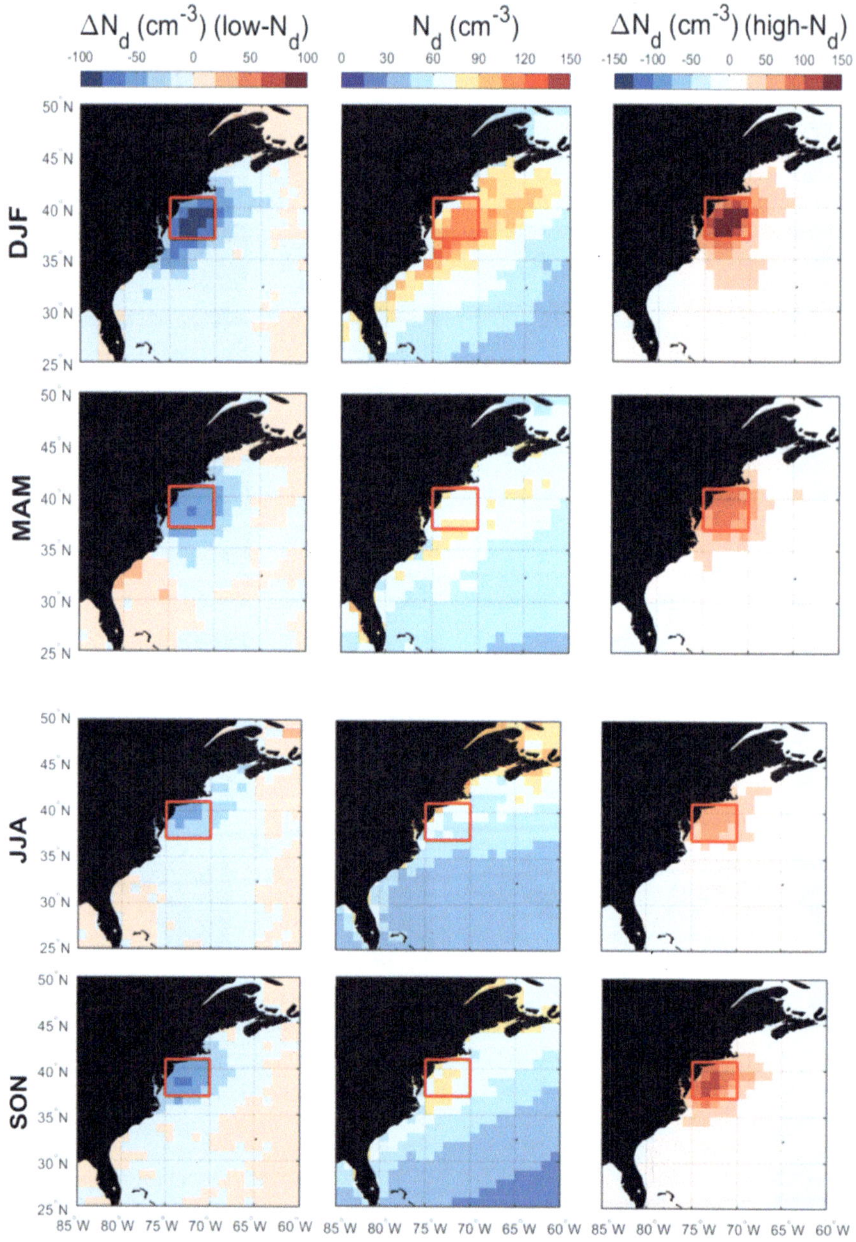

Figure S3: Seasonal climatology of N_d (middle column) and anomalies from seasonal averages for low-N_d days (left column) and high-N_d days (right column). The red box represents sub-domain C-N for which the analysis was conducted, as explained in Section 4.1.

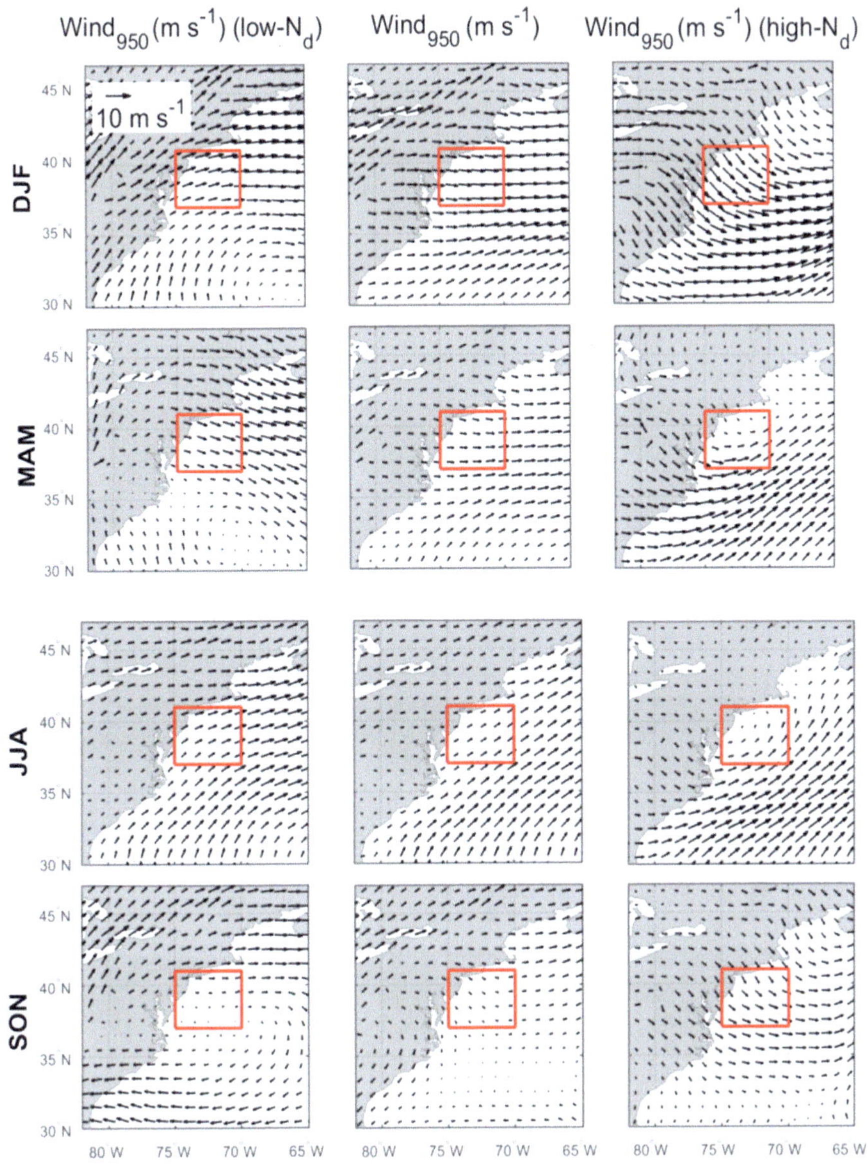

Figure S4: Seasonal climatology of winds at 950 hPa (middle column) and anomalies from seasonal averages for low-N_d days (left column) and high-N_d days (right column). The reference wind vector is shown on the top left panel. The red box represents sub-domain N for which the analysis was conducted.

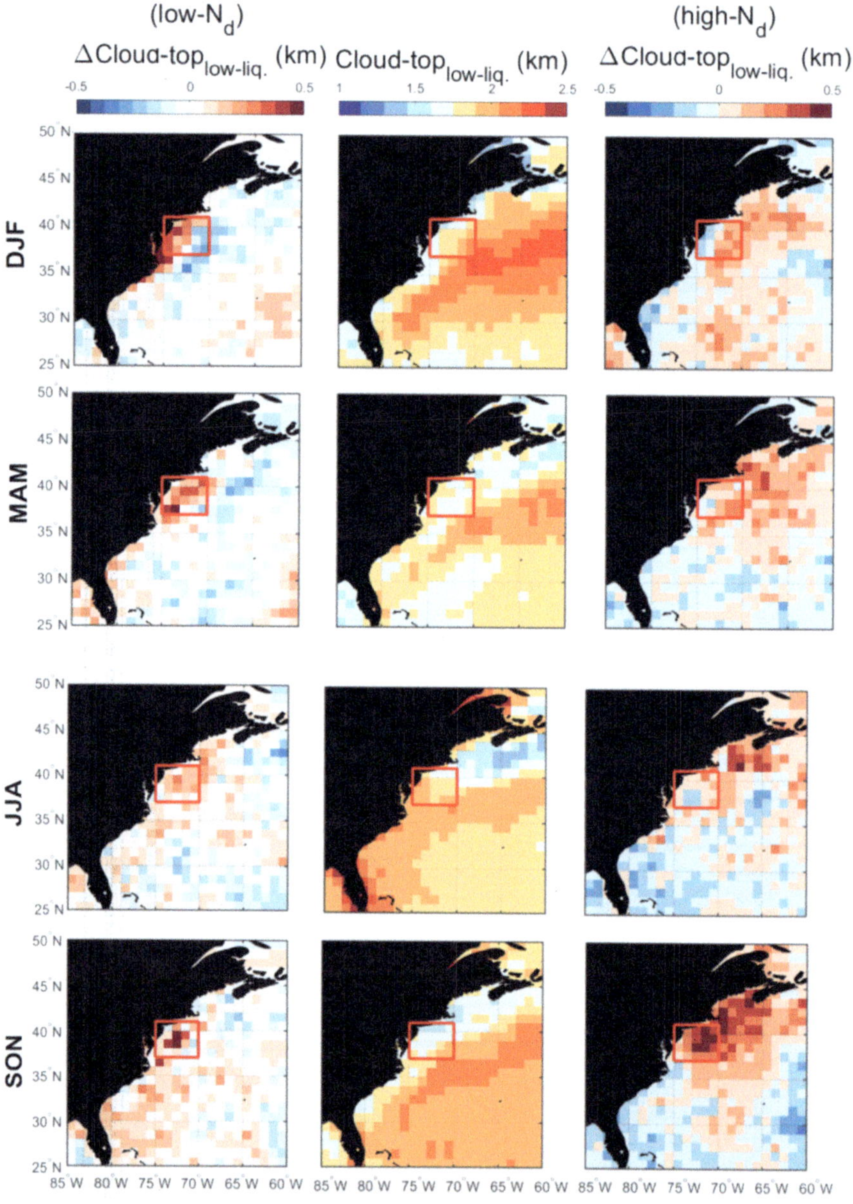

Figure S5: Seasonal averages of low-level liquid cloud-top effective height (middle column) and associated anomalies on low-Nd days (left column) and high-Nd days (right column). The red box represents sub-domain C-N for which the analysis was conducted.

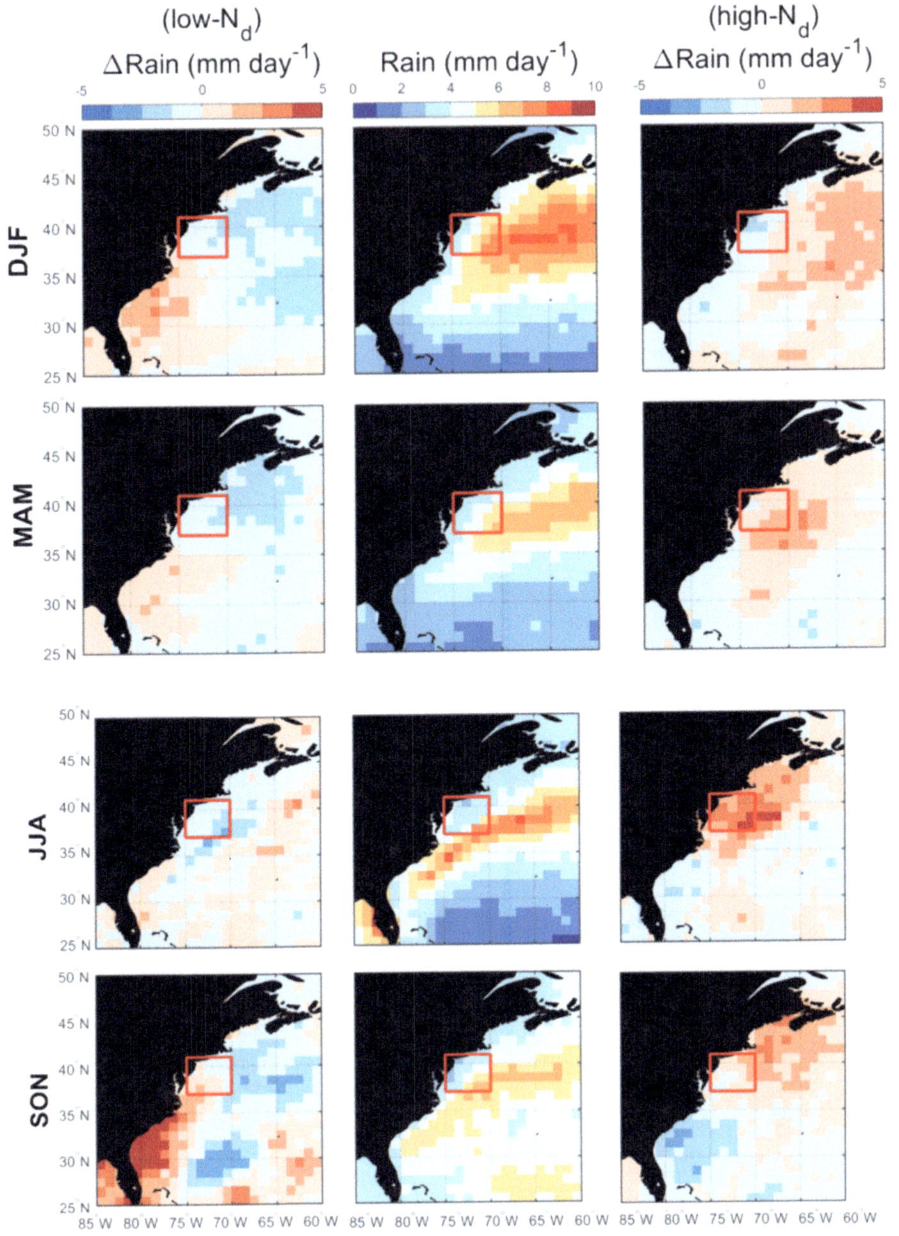

Figure S6: Seasonal averages of PERSIANN-CDR precipitation rate (middle column) and associated anomalies on low-N_d days (left column) and high-N_d days (right column). The red box represents sub-domain C-N for which the analysis was conducted.

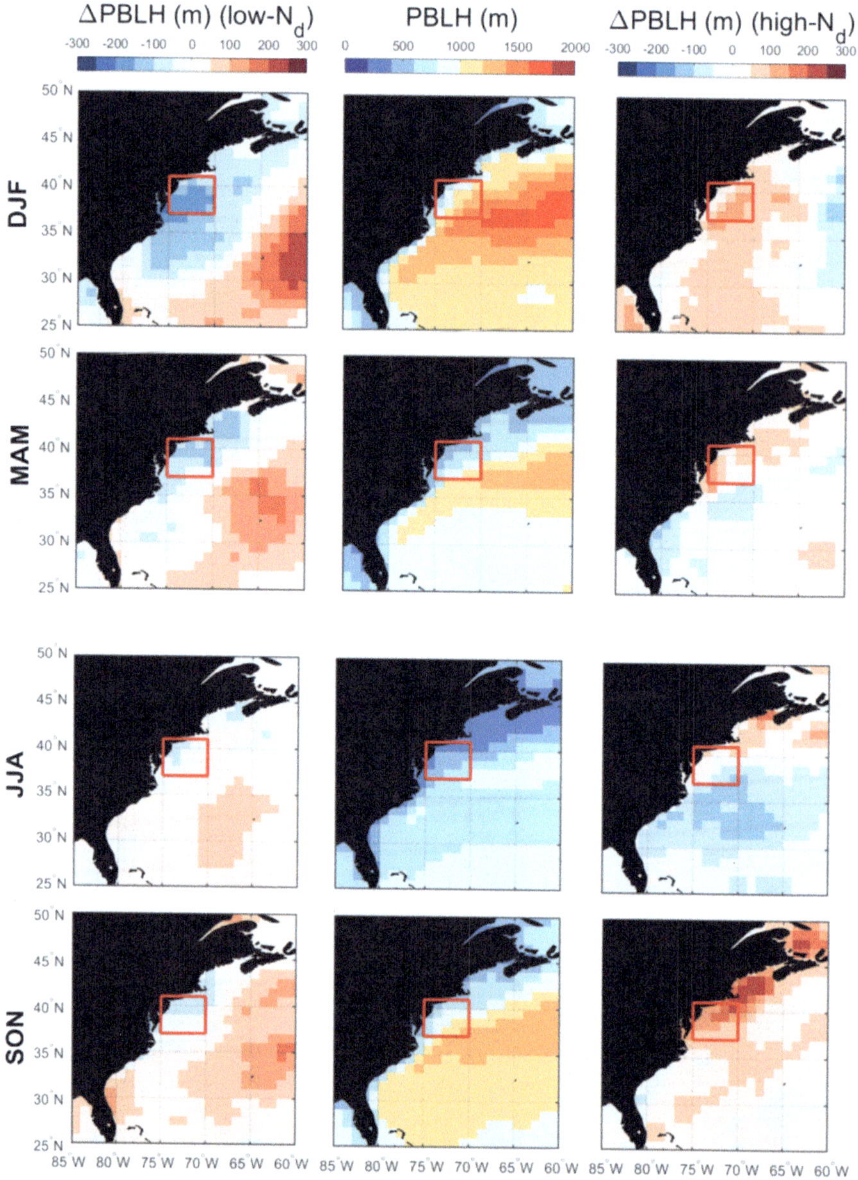

Figure S7: Seasonal averages of planetary boundary layer height (middle column) and associated anomalies on low-N_d days (left column) and high-N_d days (right column). The red box represents sub-domain C-N for which the analysis was conducted.

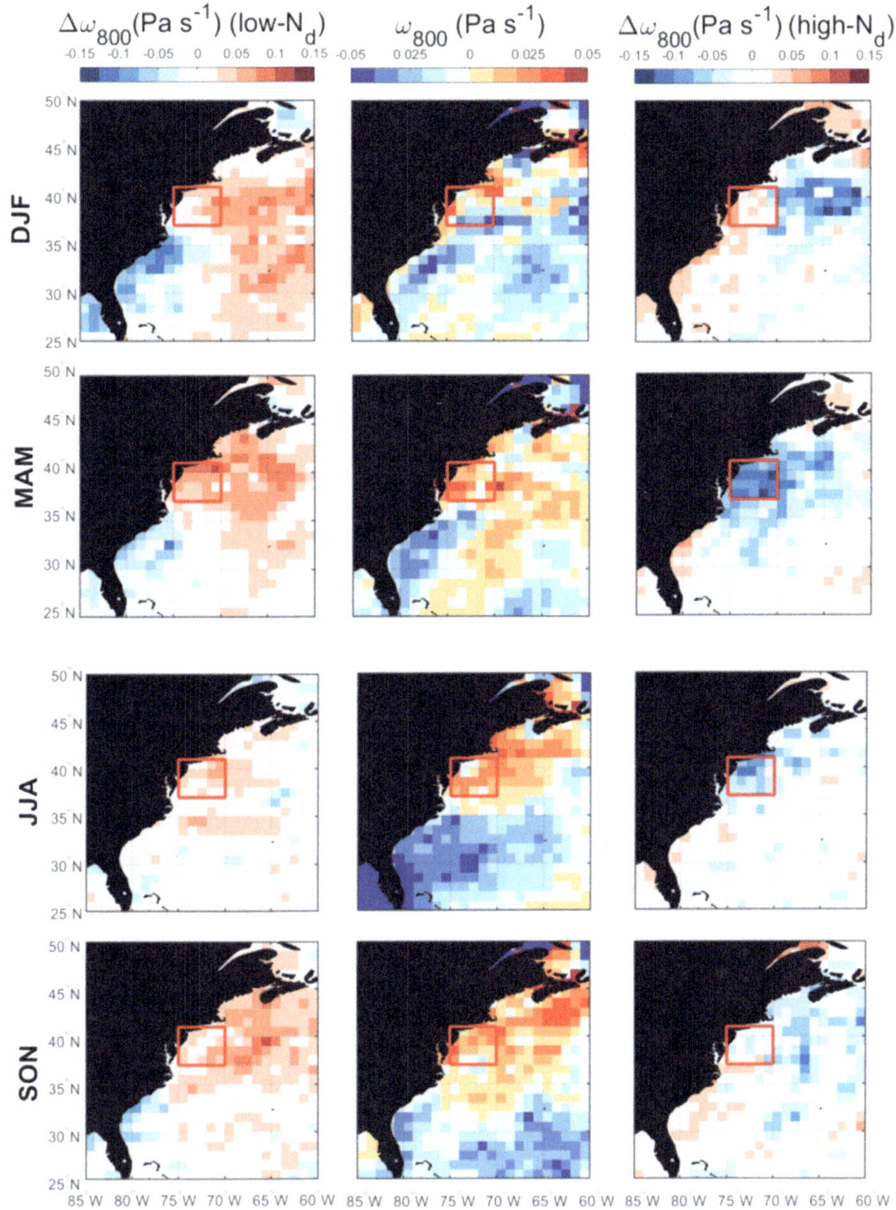

Figure S8: Seasonal averages of vertical pressure velocity at 800 hPa (middle column) and associated anomalies on low-N_d days (left column) and high-N_d days (right column). The red box represents sub-domain C-N for which the analysis was conducted.

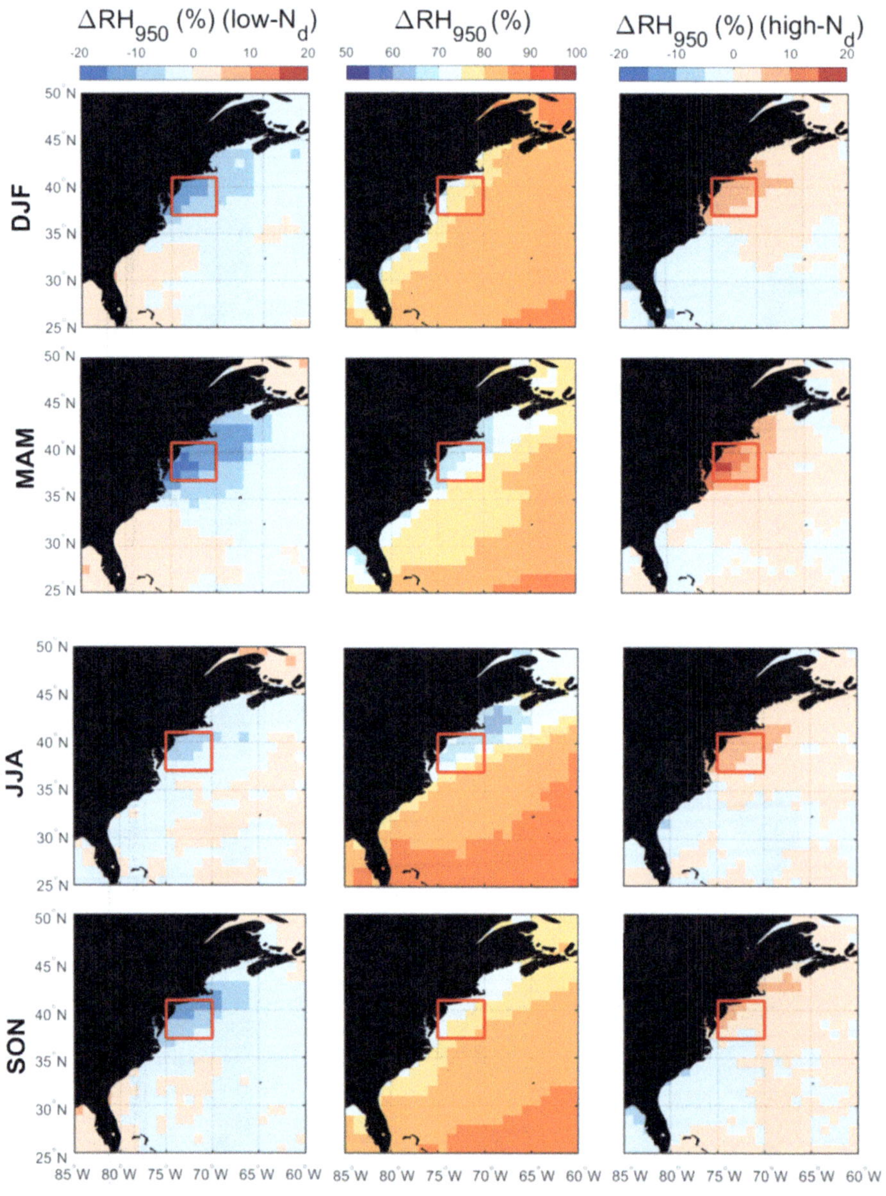

Figure S9: Seasonal averages of relative humidity at 950 hPa (middle column) and associated anomalies on low-N_d days (left column) and high-N_d days (right column). The red box represents sub-domain C-N for which the analysis was conducted.

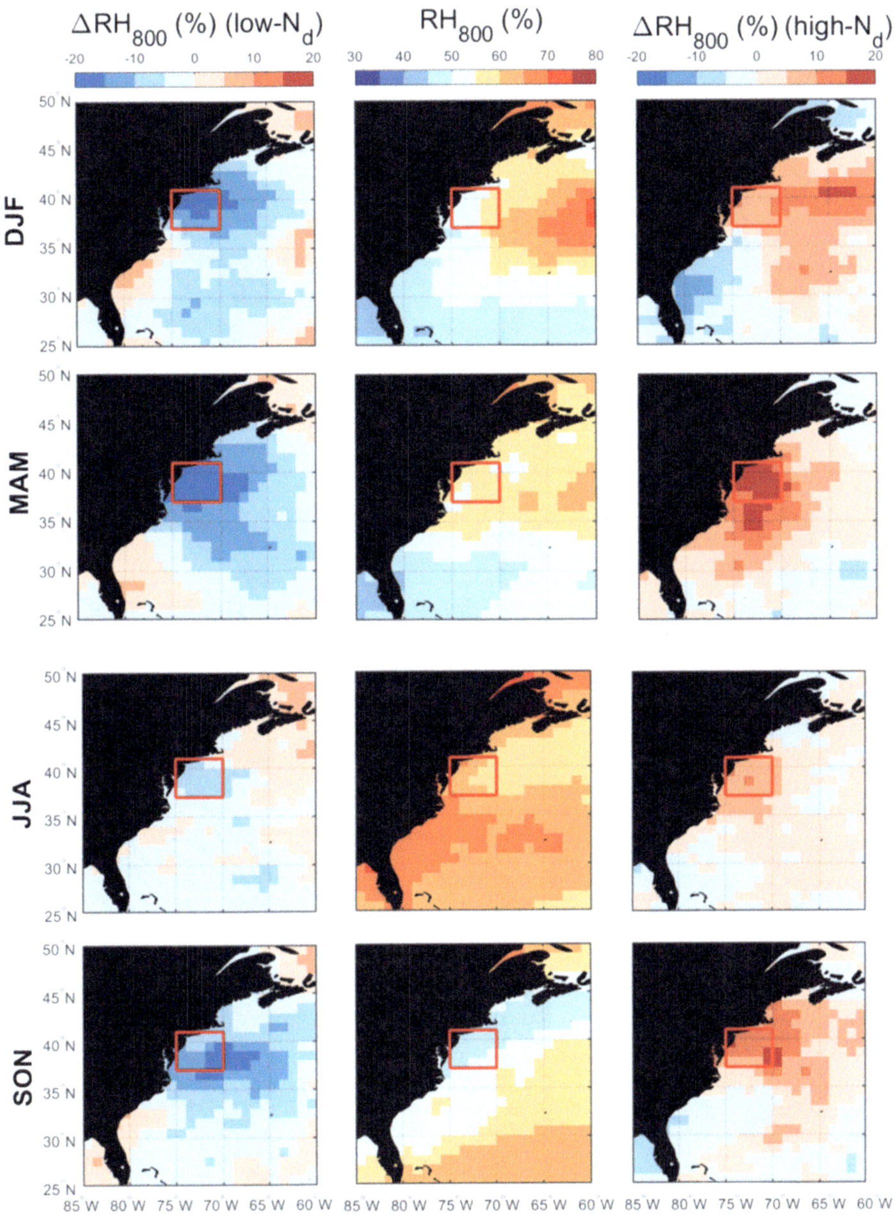

Figure S10: Seasonal averages of relative humidity at 800 hPa (middle column) and associated anomalies on low-N_d days (left column) and high-N_d days (right column). The red box represents sub-domain C-N for which the analysis was conducted.

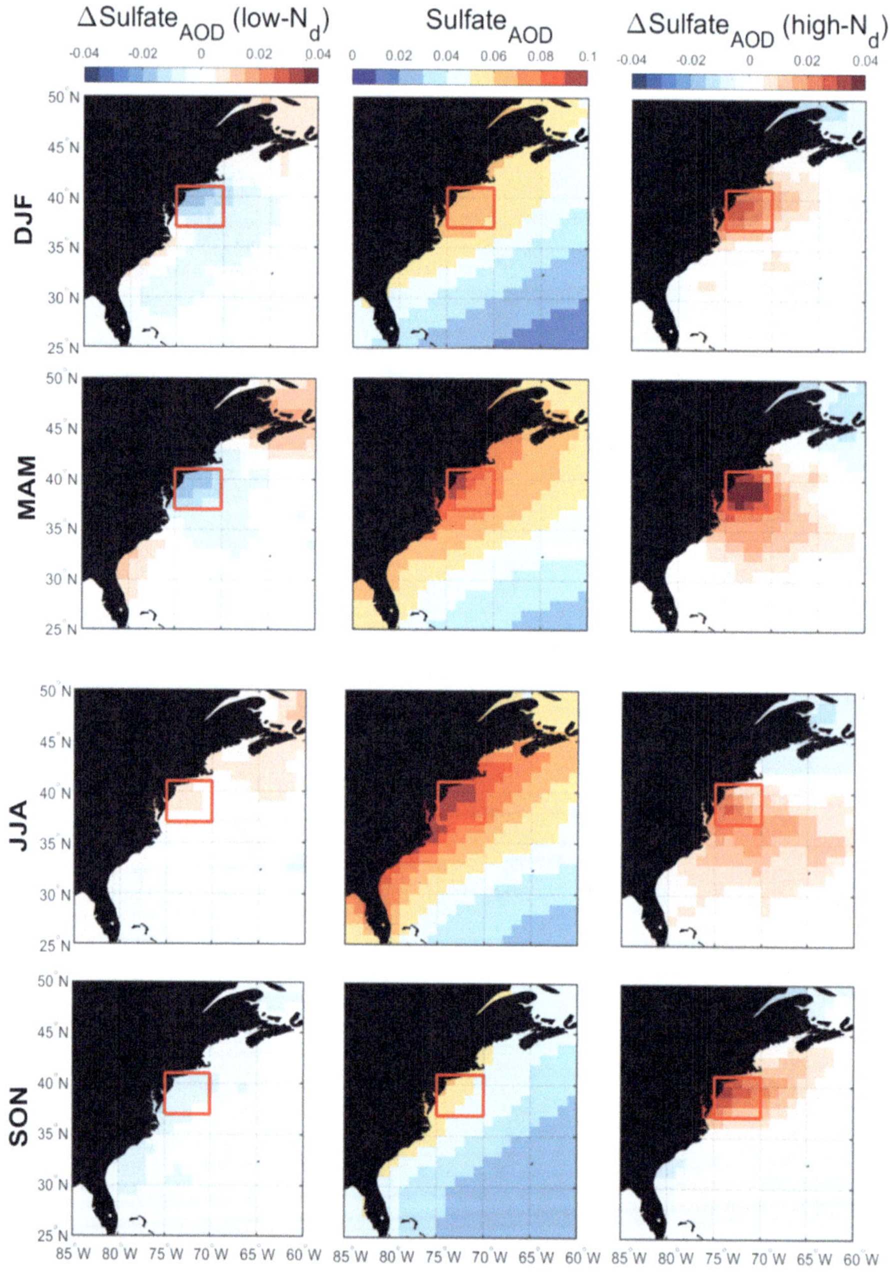

Figure S11: Seasonal averages of sulfate AOD (middle column) and associated anomalies on low-N_d days (left column) and high-N_d days (right column). The red box represents sub domain C-N for which the analysis was conducted.

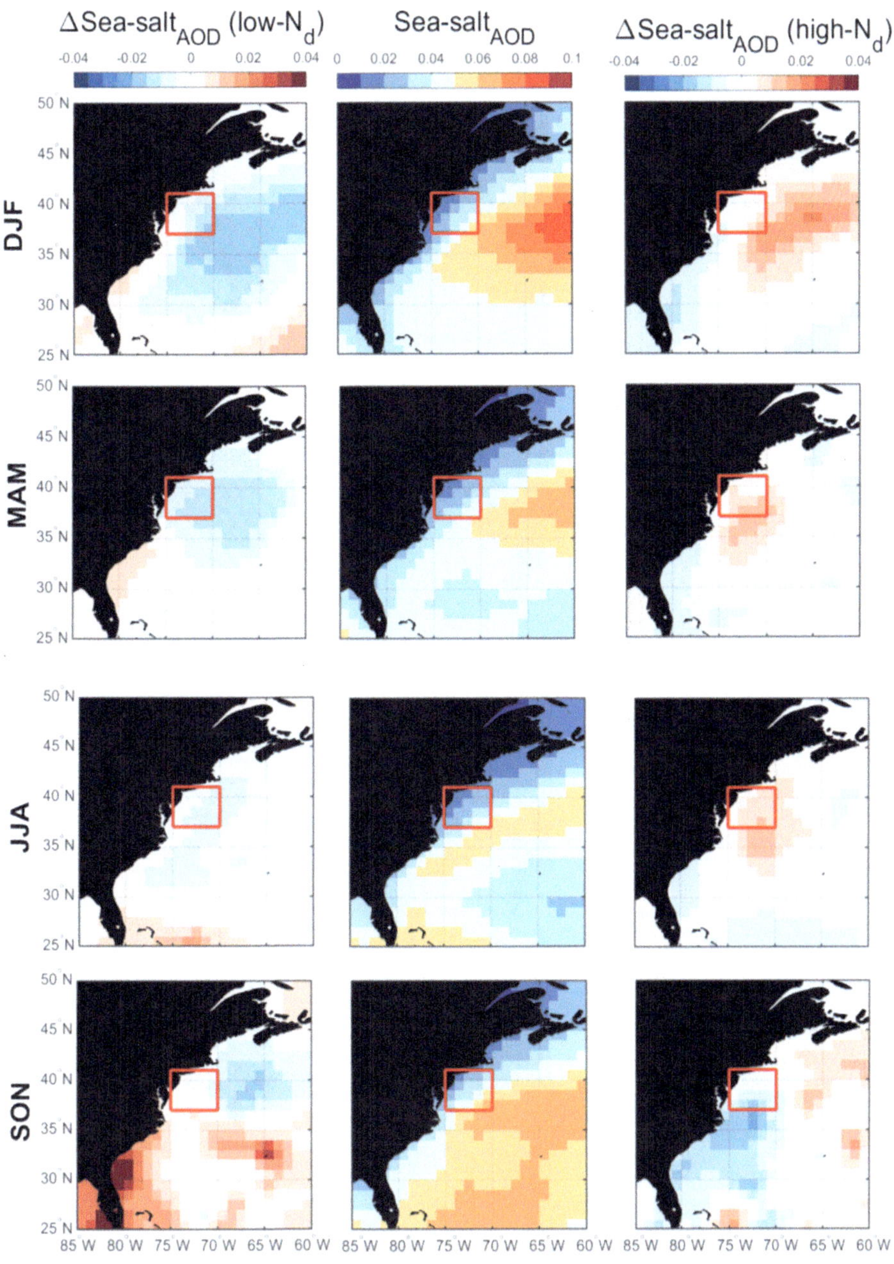

Figure S12: Seasonal averages of sea-salt AOD (middle column) and associated anomalies on low-Nd days (left column) and high-Nd days (right column). The red box represents sub domain C-N for which the analysis was conducted.

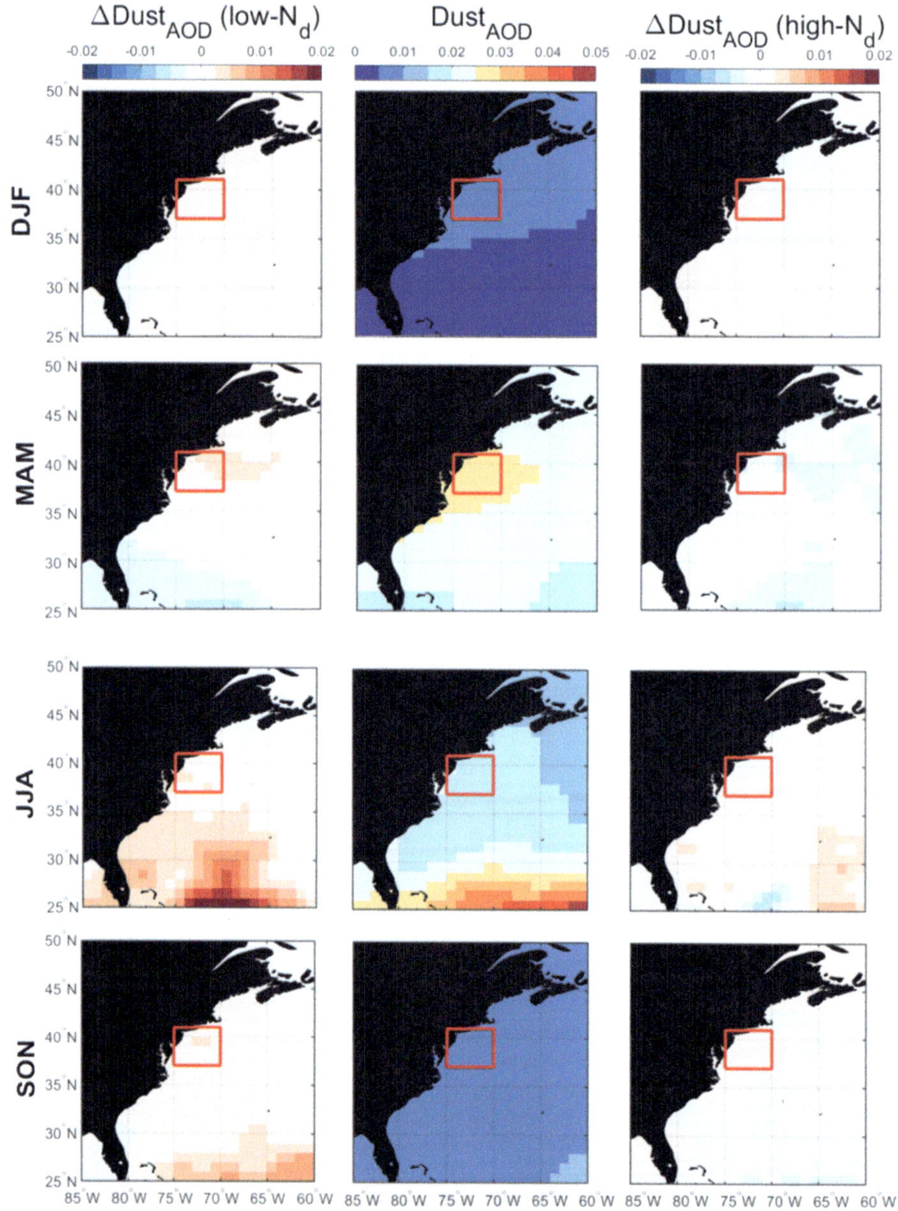

Figure S13: Seasonal averages of dust AOD (middle column) and associated anomalies on low-Nd days (left column) and high-Nd days (right column). The red box represents sub domain C-N for which the analysis was conducted.

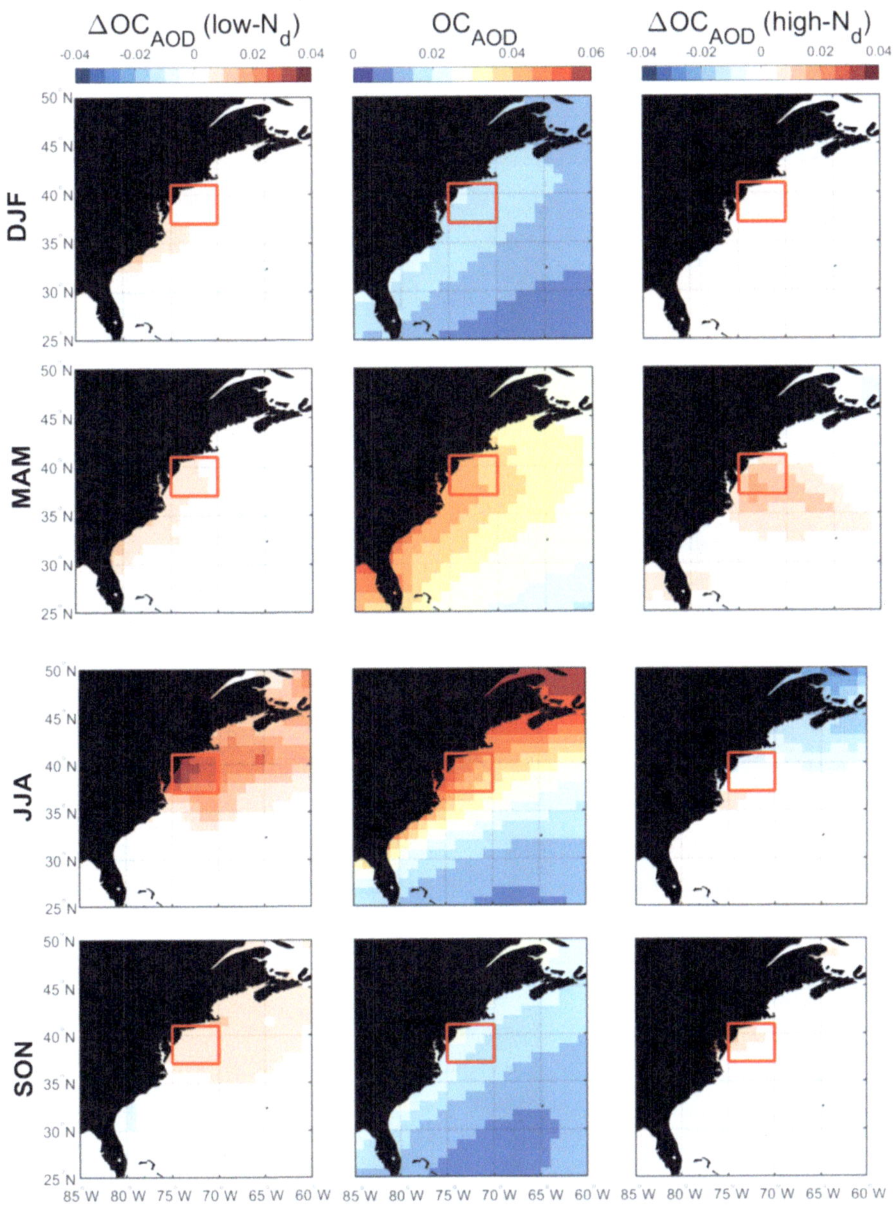

Figure S14: Seasonal averages of organic carbon AOD (middle column) and associated anomalies on low-Nd days (left column) and high-Nd days (right column). The red box represents sub-domain C-N for which the analysis was conducted.

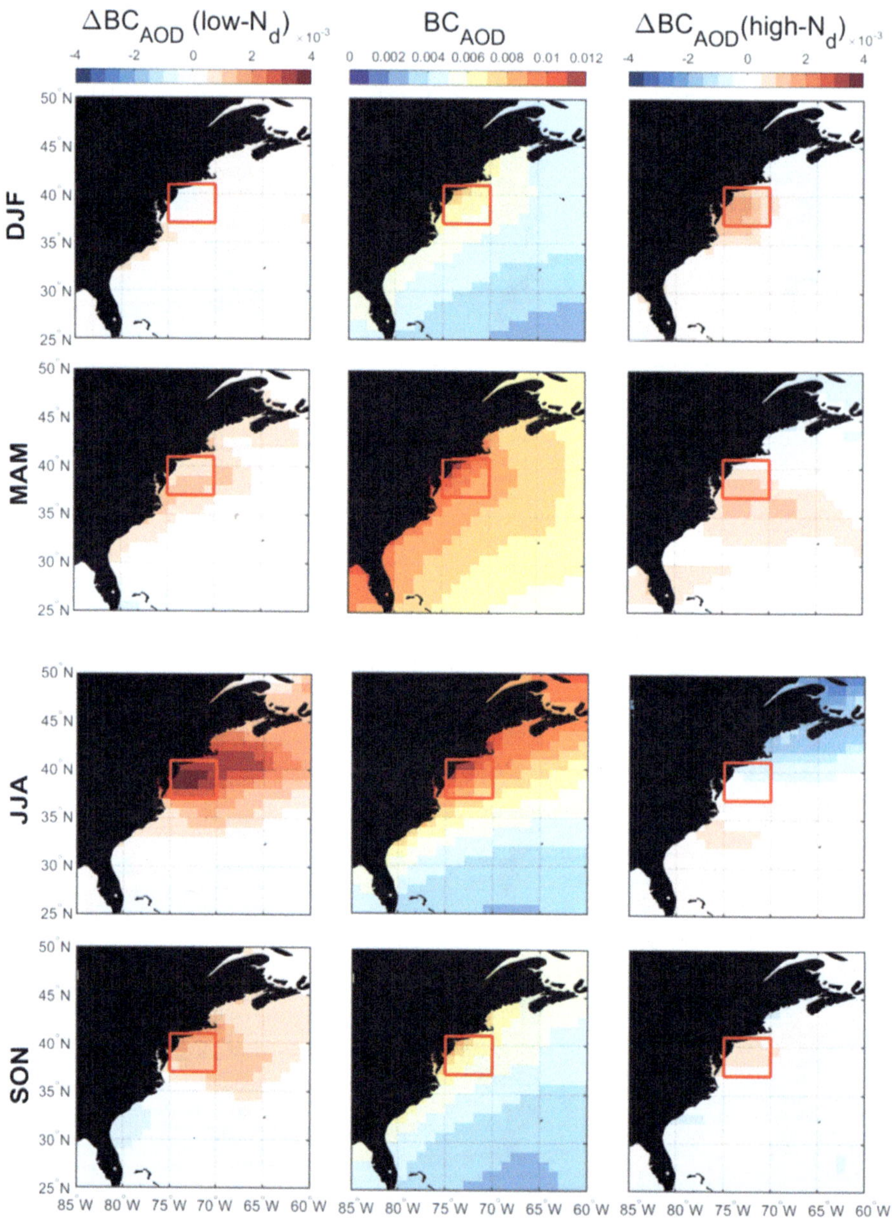

Figure S15: Seasonal averages of black carbon AOD (middle column) and associated anomalies on low-Nd days (left column) and high-Nd days (right column). The red box represents sub-domain C-N for which the analysis was conducted.

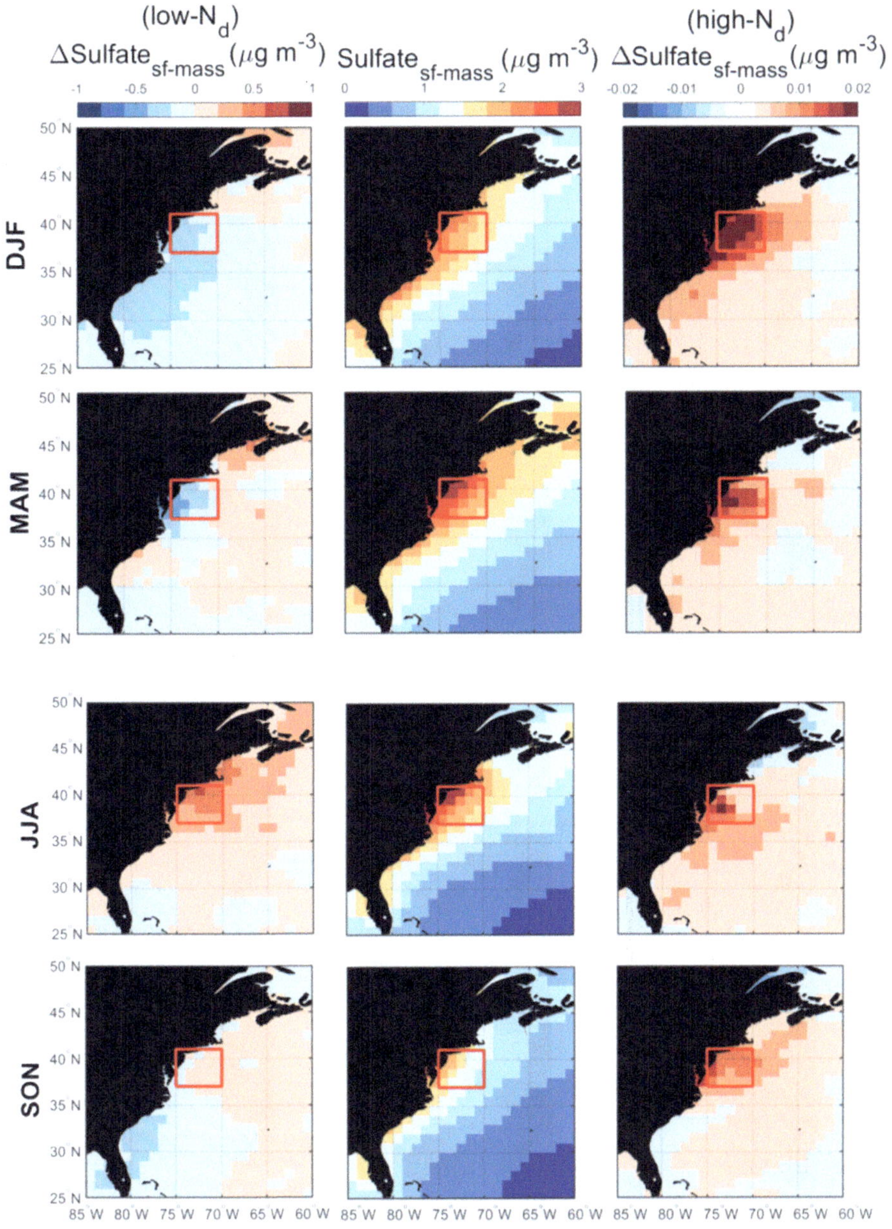

Figure S16: Seasonal averages of sulfate surface mass concentration (middle column) and associated anomalies on low-Nd days (left column) and high-Nd days (right column). The red box represents sub-domain C-N for which the analysis was conducted.

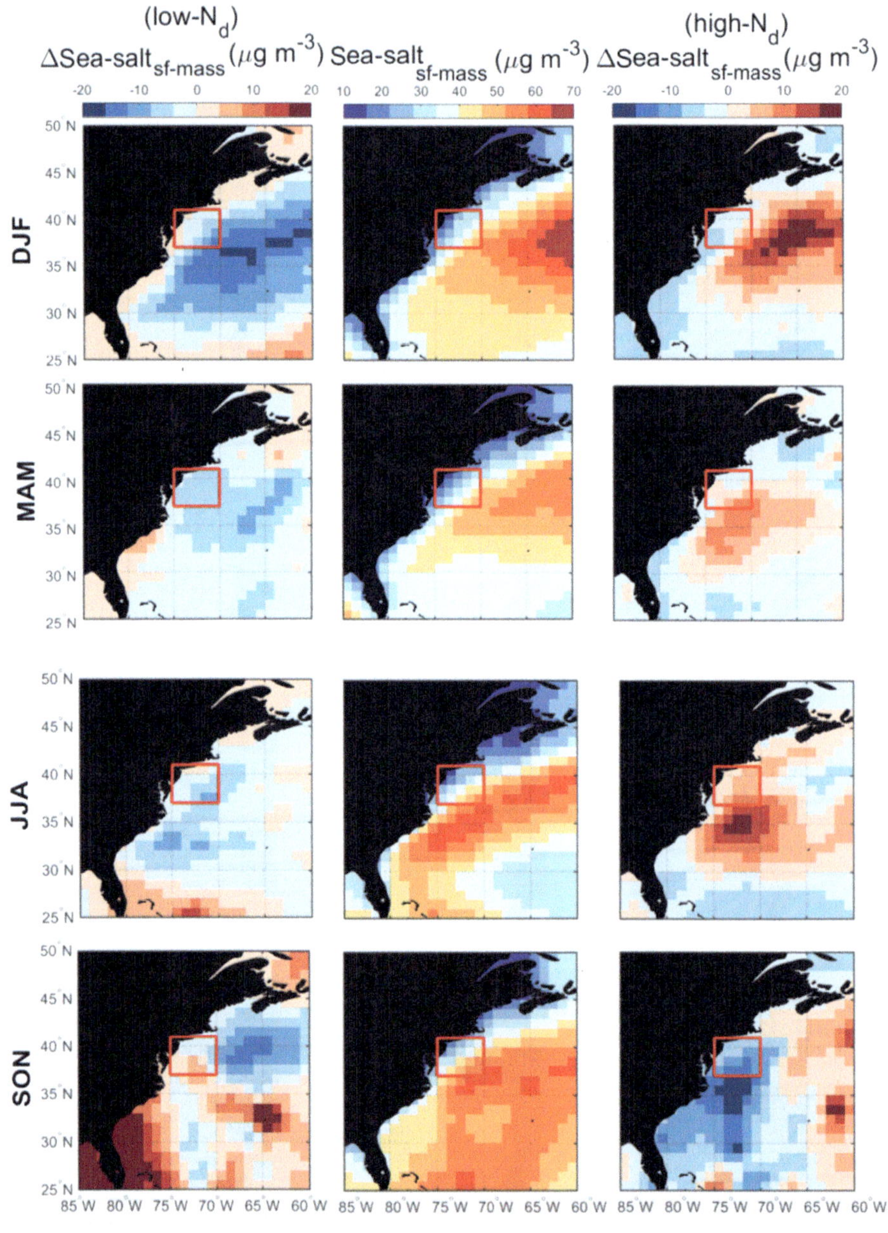

Figure S17: Seasonal averages of sea-salt surface mass concentration (middle column) and associated anomalies on low-Nd days (left column) and high-Nd days (right column). The red box represents sub-domain C-N for which the analysis was conducted.

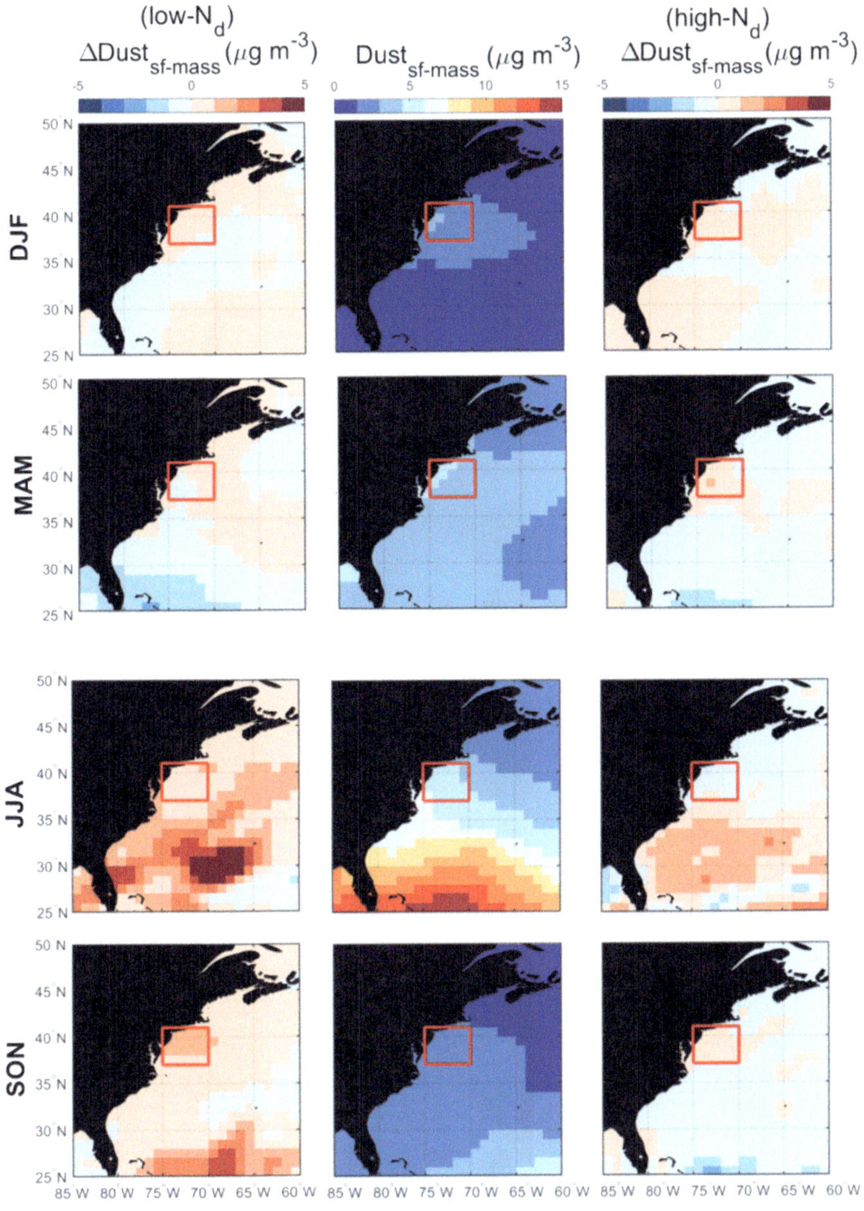

Figure S18: Seasonal averages of dust surface mass concentration (middle column) and associated anomalies on low-Nd days (left column) and high-Nd days (right column). The red box represents sub-domain C-N for which the analysis was conducted.

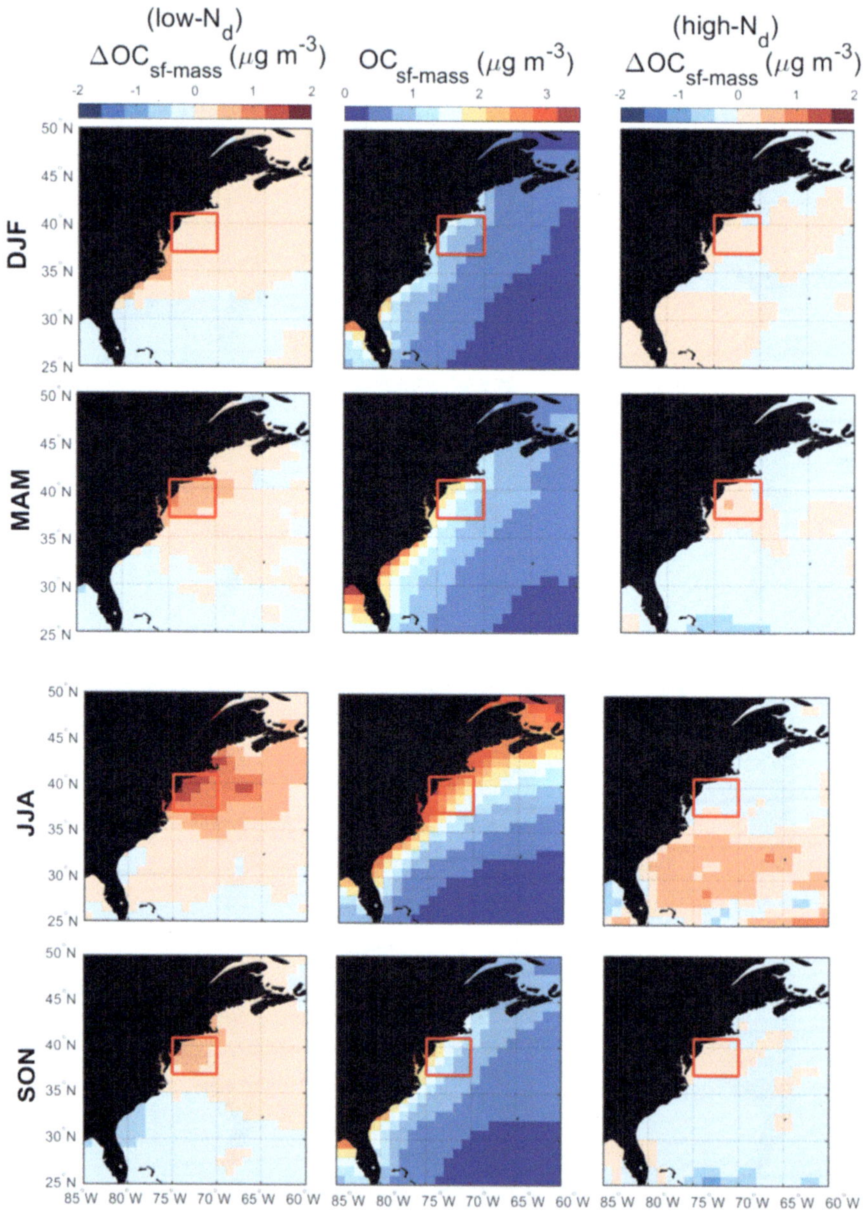

Figure S19: Seasonal averages of organic carbon (OC) surface mass concentration (middle column) and associated anomalies on low-Nd days (left column) and high-Nd days (right column). The red box represents sub-domain C-N for which the analysis was conducted.

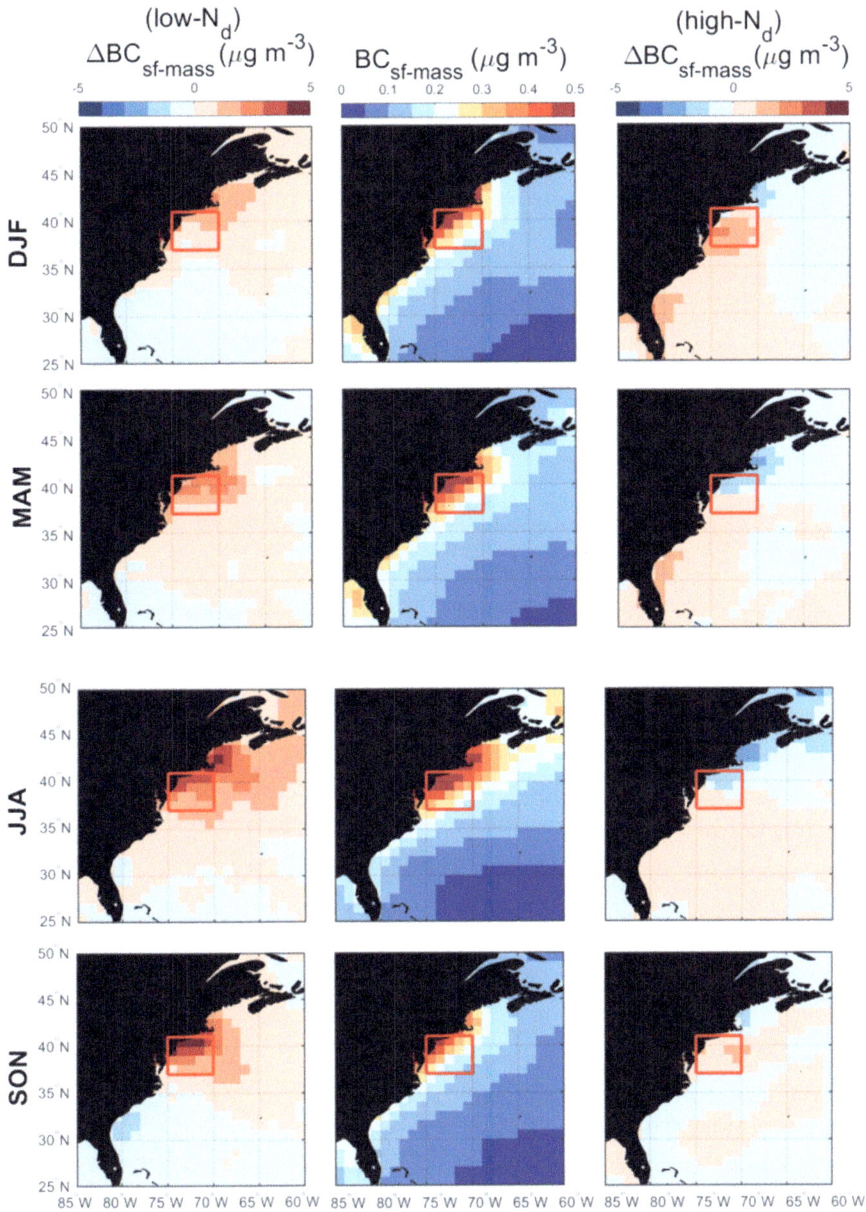

Figure S20: Seasonal averages of black carbon (BC) surface mass concentration (middle column) and associated anomalies on low-Nd days (left column) and high-Nd days (right column). The red box represents sub-domain C-N for which the analysis was conducted.

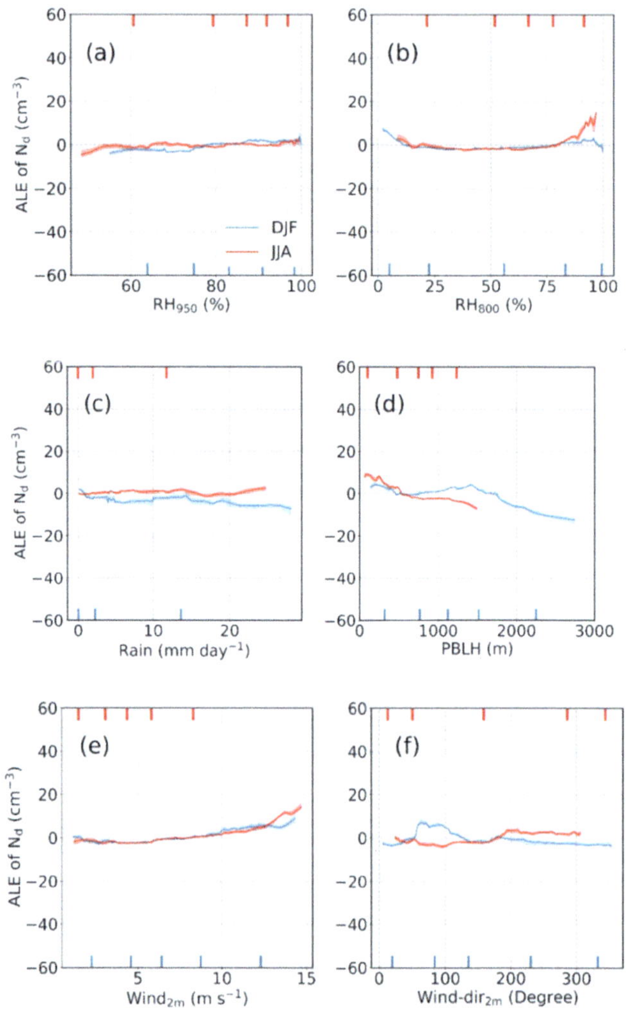

Figure S21: Average local accumulated effect (ALE) profiles based on GBRT modeling of the following parameters: (a) relative humidity at 950 hPa (RH_{950}), (b) relative humidity at 800 hPa (RH_{800}), (c) rain rate, (d) planetary boundary layer height (PBLH), (e) wind speed at 2 m ($Wind_{2m}$), and (f) wind direction at 2 m ($wind\text{-}dir_{2m}$). Blue and red profiles represent ALEs of DJF and JJA, respectively. Shaded areas show the ALE ranges stemming from the variability of the obtained models from the cross-validation resampling procedure. Markers on the bottom and top x-axes denote the values of 5th, 25th, 50th, 75th, and 95th percentiles for each input variable; note that the first three markers on the x-axes in panel (c) are very close and thus on top of each other.

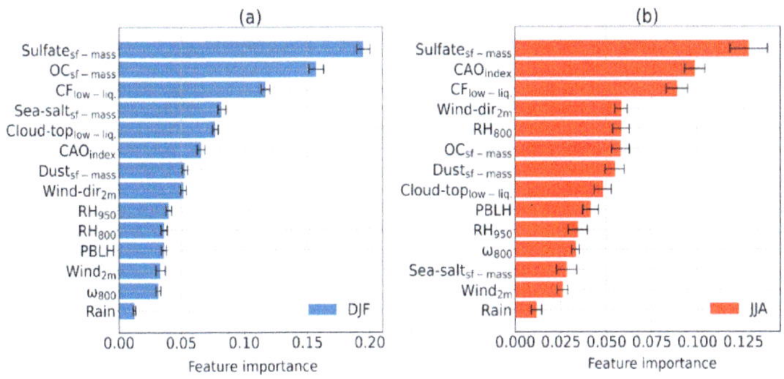

Figure S22: Average permutation feature importance of input parameters for (a) DJF and (b) JJA based on GBRT models trained in each season on subsets of data including only 158 samples with low-level liquid cloud fraction greater than or equal to 0.7 (i.e., $CF_{low\text{-}liq.} \geq 0.7$). Feature importance values were calculated based on using the test set. Error bars exhibit the range of feature importance values stemming from the variability of the obtained models from the cross-validation resampling procedure.

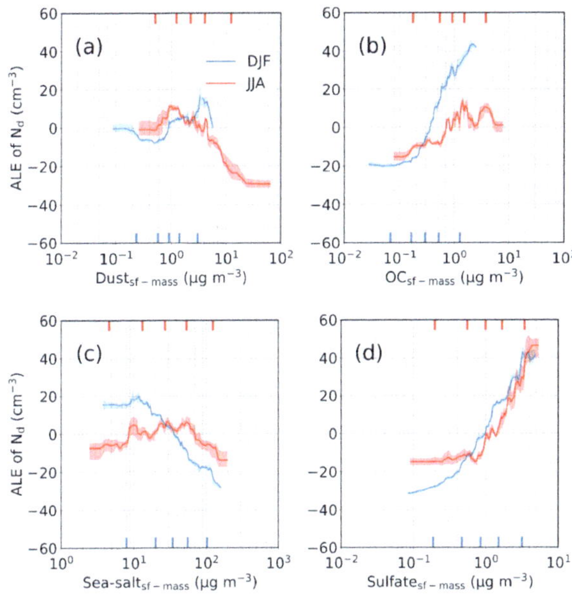

Figure S23: Average local accumulated effect (ALE) profiles based on GBRT modeling for surface mass concentrations of the following parameters: (a) dust, (b) organic carbon, (c) sea-salt, and (d) sulfate. ALE profiles were based on GBRT modeling on subsets of data including only samples with low-level liquid cloud fraction greater than or equal to 0.7 (i.e., $CF_{low\text{-}liq.} \geq 0.7$). Blue and red profiles represent ALEs of DJF and JJA, respectively. Shaded areas show the ALE ranges stemming from the variability of the obtained models from the cross-validation resampling procedure. Markers on the bottom and top x-axes denote the values of 5th, 25th, 50th, 75th, and 95th 171 percentiles for each input variable.

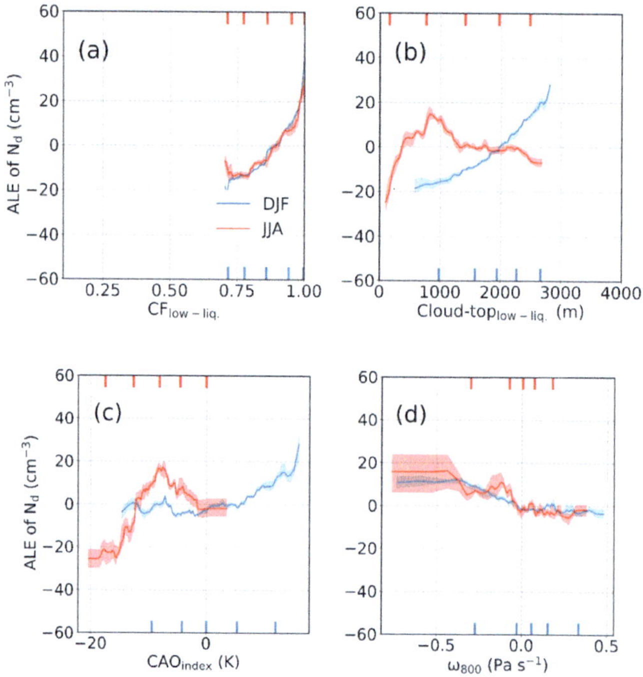

Figure S24: Same as Figure S23 but for the following input parameters: (a) low-level liquid 175 cloud fraction ($CF_{low\text{-}liq.}$), (b) cloud-top effective height of low-level liquid cloud (cloud-toplow176 liq.), (c) cold-air outbreak (CAO) index, and (d) vertical pressure velocity at 800 hPa (ω800).

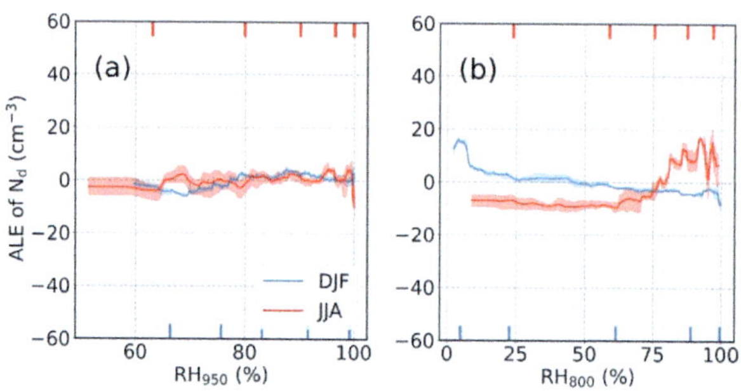

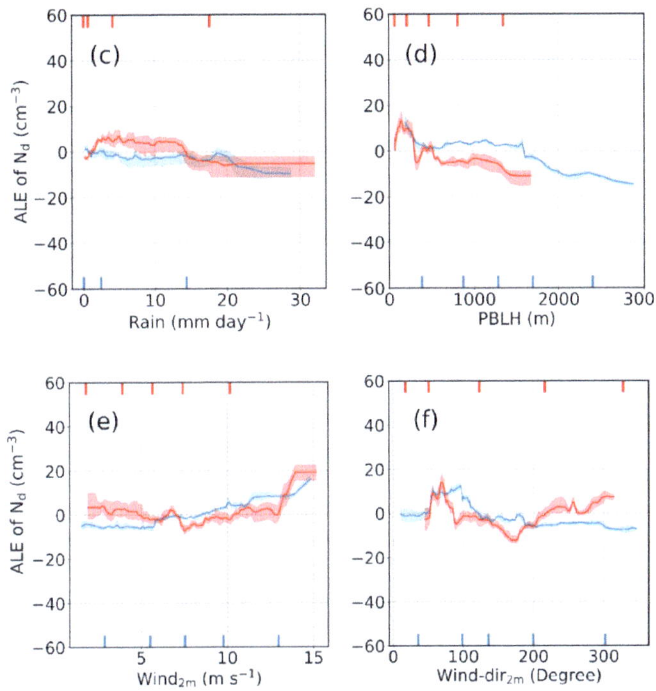

Figure S25: Same as Figure S23 but for the following input parameters: (a) relative humidity at 950 hPa (RH_{950}), (b) relative humidity at 800 hPa (RH_{800}), (c) rain rate, (d) planetary boundary layer height (PBLH), (e) wind speed at 2 m ($Wind_{2m}$), and (f) wind direction at 2 m ($wind\text{-}dir_{2m}$).

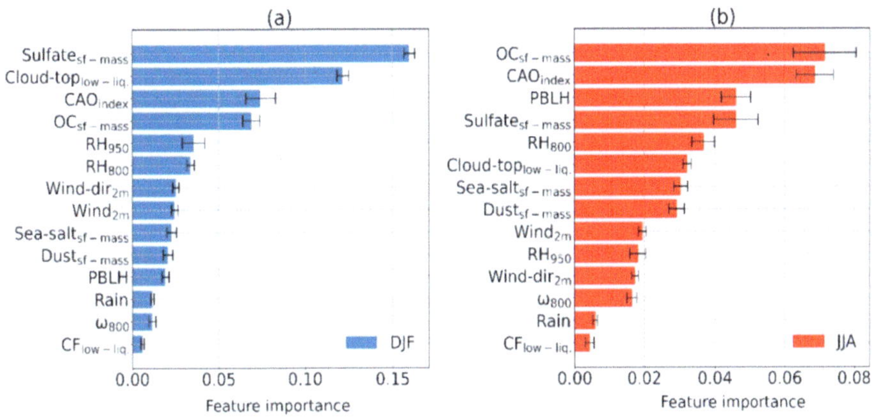

Figure S26: Average permutation feature importance of input parameters for (a) DJF and (b) JJA based on GBRT models trained in each season on subsets of data including only samples with low-level liquid cloud fraction between 0.2 and 0.4 (i.e., $0.2 \leq CF_{low\text{-}liq.} \leq 0.4$). Feature importance values were calculated based on using the test set. Error bars exhibit the range of feature importance values stemming from the variability of the obtained models from the cross-validation resampling procedure.

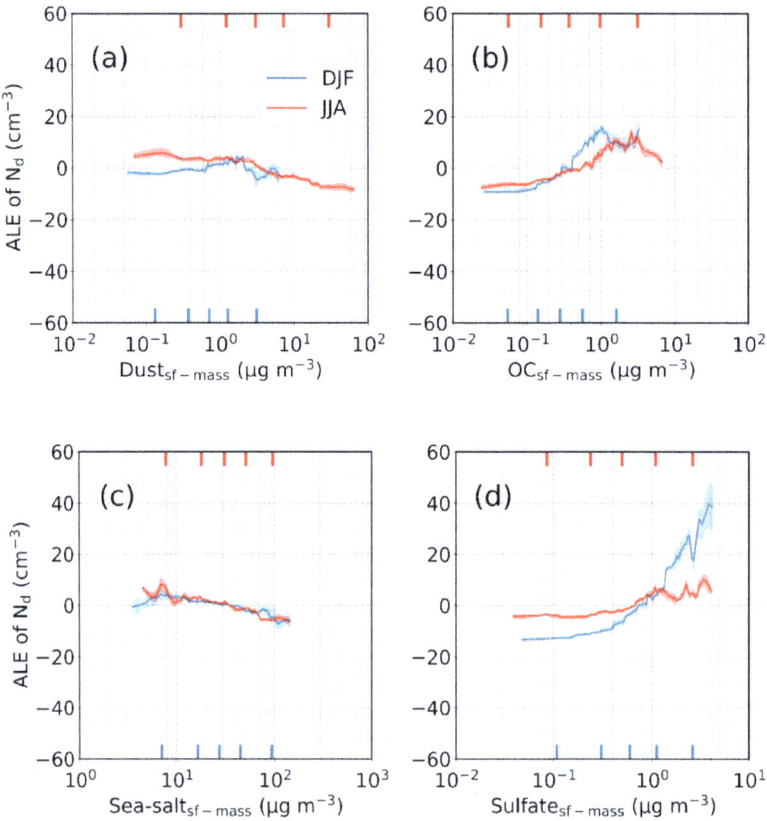

Figure S27: Average local accumulated effect (ALE) profiles based on GBRT modeling for surface mass concentrations of the following parameters: (a) dust, (b) organic carbon, (c) sea-salt, and (d) sulfate. ALE profiles were based on GBRT modeling on subsets of data including only samples with low-level liquid cloud fraction between 0.2 and 0.4 (i.e., $0.2 \leq C_{Flow\text{-}liq.} \leq 0.4$). Blue and red profiles represent ALEs of DJF and JJA, respectively. Shaded areas show the ALE ranges stemming from the variability of the obtained models from the cross-validation resampling procedure. Markers on the bottom and top x-axes denote the values of 5th, 25th, 50th, 75th, and 95th percentiles for each input variable.

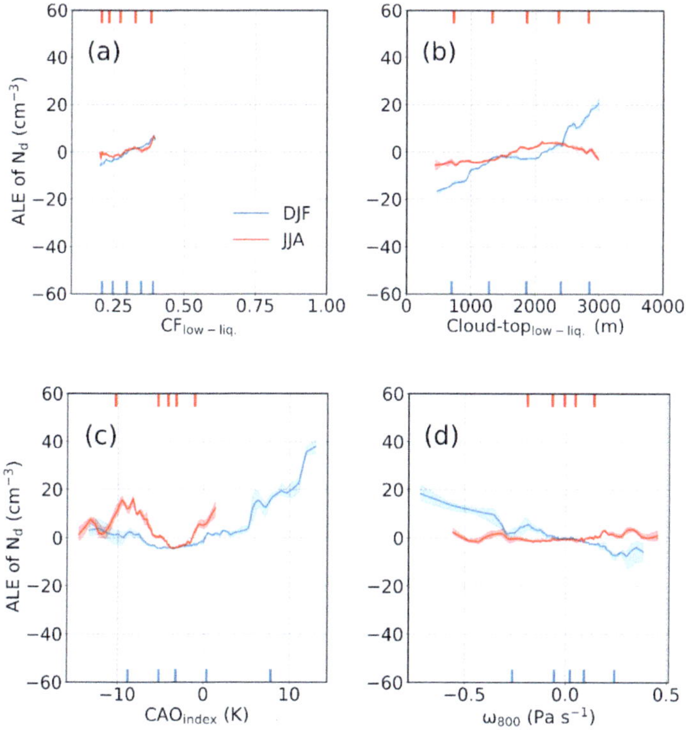

Figure S28: Same as Figure S27 but for the following input parameters: (a) low-level liquid cloud fraction ($CF_{\text{low-liq.}}$), (b) cloud-top effective height of low-level liquid cloud (cloud- liq.), (c) cold-air outbreak (CAO) index, and (d) vertical pressure velocity at 800 hPa (ω800).

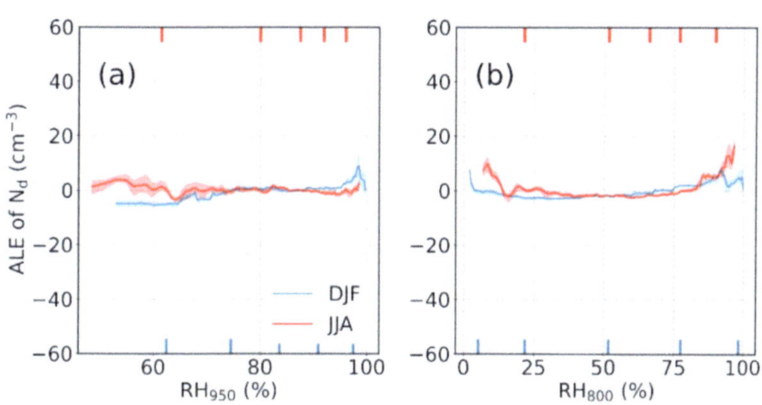

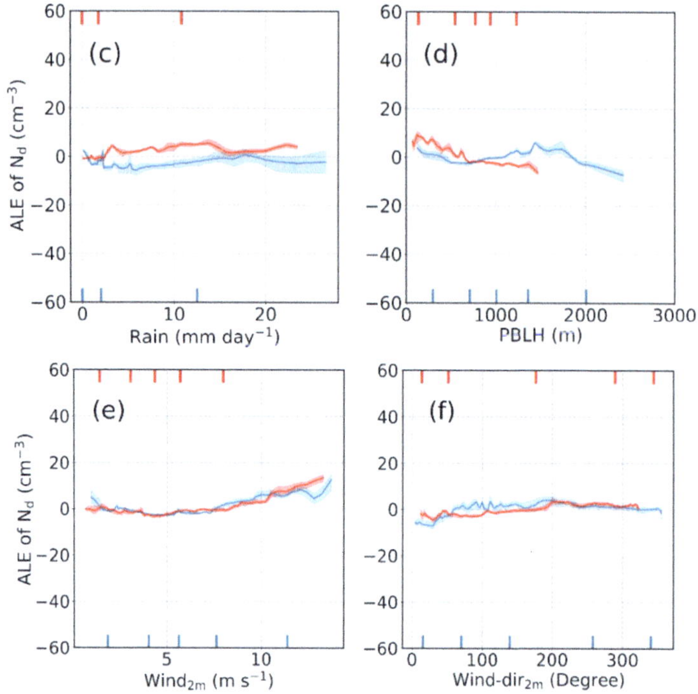

Figure S29: Same as Figure S27 but for the following input parameters: (a) relative humidity at 950 hPa (RH_{950}), (b) relative humidity at 800 hPa (RH_{800}), (c) rain rate, (d) planetary boundary layer height (PBLH), (e) wind speed at 2 m ($Wind_{2m}$), and (f) wind direction at 2 m (wind-dir2m).

ACKNOWLEDGEMENTS

The work was funded by NASA grant 80NSSC19K0442 in support of ACTIVATE, a NASA Earth Venture Suborbital-3 (EVS-3) investigation funded by NASA's Earth Science Division and managed through the Earth System Science Pathfinder Program Office. The authors acknowledge the NOAA Air Resources Laboratory (ARL) for the provision of the HYSPLIT transport and dispersion model and READY website (http://ready.arl.noaa.gov, last access: 14 June 2021) used in this work.

Financial Support

This research has been supported by the National Aeronautics and Space Administration (grant no. 80NSSC19K0442).

Footnotes

Competing interests. The authors declare that they have no conflict of interest.

Supplement. The supplement related to this article is available online at: https://doi.org/10.5194/acp-21-10499-2021-supplement.

Special issue statement. This article is part of the special issue "Marine aerosols, trace gases, and clouds over the North Atlantic (ACP/AMT inter-journal SI)". It is not associated with a conference.

Review statement. This paper was edited by Stefania Gilardoni and reviewed by two anonymous referees.

Code availability. ALE plots were calculated using the source code provided by Jumelle et al. (2021) available at https://github.com/blent-ai/ALEPython.

Data Availability

All the data used in this study are open access and can be found online as follows. CERES-MODIS data as described by Minnis et al. (2011) are available at https://ceres.larc.nasa.gov/data/; CALIPSO data are available at https://subset.larc.nasa.gov/calipso/ (CALIPSO, 2021); PERSIANN-CDR data as described by Nguyen et al. (2019) are available at https://chrsdata.eng.uci.edu/; MERRA-2 data are available at https://disc.gsfc.nasa.gov/ (GMAO, 2021); TCAP CCN data as described by Berg et al. (2016) are available at https://adc.arm.gov/discovery; ACTIVATE Airborne Data are available at https://doi.org/10.5067/ASDC/ACTIVATE_Aerosol_AircraftInSitu_Falcon_Data_1 (NASA/LARC/SD/ASDC, 2020a), https://doi.org/10.5067/ASDC/ACTIVATE_Cloud_AircraftInSitu_Falcon_Data_1 (NASA/LARC/SD/ASDC, 2020b), and https://doi.org/10.5067/ASDC/ACTIVATE_MetNav_AircraftInSitu_Falcon_Data_1 (NASA/LARC/SD/ASDC, 2020c).

REFERENCES

1. Abel SJ, Boutle IA, Waite K, Fox S, Brown PRA, Cotton R, Lloyd G, Choularton TW, and Bower KN: The Role of Precipitation in Controlling the Transition from Stratocumulus to Cumulus Clouds in a Northern Hemisphere Cold-Air Outbreak, *J. Atmos. Sci*, 74, 2293–2314, 10.1175/Jas-D-16-0362.1, 2017.
2. Ackerman AS, Kirkpatrick MP, Stevens DE, and Toon OB: The impact of humidity above stratiform clouds on indirect aerosol climate forcing, *Nature*, 432, 1014–1017, 10.1038/nature03174, 2004.
3. Albrecht BA: Aerosols, Cloud Microphysics, and Fractional Cloudiness, *Science*, 245, 1227–1230, 10.1126/science.245.4923.1227, 1989.
4. Aldhaif AM, Lopez DH, Dadashazar H, Painemal D, Peters AJ, and Sorooshian A: An Aerosol Climatology and Implications for Clouds at a Remote Marine Site: Case Study Over Bermuda, *J. Geophys. Res.-Atmos*, 126, e2020JD034038, 10.1029/2020jd034038, 2021.
5. Apley DW and Zhu JY: Visualizing the effects of predictor variables in black box supervised learning models, *J. Roy. Stat. Soc. B*, 82, 1059–1086, 10.1111/rssb.12377, 2020.
6. Ashouri H, Hsu K-L, Sorooshian S, Braithwaite DK, Knapp KR, Cecil LD, Nelson BR, and Prat OP: PERSIANN-CDR: Daily Precipitation Climate Data Record from Multisatellite Observations for Hydrological and Climate Studies, *B. Am. Meteorol. Soc*, 96, 69–83, 10.1175/BAMS-D-13-00068.1, 2015.
7. Bennartz R, Fan JW, Rausch J, Leung LR, and Heidinger AK: Pollution from China increases cloud droplet number, suppresses rain over the East China Sea, *Geophys. Res. Lett*, 38, L09704, 10.1029/2011gl047235, 2011.
8. Berg LK, Fast JD, Barnard JC, Burton SP, Cairns B, Chand D, Comstock JM, Dunagan S, Ferrare RA, Flynn CJ, Hair JW, Hostetler CA, Hubbe J, Jefferson A, Johnson R, Kassianov EI, Kluzek CD, Kollias P, Lamer K, Lantz K, Mei F, Miller MA, Michalsky J, Ortega I, Pekour M, Rogers RR, Russell PB, Redemann J, Sedlacek AJ, Segal-Rosenheimer M, Schmid B, Shilling JE, Shinozuka Y, Springston SR, Tomlinson JM, Tyrrell M, Wilson JM, Volkamer R, Zelenyuk A, and Berkowitz CM: The Two-Column Aerosol Project: Phase IOverview and impact of elevated aerosol layers on aerosol optical depth, *J. Geophys. Res.-Atmos*, 121, 336–361, 10.1002/2015jd023848, 2016. (data availalbe

at: https://adc.arm.gov/discovery, last access: 14 June 2021).

9. Berg W, L'Ecuyer T, and van den Heever S: Evidence for the impact of aerosols on the onset and microphysical properties of rainfall from a combination of satellite observations and cloud-resolving model simulations, *J. Geophys. Res.-Atmos*, 113, D14S23, 10.1029/2007jd009649, 2008.
10. Boucher O and Lohmann U: The Sulfate-Ccn-Cloud Albedo Effect – a Sensitivity Study with 2 General-Circulation Models, *Tellus B*, 47, 281–300, 10.1034/j.1600-0889.47.issue3.1.x, 1995.
11. Boucher O, Randall D, Artaxo P, Bretherton C, Feingold G, Forster P, Kerminen V-M, Kondo Y, Liao H, Lohmann U, Rasch P, Satheesh SK, Sherwood S, Stevens B, and Zhang XY: Clouds and aerosols, in: *Climate Change 2013: The Physical Science Basis, Contribution of Working Group I to the Fifth Assessment Report of the Intergovernmental Panel on Climate Change*, edited by: Stocker TF, Qin D, Plattner G-K, Tignor M, Allen SK, Doschung J, Nauels A, Xia Y, Bex V, and Midgley PM, Cambridge University Press, Cambridge, UK, 571–657, 2013.
12. Breon FM, Tanre D, and Generoso S: Aerosol effect on cloud droplet size monitored from satellite, *Science*, 295, 834–838, 10.1126/science.1066434, 2002.
13. Brummer B: Boundary-layer modification in wintertime cold-air outbreaks from the arctic sea ice, *Bound.-Lay. Meteorol*, 80, 109–125, 10.1007/Bf00119014, 1996.
14. Buchard V, Randles CA, da Silva AM, Darmenov A, Colarco PR, Govindaraju R, Ferrare R, Hair J, Beyersdorf AJ, Ziemba LD, and Yu H: The MERRA-2 Aerosol Reanalysis, 1980 Onward, Part II: Evaluation and Case Studies, *J. Climate*, 30, 6851–6872, 10.1175/Jcli-D-16-0613.1, 2017.
15. CALIPSO: CALIPSO LID L2 Standard HDF File – Version 3.30, *NASA Langley Research Center Atmospheric Science Data Center DAAC [data set]*, available at: https://subset.larc.nasa.gov/calipso/, last access: 14 June 2021.
16. Chin M, Ginoux P, Kinne S, Torres O, Holben BN, Duncan BN, Martin RV, Logan JA, Higurashi A, and Nakajima T: Tropospheric Aerosol Optical Thickness from the GOCART Model and Comparisons with Satellite and Sun Photometer Measurements, *J. Atmos. Sci*, 59, 461–483, 10.1175/1520-0469(2002)059<0461:TAOTFT>2.0.CO;2, 2002.

17. Colarco P, da Silva A, Chin M, and Diehl T: Online simulations of global aerosol distributions in the NASA GEOS-4 model and comparisons to satellite and ground-based aerosol optical depth, *J. Geophys. Res.-Atmos*, 115, D14207, 10.1029/2009JD012820, 2010.
18. Corral AF, Braun R, Cairns B, Gorooh VA, Liu H, Ma L, Mardi A, Painemal D, Stamnes S, van Diedenhoven B, Wang H, Yang Y, Zhang B, and Sorooshian A: An Overview of Atmospheric Features Over the Western North Atlantic Ocean and North American East Coast – Part 1: Analysis of Aerosols, Gases, and Wet Deposition Chemistry, *J. Geophys. Res.-Atmos*, 126, e2020JD032592, 10.1029/2020JD032592, 2021.
19. Dadashazar H, Wang Z, Crosbie E, Brunke M, Zeng XB, Jonsson H, Woods RK, Flagan RC, Seinfeld JH, and Sorooshian A: Relationships between giant sea salt particles and clouds inferred from aircraft physicochemical data, *J. Geophys. Res.-Atmos*, 122, 3421–3434, 10.1002/2016jd026019, 2017.
20. Dadashazar H, Crosbie E, Majdi MS, Panahi M, Moghaddam MA, Behrangi A, Brunke M, Zeng X, Jonsson HH, and Sorooshian A: Stratocumulus cloud clearings: statistics from satellites, reanalysis models, and airborne measurements, *Atmos. Chem. Phys*, 20, 4637–4665, 10.5194/acp-20-4637-2020, 2020.
21. DeCarlo PF, Dunlea EJ, Kimmel JR, Aiken AC, Sueper D, Crounse J, Wennberg PO, Emmons L, Shinozuka Y, Clarke A, Zhou J, Tomlinson J, Collins DR, Knapp D, Weinheimer AJ, Montzka DD, Campos T, and Jimenez JL: Fast airborne aerosol size and chemistry measurements above Mexico City and Central Mexico during the MILAGRO campaign, *Atmos. Chem. Phys*, 8, 4027–4048, 10.5194/acp-8-4027-2008, 2008.
22. Deuze JL, Breon FM, Devaux C, Goloub P, Herman M, Lafrance B, Maignan F, Marchand A, Nadal F, Perry G, and Tanre D: Remote sensing of aerosols over land surfaces from POLDER-ADEOS-1 polarized measurements, *J. Geophys. Res.-Atmos*, 106, 4913–4926, 10.1029/2000jd900364, 2001.
23. Feingold G: Modeling of the first indirect effect: Analysis of measurement requirements, *Geophys. Res. Lett*, 30, 1997, 10.1029/2003gl017967, 2003.
24. Fletcher J, Mason S, and Jakob C: The Climatology, Meteorology, and Boundary Layer Structure of Marine Cold Air Outbreaks in Both

Hemispheres, *J. Climate*, 29, 1999–2014, 10.1175/Jcli-D-15-0268.1, 2016.

25. Fuchs J, Cermak J, and Andersen H: Building a cloud in the southeast Atlantic: understanding low-cloud controls based on satellite observations with machine learning, *Atmos. Chem. Phys*, 18, 16537–16552, 10.5194/acp-18-16537-2018, 2018.
26. Gelaro R, McCarty W, Suarez MJ, Todling R, Molod A, Takacs L, Randles CA, Darmenov A, Bosilovich MG, Reichle R, Wargan K, Coy L, Cullather R, Draper C, Akella S, Buchard V, Conaty A, da Silva AM, Gu W, Kim GK, Koster R, Lucchesi R, Merkova D, Nielsen JE, Partyka G, Pawson S, Putman W, Rienecker M, Schubert SD, Sienkiewicz M, and Zhao B: The Modern-Era Retrospective Analysis for Research and Applications, Version 2 (MERRA-2), *J. Climate*, 30, 5419–5454, 10.1175/Jcli-D-16-0758.1, 2017.
27. Global Modeling and Assimilation Office (GMAO): *MERRA-2, Greenbelt, MD, USA, Goddard Earth Sciences Data and Information Services Center (GES DISC)*, available at: https://disc.gsfc.nasa.gov/, last access: 14 June 2021.
28. Grandey BS and Stier P: A critical look at spatial scale choices in satellite-based aerosol indirect effect studies, *Atmos. Chem. Phys*, 10, 11459–11470, 10.5194/acp-10-11459-2010, 2010.
29. Grosvenor DP, Sourdeval O, Zuidema P, Ackerman A, Alexandrov MD, Bennartz R, Boers R, Cairns B, Chiu JC, Christensen M, Deneke H, Diamond M, Feingold G, Fridlind A, Hunerbein A, Knist C, Kollias P, Marshak A, McCoy D, Merk D, Painemal D, Rausch J, Rosenfeld D, Russchenberg H, Seifert P, Sinclair K, Stier P, van Diedenhoven B, Wendisch M, Werner F, Wood R, Zhang ZB, and Quaas J: Remote Sensing of Droplet Number Concentration in Warm Clouds: A Review of the Current State of Knowledge and Perspectives, *Rev. Geophys*, 56, 409–453, 10.1029/2017rg000593, 2018.
30. Gryspeerdt E, Quaas J, and Bellouin N: Constraining the aerosol influence on cloud fraction, *J. Geophys. Res.-Atmos*, 121, 3566–3583, 10.1002/2015jd023744, 2016.
31. Gryspeerdt E, Quaas J, Ferrachat S, Gettelman A, Ghan S, Lohmann U, Morrison H, Neubauer D, Partridge DG, Stier P, Takemura T, Wang HL, Wang MH, and Zhang K: Constraining the instantaneous aerosol influence on cloud albedo, *P. Natl. Acad. Sci. USA*, 114, 4899–4904, 10.1073/pnas.1617765114, 2017.

32. Hasekamp OP, Gryspeerdt E, and Quaas J: Analysis of polarimetric satellite measurements suggests stronger cooling due to aerosol-cloud interactions, *Nat. Commun*, 10, 5405, 10.1038/s41467-019-13372-2, 2019.
33. Hastie T, Tibshirani R, and Friedman J: *The elements of statistical learning: data mining, inference and prediction*, 2nd edn., Springer, New York City, USA, 744 pp., 2009.
34. Jumelle M, Kuhn-Regnier A, Rajaratnam S, and Mahima A: *Python Accumulated Local Effects package, GitHub*, available at: https://github.com/blent-ai/ALEPython, last access: 14 June 2021.
35. Kolstad EW, Bracegirdle TJ, and Seierstad IA: Marine cold-air outbreaks in the North Atlantic: temporal distribution and associations with large-scale atmospheric circulation, *Clim. Dynam*, 33, 187–197, 10.1007/s00382-008-0431-5, 2009.
36. Liu J and Li Z: Aerosol properties and their influences on low warm clouds during the Two-Column Aerosol Project, *Atmos. Chem. Phys*, 19, 9515–9529, 10.5194/acp-19-9515-2019, 2019.
37. Lloyd G, Choularton TW, Bower KN, Gallagher MW, Crosier J, O'Shea S, Abel SJ, Fox S, Cotton R, and Boutle IA: In situ measurements of cloud microphysical and aerosol properties during the break-up of stratocumulus cloud layers in cold air outbreaks over the North Atlantic, *Atmos. Chem. Phys*, 18, 17191–17206, 10.5194/acp-18-17191-2018, 2018.
38. Loeb NG, Manalo-Smith N, Su WY, Shankar M, and Thomas S: CERES Top-of-Atmosphere Earth Radiation Budget Climate Data Record: Accounting for in-Orbit Changes in Instrument Calibration, *Remote Sens.-Basel*, 8, 182, 10.3390/rs8030182, 2016.
39. Lowenthal DH, Borys RD, Choularton TW, Bower KN, Flynn MJ, and Gallagher MW: Parameterization of the cloud droplet-sulfate relationship, *Atmos. Environ*, 38, 287–292, 10.1016/j.atmosenv.2003.09.046, 2004.
40. MacDonald AB, Hossein Mardi A, Dadashazar H, Azadi Aghdam M, Crosbie E, Jonsson HH, Flagan RC, Seinfeld JH, and Sorooshian A: On the relationship between cloud water composition and cloud droplet number concentration, *Atmos. Chem. Phys*, 20, 7645–7665, 10.5194/acp-20-7645-2020, 2020.
41. McComiskey A, Feingold G, Frisch AS, Turner DD, Miller MA, Chiu JC, Min QL, and Ogren JA: An assessment of aerosol-cloud interactions

in marine stratus clouds based on surface remote sensing, *J. Geophys. Res.-Atmos*, 114, D09203, 10.1029/2008jd011006, 2009.

42. McCoy DT, Bender FAM, Mohrmann JKC, Hartmann DL, Wood R, and Grosvenor DP: The global aerosol-cloud first indirect effect estimated using MODIS, MERRA, and AeroCom, *J. Geophys. Res.-Atmos*, 122, 1779–1796, 10.1002/2016jd026141, 2017.
43. McCoy DT, Bender FA-M, Grosvenor DP, Mohrmann JK, Hartmann DL, Wood R, and Field PR: Predicting decadal trends in cloud droplet number concentration using reanalysis and satellite data, *Atmos. Chem. Phys*, 18, 2035–2047, 10.5194/acp-18-2035-2018, 2018.
44. Minnis P, Sun-Mack S, Young DF, Heck PW, Garber DP, Chen Y, Spangenberg DA, Arduini RF, Trepte QZ, Smith WL, Ayers JK, Gibson SC, Miller WF, Hong G, Chakrapani V, Takano Y, Liou KN, Xie Y, and Yang P: CERES Edition-2 Cloud Property Retrievals Using TRMM VIRS and Terra and Aqua MODIS Data – Part I: Algorithms, *IEEE T. Geosci. Remote*, 49, 4374–4400, 10.1109/Tgrs.2011.2144601, 2011. (data available at: https://ceres.larc.nasa.gov/data/, last access: 14 June 2021).
45. Minnis P, Sun-Mack S, Chen Y, Chang F, Yost CR, Smith WL, Heck PW, Arduini RF, Bedka ST, Yi Y, Hong G, Jin Z, Painemal D, Palikonda R, Scarino BR, Spangenberg DA, Smith RA, Trepte QZ, Yang P, and Xie Y: CERES MODIS Cloud Product Retrievals for Edition 4 – Part I: Algorithm Changes, *IEEE T. Geosci. Remote*, 59, 1–37, 10.1109/TGRS.2020.3008866, 2020.
46. Molnar C: *Interpretable Machine Learning, A Guide for Making Black Box Models Explainable*, available at: https://christophm.github.io/interpretable-ml-book/ (last access: 14 June 2021), 2019.
47. Myhre G, Shindell D, Bréon F-M, Collins W, Fuglestvedt J, Huang J, Koch D, Lamarque J-F, Lee D, Mendoza B, Nakajima T, Robock A, Stephens G, Takemura T, and Zhang H: Anthropogenic and natural radiative forcing, in: *Climate Change 2013: The Physical Science Basis, Contribution of Working Group I to the Fifth Assessment Report of the Intergovernmental Panel on Climate Change*, edited by: Stocker TF, Qin D, Plattner G-K, Tignor M, Allen SK, Doschung J, Nauels A, Xia Y, Bex V, and Midgley PM, Cambridge University Press, Cambridge, UK, 659–740, 2013.
48. Nakajima T, Higurashi A, Kawamoto K, and Penner JE: A possible correlation between satellite-derived cloud and aerosol microphysical

parameters, *Geophys. Res. Lett*, 28, 1171–1174, 10.1029/2000gl012186, 2001.

49. NASA/LARC/SD/ASDC: *ACTIVATE Falcon In Situ Aerosol Data [data set]*, 10.5067/ASDC/ACTIVATE_Aerosol_AircraftInSitu_Falcon_Data_1, 2020a.
50. NASA/LARC/SD/ASDC: *ACTIVATE Falcon In Situ Cloud Data [data set]*, 10.5067/ASDC/ACTIVATE_Cloud_AircraftInSitu_Falcon_Data_1, 2020b.
51. NASA/LARC/SD/ASDC: *ACTIVATE Falcon In-Situ Meteorological and Navigational Data [data set]*, 10.5067/ASDC/ACTIVATE_MetNav_AircraftInSitu_Falcon_Data_1, 2020c.
52. Naud CM, Booth JF, and Lamraoui F: Post Cold Frontal Clouds at the ARM Eastern North Atlantic Site: An Examination of the Relationship Between Large-Scale Environment and Low-Level Cloud Properties, *J. Geophys. Res.-Atmos*, 123, 12117–12132, 10.1029/2018jd029015, 2018.
53. Nguyen P, Ombadi M, Sorooshian S, Hsu K, AghaKouchak A, Braithwaite D, Ashouri H, and Thorstensen AR: The PERSIANN family of global satellite precipitation data: a review and evaluation of products, *Hydrol. Earth Syst. Sci*, 22, 5801–5816, 10.5194/hess-22-5801-2018, 2018.
54. Nguyen P, Shearer EJ, Tran H, Ombadi M, Hayatbini N, Palacios T, Huynh P, Braithwaite D, Updegraff G, Hsu K, Kuligowski B, Logan WS, and Sorooshian S: The CHRS Data Portal, an easily accessible public repository for PERSIANN global satellite precipitation data, *Scientific Data*, 6, 1–10, 10.1038/sdata.2018.296, 2019. (data available at: https://chrsdata.eng.uci.edu/, last access: 14 June 2021).
55. Painemal D: Global Estimates of Changes in Shortwave Low-Cloud Albedo and Fluxes Due to Variations in Cloud Droplet Number Concentration Derived From CERES-MODIS Satellite Sensors, *Geophys. Res. Lett*, 45, 9288–9296, 10.1029/2018gl078880, 2018.
56. Painemal D, Chang F-L, Ferrare R, Burton S, Li Z, Smith WL Jr., Minnis P, Feng Y, and Clayton M: Reducing uncertainties in satellite estimates of aerosol–cloud interactions over the subtropical ocean by integrating vertically resolved aerosol observations, *Atmos. Chem. Phys*, 20, 7167–7177, 10.5194/acp-20-7167-2020, 2020.

57. Painemal D, Corral AF, Sorooshian A, Brunke MA, Chellappan S, Gorooh VA, Ham S, O'Neill L, Smith WL Jr., Tselioudis G, Wang H, Zeng X, and Zuidema P: An Overview of Atmospheric Features Over the Western North Atlantic Ocean and North American East Coast – Part 2: Circulation, Boundary Layer, and Clouds, *J. Geophys. Res.-Atmos*, 126, e2020JD033423, 10.1029/2020JD033423, 2021.
58. Papritz L, Pfahl S, Sodemann H, and Wernli H: A Climatology of Cold Air Outbreaks and Their Impact on Air-Sea Heat Fluxes in the High-Latitude South Pacific, *J. Climate*, 28, 342–364, 10.1175/Jcli-D-14-00482.1, 2015.
59. Pedregosa F, Varoquaux G, Gramfort A, Michel V, Thirion B, Grisel O, Blondel M, Prettenhofer P, Weiss R, Dubourg V, Vanderplas J, Passos A, Cournapeau D, Brucher M, Perrot M, and Duchesnay E: Scikit-learn: Machine Learning in Python, *J. Mach. Learn. Res*, 12, 2825–2830, 2011.
60. Penner JE, Xu L, and Wang MH: Satellite methods underestimate indirect climate forcing by aerosols, *P. Natl. Acad. Sci. USA*, 108, 13404–13408, 10.1073/pnas.1018526108, 2011.
61. Persson C, Bacher P, Shiga T, and Madsen H: Multi-site solar power forecasting using gradient boosted regression trees, *Sol. Energy*, 150, 423–436, 10.1016/j.solener.2017.04.066, 2017.
62. Quaas J, Boucher O, and Lohmann U: Constraining the total aerosol indirect effect in the LMDZ and ECHAM4 GCMs using MODIS satellite data, *Atmos. Chem. Phys*, 6, 947–955, 10.5194/acp-6-947-2006, 2006.
63. Quaas J, Boucher O, Bellouin N, and Kinne S: Satellite-based estimate of the direct and indirect aerosol climate forcing, *J. Geophys. Res.-Atmos*, 113, D05204, 10.1029/2007jd008962, 2008.
64. Quinn PK, Coffman DJ, Johnson JE, Upchurch LM, and Bates TS: Small fraction of marine cloud condensation nuclei made up of sea spray aerosol, *Nat. Geosci*, 10, 674–679, 10.1038/ngeo3003, 2017.
65. Randles CA, da Silva AM, Buchard V, Colarco PR, Darmenov A, Govindaraju R, Smirnov A, Holben B, Ferrare R, Hair J, Shinozuka Y, and Flynn CJ: The MERRA-2 Aerosol Reanalysis, 1980 Onward, Part I: System Description and Data Assimilation Evaluation, *J. Climate*, 30, 6823–6850, 10.1175/Jcli-D-16-0609.1, 2017.

66. Reutter P, Su H, Trentmann J, Simmel M, Rose D, Gunthe SS, Wernli H, Andreae MO, and Pöschl U: Aerosol- and updraft-limited regimes of cloud droplet formation: influence of particle number, size and hygroscopicity on the activation of cloud condensation nuclei (CCN), *Atmos. Chem. Phys*, 9, 7067–7080, 10.5194/acp-9-7067-2009, 2009.
67. Rolph G, Stein A, and Stunder B: Real-time Environmental Applications and Display sYstem: READY, *Environ. Model. Softw*, 95, 210–228, 10.1016/j.envsoft.2017.06.025, 2017.
68. Sekiguchi M, Nakajima T, Suzuki K, Kawamoto K, Higurashi A, Rosenfeld D, Sano I, and Mukai S: A study of the direct and indirect effects of aerosols using global satellite data sets of aerosol and cloud parameters, *J. Geophys. Res.-Atmos*, 108, 4699, 10.1029/2002jd003359, 2003.
69. Shingler T, Dey S, Sorooshian A, Brechtel FJ, Wang Z, Metcalf A, Coggon M, Mülmenstädt J, Russell LM, Jonsson HH, and Seinfeld JH: Characterisation and airborne deployment of a new counterflow virtual impactor inlet, *Atmos. Meas. Tech*, 5, 1259–1269, 10.5194/amt-5-1259-2012, 2012.
70. Sinclair K, van Diedenhoven B, Cairns A, Alexandrov M, Moore R, Ziemba LD, and Crosbie E: Observations of Aerosol-Cloud Interactions During the North Atlantic Aerosol and Marine Ecosystem Study, *Geophys. Res. Lett*, 47, e2019GL085851, 10.1029/2019GL085851, 2020.
71. Sorooshian A, Anderson B, Bauer SE, Braun RA, Cairns A, Crosbie E, Dadashazar H, Diskin G, Ferrare R, Flagan RC, Hair J, Hostetler C, Jonsson HH, Kleb MM, Liu HY, MacDonald AB, McComiskey A, Moore R, Painemal D, Russell LM, Seinfeld JH, Shook M, Smith WL, Thornhill K, Tselioudis G, Wang HL, Zeng XB, Zhang B, Ziemba L, and Zuidema P: Aerosol-Cloud-Meteorology Interaction Airborne Field Investigations: Using Lessons Learned from the US West Coast in the Design of ACTIVATE off the US East Coast, *B. Am. Meteorol. Soc*, 100, 1511–1528, 10.1175/Bams-D-18-0100.1, 2019.
72. Sorooshian A, Corral AF, Braun RA, Cairns B, Crosbie E, Ferrare R, Hair J, Kleb MM, Mardi AH, Maring H, McComiskey A, Moore R, Painemal D, Scarino AJ, Schlosser J, Shingler T, Shook M, Wang HL, Zeng XB, Ziemba L, and Zuidema P: Atmospheric Research Over the Western North Atlantic Ocean Region and North American East Coast:

A Review of Past Work and Challenges Ahead, *J. Geophys. Res.-Atmos*, 125, e2019JD031626, 10.1029/2019JD031626, 2020.

73. Stein AF, Draxler RR, Rolph GD, Stunder BJB, Cohen MD, and Ngan F: NOAA's HYSPLIT Atmospheric Transport and Dispersion Modeling System, *B. Am. Meteorol. Soc*, 96, 2059–2077, 10.1175/BAMS-D-14-00110.1, 2015.
74. Stier P: Limitations of passive remote sensing to constrain global cloud condensation nuclei, *Atmos. Chem. Phys*, 16, 6595–6607, 10.5194/acp-16-6595-2016, 2016.
75. Storelvmo T, Lohmann U, and Bennartz R: What governs the spread in shortwave forcings in the transient IPCC AR4 models?, *Geophys. Res. Lett*, 36, L01806, 10.1029/2008gl036069, 2009.
76. Tackett JL, Winker DM, Getzewich BJ, Vaughan MA, Young SA, and Kar J: CALIPSO lidar level 3 aerosol profile product: version 3 algorithm design, *Atmos. Meas. Tech*, 11, 4129–4152, 10.5194/amt-11-4129-2018, 2018.
77. Terai CR, Wood R, and Kubar TL: Satellite estimates of precipitation susceptibility in low-level marine stratiform clouds, *J. Geophys. Res.-Atmos*, 120, 8878–8889, 10.1002/2015jd023319, 2015.
78. Thornhill KL, Anderson BE, Barrick JDW, Bagwell DR, Friesen R, and Lenschow DH: Air motion intercomparison flights during Transport and Chemical Evolution in the Pacific (TRACE-P)/ACE-ASIA, *J. Geophys. Res.-Atmos*, 108, 8783, 10.1029/2002JD003108, 2003.
79. Twomey S: Influence of Pollution on Shortwave Albedo of Clouds, *J. Atmos. Sci*, 34, 1149–1152, 10.1175/1520-0469(1977)034<1149:Tiopot>2.0.Co;2, 1977.
80. Winker DM, Vaughan MA, Omar A, Hu YX, Powell KA, Liu ZY, Hunt WH, and Young SA: Overview of the CALIPSO Mission and CALIOP Data Processing Algorithms, *J. Atmos. Ocean. Tech*, 26, 2310–2323, 10.1175/2009jtecha1281.1, 2009.
81. Wood R: Stratocumulus Clouds, *Mon. Weather Rev*, 140, 2373–2423, 10.1175/Mwr-D-11-00121.1, 2012.
82. Zhang Z, Ackerman AS, Feingold G, Platnick S, Pincus R, and Xue H: Effects of cloud horizontal inhomogeneity and drizzle on remote sensing of cloud droplet effective radius: Case studies based on large-eddy simulations, *J. Geophys. Res.-Atmos*, 117, D19208, 10.1029/2012jd017655, 2012.

83. Zhang Z, Werner F, Cho H-M, Wind G, Platnick S, Ackerman AS, Di Girolamo L, Marshak A, and Meyer K: A framework based on 2-D Taylor expansion for quantifying the impacts of subpixel reflectance variance and covariance on cloud optical thickness and effective radius retrievals based on the bispectral method, *J. Geophys. Res.-Atmos*, 121, 7007–7025, 10.1002/2016jd024837, 2016.

CHAPTER 16

Response of Mixed-Phase Cloud Microphysical Properties to Cloud-Seeding Near Cloud Top Over Hebei, China

Yang Yang[1,2], Chuanfeng Zhao[1,3], Jiao Fu[2] , Yi Cui[2], Xiaobo Dong[2], Rong Mai[2] and Feng Xu[2]

[1]Key Laboratory of Meteorology and Ecological Environment of Hebei Province, Shijiazhuang, China

[2]Weather Modification Office of Hebei Province, Shijiazhuang, China

[3]College of Global Change and Earth System Science, Beijing Normal University, Beijing, China

On November 29, 2019, an aircraft observation during the period of cloud-seeding was carried out for a mixed-phase cloud over Xingtai, Hebei Province, China. This study investigates the response of mixed-phase cloud

Citation: Yang Y, Zhao C, Fu J, Cui Y, Dong X, Mai R and Xu F (2022) "Response of Mixed-Phase Cloud Microphysical Properties to Cloud-Seeding Near Cloud Top Over Hebei", China. Front. Environ. Sci. 10:865966. https://doi.org/10.3389/fenvs.2022.865966.

microphysical properties to cloud-seeding near cloud top. Before cloud seeding, the cloud droplet concentration from fast cloud droplet probe ($N_{C—FCDP}$) presented a multi-peak vertical distribution structure, with a maximum concentration of 192 cm^{-3} at a height of 3,322 m; the maximum concentration of ice crystals from cloud imaging probe ($N_{C—CIP}$) was 10 L^{-1}, which appeared at 4,500 m in the upper part of cloud; and the peak value of liquid water content (LWC) in the cloud also appeared at 4,500 m, with a value of 0.15 g/m^3. The coexistence of supercooled liquid water and ice crystals implies that they are particularly suitable for cloud seeding at the height of 4,550 m. About 7–8 min later after cloud seeding at this height, the average $N_{C—FCDP}$ decreased from 160.3 to 129 cm^{-3}, and the average $N_{C—CIP}$ increased from 7.1 to 10 L^{-1}. Moreover, after cloud seeding, high $N_{C—CIP}$ as well as larger and more ice crystals appeared almost in the same areas within the cloud, and LWC presented an obvious decreasing trend. In contrast, the concentration of small cloud droplets and LWC decreased obviously after seeding. The findings suggest that the cloud microphysical properties showed obvious responses to the artificial introduction of silver iodide, which is important for human weather modification.

Keywords: mixed-phase, cloud-seeding, microphysical properties, respond, cloud top

INTRODUCTION

Clouds play an important role on both Earth's weather and climate (Ramanathan et al., 1989; Norris, 2005; Zhao and Garrett, 2015; Yang et al., 2016a; Yang et al., 2016b; Chuanfeng Zhao et al., 2020). Moreover, anthropogenic activities, by changing cloud properties, have contributed to increased extreme weather and climate events (Lohmann and Feichter, 2005; Andreae and Rosenfeld, 2008; Tao et al., 2012; Li et al., 2016; Yang et al., 2018; Zhao et al., 2018; Sun et al., 2019; Xin Zhao et al., 2020), and extreme weather generally has great impact on human life. The changes of latent heat release, precipitation, and radiation during cloud development, which are essential parts in Earth's energy budget and hydrological cycle, are all affected by the microphysical processes in the cloud (Pruppacher and Klett, 2010).

The cloud microphysical properties could be further influenced by human activities. In 1946, American scientists proposed that the number of ice crystals in cold clouds could be increased by artificially introducing silver iodide (AgI) into the cloud, thus achieving the effect of rain enhancement

(Schaefer, 1946; Vonnegut, 1947). Weather modification has gradually become an important application field of cloud and precipitation physics. At present, the theory of weather modification has been able to explain how the catalyst works in the cloud—that is, under appropriate meteorological conditions, seeding silver iodide into clouds containing supercooled water droplets can cause the liquid water droplets to transform into ice crystals, and then the newly formed ice crystals grow larger through the Bergeron process (Bergeron, 1935). After decades of experiments, the catalysis of winter orographic clouds on the windward slope has achieved a high success rate (Pokharel and Geerts, 2016; Pokharel et al., 2017; Tessendorf et al., 2020). Seeding in supercooled water clouds with a thickness less than 1 km can leave a very obvious cloud groove (Dong et al., 2021). These studies provide a solid foundation for the feasibility of artificial cloud seeding. While various experiments have been carried out, sufficient evidences of seeding effects are still needed under different conditions—for example, there are relatively few experiments on artificial catalysis and studies about the changes of microphysical properties in deep layered clouds. Particularly, many existing studies evaluate the performance of cloud seeding only by investigating the changes of macroscopic properties, such as the changes of reflectivity from ground-based radars or by directly using statistical verification. These methods are usually not intuitionistic and intelligible enough to reflect how AgI works within the cloud, leading to the effect of cloud seeding still in insufficient credibility. The response of cloud microphysical properties to seeding can be observed by airborne instruments with high temporal resolution, which provides us a good opportunity with reliable basis to understand the mechanism of artificial precipitation enhancement. In this study, through a well-designed aircraft seeding experiment of a winter snowfall deep cloud system, we investigate the changes in phase and size distribution of cloud particles due to seeding. We first analyze the weather background before cloud seeding and then the structure of the cloud system. Finally, the response of the cloud microphysical properties was examined based on the changes between before and after cloud seeding.

METHODOLOGY

Instruments

In this experiment, a snow cloud system was selected on November 29, 2019 for research about the response of mixed-phase cloud microphysical properties to cloud seeding, and aircraft detection and cloud seeding were carried out over Xingtai, Hebei Province. The King Air-350 aircraft operated by Hebei Weather Modification Office was used for the observation of clouds with an airborne detection system on board, including cloud particle detection system, meteorological measurement system, Nevzorov hot wire liquid water content (LWC), total water content (TWC) probe, and global positioning system. The cloud particle detection system in this study consists of fast cloud droplet probe (FCDP), cloud imaging probe (CIP), and cloud particle imager (CPI). The FCDP probe can obtain the concentration of cloud droplets in the size range of 2–50 μm in radius, with uncertainty of approximately 20% (Faber et al., 2018; Lance et al., 2010). CIP can detect large cloud particles in the size range of 25–1,550 μm, and the CPI probe is able to obtain high-resolution images of cloud particles. The LWC and the TWC were obtained by the Nevzorov hot wire LWC/TWC probe, and the uncertainty was about 15% (Emery et al., 2004; Korolev et al., 2013). CPI uses a CCD camera to capture a high-resolution image of cloud particles, which can distinguish water drops from ice particles. Vertical profiles of temperature and humidity were obtained using the meteorology monitoring system called Aircraft-Integrated Meteorological Measurement System. Table 1 shows these main detection equipment used in this study, along with their detected cloud variables and measurable ranges. Cloud seeding and the detection of cloud microphysical properties were performed by the same aircraft. Also noted is that the silver iodide flames used in this study contain AgI of 1.8 g per stick.

Table 1. The cloud microphysical detection instruments, the main parameters observed in this study, and the measureable range of each variable by the instrument, along with the reference for instrument details.

Instrument	Detection variable	Measurable range	References
FCDP	Cloud droplets	2–50 μm	www.specinc.com
CIP	Ice crystal particles	25–1,550 μm	Baumgardner et al. (2001)
CPI	Cloud particle images	2–3,000 μm	Lawson et al. (2001)
Nevzorov – LWC/TWC	TWC/LWC	0.005–3 g/m^3	Korolev et al. (1998)

Data Processing

The cloud particle size spectra observed in this study were obtained by FCDP and CIP probes. Considering that most of the ice crystal particles in stratiform clouds are larger than 100 μm in diameter, we assume that cloud particles larger than 100 μm are ice crystals, which can be obtained by CIP probe of 100–1,550 μm, and that the particles smaller than 100 μm are cloud droplets, which can be obtained by FCDP (Korolev and Field, 2015). Considering the higher supersaturation condition of the ice surface than the liquid droplet surface in the mixed-phase cloud, ice crystals larger than 25 μm can grow rapidly in 1 min. Thus, our assumption in general that the FCDP observed particles are all droplets in the cloud examined here should be reasonable. In order to obtain the changes of cloud particle spectra between before and after catalysis in a more detailed manner, the sampling frequency of all instruments used in this study is 1 Hz. Before using the data, we performed quality control on the observed data based on the instrument operating status and true air speed values.

The cloud microphysical parameters used in this study, including the effective radius of cloud particles r_e, number concentration of cloud particles N_{ci}, and ice water content (IWC) of clouds measured by Nevzorov-LWC/TWC $IWC_{Nevzorov}$ are calculated as follows:

$$r_e = \frac{\int r^3 n(r)dr}{\int r^2 n(r)dr} = \frac{\sum N_{ci} r_i^3}{\sum N_{ci} r_i^2} \quad (1)$$

$$N_d = \int n(r)dr = \sum N_{ci} \quad (2)$$

$$IWC_{Nevzorov} = TWC_{Nevozrov} - LWC_{Nevzorov} \quad (3)$$

where $n(r)$ is the droplet size distribution with number concentration of n at size r, and N_{ci} and r_i are the droplet number concentration and the particle size radius of the i-th bin of the instrument, respectively. Nevzorov hot wire LWC/TWC probe can only obtain liquid and total water content, and $IWC_{Nevzorov}$ is calculated by subtracting $LWC_{Nevzorov}$ from $TWC_{Nevzorov}$. Due to the potential uncertainty of each instrument, the first and last bins of FCDP and the first three bins of CIP were discarded and not used, which was similar to our previous studies (Yang et al., 2019; Zhao et al., 2019; Dong et al., 2020). The whole size range of cloud particles at 3–1,550 μm was obtained by combining the observations of FCDP and CIP. The time used in this study were all converted to local time (UTC + 8).

RESULTS AND DISCUSSION

Weather Background and Flight Plan

Weather background was analyzed using 1° × 1° NCEP/NCAR reanalysis data at 14:00 LT on November 29, 2019, and the flight seeding and observation areas were marked in Figure 1. There was a zonal circulation with two troughs and a ridge at 500 hPa (Figure 1A) at mid-high latitudes over Asia. There was also an obvious westerly trough near Hetao, and the detection area was dominated by a southwest airflow in front of the trough. The trough line at 700 hPa (Figure 1B) was located from the Mongolian Plateau to the Hetao area. The flight observation area was under the control of a southwest airflow in the south of the trough with obvious warm moisture advection, making the relative humidity close to saturation. The warm–humid air from the southwest and the cold air from the north converged, causing the formation of a large cloud system at the east of Hetao and the north of Shanxi. The cloud system gradually affected Hebei during its eastward movement. In this process, the atmosphere over the observation area was relatively stable, and the cloud system was dominated by stratus clouds. On the day of November 29, 2019, the relative humidity above the Xingtai site (Figure 1C) that lies within the observation area was relatively low at below 3,000 m, and the wet layer was mainly concentrated in 3,000–5,000 m, which was consistent with the detection results of the aircraft.

Figure 1

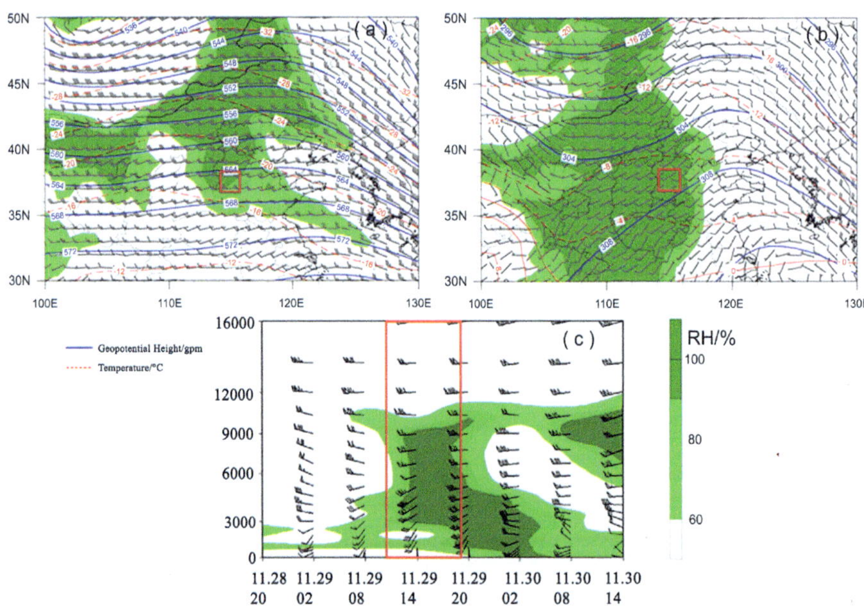

Figure 1. The weather pattern at 14:00 LT on November 29, 2019 at (A) 500 hPa and (B) 700 hPa. The blue line represents the geopotential height field, while the red dotted line represents the temperature field. (C) Relative humidity

(filled with green-colored contours) and wind field from NCEP reanalysis data above Xingtai site over the observation area in time series. The Xingtai site area was marked by a rectangle in (A,B).

The aircraft took off from Zhengding Airport at 14:50 LT, lifted to 2,100 m, and then leveled over Xingtai, which was located in south-central Hebei Province. After arriving at the detection area, the aircraft first descended to a height of 600 m and then conducted vertical detection of the macro- and micro-properties of cloud by flying from the bottom to the top during the period 15:52–16:18 LT, with an altitude range of 600–7,000 m and a hovering radius of 10 km. After knowing the vertical structure of cloud properties, the aircraft was lowered from a height of 7,000–4,550 m from 16:19 to 16:34 LT. Then, cloud seeding was carried out over Xingtai along the same circular trajectory at 16:37 LT at an altitude of 4,550 m, where the supercooled liquid water content was relatively abundant, and a total of 14 AgI bars were lighted. After cloud seeding, the flight continued to detect the cloud macro- and micro-physical properties in order to know the effect of cloud seeding.

Vertical Structure of Cloud Properties

Figure 2 shows the vertical distribution of stratus cloud microphysical properties along with the temperature and relative humidity profiles. It can be seen from Figure 2A that good humidity conditions existed above the height of 2,500 m, which was consistent with the results at around 16:00, as shown in Figure 1C.

The corresponding heights of 500 and 700 hpa are about 5,600 and 3,000 m, respectively, and the temperature values in Figure 2A are approximately the same as those in Figure 1A and Figure 1B. As shown in Figure 2A, the vertical detection area of the aircraft was all above the height of 0°C layer, and the temperature decreased with the increase of altitude.

However, there were two thin temperature inversion layers near the cloud base and the cloud top, respectively. The temperature at the lower inversion layer varied from -8.2 to −5.4°C with heights from 2,100 m to 2,355 m, and the temperature at the upper inversion layer varied from -13.5 to −10.2°C with heights from 4,600 to 4,800 m. Unlike temperature, relative humidity gradually increased with height, reached saturation in the cloud region, and gradually decreased with height above the cloud top.

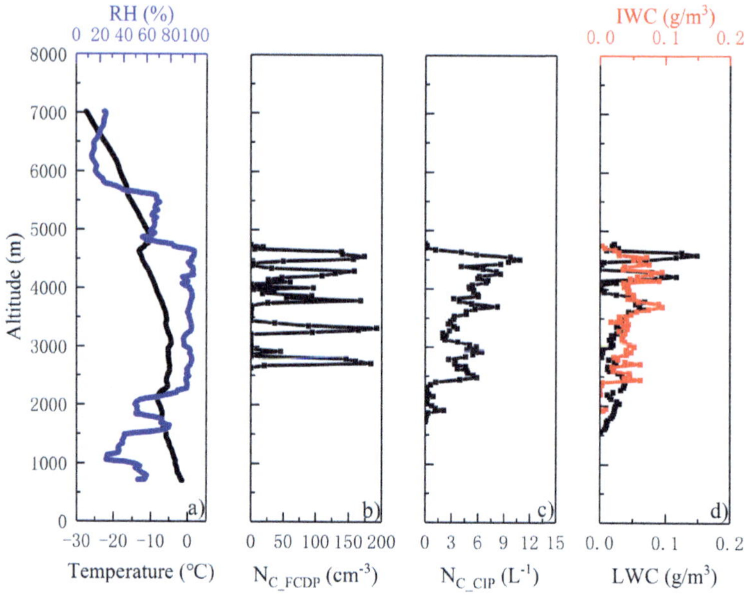

Figure 2. (A) Temperature and relative humidity profiles, (B) cloud droplet number concentration observed by fast cloud droplet probe, (C) ice crystal number concentration observed by cloud imaging probe, and (D) liquid water content and ice water content over Xingtai on November 29, 2019.

It was generally believed that when the number concentration obtained by FCDP ($N_{C\text{—}FCDP}$) was greater than 10 cm^{-3} and LWC $\geq$ 0.01 $g \cdot m^{-3}$ at the same time, the corresponding location of the aircraft was determined to be in the cloud (Hobbs, 1991). The dramatic changes of $N_{C\text{—}FCDP}$ with height, as shown in Figure 2B, indicated that the vertical distribution of stratus cloud properties varied greatly. The maximum $N_{C\text{—}FCDP}$ appeared in lower clouds at heights of about 3,220 m, with the peak value of 192 cm^{-3}. The number concentration of ice crystals ($N_{C\text{—}CIP}$) showed a growing trend with height within clouds and reached the maximum value of 10 L^{-1} at the top of the cloud, as shown in Figure 2C. The maximum LWC of 0.15 g/m^3 appeared near the top of the cloud (Figure 2D), which was similar to the results found by Zhao and Lei (2014). IWC was on the same order of magnitude as LWC, with peak values of up to 0.1 g/m^3 at heights of around 4,000 m. Figure 2 shows that both liquid droplets and ice crystals were detected at altitudes from 3,500 to 4,700 m, indicating the mixed-phase status at this height range, which was also the optimal part for cloud seeding due to the existence of supercooled liquid droplets.

Horizontal Distribution of Cloud Properties

After knowing the vertical properties of the cloud system, the flight leveled at the height of 4,550 m, which was rich of supercooled liquid water, to obtain the horizontal microphysical properties of the cloud. The corresponding flight time was 16:34-16:43 LT, and two laps were flown along the same trajectory due to the wind speed in the detection area being less than 5 m/s. We believe that the cloud properties do not change much in a short time. To make sure that the LWC obtained during this experiment was reliable, Figure 3 shows a comparison of LWC in the cloud obtained by two different instruments of FCDP and Nevzorov hot wire, which were denoted as LWC_{FCDP} and $LWC_{NEVZOROV}$, respectively. LWC_{FCDP} and $LWC_{NEVZOROV}$ had a good consistency on time series (Figure 3A), and the correlation coefficient between them reached 0.81 (Figure 3B), indicating that the LWC detection results had high reliability and accuracy. Note that the LWC from Nevzorov hot wire is slightly smaller than that from FCDP, which might be associated with the potential ice crystals with sizes smaller than 100 μm. LWC_{FCDP} was used in the following sections to represent the liquid water content during the aircraft observation period.

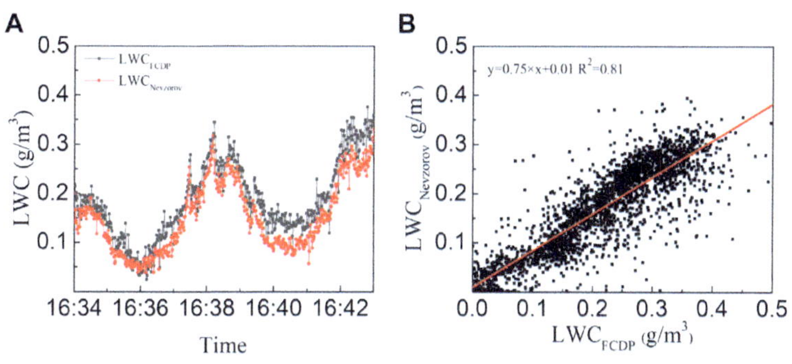

Figure 3. For a cloud layer at height 4,550 m: (A) temporal variation of LWC_{FCDP} and $LWC_{NEVZOROV}$, (B) scattered plots between LWC_{FCDP} and $LWC_{NEVZOROV}$.

Figure 4 presents the cloud droplet number concentration (N_{C_FCDP}; Figure 4A), the cloud droplet effective radius (r_e) measured by FCDP (Figure 4B), and the liquid water content (LWC_{FCDP}, Figure 4C) during the level flight process at a height of 4,550 m. The flight observation started in a counter-clockwise direction from point A. Figure 4 presents the observation results from 16:34 LT (the observation started from point A) to 16:39 LT (the flight returned to point A for the second time after one lap). The cloud seeding and detection were carried along the same circular trajectory. Even though the AgI flame bars were lighted at 16:37 LT, the aircraft still detected the unseeded area before the end of the lap. Differently from the general expectation that cloud microphysical properties vary weakly at a horizontal layer for stratus clouds (Wang et al., 2013), $N_{C—FCDP}$, r_e, and LWC_{FCDP} varied a lot horizontally. The $N_{C—FCDP}$ ranged from 70 to 248.4 cm^{-3}, corresponding to which the cloud droplet r_e ranged from 43.71 to 10.2 μm. In addition, the large amount of small cloud droplets implied that condensation, rather than collision, was likely the main way of droplet growth. By comparing the horizontal distribution of $N_{C—FCDP}$, r_e and LWC_{FCDP} along the flight track, it was clear that high $N_{C—FCDP}$ corresponded to low r_e and LWC_{FCDP} and high LWC_{FCDP} corresponded to high r_e. This suggested that while small droplets had a great contribution to number concentration, LWC in clouds was mainly contributed by large droplets.

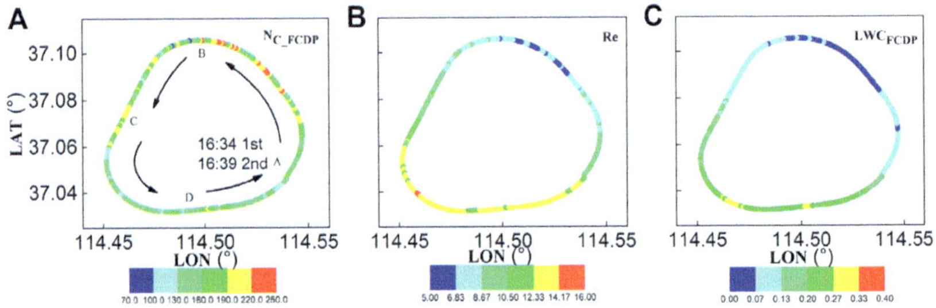

Figure 4. For a cloud layer at height 4,550 m: (A) concentration of cloud droplets ($N_{C—FCDP}$) with unit cm^{-3}, (B) effective radius (r_e) with unit μm, and (C) LWC_{FCDP} with unit g m^{-3} along the first flight observation trap.

Not only cloud microphysical properties but also the cloud particle phase varied greatly along the flight track at the selected cloud layer. From 16:34 to 16:37 LT (A-B-C in Figure 4A), the concentration of cloud droplets was relatively high, while r_e and LWC in the cloud were relatively small. The cloud particle images obtained by CPI during this period are shown in Figure 5A. In this period, the particles are mostly spherical droplets with a small amount of ice crystals. The effective radius of the particles is relatively small, so the concentration of LWC_{FCDP} was low, although $N_{C—FCDP}$ was at a high level. From 16:37-16:39 LT (C-D-A in Figure 4A), the number concentration of cloud droplets was relatively low, with an average of 152.24 cm^{-3}. The cloud particle image (Figure 5B) showed that, at this time, cloud particles were mostly in ice phase. Large ice crystals coexist with small ice crystals in irregular shapes, which might come from the shattering of large crystals. The differences of microphysical properties in the clouds at different horizontal locations were clear, which reflect the serious heterogeneity of stratus cloud properties when they were in a mixed phase.

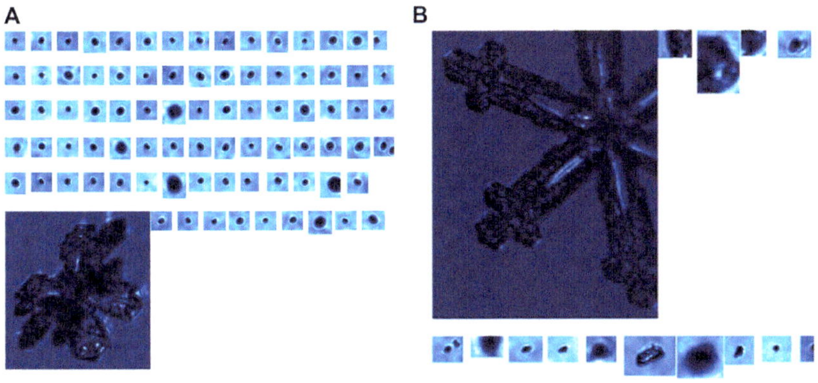

Figure 5. Cloud particle images at the height of 4,550 m in the cloud: (A) A-B-C and (B) C-D-A.

Microphysical Properties After Cloud Seeding

Figure 6 shows the temporal variation of cloud microphysical properties along with the flight height and environmental temperature. Note that the time periods shown in Figure 6 include those before and after cloud seeding. The flight made the observations of vertical profiles during the period from 15:52 to 16:18 LT at heights of 600–7,000 m. Then, the altitude of 4,550 m near cloud top, where LWC (supercooled droplets) was relatively abundant, was chosen for cloud seeding at 16:37 LT, with a total of 14 AgI bars lighted. The cloud seeding and detection were carried along the same circular trajectory. Even though the AgI flame bars were lighted at 16:37 LT, the aircraft still detected the unseeded area before the end of the lap. The aircraft returned to the position where the cloud seeding started at 16:43 LT, and the AgI flame bars got extinguished at 16:58 LT. Then, the aircraft started the observation of cloud microphysical properties after seeding along the same trajectory until 17:05 LT. The concentration of N_{C_CIP} and $N_{C—FCDP}$ were 7.1 L^{-1} and 160.3 cm^{-3}, on average, after two circles of detection in horizontal flight, which showed no significant difference from that in the vertical detection stage. As presented from the particle images, the cloud droplets and ice crystals coexisted before cloud seeding (Figure 5A and Figure 5B). The concentration of $N_{C—FCDP}$ (Figure 6B) decreased from 160.3 to 129 cm^{-3}, on average, in about 7 to 8 min after seeding. When the plane arrived at the initial seeding position after a circle of flat flight, the concentration of $N_{C\text{—}FCDP}$ reduced obviously to 88.94 cm^{-3} in minimum, and the horizontal spatial distribution became more uniform. LWC showed obvious regional differences before seeding, and these differences were no longer evident during the seeding process. After seeding (16:58–17:05 LT), LWC presented a significant decreasing trend, which indicated that the silver iodide ice nuclei were introduced into the clouds, and the supercooled water in the cloud was consumed through the Bergeron process. It can be seen from Figure 6C that the concentration of ice crystals at 4,550 m was distributed relatively uniformly before seeding, and the concentration fluctuates after seeding. The concentration of $N_{C\text{—}CIP}$ increased slightly after seeding, and several peaks appeared (for example, $N_{C\text{—}CIP}$ reached 52.2 L^{-1} at 16:53:59 LT and 149.3 L^{-1} at 17:00:52 LT). The concentration of ice crystals increased from 7.1 to 10 L^{-1} on average. The maximum $N_{C\text{—}CIP}$ reached 785.5 L^{-1} especially in some specific regions. In other studies in central and southern Hebei Province, it was also found that the concentration of ice crystals in some areas increased to hundreds after seeding. Compared with the results of this study, the concentrations of ice crystals are of a similar magnitude (Dong et al., 2020; Dong et al., 2021). Moreover, Figure 6E shows that the proportion of LWC decreased significantly from 67 to 50% after seeding, indicating that part of the LWC was consumed. As a result, the temperature decreased to a certain extent, instead of increasing, due to latent heat released by condensation as we usually thought.

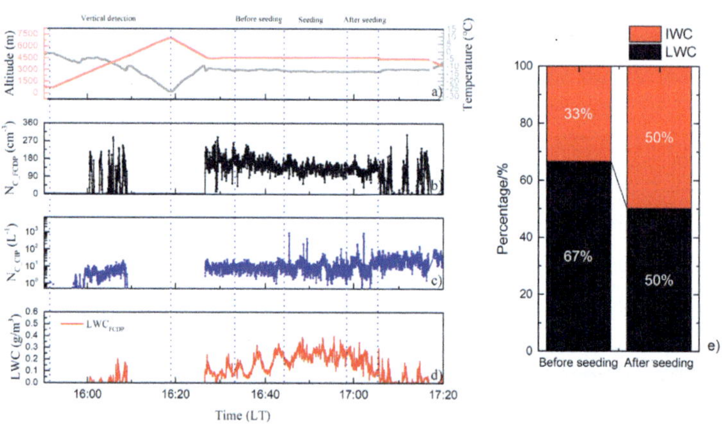

Figure 6. Changes of microphysical properties in the cloud during the observation and the seeding periods: (A) temperature and height, (B) $N_{C\text{—}FCDP}$, (C) $N_{C\text{—}CIP}$, (D) LWC_{FCDP}, and (E) percentage of liquid water content and ice water content relative to the total water content.

Figure 7 shows the cloud particle size distribution measured by FCDP and CIP in the region with abundant LWC before (16:34–16:43 LT) and after seeding (16:58–17:05 LT). Before seeding, the cloud showed a bimodal particle size distribution. The concentration of cloud particles showed a peak value of about 44.10 cm^{-3} in the size range of 4.5–6 μm (droplets) and another peak value of 0.195 cm^{-3} in the size range of 350–375 μm (ice particles). After seeding, the cloud showed a unimodal particle size distribution. The spectrum of small cloud particles measured by FCDP shifted to the right, and the peak value appeared in the range of 12–14 μm. Compared with cloud particle size distribution before seeding, the peak concentration was slightly decreased, which was about 29.28 cm^{-3}. The concentration of small cloud particles (1.5–4.5 μm) on the left side of the peak decreased by about 1 order of magnitude after seeding, but the concentration on the right side of the peak (6–50 μm) increased by about 1 order of magnitude. The changes of particle size distribution of the ice crystals measured by CIP were more obvious. The peak value of ice crystal number concentration at 350–375 μm disappeared after seeding, and the concentration of ice crystals at 100–675 μm increased significantly by about 1–2 orders of magnitude. Particularly, the ice crystal concentration at 100–125 μm increased from 2.73×10^{-4} to 7.08×10^{-2} cm^{-3}. Moreover, cloud particle images showed that flaky or columnar ice crystals appeared after seeding. From the analysis above, it can be concluded that the aircraft cloud seeding in this study presented a remarkable effect on mixed-phase cloud microphysical properties.

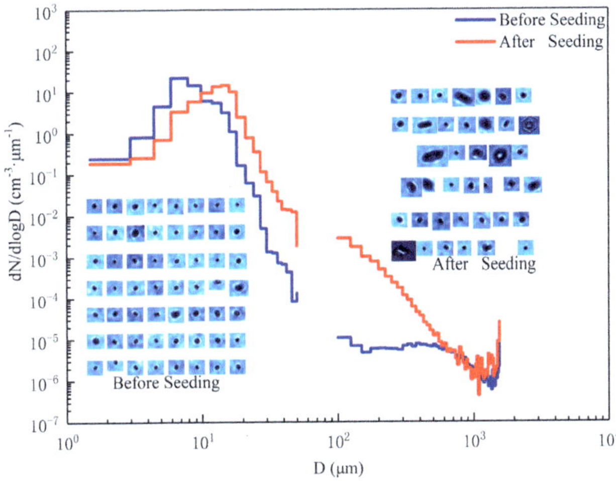

Figure 7. Size distribution and phase state of particles in the cloud with abundant liquid water content before and after seeding.

SUMMARY

Based on the aircraft observations for a mixed-phase cloud over Xingtai, Hebei Province, China, on November 29, 2019, this study investigates the response of mixed-phase cloud microphysical properties to cloud seeding near cloud top. Several valuable findings have been found.

Through the weather background analysis before cloud seeding, the suitable dynamic and humidity conditions between 700 and 500 hPa led to the abundance of supercooled water in the cloud, which was consistent with the aircraft detection results, and had a certain guiding significance for the early identification about the existence of supercooled water.

The aircraft observations showed that the vertical distribution of cloud properties was not uniform.

The vertical distribution of $N_{C\text{—}FCDP}$ presented a multi-peak structure, with the maximum value that appeared at 3,322 m. In contrast, the maximum value of $N_{C\text{—}CIP}$ appeared at 4,550 m in the upper part of the cloud. The height with peak LWC in the cloud was consistent with $N_{C\text{—}CIP}$, which was also 4,550 m. Liquid water and ice water coexisted at the altitude range from 3,500 to 4,700 m. Therefore, the middle and upper parts of this cloud were considered as mixed phase with the coexistence of supercooled liquid water and ice crystals, illustrating the suitable cloud seeding condition at the height of 4,550 m.

Cloud microphysical properties had distinct responses to cloud seeding. After seeding, the concentration of ice crystals increased on average, and the size of ice crystals increased obviously at the same time.

The horizontal variation of LWC in the cloud was no longer obvious, and LWC showed an obvious decreasing trend after seeding, which indicated that the silver iodide ice nuclei were introduced into the clouds through aircraft seeding and the supercooled water in the cloud was consumed. Similarly, the cloud particle size response to cloud seeding was also obvious with increased ice crystal sizes.

In short, the phase state, size distribution, and LWC in the cloud all had an obvious response to the artificial introduction of silver iodide, indicating that cloud seeding played a significant effect in this study. The results of this study are helpful to understand the response mechanism of cloud microphysical properties to cloud seeding and provide an effective basis for the artificial precipitation enhancement effect.

Data Availability Statement

The original contributions presented in the study are included in the article/Supplementary Material, further inquiries can be directed to the corresponding author.

AUTHOR CONTRIBUTIONS

CZ provided theoretical guidance. XD and RM carried out the flight observation mission. YC, FX, and JF helped with data analysis.

Funding

This work was supported by the National Key Research and Development Program of China (Grant 2019YFC1510301), the Hebei province Key Research and Development project (Grants 20375402D and 19275420D), the National Natural Science Foundation of China (Grant 41925022), and the Key Laboratory of Meteorology and Ecological Environment of Hebei Province Project (Grant Z202002H).

Conflict of Interest

The authors declare that the research was conducted in the absence of any commercial or financial relationships that could be construed as a potential conflict of interest.

Publisher's Note

All claims expressed in this article are solely those of the authors and do not necessarily represent those of their affiliated organizations or those of the publisher, the editors, and the reviewers. Any product that may be evaluated in this article or claim that may be made by its manufacturer is not guaranteed or endorsed by the publisher.

ACKNOWLEDGMENTS

Special appreciation is extended to the project participants for aircraft data acquisition. The NCEP/NCAR reanalysis data used for weather background analysis can be downloaded from https://psl.noaa.gov/data/gridded/data.ncep.reanalysis.pressure.html. The aircraft datasets and data description for this research can be downloaded publicly online from https://pan.bnu.edu.cn/l/KnHC01.

REFERENCES

1. Andreae, M. O., and Rosenfeld, D. (2008). Aerosol-cloud-precipitation Interactions. Part 1. The Nature and Sources of Cloud-Active Aerosols. *Earth Sci. Rev.* 89 (1-2), 13–41. doi:10.1016/j.earscirev.2008.03.001
2. Baumgardner, D., Jonsson, H., Dawson, W., O'Connor, D., and Newton, R. (2001). The Cloud, Aerosol and Precipitation Spectrometer: a New Instrument for Cloud Investigations. *Atmos. Res.* 59-60, 251–264. doi:10.1016/S0169-8095(01)00119-3
3. Bergeron, T. (1935). On the physics of cloud and precipitation. Proc 5th Assembly of IUGG, Lisbon, 156–178.
4. Dong, X., Zhao, C., Yang, Y., Wang, Y., Sun, Y., and Fan, R. (2020). Distinct Change of Supercooled Liquid Cloud Properties by Aerosols from an Aircraft-Based Seeding experiment. *Earth Space Sci.* 7 (8), e2020EA001196. doi:10.1029/2020EA001196
5. Dong, X., Zhao, C., Huang, Z., Mai, R., Lv, F., Xue, X., et al. (2021). Increase of Precipitation by Cloud Seeding Observed from a Case Study in November 2020 over Shijiazhuang, China. *Atmos. Res.* 262 (5), 105766. doi:10.1016/j.atmosres.2021.105766
6. Emery, E., Miller, D., Plaskon, S., Strapp, W., and Lillie, L. (2004). "Ice Particle Impact on Cloud Water Content Instrumentation," in 42nd AIAA Aerospace Sciences Meeting and Exhibit, Reno, Nevada, January 5–8, 2004. doi:10.2514/6.2004-731
7. Faber, S., French, J. R., and Jackson, R. (2018). Laboratory and In-Flight Evaluation of Measurement Uncertainties from a Commercial Cloud Droplet Probe (CDP). *Atmos. Meas. Tech.* 11 (6), 3645–3659. doi:10.5194/amt-11-3645-2018
8. Hobbs, P. V. (1991). Twenty Years of Airborne Research at the University of Washington. *Bull. Amer. Meteorol. Soc.* 72 (11), 1707–1716. doi:10.1175/1520-0477(1991)072<1707:tyoara>2.0.co;2
9. Korolev, A., and Field, P. R. (2015). Assessment of the Performance of the Inter-arrival Time Algorithm to Identify Ice Shattering Artifacts in Cloud Particle Probe Measurements. *Atmos. Meas. Tech.* 8 (2), 761–777. doi:10.5194/amt-8-761-2015
10. Korolev, A. V., Strapp, J. W., Isaac, G. A., and Nevzorov, A. N. (1998). The Nevzorov Airborne Hot-Wire LWC-TWC Probe: Principle of Operation and Performance Characteristics. *J. Atmos. Ocean.*

Technol. 15 (6), 1495–1510. doi:10.1175/1520-0426(1998)015<1495:tnahwl>2.0.co;2

11. Korolev, A., Strapp, J. W., Isaac, G. A., and Emery, E. (2013). Improved Airborne Hot-Wire Measurements of Ice Water Content in Clouds. *J. Atmos. Ocean. Techn.* 30 (9), 2121–2131. doi:10.1175/JTECH-D-13-00007.1
12. Lance, S., Brock, C. A., Rogers, D., and Gordon, J. A. (2010). Water droplet calibration of the Cloud Droplet Probe (CDP) and in-flight performance in liquid, ice and mixed-phase clouds during ARCPAC. *Atmos. Meas. Tech.* 3 (6), 1683–1706. doi:10.5194/amt-3-1683-2010
13. Lawson, R. P., Baker, B. A., Schmitt, C. G., and Jensen, T. L. (2001). An Overview of Microphysical Properties of Arctic Clouds Observed in May and July 1998 during FIRE ACE. *J. Geophys. Res.* 106 (D14), 14989–15014. doi:10.1029/2000JD900789
14. Li, Z., Lau, W. K. M., Ramanathan, V., Wu, G., Ding, Y., Manoj, M. G., et al. (2016). Aerosol and Monsoon Climate Interactions over Asia. *Rev. Geophys.* 54 (4), 866–929. doi:10.1002/2015RG000500
15. Lohmann, U., and Feichter, J. (2005). Global Indirect Aerosol Effects: a Review. *Atmos. Chem. Phys.* 5 (3), 715–737. doi:10.5194/acp-5-715-2005
16. Norris, J. R. (2005). Multidecadal Changes in Near-Global Cloud Cover and Estimated Cloud Cover Radiative Forcing. *J. Geophys. Res. Atmos.* 110. doi:10.1029/2004JD005600
17. Pokharel, B., and Geerts, B. (2016). A Multi-Sensor Study of the Impact of Ground-Based Glaciogenic Seeding on Clouds and Precipitation over Mountains in Wyoming. Part I: Project Description. *Atmos. Res.* 182, 269–281. doi:10.1016/j.atmosres.2016.08.008
18. Pokharel, B., Geerts, B., Jing, X., Friedrich, K., Ikeda, K., and Rasmussen, R. (2017). A Multi-Sensor Study of the Impact of Ground-Based Glaciogenic Seeding on Clouds and Precipitation over Mountains in Wyoming. Part II: Seeding Impact Analysis. *Atmos. Res.* 183, 42–57. doi:10.1016/j.atmosres.2016.08.018
19. Pruppacher, H. R., and Klett, J. D. (2010). "Microstructure of Atmospheric Clouds and Precipitation," in *Microphysics of Clouds and Precipitation* (Dordrecht: Springer), Vol. 18, 10–73. Atmospheric and Oceanographic Sciences Library. doi:10.1007/978-0-306-48100-0_2

20. Ramanathan, V., Cess, R. D., Harrison, E. F., Minnis, P., Barkstrom, B. R., Ahmad, E., et al. (1989). Cloud-radiative Forcing and Climate: Results from the Earth Radiation Budget experiment. *Science* 243 (4887), 57–63. doi:10.1126/science.243.4887.57
21. Schaefer, V. J. (1946). The Production of Ice Crystals in a Cloud of Supercooled Water Droplets. *Science* 104 (2707), 457–459. doi:10.1126/science.104.2707.457
22. Sun, Y., Zhao, C., Su, Y., Ma, Z., Li, J., Letu, H., et al. (2019). Distinct Impacts of Light and Heavy Precipitation on PM 2.5 Mass Concentration in Beijing. *Earth Space Sci.* 6 (10), 1915–1925. doi:10.1029/2019EA000717
23. Tao, W., Chen, J., Li, Z., Wang, C., and Zhang, C. (2012). Impact of Aerosols on Convective Clouds and Precipitation. *Rev. Geophys.* 50. doi:10.1029/2011RG000369
24. Tessendorf, S. A., Ikeda, K., Weeks, C., Rasmussen, R., Wolff, J., and Xue, L. (2020). An Assessment of Winter Orographic Precipitation and Cloud-Seeding Potential in Wyoming. *J. Appl. Meteorol. Climatol.* 59 (7), 1217–1238. doi:10.1175/JAMC-D-19-0219.1
25. Vonnegut, B. (1947). The Nucleation of Ice Formation by Silver Iodide. *J. Appl. Phys.* 18, 593–595. doi:10.1063/1.1697813
26. Wang, L., Yin, Y., Lunge, L. I., Wang, X., and Fugang, L. I. (2013). Analyses on Typical Autumn Multi-Layer Stratiform Clouds over the Sanjiangyuan National Nature Reserve with Airborne Observations. *Chin. J. Atmos. Sci.* 37 (28), 1038–1058. doi:10.1016/j.atmosenv.2009.04.022
27. Yang, X., Zhao, C., Guo, J., and Wang, Y. (2016a). Intensification of Aerosol Pollution Associated with its Feedback with Surface Solar Radiation and Winds in Beijing. *J. Geophys. Res. Atmos.* 121 (8), 4093–4099. doi:10.1002/2015JD024645
28. Yang, X., Zhao, C., Zhou, L., Wang, Y., and Liu, X. (2016b). Distinct Impact of Different Types of Aerosols on Surface Solar Radiation in China. *J. Geophys. Res. Atmos.* 121 (11), 6459–6471. doi:10.1002/2016JD024938
29. Yang, X., Zhou, L., Zhao, C., and Yang, J. (2018). Impact of Aerosols on Tropical Cyclone-Induced Precipitation over the mainland of China. *Climatic Change* 148, 173–185. doi:10.1007/s10584-018-2175-5

30. Yang, Y., Zhao, C., Dong, X., Fan, G., Zhou, Y., Wang, Y., et al. (2019). Toward Understanding the Process-Level Impacts of Aerosols on Microphysical Properties of Shallow Cumulus Cloud Using Aircraft Observations. *Atmos. Res.* 221, 27–33. doi:10.1016/j.atmosres.2019.01.027
31. Zhao, C., Yang, Y., Fan, H., Huang, J., Fu, Y., Zhang, X., et al. (2020). Aerosol Characteristics and Impacts on Weather and Climate over the Tibetan Plateau. *Natl. Sci. Rev.* 7 (3), 492–495. doi:10.1093/nsr/nwz184
32. Zhao, C., and Garrett, T. J. (2015). Effects of Arctic Haze on Surface Cloud Radiative Forcing. *Geophys. Res. Lett.* 42 (2), 557–564. doi:10.1002/2014GL062015
33. Zhao, X., Sun, Y., Zhao, C., and Jiang, H. (2020). Impact of Precipitation with Different Intensity on PM 2.5 over Typical Regions of China. *Atmosphere* 11 (9), 906. doi:10.3390/atmos11090906
34. Zhao, Z., and Lei, H. (2014). Aircraft Observations of Liquid and Ice in Midlatitude Mixed-phase Clouds. *Adv. Atmos. Sci.* 31, 604–610. doi:10.1007/s00376-013-3083-2
35. Zhao, C., Lin, Y., Wu, F., Wang, Y., Li, Z., Rosenfeld, D., et al. (2018). Enlarging Rainfall Area of Tropical Cyclones by Atmospheric Aerosols. *Geophys. Res. Lett.* 45 (16), 8604–8611. doi:10.1029/2018GL079427
36. Zhao, C., Zhao, L., and Dong, X. (2019). A Case Study of Stratus Cloud Properties Using In Situ Aircraft Observations over Huanghua, China. *Atmosphere* 10 (1), 19. doi:10.3390/atmos10010019

CHAPTER 17

Hygroscopicity and Ice Nucleation Properties of Dust/Salt Mixtures Originating from the Source of East Asian Dust Storms

Jun Li [1], Wanyu Liu[1], Dimitri Castarède[2], Wenjun Gu[3], Linjie Li [2], Takuji Ohigashi [4], Guiqian Zhang[1], Mingjin Tang[3], Erik S. Thomson[2], Mattias Hallquist [2], Sen Wang[1] and Xiangrui Kong[2]

[1]Shaanxi Key Laboratory of Earth Surface System and Environmental Carrying Capacity, Northwest University, Xi'an, China

[2]Department of Chemistry and Molecular Biology, Atmospheric Science, University of Gothenburg, Gothenburg, Sweden

[3]State Key Laboratory of Organic Geochemistry, Guangzhou Institute of Geochemistry, Chinese Academy of Sciences, Guangzhou, China

[4]UVSOR Synchrotron, Institute for Molecular Science, Okazaki, Japan

Citation: Li J, Liu W, Castarède D, Gu W, Li L, Ohigashi T, Zhang G, Tang M, Thomson ES, Hallquist M, Wang S and Kong X (2022) "Hygroscopicity and Ice Nucleation Properties of Dust/Salt Mixtures Originating from the Source of East Asian Dust Storms". Front. Environ. Sci. 10:897127. https://doi.org/10.3389/fenvs.2022.897127.

Dust storms are common meteorological events that occur frequently in the late spring and early summer in arid and semi-arid areas. The resulting lofted dust and salt mixtures can impact atmospheric chemistry and climate systems through the many pathways represented by aerosol-cloud-climate interactions.

In this study, dust/salt samples were collected from important sources of the East Asian dust storm, including the Badain Jaran Desert, the Tengger Desert and the Ulan Buh Desert in northwestern China. Ion chromatography (IC) measurements were performed to determine the concentrations of cations and anions.

The ionic concentrations, pH and dissolvable fractions of sand samples show a positive correlation, indicating that the dissolved content is rich in alkaline ions. A positive matrix factorization (PMF) receptor model was employed to analyze the IC results, and from the PMF solutions non-obvious connections to local geography emerge. The results of hygroscopic experiments of sand samples which were measured by a vapor sorption analyzer indicate that the hygroscopicity may be related to the soluble content of samples, and the observed hygroscopic behavior can be well described by a thermodynamic model. The morphology of individual particles was chemically mapped by the synchrotron-based scanning transmission X-ray microscopy, and needle-shaped $CaCO_3$ particles were observed to adhere to more irregular high K-containing particles. Moreover, a continuous flow diffusion chamber was used to investigate the ice nucleation abilities of typical salts, with both homogeneous freezing and deposition nucleation being observed. The results indicate that the salts primarily act as cloud condensation nuclei but can also act as ice nucleating particles at low temperatures.

Keywords: ice nucleation, hygroscopicity, AIOMFAC, ion chromatography, STXM, PMF

INTRODUCTION

Dust storms are common meteorological events which occur frequently in the late spring and early summer in arid and semi-arid areas (Barkan et al., 2005; Bou Karam et al., 2014; Liu et al., 2020). North Africa, Middle East Asia (including Western China), Western Australia and Western North America are the four most important dust-prone areas in the world (Su et al., 2016). Annually, about 1,000 to 3,000 trillion grams (Tg) of dust is emitted into the atmosphere (Huang et al., 2010), and more than 40% of the land areas on Earth are affected by aeolian processes. In Asia, about 800 Tg of dust is lofted into the atmosphere every year, of which about 30% is redeposited onto deserts, 20% is transported over regional scales, and 50% is transported to the Pacific Ocean or even further (Uno et al., 2009; Li et al., 2010; An et al., 2018). Asian dust storms follow a typical transport path across the northern regions to the eastern seas of China under the influence of strong northwest to southeast airflow at high altitude, and also affect Japan and Korea, and at long-range the United States and Europe, as well as both Arctic and equatorial regions under special weather conditions (Duce et al., 1980; Kanayama et al., 2002; Okin, 2005; Ziegler et al., 2008; Huang et al., 2015). The global character of dust has been demonstrated, for example, by assessing the contributed fractions from North Africa (15%), the Middle East (21%), the Taklimakan Desert (25%), and Northeast China/Mongolia (42%) to aerosols collected in California (Creamean et al., 2013).

Mineral dust aerosols are a primary component in the atmosphere, accounting for almost half of tropospheric aerosol mass, and they can deteriorate human living conditions (Engelstaedter et al., 2006; Prospero et al., 2014), endanger human health (Ichinose et al., 2008; Venero-Fernández, 2016), and affect ecosystems (Mahowald et al., 2014; He et al., 2016; Kok et al., 2021; Rashki et al., 2021). Dust particles can also absorb or scatter radiation to heat or cool the atmosphere, thereby affecting the radiative balance of the Earth-atmosphere system (Che et al., 2019; Zhang et al., 2019). During long-distance transportation processes, dust aerosols can participate in chemical reactions by interacting with the sulfide, nitride and black carbon in the atmosphere, and thus influence terrestrial and marine biogeochemical cycles (Costa et al., 2006; Nee et al., 2007). N_2O_5 produced by the reaction of NO_2 and O_3 can be hydrolyzed to HNO_3 on the surface of salt-containing mineral dusts (e.g., CaSO4 and Na_2SO_4), and can be further neutralized by NH_3 to NH_4NO_3 through heterogeneous reactions (Wu et al., 2020). Some studies show that, in dusty weather, nitrate, mainly $Ca(NO_3)_2$, can be formed by the reaction of HNO_3 and carbonate in dust particles under alkaline condition (Krueger et al., 2004; Tobo et al., 2010; Tang et al., 2016).

Mixtures of mineral dust and salts also impact atmospheric chemistry and the climate system due to their high hygroscopicity and their potential to act as cloud particle seeds. Hygroscopicity determines the ability of aerosol particles to grow in humid environments, which is important for optical properties and atmospheric aging processes, and the ability to act as cloud condensation nuclei (CCN) or ice nucleating particles (INPs) (Kumar et al., 2011; Karydis et al., 2017). Mixtures that have high hygroscopicity can act as

CCN and potentially INPs, to change the optical characteristics and lifetime of clouds, and to suppress or enhance precipitation under certain conditions, all of which have indirect climate effects (Rosenfeld et al., 2001; Rudich et al., 2002; Sassen, 2002; Yin et al., 2002; Huang et al., 2006; Hoose and Möhler, 2012; Huang et al., 2014; DeMott et al., 2015; Chen et al., 2017; Gui et al., 2017; Kok et al., 2018; Huang et al., 2021). Mineral dust has a significant impact on the hygroscopicity of particles, soluble inorganic ions such as Ca^{2+}, Mg^{2+}, Na^{+}, and K^{+} on dust surfaces can react heterogeneously with HNO_3 and HCl, processes that can coat the dust particles with soluble substances, further increasing the hygroscopicity and the ability to act as CCN (Kelly et al., 2007). Based on a 4-year combined CALIPSO and CloudSat dataset, Zhang et al. proposed that dust increased the effective INPs leading to glaciated mixed-phase clouds over east Asia (Zhang et al., 2015). Based on satellite retrievals, Yuan et al. analyzed a typical dust-rain storm in northwest India in 2018, and showed that under the influence of dust, a supercooled liquid water cloud was likely to freeze due to increased INPs. In this scenario, ice water path and cloud optical depth were both increased, leading to heavy rainfall in northern India (Yuan et al., 2021).

In the context of global climate change, the trend of warming and drying that is intensifying desertification in northern China is clear (Fu et al., 2008; Wang et al., 2008). The dominant regional geophysical feature, the neighboring *Third Pole*, that is the elevated region represented by the Himalayas and the Tibetan Plateau, impacts the entire planet's climate directly and *via* global teleconnections (You et al., 2021). This makes desert research in northwestern China well situated to help build a deeper understanding of driving mechanisms and consequences of climate change (You et al., 2021). This has motivated investigations of the physiochemical properties and sediment characteristics of desert lakes (Wang et al., 2013; Bogen et al., 2015; Liu et al., 2016; Schillereff et al., 2016; Li et al., 2018), but few studies thoroughly probe the physiochemical properties and hygroscopicity of the deserts' salt/sand mixtures. In this study, dust, salts and mixed samples were collected from important source regions of East Asian Dust Storms, including the Badain Jaran Desert, the Tengger Desert and the Ulan Buh Desert in northwestern China. Ion chromatography (IC) measurements were performed to determine the concentrations of cations and anions. Synchrotron-based scanning transmission X-ray microscopy (STXM) was carried out to investigate particle morphology and to chemically map representative dust/salt particles. Hygroscopic properties of the samples were measured using a vapor sorption analyzer, and a thermodynamic model was used to predict the deliquescence relative humidity (DRH) based on samples' chemical compositions. To expose links between physiochemical properties and the sample origins/types, a positive matrix factorization (PMF) receptor model was implemented to analyze the results of the IC and the thermodynamic model. Ice nucleation activity has been investigated with the Portable Ice Nucleation Chamber II (PINCii), with which both homogeneous freezing and deposition nucleation were observed.

SAMPLING SITES

The sampling was carried out between 17 May and 24 May 2021. Figure 1 shows the sampling locations in the Tengger, Badain Jaran and Ulan Buh deserts. In this study, not all but representative samples are analyzed. Different types of samples, e.g., brines, high salt-content samples and sand samples, were collected in some sites, e.g., Gurinai (sample #14, #15 and #16). The background information of the three deserts is summarized below.

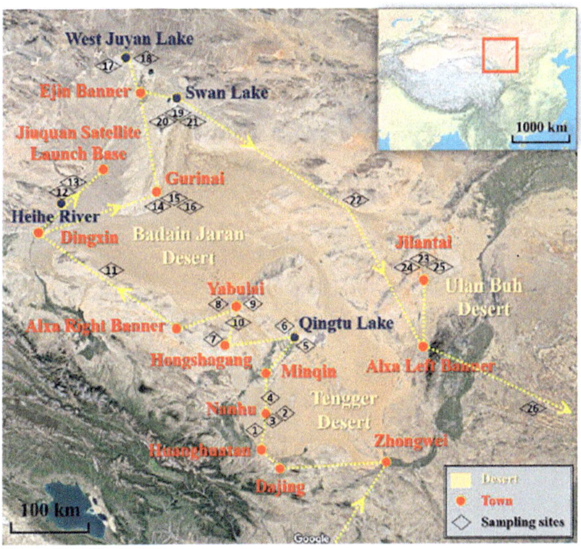

Figure 1. Sampling route, sampling sites, and aqueous environments marked in blue.

Badain Jaran Desert

The Badain Jaran Desert (39°20′~41°30′N, 99°48′~104°14′E) is located in the western Inner Mongolia Autonomous Region, with Yabulai Mountain and the Tengger Desert to the southeast, Guaizi Lake and the ancient Juyan Lake to the north, the Heihe River and Gurinai Lake to the west, and Beida Mountain to the south (Ma et al., 2014). The Badain Jaran Desert is the second largest desert in China, with an area of about 5.2 × 10⁴ km², and mobile sand dunes accounting for approximately 83% of the total desert area (Zhu et al., 2010). The southeastern part of the desert has a higher topography, including more and well-developed tall sand hills, and relatively more precipitation and lakes. The sand dunes in the Badain Jaran Desert are the highest sand hills in the world, with a relative height greater than 300 m and a maximum height of close to 500 m (Qu et al., 2003).

The Badain Jaran Desert has a temperate continental desert climate with extremely arid areas and scarce precipitation (Dong et al., 2013). The average annual precipitation in the southern part of the desert is 90–115 mm, while it is 35–43 mm in the northern area and the precipitation is mostly concentrated in June to August (Ma et al., 2011). The annual evaporation exceeds 3,500 mm, which is 40–80 times larger than the annual precipitation. The annual average temperature is 7–8°C, with the average summer temperature being 25°C and average winter temperature –9°C (Xu and Li, 2016). The annual average wind speed is 4 m/s, which is mainly a northwest wind (Hu and Yang, 2016).

Tengger Desert

The Tengger Desert (37°30′~40°00′N, 102°20′~106°00′E) is located in the southwest area of Alxa Left Banner, Inner Mongolia Autonomous Region, close to the central Gansu Province. The Great Wall lies to the south, Helan Mountain to the east and Yabulai Mountain to the west. The Tengger Desert is the fourth largest desert in China, with a total area of about 4.3 × 10^4 km^2. The desert includes two parts, Nanjiling in the north and Tengger in the south, which are collectively referred to as the Tengger Desert. Sand dunes account for 71% of the total desert area, lake basins 7%, and mountainous hills and flat lands 22%. Among the dunes, mobile dunes, semi-fixed dunes, and fixed dunes represent fractions of 67.2, 17.4, and 15.4%, respectively.

The Tengger Desert is a typical continental arid climate and is very sensitive to climate change (Zhang et al., 2000; Ji et al., 2014). The annual average temperature is 7.0–9.7°C, and the average summer temperature is 24.2°C (Gao et al., 2020). The 60-year average precipitation in the Tengger Desert is 146.5 mm, wherein summer and autumn precipitation accounts for about 80% of the annual precipitation, and more rain falls in east and south

(Wang et al., 2020). The average annual evaporation is ≈ 2,600 mm (Li et al., 2015). The annual average wind speed is 3–4 m/s, and the main wind direction is northwest wind.

Ulan Buh Desert

The Ulan Buh Desert (39°41′07″~40°31′14″N, 105°59′45″~106°41′53″E) is located in the eastern part of the Alxa Plateau in Inner Mongolia, with the Jilantai Saline Lake to the west, Langshan Mountain to the north, the Yellow River to the east, and Helan Mountain to the south. The Ulan Buh Desert is one of eight major deserts in China with a total area of about 9.991 × 10^3 km^2 (Jia et al., 2015). The desert's altitude ranges from 1,028 to 1,054 m, and its terrain slopes from south to west (Chun et al., 2009). Mobile sand dunes account for about 49.9% of the total desert area, semi-fixed sand dunes account for around 21.8%, and fixed sand dunes account for ≈23.4%. The Ulan Buh Desert has a mid-temperate arid climate, with an average annual temperature of 7.8°C, an average annual precipitation of 102.9 mm, and an average annual evaporation of 2,258.8 mm.

METHODOLOGY

For sampling the sand samples, about 300–500 g samples were collected at each site with a sampling area of around 10 × 10 cm^2 and a sampling depth of about 0–10 cm. For the high salt-content samples, about 100–150 g samples were taken at the surface of each sampling site with the sampling area of about 10 × 10 cm^2. Sand samples (#1, #6, #8, #11, #12, #15, #18, #19, #20, #22, #25, #26) and high salt-content samples (#13, #17, #21) were first dried in a Muffle furnace at 90°C.

From the dried samples 0.4 g was extracted and added to 40 ml distilled water to dissolve soluble content. The suspensions were sonicated and centrifuged, with the upper part extracted and filtered through a 0.22 μm membrane filter. Filtered sample solutions were stored in polyethylene bottles sealed with Parafilm and refrigerated at 4°C before measurements. The undissolved sands were dried and weighed to calculate the fractions of dissolvable content. The pH and conductivity of all filtered sample solutions were measured using a pH meter (FE28-Standard, Mettler Toledo, China) and a portable conductivity meter (F3-Standard, Mettler Toledo, China), respectively.

Ion Chromatography

Filtered sample solutions prepared as described above were used to determine the ionic compositions. The Dionex AQUION Ion Chromatography system (Thermo Fisher Scientific, MA, United States) was used to measure cations (Na^+, K^+, Mg^{2+}, Ca^{2+}, and NH_4^+) and anions (Cl^-, SO_4^{2-}, NO_3^-, NO_2^-, and F^-) simultaneously. High salt-content sample solutions (#13, #17, #21) were diluted 20–100 times to reduce concentrations to be within the instrumental detection limits. The sand sample solutions were measured without dilution. Cations and anions were measured with CS12A IC columns (Dionex IonPac, Thermo Fisher) with a 30 mmol L^{-1} methanesulfonic acid eluent, and AS11-HC IC columns (Dionex IonPac, Thermo Fisher) with a 20 mmol L^{-1} KOH eluent, respectively. All samples were measured in duplicate with good reproducibility.

Hygroscopicity Measurements

Hygroscopic properties of 0.5–1 mg sand samples (#12, #18, #20) were measured using a vapor sorption analyzer (Q5000SA, TA Instruments). Relative humidity (RH) was elevated in a stepwise manner (at 25 ± 0.1°C), with the sample mass change, during the RH ramp, monitored with a balance that has a 0–100 mg dynamic range and <0.1 μg measurement sensitivity. Only brief descriptions are presented here, with detailed instrumental and operational information found in previous publications (Gu et al., 2017; Guo et al., 2019; Tang et al., 2019). The mass change of an empty sample holder during the RH ramp was measured as a baseline, which was removed during data processing. The apparatus is calibrated using pure NaCl, KCl and $(NH_4)_2SO_4$ and the differences between measured results and the theoretical deliquescence RHs were within 1% (Tang et al., 2019). The samples were dried to <1% RH before each experiment, and then the RH was increased to 90% over 30 min, in 10% RH steps. At each step, the samples were equilibrated until the observed mass change ratio was less than 0.05% in 1 minute. Triplicate samples were prepared and measured, with the average values reported.

Thermodynamic Model

The Aerosol Inorganic-Organic Mixtures Functional groups Activity Coefficients (AIOMFAC) model was used to simulate the DRH of the samples based on their chemical matrix (Zuend et al., 2008; Zuend et al.,

2011). The cations and anions were balanced by scaling, and water was selected to be present in the system. From the results, the molar Ion Activity Product (IAP) is calculated based on the generated ion activities. As an example, for $MgCl_2$ the IAP is calculated as,

$$IAP = \left[a_{m\left(Mg^{2+}\right)}\right] \times \left[a_{m(Cl^{-})}\right]^2 \quad (1)$$

where a_m is the activity of the relevant ion. The IAP is a function of RH (Figure 2), and when the IAP reaches the IAP of saturated solutions (IAP_{sat}), the corresponding RH is the DRH. The IAP_{sat} is calculated based on the ionic concentrations of saturated solution from literature. Figure 2 shows one example with the IAP plotted versus RH. The horizontal dashed lines indicate the IAP_{sat} of saturated solutions of four common salts, while the vertical dashed lines point to the DRHs simulated by the AIOMFAC model.

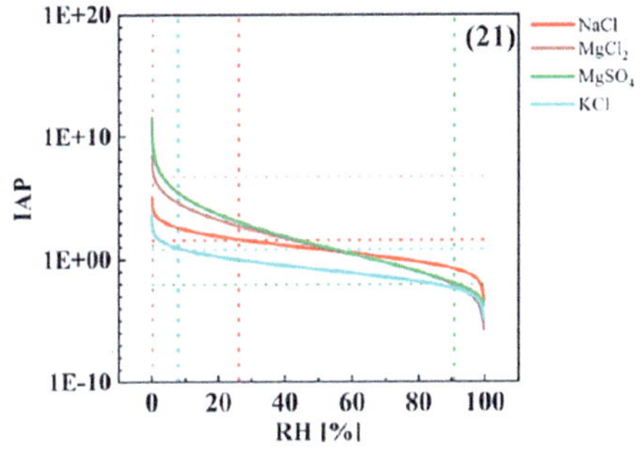

Figure 2. IAP as a function of RH, including four salt components in sample #21. The horizontal dashed lines indicate the IAP_{sat} of saturated solutions of four common salts marked on the *y*-axis. The vertical dashed lines point to the DRH simulated by the AIOMFAC model on the *x*-axis.

Positive Matrix Factorization Analysis

A Positive Matrix Factorization (PMF) receptor model was used to explore the correlations of the chemical compositions between the samples. An open-source program EPA PMF 5.0 was employed for the analysis (Norris et al., 2014). The input datasets to the PMF program are the ionic concentrations measured by the IC and the uncertainty of the measurements. A 3-factor solution is identified from a 20-iteration PMF calculation, and the interpretation of the factors is discussed, where the recognized factors are closely linked to the local characteristics of the sampling sites.

STXM Experiment

Scanning transmission X-ray microscopy (STXM) measurements were performed at the BL4U UVSOR-III beamline (Ohigashi et al., 2013). The STXM technique records high-resolution images of soft X-rays transmitted through samples at set photon energies (Kilcoyne et al., 2003). The photon energy spectra provide information on mixing states and chemical compositions. For STXM measurements sample particles were deposited on filmed TEM grids (Copper 230 mesh, Beijing Zhongjingkeyi Technology Co., Ltd.), and data was analyzed with the aXis2000 program. Example optical images (sample #20) are presented in Figure 3, where Figure 3A is an image captured at 348 eV photon energy, corresponding to calcium absorption. In this case the calcium-rich particles appear dark with a needle-like morphology. Potassium-containing particles in the same sample are highlighted in Figure 3B (297 eV) as the dark more polygonal particles.

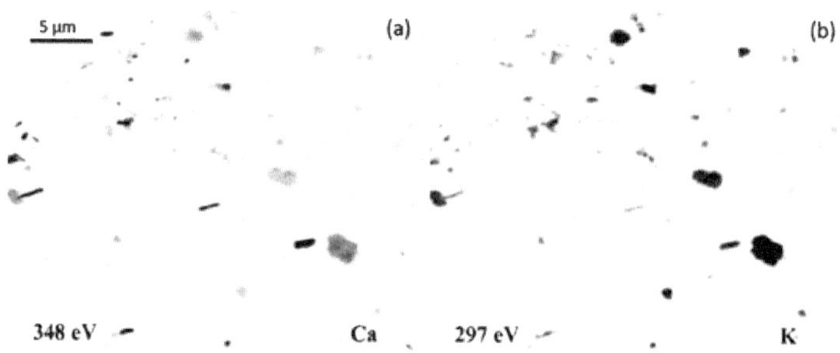

Figure 3. STXM images show the morphology of sample #20 particles. (A,B) Snapshots measured at 348 and 297 eV, which are high absorption regions for Ca 2p and K 2p, respectively.

Ice Nucleation Experiments

Ice nucleation experiments were carried out using a Portable Ice Nucleation Chamber II (PINCii), which is a newly-built Continuous Flow Diffusion Chamber (CFDC) apparatus (Brasseur et al., 2022; Castarède, 2021). Solutions of four samples with higher salt content were diluted and atomized, with the resulting aerosol flow being dried and split. One part of the resulting flow was monitored using a condensation particle counter (TSI CPC 3040) to measure input particle concentrations, while the remainder of the flow was directed through the PINCii. Experiments were run at four selected temperatures with the RH increased by 1% per min from ice saturation,

RH_{ice} = 100%. The output flow was measured using a four Channel OPC (Remote 3,104, Lighthouse Worldwide Solutions, United States) that detects ice nucleation by size discrimination. Using the input and outlet flows and particle counting the concentration of ice active particles in a given size channel (particles larger than 3 μm, $C_{>3\mu m}$) is determined.

RESULTS AND DISCUSSIONS

Dissolvable Content and Ionic Composition

Dissolvable content fractions of the sand samples are plotted together with pH and conductivity in Figure 4A. The dissolvable fractions range between 0.15 and 3.15%, with pH between 7 and 8 except for sample #20 which has pH ≈ 8.6. The dissolvable fraction, pH and conductivity are positively correlated, which indicates that the dissolved fractions are rich in alkaline ions. Samples #6 and #20 have relatively large values for all three parameters, while sample #12 has a notably large dissolvable fraction but does not exhibit increased conductivity or pH as measured. This indicates that the dissolved content of the sample #12 is not in ionic form. In general, all the solutions show slight alkalinity, agreeing with previous studies, which report that the precipitation in northwest China is commonly slightly alkaline due to the presence of alkaline dust in the atmosphere (Cao et al., 2009; Wang and Xu, 2009).

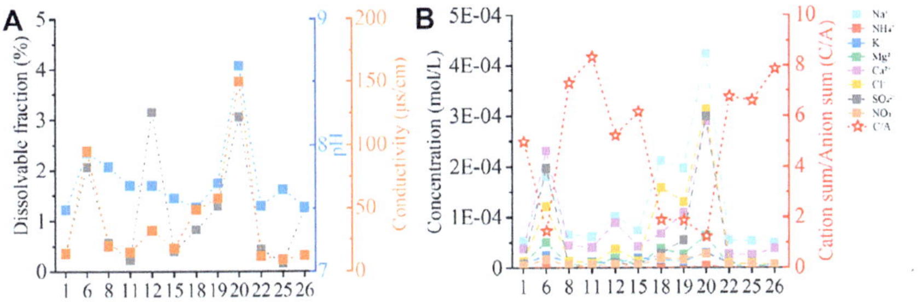

Figure 4. (A) The dissolvable fractions, pH and conductivities of the 12 sample solutions. (B) On the left vertical axis the ionic molar concentrations measured by IC are shown, while the right hand axis scaling depicts the ratio of total cationic charges to total anionic charges (C/A). The amount of each cation or anion (*n*) is multiplied by the number of charges and then added together to get the total charges of the cation or anion, i.e., cation sum = $n(Na^+) + n(NH_4^+) + n(K^+) + 2n(Mg^{2+}) + 2n(Ca^{2+})$; anion sum = $n(Cl^-) + 2n(SO_4^{2-}) + n(NO_3^-)$.

In Figure 4B the measured ionic molar concentrations are presented in addition to the ratios of total cationic charges to total anionic charges (C/A). The total cationic charges are generally greater than the total anionic charges, yielding C/A ratios ranging from approximately 1–8. For the samples with low C/A ratios, i.e., cation and anion balanced, the Cl^- and SO_4^{2-} concentrations are relatively higher than in other samples. In the high C/A ratio samples, the missing anions are probably carbonates and bicarbonates, which are not detected by IC but have been reported to be abundant in desert environments (Zhu and Yang, 2010).

To visualize the relative abundance in detail, normalized ionic molar concentrations are plotted in Figure 5. For comparison three high salt-content samples (salt crusts; #13, #17 and #21) are added (their absolute ionic concentrations are too high to be shown in Figure 4). In general, the sand sample solutions contain increased fractions of Ca^{2+} and Cl^- compared to the high salt-content samples, indicating the high content of Ca^{2+}- and Cl^-- containing species are associated with sand. In the high salt-content samples, the SO_4^{2-} fractions are commonly the greatest except for Na^+, and in two samples (#17 and #21) the Mg^{2+} levels are relatively large. These Mg^{2+} and SO_4^{2-} fractions are commonly associated with aqueous systems (Li et al., 2020), which matches the geographic features of these sampling locations: sample #13 is in the vicinity of the Heihe River, sample #17 is close to the West Juyan Lake, and sample #21 is near the Swan Lake (Figure 1).

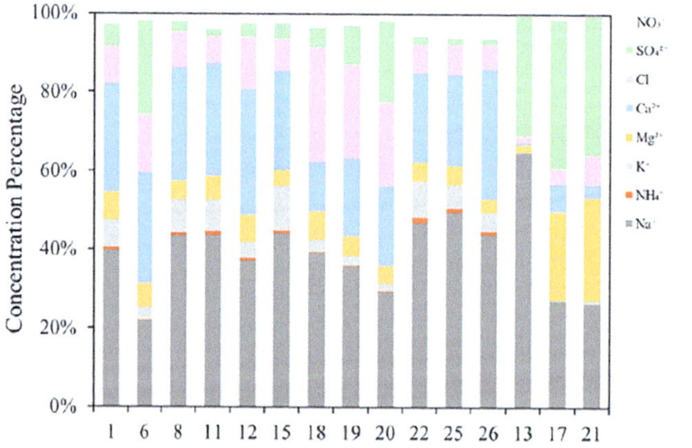

Figure 5. Normalized ionic molar concentrations as a function of species for sand sample solutions (#1, #6, #8, #11, #12, #15, #18, #19, #20, #22, #25, #26) and high salt-content sample solutions (#13, #17, #21).

To further explore the differences and connections between the chemical compositions of samples, PMF was performed to analyze ionic concentrations obtained from the IC experiments. A 3-factor PMF solution converges best, and the results are shown in Figure 6. The factor profiles, i.e., the ionic compositions of each factor, are shown in Figure 6A. It is not straightforward to directly interpret the factor profile due to its complexity and shared ionic profiles. However, the factor contributions shown in Figure 6B are relatively easy to rationalize, where the samples can be grouped based on the relative significance of factors. Color codes beneath the *x*-axis are used to illustrate this grouping, i.e., sample #18 and #19 are dominated by Factor 1 (Na^+, Ca^{2+} and Cl^-); samples #1, #8, #11, #15, #22, #25, and #26 are dominated by Factor 2 (Na^+ and Ca^{2+}); samples #13, #17 and #21 are dominated by Factor 3 (SO_4^{2-}); samples #6 and #20 are dominated by Factors 1 and 3; sample #12 is dominated by Factors 1 and 2 together.

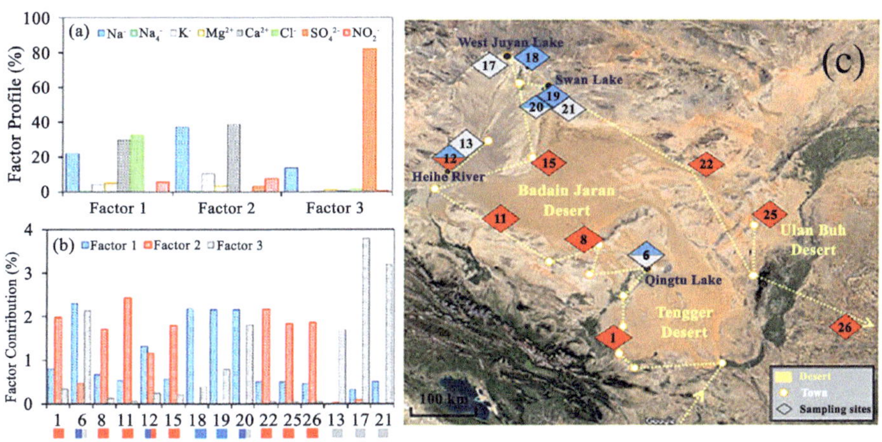

Figure 6. PMF 3-factor solution based on ionic concentrations. (A) Factor profile and (B) factor contribution of all samples, where the color codes below the *x*-axis indicate factors. (C) Sampling sites with color codes shown in map, aqueous environments marked in blue typeface. The blue code is for factor 1, the red code is for factor 2, and the grey code is for factor 3. Some samples show joint features form multiple factors.

In Figure 6C these color-coded samples are mapped, and it appears that the PMF factors correspond with local characteristics of the sampling locations. The factor 2 samples in red are close to the deserts, while the other samples are closer to aqueous sources and playa regions, i.e., the sampling location of sample #6 is near Qingtu Lake, the samples #12 and #13 are near the Heihe River, samples #17 and #18 are near West Juyan Lake, the samples #19, #20, and #21 are near Swan Lake. Thus, the PMF analysis illuminates non-obvious classification links related to the local geography that do not stand out from examining the ionic compositions alone (Figure 5). Although the calculated factor profiles are complicated in their compositions, the factor contributions provide useful information that can be used to infer geographic features based on the chemical analyses, which can be of use for analyzing samples from unknown locations or environments.

Hygroscopicity

Experimental Results

Three sand samples (#12, #18, #20) were measured for hygroscopicity using a vapor sorption analyzer and the results are shown in Figure 7. In Figure 7A growth factors (m/m_0) are plotted as a function of RH, with Figure 7B providing a rescaled *y*-axis to illuminate the low RH behavior. In all samples mass increases immediately as RH rises. Two behaviors are observed: 1) When RH $\leq$ 60%, the mass growth is gradual and limited and 2) when RH > 60%, the mass growth diverges. Among the three samples, #20 has the greatest hygroscopicity, which is expected because the sample also has more soluble content (65.80 mg/L) than sample #18 (18.74 mg/L) and sample #12 (9.62 mg/L).

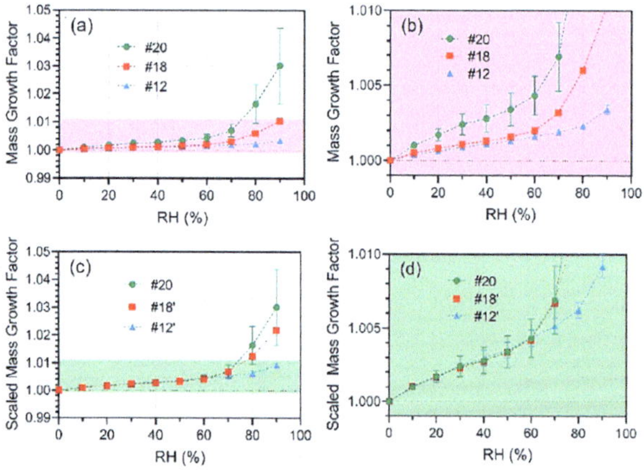

Figure 7. (A) Mass growth factor (m/m_0) as a function of RH. (B) Zoomed-in, (C) scaled, and (D) scaled and zoomed-in mass growth factors. Sample #18 is scaled by a factor of 2.1 and sample #12 is multiplied by 2.7 to overlap with sample #20.

In Figures 7C,D the hygroscopic curves are scaled, i.e., sample #18 is scaled by a factor of 2.1 and sample #12 is scaled by a factor of 2.7. The good overlap below RH = 60% indicates similar hygroscopicity in this RH range, although the absolute values vary, which is due to the variation in hygroscopic contents. Based on the dissolvable fractions, sample #20 has a higher soluble content than sample #18 by a factor of 3.5, and higher than sample #12 by 6.8 times. In other words, the variations in soluble contents are higher than the variations in mass growth factors, meaning that the hygroscopic fractions in the three samples are different. Such differences are reflected above RH = 60%, where the three scaled hygroscopic curves diverge. Samples #20 and #18 are more hygroscopic than sample #12, and the hygroscopicity is stronger at higher RH.

Thermodynamic Model

The AIOMFAC thermodynamic model was used to formulate predictions for the hygroscopic behavior based on chemical compositions. In Figure 8 modeled Ion Activity Products (IAP) are presented as a function of RH for three sand samples based on their ionic components, and complete results of all samples can be found in Supplementary Figure S1 in the Supplementary Information (SI). The corresponding RHs where the IAP curves intersect with the horizontal dashed lines (IAP of saturated solutions; IAP_{sat}) are the predicted DRHs of individual salts in the salt mixtures.

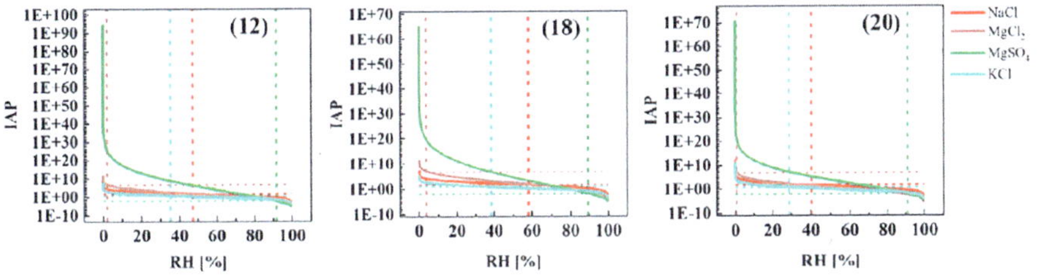

Figure 8. AIOMFAC modeled IAP as a function of RH. The horizontal dashed lines indicate the IAP_{sat} of saturated solutions of four common salts (NaCl, $MgCl_2$, $MgSO_4$, KCl). The vertical dashed lines are the model predicted DRH for individual salts in the corresponding salt mixtures.

The DRHs of four common deliquescing salts ($MgCl_2$, KCl, NaCl and $MgSO_4$) predicted by the AIOMFAC model are 0–5%, 30–40%, 40–60%, ≈90%, respectively. Among these four components $MgCl_2$ has the lowest DRH value, which determines the earliest bulk water dissolution. NaCl and $MgSO_4$ are major components in most samples, and thus these two components strongly influence the mass growth of the mixtures. The results predicted by the thermodynamic model are consistent with the results of the hygroscopicity experiments, showing that the observed hygroscopicity is thermodynamically well described. The PMF 2-factor solution based on the DRHs predicted by the AIOMFAC model can be found in Supplementary Figure S2 in the Supplementary Information.

Individual Particle Chemical Analysis

Individual particles were chemically mapped using the STXM technique. Particle images are shown in Figure 3 and a zoomed-in view in Figure 9A. Figure 9A shows the two typically observed particle morphologies, needle-like and more cubic or polygonal. We observe distinct chemical characteristics of the two morphologies, as illustrated in Figures 9B–D, which include the carbon K-edge and potassium L-edge spectra (Figure 9B), the calcium L-edge spectra (Figure 9C) and the oxygen K-edge spectra (Figure 9D).

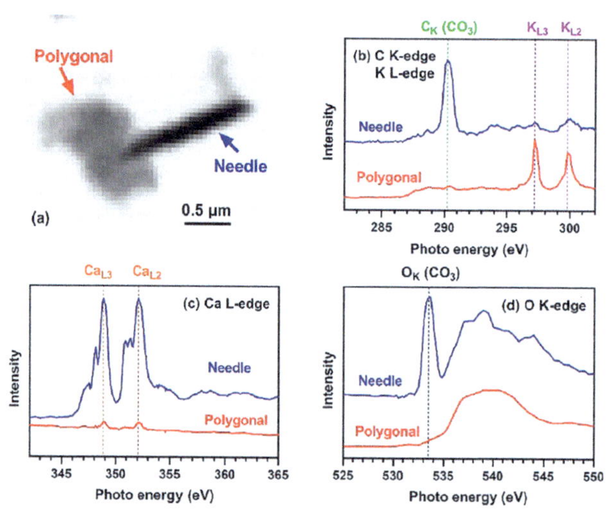

Figure 9. STXM image and X-ray spectra of representative particles. (A) Optical image, (B) spectra scanned at C K-edge and K L-edge, (C) Ca L-edge spectra, (D) O K-edge spectra. Blue lines and red lines represent the mean spectra for needle-like particles and polygonal particles, respectively.

The carbon K-edge includes the presence of a carbonate peak (at 290.2 eV) for needle particles that is minimally present for the polygonal particles (Benzerara et al., 2004). Given the region around 284–290 eV is flat, these particles cannot include significant black or organic carbon components. At the potassium L-edge, a doublet at 297.2 and 299.8 eV indicates the presence of potassium in the polygonal particle (Moffet et al., 2013). In Figure 9C, the calcium L-edge shows the presence of calcium (at 348.8 and 352.1 eV) in the needle, suggesting the needle is likely $CaCO_3$ (Benzerara et al., 2004). At the oxygen K-edge (Figure 9D), one sharp peak is observed at 533.6 eV in the needle, which likely corresponds to the oxygen in the carbonate ions, which further supports the supposition that the needle-like particles are dominated by CO_3 and Ca. The broad components in the oxygen K-edge spectra are associated with other oxygen species. These results show the mixing state of the $CaCO_3$ and K/O salts in sand samples, and these two components most likely have different solubilities, i.e., $CaCO_3$ has low solubility in water but the K/O salts may be highly dissolvable. This indicates that IC measurements can only detect low level Ca^{2+}, though there are abundant $CaCO_3$ needles in the sand.

Ice Nucleation

Four samples with higher salt content (#13, #17, #20 and #21, for chemical compositions see Section 4.1) were tested using PINCii for their ice nucleation ability, and results are shown in Figure 10. The four samples were used to produce salt solutions, which were atomized and dried before being introduced into PINCii. These samples were taken from locations close to samples (#12, #18 and #20) that were studied for hygroscopicity (Section 4.2). In Figures 10A–D, the concentrations of activated ice particles are plotted against RH_{ice} at different temperatures. During experiments, the RH was increasing isothermally. In all subplots the black solid line represents water saturation, and the dashed line is the Koop *et al.* (Koop et al., 2000) prediction for homogeneous freezing of a solution (nucleation rate; $J = 1.7 \times 10^4$). The heat map of concentration is scaled to make the ice nucleation onset stand out at the transition to green. Onset values are quantified using a cumulative sum algorithm for particle counts in the selected size channel for ice (>3 μm) and are replotted in Figure 10E, which summarizes the critical RH_{ice} for nucleation onsets. For most temperatures and samples, the observed nucleation matches what is predicted for homogeneous freezing. However, at the lowest temperature (−50°C) ice nucleation activation is also observed well below the homogeneous freezing line, which strongly suggests that some of these samples initiate deposition nucleation, which may occur at RHs between ice saturation and homogeneous freezing (Kong et al., 2018; Wagner et al., 2018).

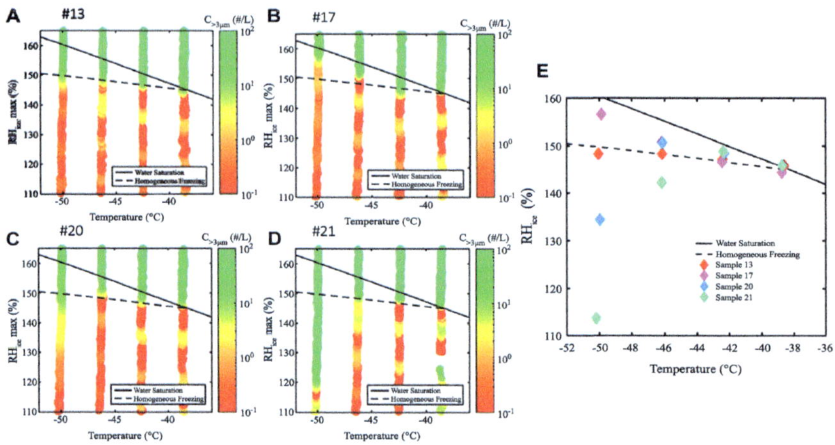

Figure 10. (A–D) Particle concentrations observed in the OPC >3 μm size channel, indicative of ice active particles, plotted against maximum lamina supersaturations with respect to RH_{ice} at different temperatures. During experiments, the RH was increased isothermally at 1% per minute. The heat map is used to emphasize ice nucleation onset—indicated by the transition to green. (E) Critical RH_{ice} of ice nucleation onset. Water saturation is represented by the solid black line, while the dashed line represents homogeneous freezing of a solution ($J = 1.7 \times 10^4$) as given by Koop et al. (Koop et al., 2000).

Deposition nucleation primarily occurs on solid surfaces or within micropores (Marcolli, 2014; David et al., 2019), but these salt mixtures are very hygroscopic and start to take up water at low RH, as shown in Figure 7. This raises the question as to whether such salt mixture particles can serve as both CCN and INP. The parallel CCN and INP behaviors have been observed on sea salt particles in previous studies (Kong et al., 2018; Wagner et al., 2018). However, in previous studies, deposition nucleation was primarily observed below the DRH. In this study, the salt mixtures may also contain small less-soluble components, even though they are observed to be very hygroscopic. Therefore, the specific behavior may reflect how well the less-soluble components facilitate ice nucleation, and simultaneously how much water uptake is promoted. The balance between water uptake, solvation, and ice nucleation is an area of active investigation in the cloud micro-physics community (Hoose and Möhler, 2012; DeMott et al., 2015; Brasseur et al., 2022). There may also be other important properties of the solvating salt surfaces, such as surface-promoted redox chemistry, that result in important physicochemical phenomena (Kong et al., 2021).

CONCLUSION

Dust/salt mixture samples were collected from important source areas of East Asian Dust Storms, including the Badain Jaran Desert, the Tengger Desert and the Ulan Buh Desert in northwestern China.

The concentrations of cations and anions were determined by IC measurements. The ionic concentrations, pH and dissolvable fractions of sand samples are positively correlated, indicating that the dissolved content is rich in alkaline ions.

A PMF receptor model was implemented to analyze the IC results and the PMF solution seemingly connects with the location of sampling sites, i.e., whether there are or were aqueous environments nearby. The hygroscopic properties of the samples were measured by a vapor sorption analyzer, and a thermodynamic model was used to well describe the DRH based on chemical compositions. Synchrotron-based STXM was carried out to examine the sample morphology and chemically map typical dust/salt particles. Needle-shaped $CaCO_3$ particles are observed as are more irregularly shaped high K-containing particles.

Ice nucleation activity was measured for typical salts, with both homogeneous freezing and deposition nucleation observed from atomized and dried salt solutions/particles. The results indicate that the salts primarily act as cloud condensation nuclei that freeze at low temperatures but can also act as ice nucleating particles at lower vapor saturations and at low temperatures. We hypothesize that this is due the co-existence of high- and low-hygroscopic materials within the salt mixtures.

Data Availability Statement

The raw data supporting the conclusion of this article will be made available by the authors, without undue reservation.

Author Contributions

XK and SW contributed to the conception and the design of the study. JL, WL, and GZ carried out the sampling campaign. JL, WL, DC, WG, and TO conducted the experiments. JL, WL, DC, WG, and LL analyzed the data. MT, ET, MH, SW, and XK supervised the project. JL, LL, and XK wrote the first draft of the manuscript. All authors contributed to manuscript revision, read, and approved the submitted version.

Funding

This work is supported by the National Natural Science Foundation of China (41975160) and the Swedish Research Council (2021-04042). It is a contribution to the Swedish Strategic Research Area Modelling the Regional and Global Earth System (MERGE). EST and DC have been supported by the Swedish Research Councils, FORMAS (2017-00564) and VR (2020-03497).

Conflict of Interest

The authors declare that the research was conducted in the absence of any commercial or financial relationships that could be construed as a potential conflict of interest.

Publisher's Note

All claims expressed in this article are solely those of the authors and do not necessarily represent those of their affiliated organizations, or those of the publisher, the editors and the reviewers. Any product that may be evaluated in this article, or claim that may be made by its manufacturer, is not guaranteed or endorsed by the publisher.

ACKNOWLEDGMENTS

A part of this work was performed at the BL4U of UVSOR Synchrotron Facility, Institute for Molecular Science (IMS program 21-840).

SUPPLEMENTARY MATERIAL

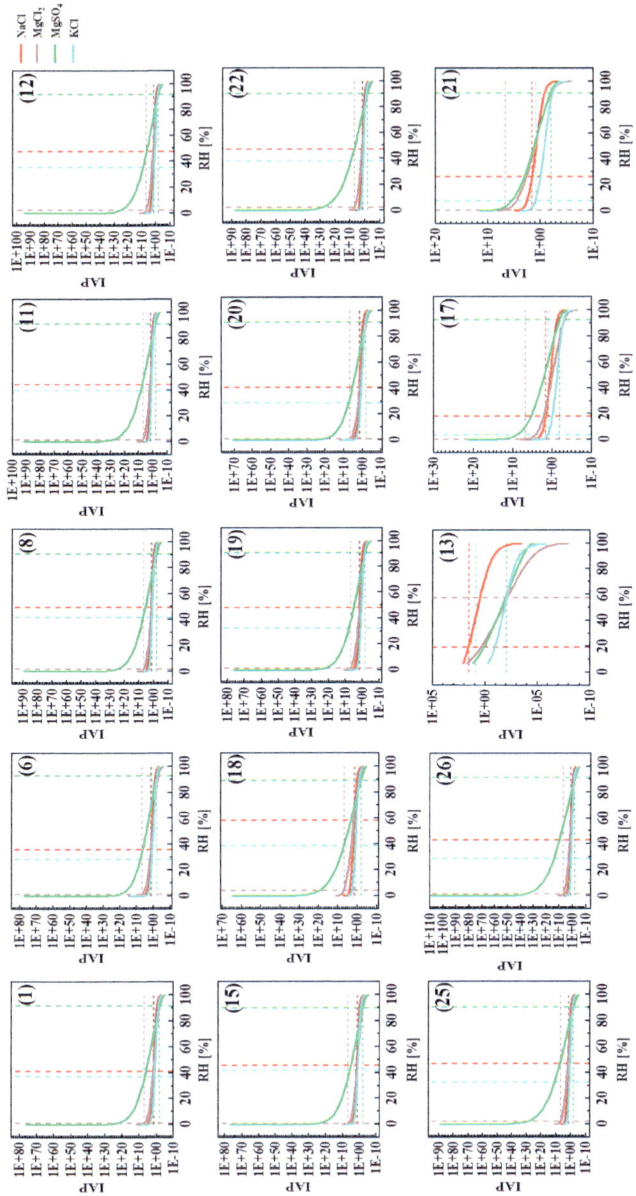

Figure S1. AIOMFAC modeled IAP as a function of RH. The horizontal dashed lines indicate the IAPsat of saturated solutions of four common salts (NaCl, $MgCl_2$, $MgSO_4$, KCl) marked on the y-axis. The vertical dashed lines point to the DRH simulated by the AIOMFAC model on the x-axis.

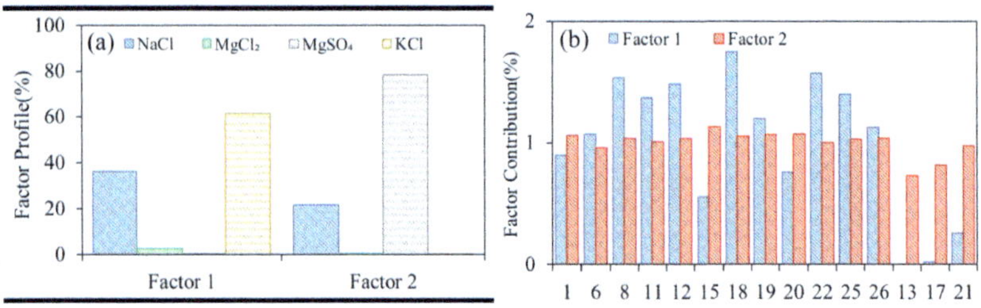

Figure S2. PMF 2-factor solution based on the DRHs predicted by the model. (a) Factor profile and (b) factor contribution of all the 15 samples.

REFERENCES

1. An, L., Che, H., Xue, M., Zhang, T., Wang, H., Wang, Y., et al. (2018). Temporal and Spatial Variations in Sand and Dust Storm Events in East Asia from 2007 to 2016: Relationships with Surface Conditions and Climate Change. *Sci. Total Environ.* 633, 452–462. doi:10.1016/j.scitotenv.2018.03.068
2. Barkan, J., Alpert, P., Kutiel, H., and Kishcha, P. (2005). Synoptics of Dust Transportation Days from Africa toward Italy and Central Europe. *J. Geophys. Res.-Atmos.* 110, D07208. doi:10.1029/2004jd005222
3. Benzerara, K., Yoon, T. H., Tyliszczak, T., Constantz, B., Spormann, A. M., and Brown, G. E. (2004). Scanning Transmission X-Ray Microscopy Study of Microbial Calcification. *Geobiology* 2, 249–259. doi:10.1111/j.1472-4677.2004.00039.x
4. Bogen, J., Xu, M., and Kennie, P. (2015). The Impact of Pro-glacial Lakes on Downstream Sediment Delivery in Norway. *Earth Surf. Proc. Land.* 40, 942–952. doi:10.1002/esp.3669
5. Bou Karam, D., Williams, E., Janiga, M., Flamant, C., McGraw-Herdeg, M., Cuesta, J., et al. (2014). Synoptic-scale Dust Emissions over the Sahara Desert Initiated by a Moist Convective Cold Pool in Early August 2006. *Q.J.R. Meteorol. Soc.* 140, 2591–2607. doi:10.1002/qj.2326
6. Brasseur, Z., Castarède, D., Thomson, E. S., Adams, M. P., Drossaart van Dusseldorp, S., Heikkilä, P., et al. (2022). Measurement Report: Introduction to the HyICE-2018 Campaign for Measurements of Ice-Nucleating Particles and Instrument Inter-comparison in the Hyytiälä Boreal Forest. *Atmos. Chem. Phys.* 22, 5117–5145. doi:10.5194/acp-22-5117-2022
7. Cao, Y.-Z., Wang, S., Zhang, G., Luo, J., and Lu, S. (2009). Chemical Characteristics of Wet Precipitation at an Urban Site of Guangzhou, South China. *Atmos. Res.* 94, 462–469. doi:10.1016/j.atmosres.2009.07.004
8. Castarède, D. (2021). *Water Condensation and Freezing in the Atmosphere: Exploring Deliquescence and Ice Nucleation.* Gothenburg: Department of Chemistry and Molecular Biology, University of Gothenburg.
9. Che, H., Xia, X., Zhao, H., Dubovik, O., Holben, B. N., Goloub, P., et al. (2019). Spatial Distribution of Aerosol Microphysical and

Optical Properties and Direct Radiative Effect from the China Aerosol Remote Sensing Network. *Atmos. Chem. Phys.* 19, 11843–11864. doi:10.5194/acp-19-11843-2019

10. Chen, S., Huang, J., Kang, L., Wang, H., Ma, X., He, Y., et al. (2017). Emission, Transport, and Radiative Effects of Mineral Dust from the Taklimakan and Gobi Deserts: Comparison of Measurements and Model Results. *Atmos. Chem. Phys.* 17, 2401–2421. doi:10.5194/acp-17-2401-2017
11. Chun, X., Chen, F., Fan, Y., Xia, D., and Zhao, H. (2009). Evidence of Palaeolake Existence in Ulan Buh Desert and its Environmental Evolution (In Chinese). *Acta Geogr. Sin.* 64, 339–348. doi:10.11821/xb200903009
12. Costa, M. J., Sohn, B.-J., Levizzani, V., and Silva, A. M. (2006). Radiative Forcing of Asian Dust Determined from the Synergized GOME and GMS Satellite Data-A Case Study. *J. Meteorological Soc. Jpn.* 84, 85–95. doi:10.2151/jmsj.84.85
13. Creamean, J. M., Suski, K. J., Rosenfeld, D., Cazorla, A., DeMott, P. J., Sullivan, R. C., et al. (2013). Dust and Biological Aerosols from the Sahara and Asia Influence Precipitation in the Western U.S. *Science* 339, 1572–1578. doi:10.1126/science.1227279
14. David, R. O., Marcolli, C., Fahrni, J., Qiu, Y., Perez Sirkin, Y. A., Molinero, V., et al. (2019). Pore Condensation and Freezing Is Responsible for Ice Formation below Water Saturation for Porous Particles. *Proc. Natl. Acad. Sci. U.S.A.* 116, 8184–8189. doi:10.1073/pnas.1813647116
15. DeMott, P. J., Prenni, A. J., McMeeking, G. R., Sullivan, R. C., Petters, M. D., Tobo, Y., et al. (2015). Integrating Laboratory and Field Data to Quantify the Immersion Freezing Ice Nucleation Activity of Mineral Dust Particles. *Atmos. Chem. Phys.* 15, 393–409. doi:10.5194/acp-15-393-2015
16. Dong, Z., Qian, G., Lv, P., and Hu, G. (2013). Investigation of the Sand Sea with the Tallest Dunes on Earth: China's Badain Jaran Sand Sea. *Earth-Science Rev.* 120, 20–39. doi:10.1016/j.earscirev.2013.02.003
17. Duce, R. A., Unni, C. K., Ray, B. J., Prospero, J. M., and Merrill, J. T. (1980). Long-range Atmospheric Transport of Soil Dust from Asia to the Tropical North Pacific: Temporal Variability. *Science* 209, 1522–1524. doi:10.1126/science.209.4464.1522

18. Engelstaedter, S., Tegen, I., and Washington, R. (2006). North African Dust Emissions and Transport. *Earth-Science Rev.* 79, 73–100. doi:10.1016/j.earscirev.2006.06.004
19. Fu, P., Huang, J., Li, C., and Zhong, S. (2008). The Properties of Dust Aerosol and Reducing Tendency of the Dust Storms in Northwest China. *Atmos. Environ.* 42, 5896–5904. doi:10.1016/j.atmosenv.2008.03.041
20. Gao, Y. H., Li, Z. L., Zhu, R. X., and Liao, H. (2020). Stable Isotope Compositions, Sources and Paleoenvironmental Significance of Holocene Calcareous Root Tubes in the Tengger Desert, Northwest China. *Catena* 195, 104846. doi:10.1016/j.catena.2020.104846
21. Gu, W., Li, Y., Zhu, J., Jia, X., Lin, Q., Zhang, G., et al. (2017). Investigation of Water Adsorption and Hygroscopicity of Atmospherically Relevant Particles Using a Commercial Vapor Sorption Analyzer. *Atmos. Meas. Tech.* 10, 3821–3832. doi:10.5194/amt-10-3821-2017
22. Gui, K., Che, H., Chen, Q., Zeng, Z., Zheng, Y., Long, Q., et al. (2017). Water Vapor Variation and the Effect of Aerosols in China. *Atmos. Environ.* 165, 322–335. doi:10.1016/j.atmosenv.2017.07.005
23. Guo, L., Gu, W., Peng, C., Wang, W., Li, Y. J., Zong, T., et al. (2019). A Comprehensive Study of Hygroscopic Properties of Calcium- and Magnesium-Containing Salts: Implication for Hygroscopicity of Mineral Dust and Sea Salt Aerosols. *Atmos. Chem. Phys.* 19, 2115–2133. doi:10.5194/acp-19-2115-2019
24. He, M., Ichinose, T., Song, Y., Yoshida, Y., Bekki, K., Arashidani, K., et al. (2016). Desert Dust Induces TLR Signaling to Trigger Th2-Dominant Lung Allergic Inflammation *via* a MyD88-dependent Signaling Pathway. *Toxicol. Appl. Pharmacol.* 296, 61–72. doi:10.1016/j.taap.2016.02.011
25. Hoose, C., and Möhler, O. (2012). Heterogeneous Ice Nucleation on Atmospheric Aerosols: A Review of Results from Laboratory Experiments. *Atmos. Chem. Phys.* 12, 9817–9854. doi:10.5194/acp-12-9817-2012
26. Hu, F., and Yang, X. (2016). Geochemical and Geomorphological Evidence for the Provenance of Aeolian Deposits in the Badain Jaran Desert, Northwestern China. *Quat. Sci. Rev.* 131, 179–192. doi:10.1016/j.quascirev.2015.10.039

27. Huang, J. F., Zhang, C. D., and Prospero, J. M. (2010). African Dust Outbreaks: A Satellite Perspective of Temporal and Spatial Variability over the Tropical Atlantic Ocean. *J. Geophys. Res.-Atmos.* 115, D05202. doi:10.1029/2009jd012516
28. Huang, J. P., Minnis, P., Lin, B., Wang, T. H., Yi, Y. H., Hu, Y. X., et al. (2006). Possible Influences of Asian Dust Aerosols on Cloud Properties and Radiative Forcing Observed from MODIS and CERES. *Geophys. Res. Lett.* 33, L06824. doi:10.1029/2005gl024724
29. Huang, J. P., Wang, T. H., Wang, W. C., Li, Z. Q., and Yan, H. R. (2014). Climate Effects of Dust Aerosols over East Asian Arid and Semiarid Regions. *J. Geophys. Res.-Atmos.* 119, 11398–11416. doi:10.1002/2014jd021796
30. Huang, S., Hu, W., Chen, J., Wu, Z., Zhang, D., and Fu, P. (2021). Overview of Biological Ice Nucleating Particles in the Atmosphere. *Environ. Int.* 146, 106197. doi:10.1016/j.envint.2020.106197
31. Huang, Z. W., Huang, J. P., Hayasaka, T., Wang, S. S., Zhou, T., and Jin, H. C. (2015). Short-cut Transport Path for Asian Dust Directly to the Arctic: A Case Study. *Environ. Res. Lett.* 10, 11. doi:10.1088/1748-9326/10/11/114018
32. Ichinose, T., Yoshida, S., Hiyoshi, K., Sadakane, K., Takano, H., Nishikawa, M., et al. (2008). The Effects of Microbial Materials Adhered to Asian Sand Dust on Allergic Lung Inflammation. *Arch. Environ. Contam. Toxicol.* 55, 348–357. doi:10.1007/s00244-007-9128-8
33. Ji, F., Wu, Z., Huang, J., and Chassignet, E. P. (2014). Evolution of Land Surface Air Temperature Trend. *Nat. Clim. Change* 4, 462–466. doi:10.1038/nclimate2223
34. Jia, P., Wang, N. a., Cheng, H., Li, Z., Ning, K., Zhang, X., et al. (2015). A Study on the Range and Area of Ulan Buh Desert Based on 3S Technology (In Chinese). *J. Arid Land Resour. Environ.* 29, 131–138. doi:10.13448/j.cnki.jalre.2015.413
35. Kanayama, S., Yabuki, S., Yanagisawa, F., and Motoyama, R. (2002). The Chemical and Strontium Isotope Composition of Atmospheric Aerosols over Japan: The Contribution of Long-Range-Transported Asian Dust (Kosa). *Atmos. Environ.* 36, 5159–5175. doi:10.1016/s1352-2310(02)00587-3

36. Karydis, V. A., Tsimpidi, A. P., Bacer, S., Pozzer, A., Nenes, A., and Lelieveld, J. (2017). Global Impact of Mineral Dust on Cloud Droplet Number Concentration. *Atmos. Chem. Phys.* 17, 5601–5621. doi:10.5194/acp-17-5601-2017
37. Kelly, J. T., Chuang, C. C., and Wexler, A. S. (2007). Influence of Dust Composition on Cloud Droplet Formation. *Atmos. Environ.* 41, 2904–2916. doi:10.1016/j.atmosenv.2006.12.008
38. Kilcoyne, A. L. D., Tyliszczak, T., Steele, W. F., Fakra, S., Hitchcock, P., Franck, K., et al. (2003). Interferometer-controlled Scanning Transmission X-Ray Microscopes at the Advanced Light Source. *J. Synchrotron Radiat.* 10, 125–136. doi:10.1107/s0909049502017739
39. Kok, J. F., Ward, D. S., Mahowald, N. M., and Evan, A. T. (2018). Global and Regional Importance of the Direct Dust-Climate Feedback. *Nat. Commun.* 9, 241. doi:10.1038/s41467-017-02620-y
40. Kok, J. F., Adebiyi, A. A., Albani, S., Balkanski, Y., Checa-Garcia, R., Chin, M., et al. (2021). Contribution of the World's Main Dust Source Regions to the Global Cycle of Desert Dust. *Atmos. Chem. Phys.* 21, 8169–8193. doi:10.5194/acp-21-8169-2021
41. Kong, X., Castarède, D., Thomson, E. S., Boucly, A., Artiglia, L., Ammann, M., et al. (2021). A Surface-Promoted Redox Reaction Occurs Spontaneously on Solvating Inorganic Aerosol Surfaces. *Science* 374, 747–752. doi:10.1126/science.abc5311
42. Kong, X., Wolf, M. J., Roesch, M., Thomson, E. S., Bartels-Rausch, T., Alpert, P. A., et al. (2018). A Continuous Flow Diffusion Chamber Study of Sea Salt Particles Acting as Cloud Nuclei: Deliquescence and Ice Nucleation. *Tellus B Chem. Phys. Meteorology* 70, 1–11. doi:10.1080/16000889.2018.1463806
43. Koop, T., Kapilashrami, A., Molina, L. T., and Molina, M. J. (2000). Phase Transitions of Sea-Salt/water Mixtures at Low Temperatures: Implications for Ozone Chemistry in the Polar Marine Boundary Layer. *J. Geophys. Res.* 105, 26393–26402. doi:10.1029/2000jd900413
44. Krueger, B. J., Grassian, V. H., Cowin, J. P., and Laskin, A. (2004). Heterogeneous Chemistry of Individual Mineral Dust Particles from Different Dust Source Regions: the Importance of Particle Mineralogy. *Atmos. Environ.* 38, 6253–6261. doi:10.1016/j.atmosenv.2004.07.010

45. Kumar, P., Sokolik, I. N., and Nenes, A. (2011). Measurements of Cloud Condensation Nuclei Activity and Droplet Activation Kinetics of Fresh Unprocessed Regional Dust Samples and Minerals. *Atmos. Chem. Phys.* 11, 3527–3541. doi:10.5194/acp-11-3527-2011
46. Li, C., Krotkov, N. A., Dickerson, R. R., Li, Z. Q., Yang, K., and Chin, M. (2010). Transport and Evolution of a Pollution Plume from Northern China: A Satellite-Based Case Study. *J. Geophys. Res.-Atmos.* 115, D00K03. doi:10.1029/2009jd012245
47. Li, Q., Fan, Q., Wei, H., Qin, Z., Zhang, X., Du, Y., et al. (2020). Sulfur Isotope Constraints on the Formation of MgSO4-Deficient Evaporites in the Qarhan Salt Lake, Western China. *J. Asian Earth Sci.* 189, 104160. doi:10.1016/j.jseaes.2019.104160
48. Li, Z., Wang, N. a., Cheng, H., Ning, K., Zhao, L., and Li, R. (2015). Formation and Environmental Significance of Late Quaternary Calcareous Root Tubes in the Deserts of the Alashan Plateau, Northwest China. *Quat. Int.* 372, 167–174. doi:10.1016/j.quaint.2014.11.021
49. Li, Z., Wei, Z., Dong, S., and Chen, Q. (2018). The Paleoenvironmental Significance of Spatial Distributions of Grain Size in Groundwater-Recharged Lakes: A Case Study in the Hinterland of the Badain Jaran Desert, Northwest China. *Earth Surf. Process. Landforms* 43, 363–372. doi:10.1002/esp.4248
50. Liu, C., Zhang, J., Jiao, P., and Mischke, S. (2016). The Holocene History of Lop Nur and its Palaeoclimate Implications. *Quat. Sci. Rev.* 148, 163–175. doi:10.1016/j.quascirev.2016.07.016
51. Liu, Y., Wang, G. P., Hu, Z. Y., Shi, P. J., Lyu, Y. L., Zhang, G. M., et al. (2020). Dust Storm Susceptibility on Different Land Surface Types in Arid and Semiarid Regions of Northern China. *Atmos. Res.* 243, 105031. doi:10.1016/j.atmosres.2020.105031
52. Ma, N., Wang, N., Li, Z., Chen, X., Zhu, J., and Dong, C. (2011). Analysis on Climate Change in the Northern and Southern Marginal Zones of the Badain Juran Desert during the Period 1960–2009 (In Chinese). *Arid Zone Res.* 28, 242–250. doi:10.13866/j.azr.2011.02.020
53. Ma, N., Wang, N., Zhao, L., Zhang, Z., Dong, C., and Shen, S. (2014). Observation of Mega-Dune Evaporation after Various Rain Events in the Hinterland of Badain Jaran Desert, China. *Chin. Sci. Bull.* 59, 162–170. doi:10.1007/s11434-013-0050-3
54. Mahowald, N., Albani, S., Kok, J. F., Engelstaeder, S., Scanza, R., Ward, D. S., et al. (2014). The Size Distribution of Desert Dust Aerosols and

its Impact on the Earth System. *Aeolian Res.* 15, 53–71. doi:10.1016/j.aeolia.2013.09.002

55. Marcolli, C. (2014). Deposition Nucleation Viewed as Homogeneous or Immersion Freezing in Pores and Cavities. *Atmos. Chem. Phys.* 14, 2071–2104. doi:10.5194/acp-14-2071-2014
56. Moffet, R. C., Rödel, T. C., Kelly, S. T., Yu, X. Y., Carroll, G. T., Fast, J., et al. (2013). Spectro-microscopic Measurements of Carbonaceous Aerosol Aging in Central California. *Atmos. Chem. Phys.* 13, 10445–10459. doi:10.5194/acp-13-10445-2013
57. Nee, J. B., Chiang, C. W., Hu, H. L., Hu, S. X., and Yu, J. Y. (2007). Lidar Measurements of Asian Dust Storms and Dust Cloud Interactions. *J. Geophys. Res.-Atmos.* 112, D15202. doi:10.1029/2007jd008476
58. Norris, G., Duvall, R., Brown, S., and Bai, S. (2014). *EPA Positive Matrix Factorization (PMF) 5.0 Fundamentals and User Guide.* Washington, DC: US Environmental Protection Agency.
59. Ohigashi, T., Arai, H., Araki, T., Kondo, N., Shigemasa, E., Ito, A., et al. (2013). Construction of the Scanning Transmission X-Ray Microscope Beamline at UVSOR. *J. Phys. Conf. Ser.* 463, 012006. doi:10.1088/1742-6596/463/1/012006
60. Okin, G. S. (2005). Dependence of Wind Erosion and Dust Emission on Surface Heterogeneity: Stochastic Modeling. *J. Geophys. Res.-Atmos.* 110, 0148–0227. doi:10.1029/2004jd005288
61. Prospero, J. M., Collard, F.-X., Molinié, J., and Jeannot, A. (2014). Characterizing the Annual Cycle of African Dust Transport to the Caribbean Basin and South America and its Impact on the Environment and Air Quality. *Glob. Biogeochem. Cycles* 28, 757–773. doi:10.1002/2013gb004802
62. Qu, J., Chang, X., Dong, G., Wang, X., Lu, J., and Zhong, D. (2003). Fractal Behavior of Aeolian Sand Landform in Typical Megadune Area of Badin Jaran Desert (In Chinese). *J. Desert Res.* 23, 361–365.
63. Rashki, A., Middleton, N. J., and Goudie, A. S. (2021). Dust Storms in Iran - Distribution, Causes, Frequencies and Impacts. *Aeolian Res.* 48, 100655. doi:10.1016/j.aeolia.2020.100655
64. Rosenfeld, D., Rudich, Y., and Lahav, R. (2001). Desert Dust Suppressing Precipitation: A Possible Desertification Feedback Loop. *Proc. Natl. Acad. Sci. U.S.A.* 98, 5975–5980. doi:10.1073/pnas.101122798

65. Rudich, Y., Khersonsky, O., and Rosenfeld, D. (2002). Treating Clouds with a Grain of Salt. *Geophys. Res. Lett.* 29, 17/1–17/4. doi:10.1029/2002gl016055

66. Sassen, K. (2002). Indirect Climate Forcing over the Western US from Asian Dust Storms. *Geophys. Res. Lett.* 29, 136–142. doi:10.1029/2001gl014051

67. Schillereff, D. N., Chiverrell, R. C., Macdonald, N., and Hooke, J. M. (2016). Hydrological Thresholds and Basin Control over Paleoflood Records in Lakes. *Geology* 44, 43–46. doi:10.1130/g37261.1

68. Su, X., Xu, L., Wei, Q., Zhang, Z., Wang, S., and Wu, H. (2016). Study of Impacts of Dust Aerosol on Precipitation over East Asia (In Chinese). *Plateau Meteorol.* 35, 211–219. doi:10.7522/j.issn.1000-0534.2014.00091

69. Tang, M., Cziczo, D. J., and Grassian, V. H. (2016). Interactions of Water with Mineral Dust Aerosol: Water Adsorption, Hygroscopicity, Cloud Condensation, and Ice Nucleation. *Chem. Rev.* 116, 4205–4259. doi:10.1021/acs.chemrev.5b00529

70. Tang, M., Gu, W., Ma, Q., Li, Y. J., Zhong, C., Li, S., et al. (2019). Water Adsorption and Hygroscopic Growth of Six Anemophilous Pollen Species: The Effect of Temperature. *Atmos. Chem. Phys.* 19, 2247–2258. doi:10.5194/acp-19-2247-2019

71. Tobo, Y., Zhang, D., Matsuki, A., and Iwasaka, Y. (2010). Asian Dust Particles Converted into Aqueous Droplets under Remote Marine Atmospheric Conditions. *Proc. Natl. Acad. Sci. U.S.A.* 107, 17905–17910. doi:10.1073/pnas.1008235107

72. Uno, I., Eguchi, K., Yumimoto, K., Takemura, T., Shimizu, A., Uematsu, M., et al. (2009). Asian Dust Transported One Full Circuit Around the Globe. *Nat. Geosci.* 2, 557–560. doi:10.1038/ngeo583

73. Venero-Fernández, S. J. (2016). Saharan Dust Effects on Human Health: A Challenge for Cuba's Researchers. *Medicc Rev.* 18, 32–34. doi:10.37757/MR2016.V18.N3.8

74. Wagner, R., Kaufmann, J., Möhler, O., Saathoff, H., Schnaiter, M., Ullrich, R., et al. (2018). Heterogeneous Ice Nucleation Ability of NaCl and Sea Salt Aerosol Particles at Cirrus Temperatures. *J. Geophys. Res. Atmos.* 123, 2841–2860. doi:10.1002/2017jd027864

75. Wang, B., Liu, X., and Liu, L. (2020). Characteristics of Precipitation in the Surrounding Area of Tengger Desert in 1957–2017 (In Chinese). *J. Desert Res.* 40, 163–170. doi:10.7522/j.issn.1000-694X.2020.00047

76. Wang, N. a., Li, Z., Li, Y., and Cheng, H. (2013). Millennial-scale Environmental Changes in the Asian Monsoon Margin during the Holocene, Implicated by the Lake Evolution of Huahai Lake in the Hexi Corridor of Northwest China. *Quat. Int.* 313-314, 100–109. doi:10.1016/j.quaint.2013.08.039
77. Wang, W. X., and Xu, P. J. (2009). Research Progress in Precipitation Chemistry in China. *Prog. Chem.* 21, 266–281.
78. Wang, X., Huang, J., Ji, M., and Higuchi, K. (2008). Variability of East Asia Dust Events and Their Long-Term Trend. *Atmos. Environ.* 42, 3156–3165. doi:10.1016/j.atmosenv.2007.07.046
79. Wu, C., Zhang, S., Wang, G., Lv, S., Li, D., Liu, L., et al. (2020). Efficient Heterogeneous Formation of Ammonium Nitrate on the Saline Mineral Particle Surface in the Atmosphere of East Asia during Dust Storm Periods. *Environ. Sci. Technol.* 54, 15622–15630. doi:10.1021/acs.est.0c04544
80. Xu, M., and Li, Z. (2016). Accumulated Temperature Changes in Desert Region and Surrounding Area during 1960–2013: A Case Study in the Alxa Plateau, Northwest China. *Environ. Earth. Sci.* 75, 1276. doi:10.1007/s12665-016-6084-5
81. Yin, Y., Wurzler, S., Levin, Z., and Reisin, T. G. (2002). Interactions of Mineral Dust Particles and Clouds: Effects on Precipitation and Cloud Optical Properties. *J. Geophys. Res.-Atmos.* 107, AAC 19/1-19/14. doi:10.1029/2001jd001544
82. You, Q., Cai, Z., Pepin, N., Chen, D., Ahrens, B., Jiang, Z., et al. (2021). Warming Amplification over the Arctic Pole and Third Pole: Trends, Mechanisms and Consequences. *Earth-Science Rev.* 217, 103625. doi:10.1016/j.earscirev.2021.103625
83. Yuan, T. G., Huang, J. P., Cao, J. H., Zhang, G. L., and Ma, X. J. (2021). Indian Dust-Rain Storm: Possible Influences of Dust Ice Nuclei on Deep Convective Clouds. *Sci. Total Environ.* 779, 146439. doi:10.1016/j.scitotenv.2021.146439
84. Zhang, C., Liu, C., Hu, Q., Cai, Z., Su, W., Xia, C., et al. (2019). Satellite UV-Vis Spectroscopy: Implications for Air Quality Trends and Their Driving Forces in China during 2005-2017. *Light Sci. Appl.* 8, 100. doi:10.1038/s41377-019-0210-6
85. Zhang, D., Liu, D., Luo, T., Wang, Z., and Yin, Y. (2015). Aerosol Impacts on Cloud Thermodynamic Phase Change over East Asia

Observed with CALIPSO and CloudSat Measurements. *J. Geophys. Res. Atmos.* 120, 1490–1501. doi:10.1002/2014jd022630

86. Zhang, H. C., Ma, Y. Z., Wünnemann, B., and Pachur, H.-J. (2000). A Holocene Climatic Record from Arid Northwestern China. *Palaeogeogr. Palaeoclimatol. Palaeoecol.* 162, 389–401. doi:10.1016/s0031-0182(00)00139-5
87. Zhu, B., and Yang, X. (2010). The Origin and Distribution of Soluble Salts in the Sand Seas of Northern China. *Geomorphology* 123, 232–242. doi:10.1016/j.geomorph.2010.07.001
88. Zhu, J., Wang, N., Chen, H., Dong, C., and Zhang, H. (2010). Study on the Boundary and the Area of Badain Jaran Desert Based on Remote Sensing Imagery (In Chinese). *Prog. Geogr.* 29, 1087–1094. doi:10.3724/SP.J.1084.2010.00199
89. Ziegler, C. L., Murray, R. W., Plank, T., and Hemming, S. R. (2008). Sources of Fe to the Equatorial Pacific Ocean from the Holocene to Miocene. *Earth Planet. Sci. Lett.* 270, 258–270. doi:10.1016/j.epsl.2008.03.044
90. Zuend, A., Marcolli, C., Booth, A. M., Lienhard, D. M., Soonsin, V., Krieger, U. K., et al. (2011). New and Extended Parameterization of the Thermodynamic Model AIOMFAC: Calculation of Activity Coefficients for Organic-Inorganic Mixtures Containing Carboxyl, Hydroxyl, Carbonyl, Ether, Ester, Alkenyl, Alkyl, and Aromatic Functional Groups. *Atmos. Chem. Phys.* 11, 9155–9206. doi:10.5194/acp-11-9155-2011
91. Zuend, A., Marcolli, C., Luo, B. P., and Peter, T. (2008). A Thermodynamic Model of Mixed Organic-Inorganic Aerosols to Predict Activity Coefficients. *Atmos. Chem. Phys.* 8, 4559–4593. doi:10.5194/acp-8-4559-2008

SECTION 5
THERMAL RADIATION

CHAPTER 18

Disquisitions Relating to Principles of Thermodynamic Equilibrium in Climate Modelling

Leslie V. Woodcock

Department of Physics, University of Algarve, 8005-139 Faro, Portugal

ABSTRACT

We revisit the fundamental principles of thermodynamic equilibrium in relation to heat transfer processes within the Earth's atmosphere. A knowledge of equilibrium states at ambient temperatures (T) and pressures (p) and deviations for these p-T states due to various transport 'forces' and flux events give rise to gradients (dT/dz) and (dp/dz) of height z throughout the atmosphere. Fluctuations about these troposphere averages determine weather and climates. Concentric and time-span average values $<T>$ (z, Δt)) and its gradients known as the lapse rate = d $<$ T(z) $>$/dz have hitherto

Citation: Woodcock, Leslie V. 2022. "Disquisitions Relating to Principles of Thermodynamic Equilibrium in Climate Modelling" Entropy 24, no. 4: 459. https://doi.org/10.3390/e24040459.

been assumed in climate models to be determined by a closed, reversible, and adiabatic expansion process against the constant gravitational force of acceleration (g). Thermodynamics tells us nothing about the process mechanisms, but adiabatic-expansion hypothesis is deemed in climate computer models to be convection rather than conduction or radiation. This prevailing climate modelling hypothesis violates the 2nd law of thermodynamics. This idealized hypothetical process cannot be the causal explanation of the experimentally observed mean lapse rate (approx.−6.5 K/km) in the troposphere. Rather, the troposphere lapse rate is primarily determined by the radiation heat-transfer processes between black-body or IR emissivity and IR and sunlight absorption. When the effect of transducer gases (H_2O and CO_2) is added to the Earth's emission radiation balance in a 1D-2level primitive model, a linear lapse rate is obtained. This rigorous result for a perturbing cooling effect of transducer ('greenhouse') gases on an otherwise sunlight-transducer gas-free troposphere has profound implications. One corollary is the conclusion that increasing the concentration of an existing weak transducer, i.e., CO_2, could only have a net cooling effect, if any, on the concentric average <T> ($z = 0$) at sea level and lower troposphere ($z < 1$ km). A more plausible explanation of global warming is the enthalpy emission 'footprint' of all fuels, including nuclear.

Keywords: climate modelling, thermal equilibrium, atmospheric thermodynamics, adiabatic expansion, lapse rate, troposphere, radiation balance

INTRODUCTION

The subject of classical thermodynamics provides a description of the properties of all pure elements and compounds, multicomponent mixtures, and colloidal fluids.

The earth's lower atmosphere, the troposphere, that determines climates is a multicomponent mixture of gases, and in part, a colloid of water in air. In an idealized system state known as the thermodynamic limit, the system is deemed to be infinitely large so that there are no surface effects or finite size effects, and all molecular fluctuation-correlation timescales are deemed to be so short that all such systems in this limit are truly homogeneous. Systems that are inhomogeneous in the thermodynamic limit either in space or time are generally described as being thermodynamically "small". In this context, the atmosphere is a "small" system.

In the thermodynamic limit, before the first and second laws of thermodynamics can be formulated, there needs to be a formal definition of the two principles of mechanical equilibrium and thermal equilibrium proposed originally by Isaac Newton in 1687 and Joseph Black in 1850, respectively. Only then, can we define state functions of p-T that determine the first and second laws of thermodynamics, simply stated Q_{rev} (=ΔH enthalpy change) and Q_{rev}/T (=ΔS entropy change), respectively [1]. Q_{rev} is reversible heat transferred to or from a system in a reversible process. Whilst real systems and real processes all involve deviations from these well-defined equilibrium states, a proper understanding of the underlying thermodynamic equilibria is fundamental to an understanding and description of real systems and irreversible processes, which, to some degree, are thermodynamically small in a gravitational field.

There are various reasons for thermodynamic "smallness", colloidal particles with long time scales, multiphase systems in weak field, systems of low dimensionality, and many different types of thermodynamically small systems of interest in nature. When a gravitational field is applied to a thermodynamic system such as a molecular fluid, it is trivial to show that the gravitational force on each individual molecule is many orders of magnitude less than the root mean square fluctuating intermolecular force and is hence negligible.

As a result, even though the gravity mainly determines the pressure profile of the Earth's atmosphere as a function of height and the pressure of the oceans as a function of depth, the thermodynamic properties of the atmosphere at a particular height or the ocean pressure as a function of depth are reasonably accurate from the known equilibrium equations of state. Thus, the effect of smallness arising from an external field in a condensing multicomponent system is predictable in the limits of high and low densities. However, system smallness can also arise because of finite extent of the system in the direction of a strong external field.

We can describe the Earth's atmosphere as a thermodynamically small system because characteristic fluctuation lengths ~ km+ are of the same order as the characteristic length scales of the atmosphere under gravity. On the other hand, ironically, a one-liter sample of pure air, say, in both thermal and mechanical equilibrium with uniform T and p, a gradient in T or p of the atmospheric lapse rates 0.0001 K/cm may be negligible, i.e., sufficiently close to the thermodynamic limit and not thermodynamically "small". In a highly inhomogeneous non-equilibrium system as complex as the Earth's atmosphere, local processes with length scales of kilometers such as wind and rain are predictable. Clausius law [1] for irreversible processes tells us that for all spontaneous processes state functions change in the direction towards thermodynamic equilibrium of increasing entropy to a maximum. Thermodynamics, moreover, given the thermodynamic state functions ΔH and ΔS of all the components, can tell us what the overall direction of a spontaneous physical or chemical change is and whether a transformation occurs.

Given the state functions of its components, thermodynamic laws can tell us what the equilibrium state of the Earth's atmosphere would be if the Earth stopped rotating and orbiting, the sun stopped shining, volcanoes stopped erupting, the oceans stopped evaporating, radioactive active material stopped decomposing, etc. One must be cautious in drawing conclusions from any predictive computer model when all these external and internal processes conspire to create a truly chaotic system. The local state variables, T and p, of space and time are literally fluctuating on all length scales from millimeters to kilometers to time scales from minutes to millions of years. Yet, we can look out the window and predict the weather for the next hour or so. Given modern technology, instantaneous knowledge of several thousand thermometers and barometers strategically placed around the Earth's surface, and the most powerful computers imaginable, computer weather forecast modelling cannot extend this prediction with any certainty beyond a few days because the atmosphere is a chaotic, stochastic, and multivariate system.

The application of principles of thermodynamics to "small systems" began in the 1960s when Hill derived general equations applicable to non-macroscopic systems, which in the limiting case of an infinite system would reduce to "ordinary thermodynamics" [2,3]. Instead of deriving the fundamental relations between thermodynamic quantities in the various statistical mechanical ensembles for application to small systems, an alternative more useful and modern approach is to adopt a quasi-thermodynamic approximation and to use density functional theory (DFT) [4,5,6].

Here, we investigate hypotheses of thermodynamic equilibrium processes in the predictions of thermodynamic state averages of the Earth's atmosphere which are commonly used in climate change science modelling [7,8] to estimate the effect on average temperatures of increased concentrations of (so-called) greenhouse gases that transduce light into heat.

These open-system sub-states are thermodynamically small because their thermodynamic T-p state profiles are determined by externally imposed gradients of pressure by gravity and temperature by radiation energy transfer.

EXPERIMENTAL CRITERIA

Global Average Temperatures

The atmosphere can be treated as a multi-component single Gibbs phase of N_2, O_2, CO_2, Ar, and homogeneous mixture with variable water concentration (H_2O) at states up to a small humidity that exists at pressures from 1 atmosphere at the Earth's surface to vanishingly small beyond the thermosphere (100 km). The thermodynamic state variables are permanently in a state of flux at all points in the system. One can define averages of temperature (T) at any point in space as a function of latitude (φ), longitude (θ), height (z), and time (t), i.e., $\langle T \rangle (\varphi, \theta, z, \Delta t)$.

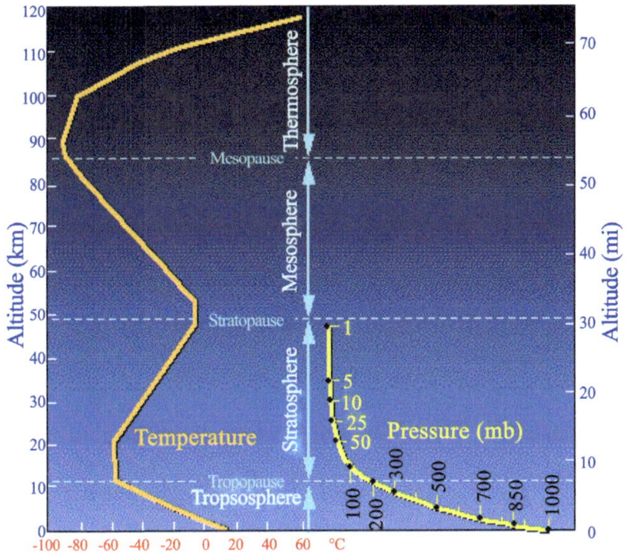

Figure 1: Experimental mean temperature (T) and pressure (p) profiles of the Earth's atmosphere; the density functional profile $< \rho(T, p)_z >$ is indicated by the increase in darkness with height; yellow lines for temperature (T) and pressure (p) are concentric averages denoted by angular brackets in the text. https://okfirst.mesonet.org/train/meteorology/VertStructure2.html (accessed on 20 December 2021).

The angular brackets denote an average over a time span Δt that, without a seasonal specification, is not less than one year to include all fluctuations arising from the Earth's daily rotation and annual orbit. "Global average temperatures", a concept widely used in climate change theory, are variously defined operationally as an average of measurements over the Earth's land and sea surfaces.

A basic objective of any climate model based upon atmospheric thermodynamics is to predict the experimental profiles of the global concentric time averages of the temperature $< T>$ (z, Δt) and $<p >$ (z, Δt). Typical profiles of long-time averaged T(z) and p(z) from meteorological research are shown in Figure 1.

Barometric Pressures

Since the atmosphere is not contained, except for gravity, the pressure profile is predicted accurately using the ideal gas equation of state (pV = RT where R is the gas constant and T is absolute K). The maximum ground-zero mean sea-level air pressure of 1 atmosphere is exactly known, and the ideal gas equation of state p = RT/V holds at that and all lower pressures. Figure 1 also shows that the average temperature up to around 100 km is everywhere below the 300 K (27 °C) but on average only by around 10%. To a first approximation, T(K) is constant assumed T = $<T>$ (z) where the local density approximation of DFT applies, and both the pressure and the density profiles are obtained from the ideal gas equation of state. The result for the pressure profile is generally called the *barometric formula*; it is in quite good agreement with the experimental logarithmic curve shown in Figure 1.

$$RT \log_e (p/p_0) = RT \log_e (\rho/\rho_0) = -gmz \tag{1}$$

where m is the molar mass of air and ρ is a density.

Substituting for RT = pV and differentiating with z gives an approximate atmospheric density profile.

$$\rho(z)\mu = -\{(dp/dz)_T\}(z)/g \tag{2}$$

We note here that the mean air density decreases approximately linearly with z in the troposphere as does the concentration (denoted by square brackets: mol/L) of its main components $[N_2]$, $[O_2]$, $[H_2O]$, [Ar] and $[CO_2]$ to a first approximation. A corollary of this observation is that the z-dependence of any absorption or emission of electromagnetic radiation, being proportional to concentration, also varies linearly with z.

Experimental Observations

The experimentally measured decrease in <T>(z) of the troposphere is approximately constant around −6.5 K/km. This quantity plays a central role in climate modelling [7,8] but its scientific origin appears to be the subject of a misrepresentation of the principles of equilibrium thermodynamics. The theory giving rise to its calculation in the troposphere appears to be based upon the misconception, that its existence is determined by a thermodynamic reversible adiabatic expansion or compression of volume elements of the Earth's atmosphere. There can be no thermodynamically reversible adiabatic expansions or contractions occurring in the atmosphere since it coexists with the Earth's surfaces with no adiabatic or closed boundaries.

The experimental evidence suggests the atmospheric lapse rates in the atmosphere are not a consequence of thermal equilibrium. A thermal equilibrium state of a system under gravity must have a uniform temperature with the lapse rate d <T>/dz = zero. The observed oscillations of the sign of the mean temperature gradient in Figure 1 must be determined primarily from non-equilibrium heat balance radiation absorptions and emissions in the first instance. The concept of a "parcel of air" in the atmosphere undergoing a reversible adiabatic expansion to create the temperature gradient, which can only occur in a closed system, may account for fluctuations around <T> (z) and affect weather events that we consider later. First, we reviewed the prevailing assumption [9] that the lapse rate is determined by adiabatic expansion of an otherwise equilibrium isothermal fluid state of air, say, a surface temperature T_0.

Lithosphere Temperature Profile

To explain the experimental troposphere lapse rate up to 10 km, consider the global lapse rate of the whole Earth. Its crust, the lithosphere with a depth of ~ 400 km, and atmosphere with a height of 100 km (Figure 2) shows that the Earth itself is cooling from its core by around 1 K/km but that the cooling rate suddenly increases to around –5 k/km on average for the Earth's lithosphere to a depth of around 400 km. At sea level (T_0), the Earth's solid crust lapse rate is closer to the experimental atmospheric lapse rate of –6.5 K/km than the DARL value –9.8 K/km. If we extend the outer limit of the average lithosphere lapse rate to the global mountainous extremities at z ~ 9 km, we see from Figure 2 that the linear constant troposphere lapse rate seen in Figure 1 up to the tropopause at 10 km can explain to a degree the experimental average lapse rate. The ground temperature at z ~ 9 km is roughly 50 K below the mean sea-level global average ground temperature, as also is the ambient atmosphere at that level. Heat transfer by transverse convection (winds) have a moderating effect to some extent upon the mean lapse rate < dT/dz > (z) in the troposphere.

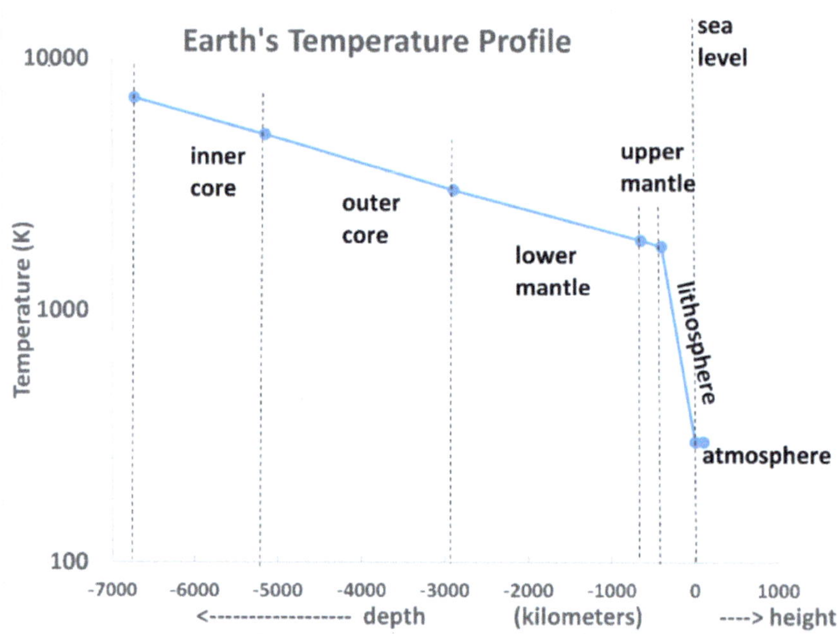

Figure 2: Mean temperature profile of the Earth and its atmosphere; the average temperature of the atmosphere is around 300 K: the average gradient <dT/dz> for the lithosphere is around –5 K/km, i.e., slightly less than the troposphere average up to the tropopause at 10 km (Figure 1): the atmosphere outer limit is 100 km.

Whereas the solid Earth's lapse rate is monotonic everywhere, the atmospheric lapse rate is oscillatory in sign and the changeovers are used to classify the concentric regions of different thermodynamic behavior. <dT/ dz> changes from negative to positive from troposphere to stratospheres and back to negative again for the mesosphere, then back to positive for the thermosphere. In the thermosphere around 100 km and higher, the temperature appears to be ever-increasing as it approaches p ~ 0 in outer space. This begs the question: what causes the reversals in sign of (d <T>/dz) z on climbing from troposphere to stratosphere, stratosphere to mesosphere, and mesosphere to thermosphere?

LAPSE RATE IN CLIMATE MODELLING

Adiabatic Lapse Rate

To calculate a dry adiabatic lapse rate (DARL), defined using the heat capacity of dry air, climate modelers hypothesize [7,8,9] that for a "parcel of air, within a still vertical column at equilibrium" with just the gravitational hydrostatic pressure, zero at the top, and height z → infinity, the pressure at the base z = 0 is equal to gm/A with the thermodynamic limit in x,y planes being open and in thermodynamic limit. However, if this hypothetical model atmosphere existed (see Figure 3) at the ground temperature T_0 at thermodynamic equilibrium, the temperature of the column would be uniform with $T(z) = T_0$ everywhere and the lapse rate (dT/dz) would equal zero for all values of z.

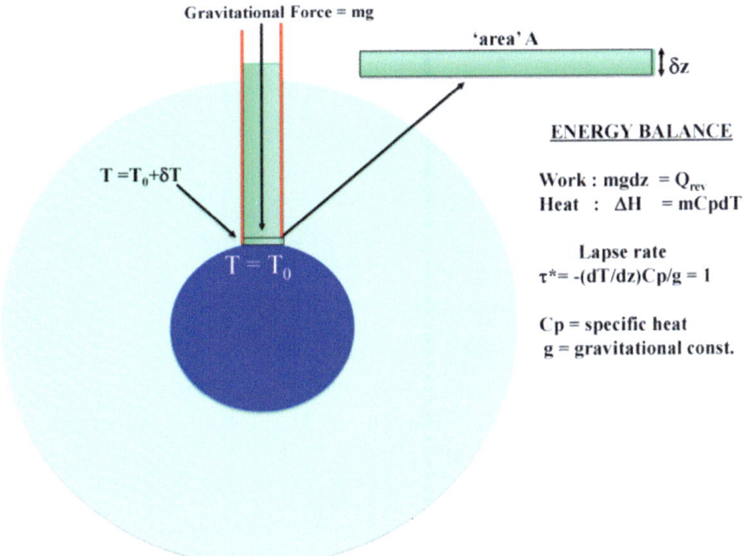

Figure 3: Schematic 2D illustration of an energy balance for a hypothetical reversible adiabatic expansion process for an infinitesimal expansion of a column volume element of the Earth's atmosphere, initially at equilibrium at temperature T, by an amount Adz (=dV) at the local gravitational pressure p(z): the hypothetical expansion in isolated closed system indicated by red lines in equilibrium with T_0 at ground zero with no roof under gravity would predict a non-equilibrium constant gradient in temperature (dT/dZ) known in climate modelling as the "dry adiabatic lapse rate" (DALR) = –9.8 K/km [7,8,9].

The lapse rate adiabatic hypothesis considers this representative sample column of gas to change states in uniform adiabatic expansion under the force of gravity such that the volume occupied by an infinitesimal 'pizza' slice of height z and width dz increases by an amount dV. The work done by this adiabatic expansion is simply pdV, where p is the pressure at level z. The pressure at level z is the force of gravity x weight (for an ideal gas, one can use the barometric formula), and since the process obeys Joules law [1] for all processes, reversible and irreversible), work performed against surroundings = enthalpy change; for a simple energy balance for an infinitesimal change dp, we have for the enthalpy of work performed (by definition and/or Joule's law [1]):

$$H = -Q_{rev} = W_{rev} = mC_p dT \qquad (3)$$

where C_p is the specific heat per unit mass (m), W is work, and Q is heat; work to change pressure is:

$$W_{rev} = Vdp \qquad (4)$$

then, using barometric Equation (2) $dp = g\rho dz$ where density $\rho = m/V$, adiabatic lapse rate is:

$$(dT/dz) = g/C_p \qquad (5)$$

The result can equally be obtained as shown schematically in Figure 3: since for an ideal gas, $pdV = -Vdp$ for any infinitesimal expansion. This simple energy balance produces a non-equilibrium temperature gradient for a mechanical equilibrium of a uniaxial expansion of a subvolume Adz of air that is isolated and closed to the surrounding atmosphere. The result is that $(dT/dz)_z$ for any isolated gas expanding reversibly and adiabatically under gravity, if the process were possible, would be a material constant that depends only upon its heat capacity C_p.

Troposphere Lapse Rate Hypothesis

There is a well-documented belief, especially amongst the climate modelling community [7,8], that because of this result, g/C_p for dry air = –9.8 K/km appears to agree with the experimental observation of a linear negative lapse rate in the troposphere of the same order. Figure 1 shows that $\langle T \rangle (z)$ varies from 15 °C (288 K) at z = 0 to –50 °C (223 K) at z = 10 km, giving an experimental lapse rate $\langle (dT/dz) \rangle (z) = -6.5$ K/km, i.e., somewhat less than the adiabatic hypothesis constant. Because of this rather coincidental result, it has been generally assumed that an average temperature $\langle T \rangle (dz)$, say, the global concentric average near sea-level, is a result of natural convection from a radiation-heated Earth's surface (~70% oceans and 30% land). Sunlight hits the surface of the Earth (land and sea) and heats them. They then heat the air above the surface. This explanation for the negative troposphere temperature gradient is summarized: (from Wikipedia [9], see also references [10,11,12] for further detailed analysis).

QUOTE: "However, when air is hot, it tends to expand, which lowers its density. Thus, hot air tends to rise and carry internal energy upward. This is the process of convection. Vertical convective motion stops when a parcel of air at a given altitude has the same density as the other air at the same elevation. When a parcel of air expands, it pushes on the air around it, doing thermodynamic work. An expansion or contraction of an air parcel without inward or outward heat transfer is an adiabatic process. Air has low thermal conductivity, and the bodies of air involved are very large, so transfer of heat by conduction is negligibly small. Also, in such expansion and contraction, intra-atmospheric radiative heat transfer is relatively slow and so negligible. Since the upward-moving and expanding parcel does work but gains no heat, it loses internal energy so that its temperature decreases. The adiabatic process for air has a characteristic temperature-pressure curve, so the process determines the lapse rate. When the air contains little water, this lapse rate is known as the dry adiabatic lapse rate: the rate of temperature decrease is 9.8 °C/km The reverse occurs for a sinking parcel of air" [11,12].

The Experimental Evidence

Below, we list some empirical objections inter alia why the foregoing prevailing hypothetical explanation for the troposphere lapse rate is inconsistent with the principles of thermodynamic equilibrium. The fundamental objection, however, is that gravity cannot determine a temperature gradient via adiabatic expansion or contraction processes without violating the second law of thermodynamics. A formal proof is given in Appendix A.

1. The first obvious indication that the climate modelers [7,8,9] adiabatic hypothesis cannot be a causal reason for <T> z in the troposphere is that the average temperature <T> (z = 9 km) above Mount Everest or the Himalayas area for example, is approximately the same as the collateral air temperatures 9 km above the sea level [13]. Where does the upward moving parcel of hot, isolated, insulated, and expanding air come from? The ground temperature at the extremity of the lithosphere (Figure 2)

is almost the same as the mean air temperature above the oceans (–50 °C at 9 km).

2. A thermodynamically reversible uniaxial adiabatic expansion of air cannot occur in an open system such as the atmosphere. The mean air pressure at any level z, whatever the ground level is below, is primarily determined by the weight above; the temperature variation is caused by radiation balance effects determined in part by transducer gas concentrations and hence the pressures. Mean temperatures <T> (z) are maintained by prevailing transverse convection fluxes, e.g., between land and sea and/or night and day, i.e., winds, and not uniaxial upward convection.
3. The adiabatic expansion hypothesis cannot explain why the lapse rate is negative in the troposphere then changes to positive values in the stratosphere and negative again in the mesosphere [14]. If the lapse rate were to be a phenomenon of equilibrium thermodynamics of the atmosphere associated with reversible adiabatic expansions, the effect would be present throughout. All adiabatic expansions of ideal gases are associated with a cooling effect; the lower the density, the closer to the ideal gas behavior. Figure 1 shows that the lapse rate in the stratosphere is a positive heating effect, yet p(z) decreases with height (z) in accordance with the barometric formula and in contradiction to any adiabatic contraction hypothesis.
4. No thermodynamic fluid in a uniform external gravitational field can be in a state of thermal equilibrium if there exists within it any gradients of temperature. Such a scenario violates the second law (Appendix A). This is Black's principle of thermodynamic equilibrium [1] (referred to also as the 'zeroth law'). It is inviolable and determines the direction of heat flow in all non-equilibrium spontaneous processes. Newton's principle of mechanical equilibrium of no unbalanced forces [1] is equally inviolable, so if the (dT/dz) in the troposphere (Figure 1) is a steady state non-equilibrium phenomenon, there would also be non-equilibrium gradients in pressure, hence the existence of transverse heat and mass transfer by convection everywhere.
5. Around 70% of the Earth's surface is water with a surface temperature that is near to contact air temperature T_0 for the most

part. The suggestions [8,9,10] that the altitude cooling rate is determined by a thermodynamic equilibrium phenomenon of a reversible adiabatic expansion of a contained and closed column of air against gravity lacks a driving force on most of the Earth's surface for most of the time.

6. The statement in the above Wikipedia quote from the climate modelling literature [7,8,9] "intra-atmospheric radiative heat transfer is relatively slow" cannot be further from the truth. All heat transfer by radiation occurs effectively and instantaneously at the speed of light. When the sun sets on a clear day with a blue sky, there is simultaneous change in the ground, sea, and air temperatures of the biosphere. The various radiation effects combine to create, in effect, an instantaneous z-dependent average thermostat that maintains T-gradients. The local density approximation of DFT, given the local state variables p and T, determine the dependent variable density ρ(z) using the ideal gas equation of state = p(z)/RT(z). Hence, there can be coincidental similarity with a sequence of temperature drops with a hypothetical column of reversible adiabatic expansion processes.

Radiation Balance Atmospheric Thermostat

The experimental slope of <T> (z) in the troposphere around –6.5 K/km in Figure 1 has no direct causal relationship with the DALR constant (–9.8 K/km) for the above reversible thermodynamic expansion of a closed gas system of dry air. Accordingly, the steady state heat balances can only be explained by the plethora of radiation, absorption and emission, and heating and cooling effects from both the Earth surface, the atmosphere itself, and sun to maintain the non-equilibrium steady state thermostat T(z). This must then include all the effects of transducer gases leading to the variable mean experimental average temperature gradients shown in Figure 1 of oscillating signs that define the concentric regions.

A temperature at any time of any volume element of the atmosphere T(z, θ, φ) is determined essentially instantaneously by the radiation balance at that radius (z) (Figure 4). We can regard the outcome of the radiation balance as a z-dependent "thermostat" that is effective on a time scale which is much shorter that the characteristic relaxation time for all other conductive and convective heat transfer processes. Accordingly, the LDA of density functional theory (DFT) [2,3] then applies to a volume element at the local equilibrium where the pressure, on average, is given by the barometric formula Equation (1) recalculated at a local <T> (z) and the density ρ(z) is given by Equation (2) as a function of T.

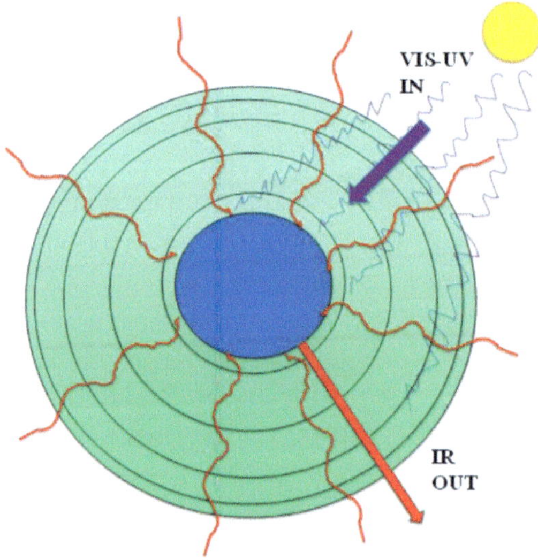

Figure 4: Schematic of a magnified 2D illustration of the atmospheric radiation balance 'thermostat' that determines the variant concentric average temperature as a function of height <T> (z) and its gradient <dT/dz> (z) of the Earth's atmosphere: the heat transfer balance of the radiation is essentially instantaneous relative to the convective response relaxation processes, resulting in the near-constant negative lapse rate (d <T>/dz) in the troposphere (z = 0 to 10 km).

This simplified scenario can explain the similarity between the mean lapse rate determined by the radiation balance and the subsequent adjustment of the atmospheric steady state temperature that is same sign and of same order as the DALR constant. In the troposphere, the intrinsic decrease in density is roughly the same as one expects for an adiabatic expansion since the local density approximation of density functional theory (DFT) [2,3] predicts a steady state change of T(z) with ρ(z) for an expansion process from air at T to T−dT and height z to z + dz. It has been suggested that the reason why the experimental mean lapse rate in the troposphere is 40% less than the DALR constant is the presence of water and its associated local heat and mass transfer effects such as enthalpy of condensation in cloudy skies and precipitation [15,16,17]. This has resulted in a revised and more general definition of a moist adiabatic lapse rate (MARL) that takes account of the change in properties of humid air. Water in the troposphere does indeed have a moderating and cooling effect due to its transducer properties such as the absorption and scattering of sunlight; these effects, however, have nothing to do with reversible adiabatic expansions.

The DALR dimensionless scaling variable $(dT/dz)g/C_p$, which we denote by τ*, can be regarded as a local or transient consequence of fluctuations around the experimental lapse rate in Figure 1 but not the cause of it. Evidently, its value [7,8,9] is a stability limit that plays a similar role in atmospheric heat transfer by convective flow as do other transport stability limits, for example, Reynolds number in fluid flow in prediction the transition to turbulence.

LAPSE RATE INSTABILITY INDICATOR

In DFT for small systems under gravity [2,3,4,5,6], temperature (T) is the equilibrium-independent state variable that determines the density (ρ) instead of the other way around. The adiabatic expansion hypothesis for the lapse rate in troposphere is based upon this misconception.

This raises the question, is the adiabatic lapse rate constant and a material and physical constant scaling law that depends only on the heat capacity with any relevance in computer modelling or the prediction of events? The answer appears to be yes—it is a natural criterion for a gradient stability limit to be predicting the transition from laminar to turbulence for heat and momentum transfer instabilities in weather forecasting. However, it can be highly misleading when one asks the question of which factors, especially CO_2, are influential in determining long-term variations in <T> (z).

There is abundant experimental evidence that the DALR constant 9.8 K/km is an instability limit wherever and whenever τ* exceeds a critical value. A large fluctuation from the mean is accompanied by fluctuations in density leading to an accompanying huge gradient in chemical potential unstable flow and mass transfer processes, leading to turbulence. It appears from an extensive catalogue of anecdotal evidence that this critical value is close to the DALR constant τ* = 1. Is this a coincidence? The evidence for the usefulness of τ* and its various modifications is in moist atmospheres that reduce its value by up to 50% at saturation [9]. This knowledge applied to modelling unstable weather phenomena such as turbulence is summarized below but there is no evidence that it is relevant to very small variations in <T> (z) with concentrations of greenhouse gases, e.g., CO_2 (z) or H_2O (z) in the troposphere.

Further compelling evidence that the lapse rate is a consequence of radiation thermostat and not uniaxial adiabatic convection processes is the longstanding observation that turbulence in the atmosphere is only found to occur in the troposphere ~0–10 km and not in the stratosphere ~10–50 km. Wherever the experimental lapse rate is lower than the τ^* prediction, i.e., for all z higher than the tropopause height which is rather sharply defined by the change in lapse rate, atmosphere shows no turbulence. By contrast, the change from −7 K/km in the troposphere and stratosphere with a positive small temperature gradient (Figure 1) shows no turbulence [14]. The presence of water within the troposphere requires modification to the experimental interpretation of unstable events. If air rises due to the thermal-gradient-induced convection processes, it cools even more due to the heat of condensation. Eventually, clouds form, releasing the heat of condensation that further changes the local lapse rate as the clouds are powerful transducers of sunlight to heat. Before saturation, the rising air follows the dry adiabatic lapse rate. After saturation, the rising air follows the moist adiabatic lapse rate [15].

The exchanges of enthalpy and radiation energy balance fluctuations on cloud formation explains the occurrence of thunderstorms. Modified moist adiabatic lapse rates can be calculated with a lower value around 5 K/km [16,17]. Thus, for moist or saturated air in the troposphere, the threshold for instability where there is cloud formation is reduced by around 50% and the actual local lapse exceeds the moist adiabatic lapse rate (MALR), giving rise to turbulent weather patterns of wind and rain. Here, however, we were concerned with questions about changes in average concentrations of greenhouse gases, notably CO_2 and H_2O, on the concentric averages <T> (z) and d < T) >/dz (z). An analysis of time-dependent phase relationships between [CO_2] (t) concentration and <T> (t, z = 0) showed a correlation over many decades with CO_2 following <T> in the phase shift [18] rather than the reverse, as suggested by IPCC climate change models [7,8].

CLIMATE MODELLING

Computer Models of Atmospheric Processes

In the "dark ages" of science before the advent of computer modelling in the 1950s, scientific research was just theory and experiment. There are two types of approximation in theories, physical and mathematical. The advent of computer simulation enabled the numerical solution of multidimensional integrals, or high-order multi-coupled differential equations of otherwise intractable physical models. Building a physical computer model with superimposed mathematical approximations and hundreds of variable parameters is not scientific research. Computer modelling may have educational applications and engineering design uses or transient predictive applications such as weather forecasting, but such models cannot be a tool of scientific research. Computer "animation" of complex stochastic systems with hundreds of adjustable input parameters tells us nothing.

In modelling the thermophysics of real systems, whatever the level—astronomical, phenomenological, molecular, electronic, or nuclear—in order to obtain new scientific information to test a hypothesis, we must perform what is known as Hamiltonian surgery. Hamiltonian is the mathematician whose name is given to the model definition of the energetics of the system. As in the laboratory, one can do an experiment on a computer model, but it is not what we put into a computer experiment that enables access to indisputable new knowledge, it is what we leave out. The simplest imaginable model of the thermodynamics of the Earth's atmosphere is a gas, surface, and gravitational field that predicts a uniform temperature. This fails because it cannot predict thermal gradients as observed (Figure 1); the surgery is too drastic. Yet, the model has told us something.

The experimental lapse rate is evidence that conduction and convection, especially transverse convection, whilst important in moderating the natural fluctuations of T and p and all spontaneous weather processes, may not be significant factors in determining the experimental non-equilibrium concentric average $< T(z) >$ lapse rate. If there were to be no radiation absorption or emission anywhere of any kind, then there would be no variations in ground temperature from poles to equator, from day to night, or summer to winter, and the entire atmosphere would come to thermal equilibrium at a mean temperature of the Earth's surface <T> (z = 0) under gravity. Then, the simple barometric formula that assumes a constant T throughout would be precise for the pressure. Therefore, the next research step must include the simplest imaginable one-dimensional (z) homogeneous in x,y, model atmosphere with only the salient basic radiation transfer effects.

Effects of H_2O and CO_2 on <T> (z)

The most primitive model only considers Planck radiation exchanges between Earth's surface, atmosphere, and sun in the form of radiation balance as summarized in Figure 5a showing the energy fluxes to and from an atmospheric volume element. The radiative flux F comes from the sun's surface as visible or UV radiation The radiative flux F_s is emitted from the Earth's surface and the atmosphere is also a black-body emitter which we call F_a. We denote as F_u (upward flux) and F_d (downward flux) to describe the energy exchange between the Earth's surface and the atmosphere occurring by all radiation processes. A more detailed description of a one-dimensional radiation balance model with a specification of parameters is given on page 211 of McGuffie and Henderson-Sellers, Section 4.2.1 [8].

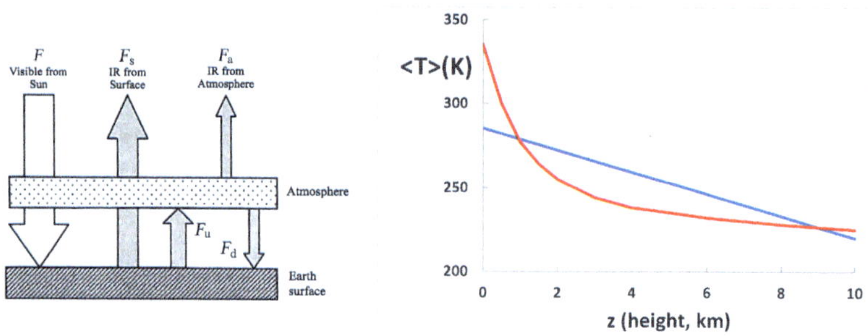

(a) (b)

Figure 5: (**a**) Energy exchange in a primitive surface plus atmosphere one-dimensional two-level radiation balance model. (**b**) profiles of the model temperature from radiation balance in an atmosphere with no transducer gases (red line) compared with the real-world experimental linear lapse rate <T> (z) (blue line): note that <T> z = 0, i.e., at sea level is reduced from 335 K (62 °C) to 285 K (12 °C), and in the troposphere up to around 1.5 km by H_2O absorption and, to a much lesser extent, CO_2.

Simply stated, the total emission from earth and atmosphere is (Figure 5a):

$$F = F_s + F_a \quad (6)$$

The energy received from the sun according to the steady state radiation balance and lost by the Earth is constant total energy, as the long-term global temperature fluctuations are less than a fraction of 1K over hundreds of years. The Stefan–Boltzmann law determines the energies in radiation energy scales of Equation (6) as the fourth power of temperature [8,19]. This has a stabilizing effect on the exchange, and the flux emitted by Earth tends to be equal to the flux absorbed, close to the steady state. This result greatly simplifies the research question: what is the effect of changing the concentration of one or more of the atmospheric components on the concentric thermal average temperature <T> (z)? The likely effect, if any, on the steady state difference <ΔT(z)> due to a presence of CO_2, for example, with all other things being equal, can be analytically obtained from the primitive model in Figure 5a. CO_2 both absorbs and emits UV-visible radiation, so the additional CO_2 warm up T(z) with the absorbed radiation, making the lapse rate smaller and hence <T> higher, maintained by radiation flux F_u and F_d balance. However, both F_s and F_a become larger than before according to the Stefan–Boltzmann radiation law [8,19] due to higher surface and atmospheric temperatures upsetting the steady state Equation (6). Then, both the surface and lower atmosphere cool down until they reach a new state of radiation balance, with Equation (6) once more satisfied. Likewise, UV or visible radiation from the sun is diminished by depletion of intensity at the absorption frequencies on passing through the atmosphere. This transducer (or greenhouse) gas absorption mainly by [H_2O], but also [CO_2], creates a heating effect at higher levels but a cooling effect of the lower troposphere and at the Earth's surface (Figure 5).

If [CO_2] increases, then it both absorbs and emits more radiation energy by amounts given by the Einstein coefficients [20]. At thermodynamic equilibrium, we have a simple balancing in which the net change in the number of any excited atoms is zero, being balanced by loss and gain due to all processes. This local thermodynamic equilibrium isothermal condition [20] requires that the net exchange of energy between different energy level states of the same component at the same temperature is zero.

This is because the probabilities of transition cannot be affected by the presence or absence of other excited atoms, other components present, or the thermodynamic state. With the more emissive atmosphere at the higher T, the same total flux $F_s + F_a$ occurs at lower temperatures; therefore, the concentric averages of <T> (z) can only be the same or lower with increasing CO_2. This simple observation is not dependent on the absorption spectrum of CO_2 but the magnitude of the effect is.

Stronger atmospheric absorption of CO_2 results in enhanced emission of the same component with everything else being equal and leads to a reduction of T(z) with decreasing z and of <T> (z = 0). The difference between the primitive model T profile, and the real-world lapse rate in the troposphere represents the total transduction effect of greenhouse gases, which is mainly due to water and CO_2 at the lower levels of the troposphere.

We also note that the extensive specific heat C_p of air and CO_2 that enters the estimates of DALR and MALR cannot affect the lapse rate effects, if any, due to adiabatic expansion. The change ΔC_p of one mole of air (all components) is due to an additional 0.0002 moles of CO_2 (increase from 0.02 to 0.04%).

$$\Delta C_p = \Delta[CO_2] \{C_p (CO_2)\text{-}C_p (air)\} = 0.00005\ C_p(air) \qquad (7)$$

A higher C_p of air would correspond to a lower adiabatic lapse rate and an increase from T_0 by 0.000005% according to Equation (5). This is clearly also a completely negligible effect.

CO_2 -Steam and Enthalpy 'Footprints'

When a fossil fuel is burned, a hydrocarbon such as gasoline (octane) is replaced by 16 moles of CO_2 and 18 moles of steam into the atmosphere roughly to every 25 moles of oxygen burned up. What would be the effect of this "CO_2-H_2O" footprint on <T> (z) if the exhausts were uniformly distributed in the troposphere? The balanced combustion chemical equation for octane is, showing a large enthalpy 'footprint' for every 0.114 kg octane.

$$C_8H_{18} + (25/2)\ O_2 = 8\ CO_2 + 9\ H_2O - 5509\ kJ\ (=\Delta H) \qquad (8)$$

If we consider two identical-twin Earths, I and II: Earth I = no gasoline engines. Earth II = with gasoline engines, converting massive moles of fuel to [CO_2 + steam]. What is the effect of gasoline engines on Earth II compared to Earth I, all other physical aspects of the solar system, atmosphere, biosphere, etc. being equal?

It can be argued that the steam footprint is negligible compared to the water already present in the atmosphere at steady state and therefore its equilibrium with the clouds, oceans, biosphere photosynthesis, etc. all remain undisturbed by gasoline emissions over several decades. Adding more water vapor to the atmosphere, albeit relatively small, can produce a 'steam footprint' global cooling effect. This would happen because more water vapor leads to more extensive saturation in the troposphere, hence more cloud formation. Clouds reflect sunlight and reduce the amount of radiation heat that reaches the Earth's surface to warm it. If the amount of solar warming decreases, then the temperature of the Earth's surface and lower troposphere would decrease. In that case, the effect on average temperatures, if any, of adding more water vapor via the industrial 'steam footprint' that invariably accompanies the industrial 'carbon footprint' would more likely be global cooling rather than warming, but is probably a negligible effect.

Similar considerations, however, apply to [CO_2]. A corollary of such an argument, therefore, is that if the steam footprint is negligible, then the

CO_2 footprint from gasoline engines is also negligible. Moreover, for $[CO_2]$, there is a balancing effect in the biosphere. Le Chatellier's principle of equilibrium chemical thermodynamics [21] works toward restoration of the chemical equilibria for the photosynthesis reaction that removes CO_2 from the atmosphere for plant growth, replacing the oxygen. If the concentration of $[CO_2]$ increases, everything grows faster. If the temperature of the biosphere (land surface) increases, everything grows faster. Both these effects, together with the CO_2 output from humans and animals, eventually maintain a $[CO_2]$ balance.

A detailed analysis of the IR microwave absorption spectrum of CO_2 by Witteman [22] was reported recently in relation to atmospheric absorption of Earth's emission. The main conclusion was that since only 10% of its spectrum is active, the thermal radiation that falls within the frequency region of the Earth's emission is fully absorbed. This is because there are only three fairly narrow absorption bands in CO_2. This is not only the case for 400 ppm of CO_2 in the atmosphere but also for much smaller concentration values. If computer model conclusions for global warming are dependent to any appreciable extent on IR thermal emission from the Earth's surface, the heating effect is practically not affected by the change in the CO_2 concentration since it is wholly absorbed at much lower concentrations in just the lower atmosphere layers.

CONCLUSIONS

Thermodynamic Equilibria Summary

We found that *Black's principle of thermal equilibrium*, alternatively known as the zeroth law of thermodynamics, plays a central role in determining the average temperatures and gradients in the various height spheres. It does not, however, determine the experimental lapse rates $<dT(z)/dz>$ that are deviations from thermodynamic equilibrium due to the radiation balance which acts at a z-dependent thermostat, causing deviations from any global average temperature $<T_G>$ that would prevail if there was no radiation balance effect to create a non-zero experimental reduced lapse rate t * throughout the atmosphere. The $\tau * = 1$ results for an idealized uniaxial adiabatic expansion against gravity is given by a simple energy balance between the gravitational force mg/A as there is no intrinsic pressure, and the enthalpy loss is caused by drop in temperature $dH = -dQ_{rev} = pdV = mgdz = C_p dT$. This simple analytic result for an idealized hypothetical process does not determine the experimental averages of temperature gradients in the atmosphere. Likewise, *Newton's principle of mechanical equilibrium*, i.e., the pressure of an equilibrium fluid in an external field, is the intrinsic pressure plus the weight above with no unbalanced forces. It is central to determining the atmospheric pressure profile that would be given accurately by the barometric pressure (p_B) but with significant deviations $\Delta P(z) = p - p_B$ especially in the troposphere, also determined primarily by the variation in temperatures T(z) that can only be a consequence of the radiation balance thermostat.

To complete this preliminary analysis, it is *Gibbs principle of chemical equilibrium* that is the driving force throughout the atmosphere for the direction of all spontaneous changes for all mass transfer events. The Gibbs principle requires uniform chemical potential for every component in the atmosphere everywhere. At pressures of 1 atmosphere and lower, however, the Gibbs energies of nitrogen, oxygen, argon, CO_2, and other minor gases are essentially the same depending only on the ideal gas concentrations. This is certainly not the case for water and possibly also CO_2 to a slight extent if there are concentration gradients $d[CO_2]/d_Z$ resulting from its solubility on water.

Mass transfer of water in the atmosphere is the basis of all catastrophic climate change events, floods, and droughts. Whilst one can approximate the driving force for the mass transfer of water in nearly dry unsaturated air by its concentration gradient $d[H_2O]/dz$ in weather forecast models using Dalton's law of partial pressures, this is certainly not true for the state changes in atmospheric water between its different states including condensations to its colloidal state in clouds, liquid state in rain, and solid state in snow and ice.

All these state changes of water, with latent heat changes, are driven by gradients in T, p, and chemical potential (μ). Non-zero gradients of μ_i is the partial molar Gibbs energy ($=dG/dn_i)_T$ driving force for mass transfer of any single component i of a multicomponent fluid. This essential driving force $d\mu(H_2O)/d(z, \theta, \varphi)$ for extreme weather patterns appears to have been neglected or overlooked as the true driving force in climate change modelling. In a nine-page index to a 453-page treatise on climate change modelling spanning 50 years [8], there is no entry for Gibbs "chemical potential"—the driving force for all spontaneous mass transfer and correlated processes that determine climate and any climate change at the Earth's lithosphere atmosphere interface.

Thermosphere Measurements

There are two more paradoxical observations in the atmosphere relating to thermodynamic equilibria in the thermosphere. An ever-increasing temperature in the thermosphere and a non-zero wind speed reveals that the outer limit of the thermosphere is apparently rotating faster than the Earth itself.

At outer limit levels of the thermosphere, however, the rarity of the atmosphere makes measurement of an experimental T(z) by conventional means more difficult. It is necessary to bring the air into thermal equilibrium with the surface of a thermometer that is calibrated to a specific scale [23]. This process takes time.

At around an altitude of 100 m, the pressure is so small that the time taken for the gas to equilibrate with the thermometer begins to exceed the observation time for equilibration between gas and thermometer. Recorded temperature apparently higher than the true temperature of thermal equilibrium can be obtained. The ever-increasing T(z) shown in Figure 1 for the thermosphere may be spurious.

A similar effect apparently occurs in any attempt at this level, ~ 100 km, to measure a long-time average wind velocity which is zero for most of the Earth's atmosphere where there is no "slip" as the atmosphere rotates with the Earth due to gravity.

The equipartition of the energy principle of thermal equilibrium explains both these non-equilibrium effects in the thermosphere. The distribution of molecular velocities in all directions is Maxwellian with a high velocity tail. At extremely low densities when experiments are measured on a finite time scale, the molecular collision frequencies of high velocity molecules are more prominent than low velocity at the thermometer or pressure detecting surface, and fluctuation times diverge. Non-zero 'average' wind fluxes from transient finite time measurements become misleading.

Gravity and Thermal Equilibrium

There appears to be a widespread misapprehension in the modelling literature that a gravitational field *per se* can produce a thermal gradient in an otherwise fluid thermodynamic equilibrium. Classical thermodynamics tells us, however, that the variations in the atmospheric lapse rates cannot be caused by a gravitational potential alone. To confirm this, we appended a formal proof (Appendix A) that a thermal gradient in a fluid in equilibrium in a gravitational field would be a violation of the second law of thermodynamics. Non-zero lapse rates in the atmosphere and temperature gradients were reported in quasi-equilibrium laboratory experiments [24]. As with the more circumspect analysis of atmospheric mean gradients, these laboratory experimental results could be explained by radiation effects.

Changes in the sign of lapse rates are mainly due to variations with z in chemical composition and concentration of the transducer gases. The crossover from a negative to positive lapse rate at the tropopause, for example, is mainly caused by a reduction in the concentration of water above the tropopause and the concentration of ozone of the ozone layer in the stratosphere.

Finally, we return to the enthalpy 'footprint' of equation (8). If all the enthalpy from all the fuels (coal, oil, gas, nuclear) burned in the last 100 years emitted into the Earths' atmosphere at once, what is the temperature rise? The heat capacity C_p of the whole atmosphere, using ideal gas value, is 5.6×10^{21} J/K; the total ΔH produced by burning fuels from 1920 to 2020 is 3.5×10^{22} J. This would increase the global mean $<T>$ by $\Delta H/C_p = 6K$ (6 degrees centigrade). This is a starting point for the real explanation of global warming!

ACKNOWLEDGMENTS

Research performed under the auspices of the Ossonoba Philosophical Society: we wish to thank Igor Khmelinskii and Peter Stallinga for helpful discussion, advice, and some corrections to first draft.

APPENDIX

Appendix A. Proof That Gravity Cannot Determine Lapse Rates by Reversible Adiabatic Expansion or Any Other Equilibrium Thermodynamic Process

Given the definition of a T-scale [23], and the principles of thermal (T) and mechanical (p) equilibrium:

1. define reversible heat $Q_{rev} = C_p \Delta T$ (heat to raise T is then unavailable for work)
2. define enthalpy $\Delta H = Q_{rev}$
3. define entropy $\Delta S = Q_{rev}/T = Cp\ \Delta \log_e T$ ("unavailable" energy per degree absolute)
4. 1st Law: Enthalpy is a state function (Hess's law)
5. 2nd Law: Entropy is a state function (Carnot's law)

Consider two adjacent segments within a column of air with a non-zero temperature gradient (dT/dz)
(e.g. following a hypothetical adiabatic expansion against gravity,
but for equilibrium state functions the history doesn't matter)

Take two adjacent sub volumes: T1 < T2 and V1 > V2

Imagine a "hypothetical " adiabatic process between states 1-3 in gravitational field (g)

T1-V1 (m_1) and T2-V2(m_2): mass $m = m_1 = m_2$

$T1 = T3 - \Delta T$; $T2 = T3 + \Delta T$; $T3 = (T2+T1)/2$

Check 1st Law $\Delta H = Q_{rev}$ (contraction) + Q_{rev} (expansion)
$= m\ Cp\ \Delta T - m\ Cp\ \Delta T = 0$ **NO VIOLATION**

[Entropy (S) is a state function: S must remain constant in adiabatic equilibrium process].

Funding

This research received no external funding.

Conflicts of Interest

The author declares no conflict of interest.

REFERENCES

1. Sandler S.I., Woodcock L.V. Historical observations of the laws of thermodynamics. *J. Chem. Eng. Data.* 2010;55:4485–4490. doi: 10.1021/je1006828.
2. Hill T.L. Thermodynamics of small systems. *J. Chem. Phys.* 1962;36:3182. doi: 10.1063/1.1732447.
3. Hill T.L. *Thermodynamics of Small Systems.* Wiley; New York, NY, USA: 1965.
4. Rowlinson J.S. Molecular theory of thermodynamically small systems. *Chem. Soc. Rev.* 1983;12:251. doi: 10.1039/cs9831200251.
5. Rowlinson J.S. Thermodynamics of inhomogeneous systems. *J. Chem. Thermodyn.* 1993;24:449. doi: 10.1006/jcht.1993.1154.
6. Saxena R., Woodcock L.V. Density functional approach to the thermodynamics of self-assembly. *J. Chem. Phys.* 2005;122:164501. doi: 10.1063/1.1858439. [PubMed]
7. Jacobson M.Z. *Fundamentals of Atmospheric Modeling.* 2nd ed. Cambridge University Press; Cambridge, UK: 2005.
8. McGuffie K., Henderson-Sellers A. *The Climate Modelling Primer.* 4th ed. John Wiley and Sons Ltd.; London, UK: 2014.
9. Wikipedia. [(accessed on 20 December 2021)]. Available online: https://en.wikipedia.org/wiki/Lapse_rate
10. Stallinga P. Comprehensive analytical study of the greenhouse effect of the Earth's atmosphere. *Atmos. Clim. Sci.* 2020;10:40–80.
11. Goody R.M., James C.G., Walker J.C.G. *Atmospheres.* Prentice-Hall; Hoboken, NJ, USA: 1972. p. 60. (Foundations of Earth Science Series).
12. Danielson E.W., Levin J., Abrams E. *Meteorology.* 2nd ed. McGraw Hill; New York, NY, USA: 2003.
13. Whiteman C.D. *Mountain Meteorology: Fundamentals and Applications.* Oxford University Press; Oxford, UK: 2000.
14. UCAR The Stratosphere. [(accessed on 20 December 2021)]. Available online: https://scied.ucar.edu/learning-zone/atmosphere/stratosphere
15. DALR. [(accessed on 20 December 2021)]. Available online: https://web.archive.org/web/20160603041448/http:/meteorologytraining.tpub.com/14312/css/14312_47.htm

16. Minder J.R., Mote P.W., Lundquist J.D. Surface temperature lapse rates over complex terrain: Lessons from the Cascade Mountains. *J. Geophys. Res.* 2010;115:D14122. doi: 10.1029/2009JD013493.
17. American Meteorological Society: 2021 Saturated Adiabatic Lapse Rate, Glossary of Meteorology. [(accessed on 20 December 2021)]. Available online: https://glossary.ametsoc.org/w/index.php?title=Moist-adiabatic_lapse_rate
18. Stallinga P., Khmelinskii I. Phase Relation between Global Temperatures and Atmospheric Carbon Dioxide. [(accessed on 20 December 2021)];*arXiv.* 2013 Available online: https://arxiv.org/abs/1311.21651311.2165
19. Wikipedia. [(accessed on 28 December 2021)]. Available online: https://en.wikipedia.org/wiki/Stefan%E2%80%93Boltzmann_law
20. Einstein Coefficients. [(accessed on 28 December 2021)]. Available online: https://en.wikipedia.org/wiki/Einstein_coefficients
21. Clark J. The Effect of Changing Conditions. 2020. [(accessed on 20 December 2021)]. Available online: https://chem.libretexts.org/Bookshelves/Physical_and_Theoretical_Chemistry
22. Witteman W.J. Global Warming by Thermal Absorption of CO2, 2021 University of Twente (NL: Preprint Stefan–Boltzmann Law. [(accessed on 28 December 2021)]. Available online: https://en.wikipedia.org/wiki/Stefan-Boltzmann_law
23. Harvey A.H. What the thermophysical property community should know about temperature scales. *Int. J. Thermophys.* 2021;42:165. doi: 10.1007/s10765-021-02915-9.
24. Kigoshi K. Thermal steady states of gases in a gravitational field. *Z. Naturforsch. A.* 2011;66:123–133. doi: 10.1515/zna-2011-1-217.

CHAPTER 19

Radiation and Heat Transfer in the Atmosphere: A Comprehensive Approach on a Molecular Basis

Hermann Harde

Laser Engineering and Materials Science, Helmut-Schmidt-University Hamburg, Holstenhofweg 85, 22043 Hamburg, Germany

ABSTRACT

We investigate the interaction of infrared active molecules in the atmosphere with their own thermal background radiation as well as with radiation from an external blackbody radiator. We show that the background radiation can be well understood only in terms of the spontaneous emission of the molecules. The radiation and heat transfer processes in the atmosphere are described by rate equations which are solved numerically for typical conditions as found in the troposphere and stratosphere, showing the conversion of heat

Citation: Hermann Harde, "Radiation and Heat Transfer in the Atmosphere: A Comprehensive Approach on a Molecular Basis", International Journal of Atmospheric Sciences, vol. 2013, Article ID 503727, 26 pages, 2013. https://doi.org/10.1155/2013/503727.

to radiation and vice versa. Consideration of the interaction processes on a molecular scale allows to develop a comprehensive theoretical concept for the description of the radiation transfer in the atmosphere. A generalized form of the radiation transfer equation is presented, which covers both limiting cases of thin and dense atmospheres and allows a continuous transition from low to high densities, controlled by a density dependent parameter. Simulations of the up- and down-welling radiation and its interaction with the most prominent greenhouse gases water vapour, carbon dioxide, methane, and ozone in the atmosphere are presented. The radiative forcing at doubled CO_2 concentration is found to be 30% smaller than the IPCC-value.

INTRODUCTION

Radiation processes in the atmosphere play a major role in the energy and radiation balance of the earth-atmosphere system. Downwelling radiation causes heating of the earth's surface due to direct sunlight absorption and also due to the back radiation from the atmosphere, which is the source term of the so heavily discussed atmospheric greenhouse or atmospheric heating effect. Upward radiation contributes to cooling and ensures that the absorbed energy from the sun and the terrestrial radiation can be rendered back to space and the earth's temperature can be stabilized.

For all these processes, particularly, the interaction of radiation with infrared active molecules is of importance. These molecules strongly absorb terrestrial radiation, emitted from the earth's surface, and they can also be excited by some heat transfer in the atmosphere. The absorbed energy is reradiated uniformly into the full solid angle but to some degree also re-absorbed in the atmosphere, so that the radiation underlies a continuous interaction and modification process over the propagation distance.

Although the basic relations for this interaction of radiation with molecules are already well known since the beginning of the previous century, up to now the correct application of these relations, their importance, and their consequences for the atmospheric system are discussed quite contradictorily in the community of climate sciences.

Therefore, it seems necessary and worthwhile to give a brief review of the main physical relations and to present on this basis a new approach for the description of the radiation transfer in the atmosphere.

In Section 2, we start from Einstein's basic quantum-theoretical considerations of radiation [1] and Planck's radiation law [2] to investigate the interaction of molecules with their own thermal background radiation under the influence of molecular collisions and at thermodynamic equilibrium [3, 4]. We show that the thermal radiation of a gas can be well understood only in terms of the spontaneous emission of the molecules. This is valid at low pressures with only few molecular collisions as well as at higher pressures and high collision rates.

In Section 3, also the influence of radiation from an external blackbody radiator and an additional excitation by a heat source is studied. The radiation and heat transfer processes originating from the sun and/or the earth's surface are described by rate equations, which are solved numerically for typical conditions as they exist in the troposphere and stratosphere. These examples right away illustrate the conversion of heat to radiation and vice versa.

In Section 4, we derive the Schwarzschild equation [5–11] as the fundamental relation for the radiation transfer in the atmosphere. This equation is deduced from pure considerations on a molecular basis, describing the thermal radiation of a gas as spontaneous emission of the molecules. This equation is investigated under conditions of only few intermolecular collisions as found in the upper mesosphere or mesopause as well as at high collision rates as observed in the troposphere. Following some modified considerations of Milne [12], a generalized form of the radiation transfer equation is presented, which covers both limiting cases of thin and dense atmospheres and allows a continuous transition from low to high densities, controlled by a density dependent parameter. This equation is derived for the spectral radiance as well as for the spectral flux density (spectral intensity) as the solid angular integral of the radiance.

In Section 5, the generalized radiation transfer equation is applied to simulate the up- and downwelling radiation and its interaction with the most prominent greenhouse gases water vapour, carbon dioxide, methane, and ozone in the atmosphere. From these calculations, a detailed energy and radiation balance can be derived, reflecting the different contributions of these gases under quite realistic conditions in the atmosphere. In particular, they show the dominant influence of water vapour over the full infrared spectrum, and they explain why a further increase in the CO_2 concentration only gives marginal corrections in the radiation budget.

It is not the objective of this paper to explain the fundamentals of the atmospheric greenhouse effect or to prove its existence within this framework. Nevertheless, the basic considerations and derived relations for the molecular interaction with radiation have some direct significance for the understanding and interpretation of this effect, and they give the theoretical background for its general calculation.

INTERACTION OF MOLECULES WITH THERMAL BATH

When a gas is in thermodynamic equilibrium with its environment it can be described by an average temperature T_G. Like any matter at a given temperature, which is in unison with its surrounding, it is also a source of gray or blackbody radiation as part of the environmental thermal bath. At the same time, this gas is interacting with its own radiation, causing some kind of self-excitation of the molecules which finally results in a population of the molecular states, given by Boltzmann's distribution.

Such interaction first considered by Einstein [1] is replicated in the first part of this section with some smaller modifications, but following the main thoughts. In the second part of this section, also collisions between the molecules are included and some basic consequences for the description of the thermal bath are derived.

Einstein's Derivation of Thermal Radiation

The molecules are characterized by a transition between the energy states E_m and E_n with the transition energy

$$\Delta E = E_m - E_n = h\nu_{mn} = \frac{hc}{\lambda_{mn}}, \tag{1}$$

where h is Planck's constant, c the vacuum speed of light,]mn the transition frequency, and λ_{mn} is the transition wavelength (see Figure 1).

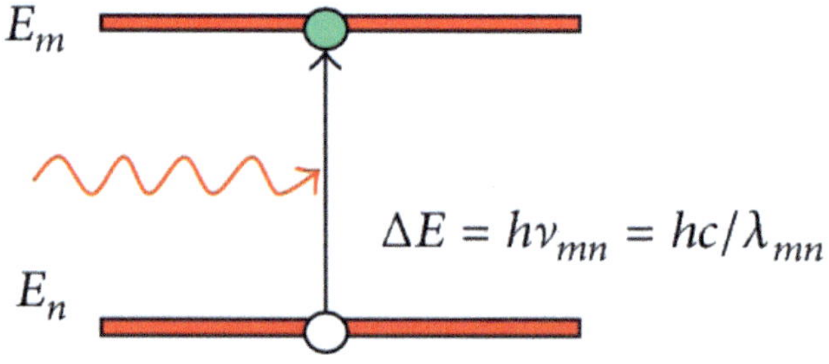

Figure 1: Two-level system with transition between states m and n.

Planckian Radiation. The cavity radiation of a black body at temperature T_G can be represented by its spectral energy density $u\lambda$ (units: J/m³ /μm), which obeys Planck's radiation law [2] with

$$u_\lambda = \frac{8\pi n^4 hc}{\lambda^5} \frac{1}{e^{hc/kT_G\lambda} - 1}, \tag{2}$$

or as a function of frequency assumes the form

$$u_\nu = u_\lambda \frac{\lambda^2}{cn} = \frac{8\pi n^3 h\nu^3}{c^3} \frac{1}{e^{h\nu/kT_G} - 1}. \tag{3}$$

This distribution is shown in Figure 2 for three different temperatures as a function of wavelength. n is the refractive index of the gas.

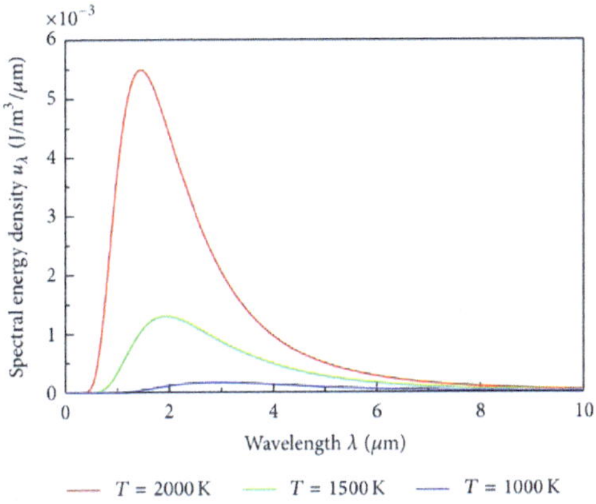

Figure 2: Spectral energy density of blackbody radiation at different temperatures.

Boltzmann's Relation. Due to Boltzmann's principle, the relative population of the states m and n in thermal equilibrium is [3, 4]

$$\frac{N_m}{N_n} = \frac{g_m}{g_n} e^{-h\nu_{mn}/kT_G}, \tag{4}$$

with N_m and N_n as the population densities of the upper and lower state, g_m and g_n the statistical weights representing the degeneracy of these states, k as the Boltzmann constant, and T_G as the temperature of the gas.

For the moment, neglecting any collisions of the molecules, between these states three different transitions can take place.

Spontaneous Emission. The spontaneous emission occurs independent of any external field from $m \rightarrow n$ and is characterized by emitting statistically a photon of energy $h\nu_{mn}$ into the solid angle 4π with the probability $dWs\ mn$ within the time interval dt

$$dW_{mn}^{s} = A_{mn} dt. \tag{5}$$

A_{mn} is the Einstein coefficient of spontaneous emission, sometimes also called the spontaneous emission probability (units: s^{-1}).

Induced Absorption. With the molecules subjected to an electromagnetic field, the energy of the molecules can change in that way, that due to a resonant interaction with the radiation the molecules can be excited or de-excited. When a molecule changes from $n \rightarrow m$, it absorbs a photon of energy $h\nu_{mn}$ and increases its internal energy by this amount, while the radiation energy is decreasing by the same amount. The probability for this process is found by integrating over all frequency components within the interval $\Delta\nu$, contributing to an interaction with the molecules:

$$dW_{nm}^{i} = B_{nm} \int_{\Delta\nu} u_{\nu} g(\nu)\, d\nu \cdot dt. \tag{6}$$

This process is known as induced absorption with B_{nm} as Einstein's coefficient of induced absorption (units: $m^3 \cdot Hz/J/s$), $u]$ as the spectral energy density of the radiation (units: J/m^3 /Hz) and $g(\nu)$ as a normalized lineshape function which describes the frequency dependent interaction of the radiation with the molecules and generally satisfies the relation:

$$\int_{0}^{\infty} g(\nu)\, d\nu = 1. \tag{7}$$

Since u_{ν} is much broader than $g(\nu)$, it can be assumed to be constant over the linewidth, and with (7), the integral in (6) can be replaced by the spectral energy density on the transition frequency:

$$\int_{\Delta\nu} u_{\nu} g(\nu)\, d\nu = u_{\nu_{mn}}. \tag{8}$$

Then the probability for induced absorption processes simply becomes

$$dW_{nm}^{i} = B_{nm} u_{\nu_{mn}} dt. \tag{9}$$

Induced Emission. A transition from $m \rightarrow n$ caused by the radiation is called the induced or stimulated emission. The probability for this transition is

$$dW^{i}_{mn} = B_{mn} \int_{\Delta\nu} u_{\nu} g(\nu)\, d\nu \cdot dt = B_{mn} u_{\nu_{mn}} dt \tag{10}$$

with B_{mn} as the Einstein coefficient of induced emission.

Total Transition Rates. Under thermodynamic equilibrium, the total number of absorbing transitions must be the same as the number of emissions. These numbers depend on the population of a state and the probabilities for a transition to the other state. According to (5)–(10), this can be expressed by

$$B_{nm} u_{\nu_{mn}} N_n = \left(B_{mn} u_{\nu_{mn}} + A_{mn} \right) N_m, \tag{11}$$

or more universally described by rate equations:

$$\frac{dN_m}{dt} = +B_{nm} u_{\nu_{mn}} N_n - \left(B_{mn} u_{\nu_{mn}} + A_{mn} \right) N_m,$$

$$\frac{dN_n}{dt} = -B_{nm} u_{\nu_{mn}} N_n + \left(B_{mn} u_{\nu_{mn}} + A_{mn} \right) N_m. \tag{12}$$

For $dN_m/dt = dN_n/dt = 0$, (12) gets identical to (11). Using (4) in (11) gives

$$g_n B_{nm} u_{\nu_{mn}} = g_m \left(B_{mn} u_{\nu_{mn}} + A_{mn} \right) e^{-h\nu_{mn}/kT_G}. \tag{13}$$

Assuming that with T_G also $u_{\nu_{mn}}$ gets infinite, B_{mn} and B_{nm} must satisfy the relation:

$$g_n B_{nm} = g_m B_{mn}. \tag{14}$$

Then (13) becomes

$$B_{mn} u_{\nu_{mn}} e^{h\nu_{mn}/kT_G} = B_{mn} u_{\nu_{mn}} + A_{mn}, \tag{15}$$

or resolving to $u_{\nu_{mn}}$ gives

$$u_{\nu_{mn}} = \frac{A_{mn}}{B_{mn}} \frac{1}{e^{h\nu_{mn}/kT_G} - 1}. \tag{16}$$

This expression in Einstein's consideration is of the same type as the Planck distribution for the spectral energy density. Therefore, comparison of (3) and (16) at ν_{mn} gives for B_{mn}:

$$B_{mn} = \frac{g_n}{g_m} B_{nm} = \frac{A_{mn} c^3}{8\pi n^3 h\nu_{mn}^3} = \frac{c}{n} \frac{1}{h\nu_{mn}} \frac{A_{mn}\lambda_{mn}^2}{8\pi n^2}, \tag{17}$$

showing that the induced transition probabilities are also proportional to the spontaneous emission rate A_{mn}, and in units of the photon energy, $h\nu_{mn}$, are scaling with λ^2 .

Relationship to Other Spectroscopic Quantities

The Einstein coefficients for induced absorption and emission are directly related to some other well-established quantities in spectroscopy, the cross sections for induced transitions, and the absorption and gain coefficient of a sample.

Cross Section and Absorption Coefficient

Radiation propagating in -direction through an absorbing sample is attenuated due to the interaction with the molecules. The decay of the spectral energy obeys Lambert-Beer's law, here given in its differential form:

$$\frac{du_\nu}{dr} = -\alpha_{nm}(\nu)\, u_\nu = -\sigma_{nm}(\nu)\, N_n u_\nu, \tag{18}$$

where $\alpha_{nm}(\nu)$ is the absorption coefficient (units: cm^{-1}) and $\sigma_{nm}(\nu)$ is the cross section (units: cm^2) for induced absorption.

For a more general analysis, however, also emission processes have to be considered, which partly or completely compensate for the absorption losses. Then two cases have to be distinguished, the situation we discuss in this section, where the molecules are part of an environmental thermal bath, and on the other hand, the case where a directed external radiation prevails upon a gas cloud, which will be considered in the next section.

In the actual case, (18) has to be expanded by two terms representing the induced and also the spontaneous emission. Quite similar to the absorption, the induced emission is given by the cross section of induced emission $\sigma_{mn}(\nu)$, the population of the upper state $_{Nm}$, and the spectral radiation density u_ν. The product $\sigma_{mn}(\nu)\cdot N_m$ now describes an amplification of u_ν and is known as gain coefficient.

Additionally, spontaneously emitted photons within a considered volume element and time interval $dt = dr \cdot n/c$ contribute to the spectral energy density of the thermal background radiation with

$$du_\nu = h\nu_{mn} A_{mn} dt N_m g(\nu) = \frac{h\nu_{mn}}{c/n} A_{mn} dr N_m g(\nu). \tag{19}$$

Then altogether this gives

$$\frac{du_\nu}{dr} = -\sigma_{nm}(\nu) N_n u_\nu + \sigma_{mn}(\nu) N_m u_\nu + \frac{h\nu_{mn}}{c/n} A_{mn} N_m g(\nu). \tag{20}$$

As we will see in Section 2.5 and later in Section 3.3 or Section 4, (19) is the source term of the thermal background radiation in a gas, and (20) already represents the theoretical basis for calculating the radiation transfer of thermal radiation in the atmosphere.

The frequency dependence of $\sigma_{nm}(\nu)$ and $\sigma_{mn}(\nu)$, and thus the resonant interaction of radiation with a molecular transition can explicitly be expressed by the normalized lineshape function $g(\nu)$ as

$$\sigma_{mn}(\nu) = \sigma^0_{mn} g(\nu), \qquad \sigma_{nm}(\nu) = \sigma^0_{nm} g(\nu).$$

Equation (20) may be transformed into the time domain by $dr = c/n \cdot dt$ and additionally integrated over the lineshape $g(\nu)$. When u_ν can be assumed to be broad compared to $g(\nu)$,the energy density u as the integral over the lineshape of width $\Delta\nu$ becomes

$$\int_{\Delta\nu} \frac{du_\nu}{dt} d\nu = \frac{du_{\nu_{mn}}}{dt} \Delta\nu = \frac{du}{dt}$$
$$= -\frac{c}{n}\left(\sigma^0_{nm} N_n - \sigma^0_{mn} N_m\right) u_{\nu_{mn}} + h\nu_{mn} A_{mn} N_m. \tag{22}$$

This energy density of the thermal radiation u (units: J/m^3) can also be expressed in terms of a photon density ρG [m^{-3}] in the gas, multiplied with the photon energy $h \cdot \nu_{mn}$ with

$$u = u_{\nu_{mn}} \Delta\nu = \rho_G \cdot h\nu_{mn}. \tag{23}$$

Since each absorption of a photon reduces the population of state n and increases m by the same amount—for an emission it is just opposite—this yields

$$\frac{d\rho_G}{dt} = \frac{dN_n}{dt} = -\frac{dN_m}{dt}$$
$$= -\frac{c/n}{h\nu_{mn}}\left(\sigma^0_{nm} N_n - \sigma^0_{mn} N_m\right) u_{\nu_{mn}} + A_{mn} N_m \tag{24}$$

which is identical with the balance in (12). Comparison of the first terms on the right side and applying (17) gives the identity

$$B_{nm} u_{\nu_{mn}} N_n = \frac{g_m}{g_n} \frac{c}{n} \frac{1}{h\nu_{mn}} \frac{A_{mn}\lambda_{mn}^2}{8\pi n^2} u_{\nu_{mn}} N_n$$

$$= \frac{c}{n} \frac{1}{h\nu_{mn}} \sigma_{nm}^0 N_n u_{\nu_{mn}}, \tag{25}$$

and therefore

$$\sigma_{nm}^0 = \frac{g_m}{g_n} \frac{A_{mn}\lambda_{mn}^2}{8\pi n^2}. \tag{26}$$

Comparing the second terms in (12) and (24) results in

$$\sigma_{mn}^0 = \frac{A_{mn}\lambda_{mn}^2}{8\pi n^2} = \frac{g_n}{g_m} \sigma_{nm}^0. \tag{27}$$

So, together with (21) and (27), we derive as the final expressions for $\sigma_{nm}(\nu)$ and $\sigma_{mn}(\nu)$:

$$\sigma_{mn}(\nu) = \frac{g_n}{g_m} \sigma_{nm}(\nu) = \frac{A_{mn}\lambda_{mn}^2}{8\pi n^2} g(\nu),$$

$$\sigma_{mn}(\nu) = \frac{h\nu_{mn}}{c/n} B_{mn} g(\nu), \qquad \sigma_{nm}(\nu) = \frac{h\nu_{mn}}{c/n} B_{nm} g(\nu). \tag{28}$$

Effective Cross Section and Spectral Line Intensity

Often the first two terms on the right side of (20) are unified and represented by an effective cross section $\overline{\sigma}_{nm}(\nu)$. Further relating the interaction to the total number density N of the molecules, it applies

$$\overline{\sigma}_{nm}(\nu) N = \sigma_{nm}(\nu) N_n - \sigma_{mn}(\nu) N_m$$

$$= \sigma_{nm}(\nu) N_n \left(1 - \frac{g_n}{g_m} \frac{N_m}{N_n}\right)$$

$$= \frac{h\nu_{mn}}{c/n} B_{nm} g(\nu) N_n \left(1 - \frac{g_n}{g_m} \frac{N_m}{N_n}\right), \tag{29}$$

and $\overline{\sigma}_{nm}(\nu)$ becomes

$$\overline{\sigma}_{nm}(\nu) = \frac{h\nu_{mn}}{c/n} B_{nm} \frac{N_n}{N}\left(1 - \frac{g_n}{g_m}\frac{N_m}{N_n}\right) g(\nu). \tag{30}$$

Integration of (30) over the linewidth gives the spectral line intensity S_{nm} of a transition (Figure 3):

$$S_{nm} = \int_{\Delta\nu} \overline{\sigma}_{nm}(\nu)\, d\nu = \frac{h\nu_{mn}}{c/n} B_{nm} \frac{N_n}{N}\left(1 - \frac{g_n}{g_m}\frac{N_m}{N_n}\right), \tag{31}$$

as it is used and tabulated in data bases [13, 14] to characterize the absorption strength on a transition.

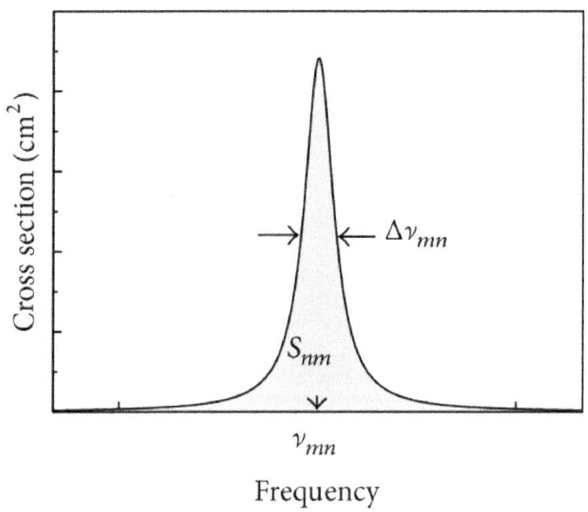

Figure 3: For explanation of the spectral line intensity.

Effective Absorption Coefficient

Similar to $\overline{\sigma}_{nm}(\nu)$, with (30) and (31), an effective absorption coefficient on a transition can be defined as

$$\begin{aligned}\overline{\alpha}_{nm}(\nu) &= \overline{\sigma}_{nm}(\nu) N \\ &= \frac{h\nu_{mn}}{c/n} B_{nm} N_n \left(1 - \frac{g_n}{g_m}\frac{N_m}{N_n}\right) g(\nu) \\ &= S_{nm} N g(\nu) = \overline{\alpha}^{0}_{nm} g(\nu),\end{aligned} \tag{32}$$

which after replacing B_{nm} from (17) assumes the more common form:

$$\overline{\alpha}_{nm}(\nu) = \frac{A_{mn}c^2}{8\pi n^2 \nu^2} N_n \left(\frac{g_m}{g_n} - \frac{N_m}{N_n} \right) g(\nu)$$

$$= \frac{A_{mn}\lambda^2}{8\pi n^2} N_n \left(\frac{g_m}{g_n} - \frac{N_m}{N_n} \right) g(\nu). \tag{33}$$

Collisions

Generally the molecules of a gas underlie collisions, which may perturb the phase of a radiating molecule, and additionally cause transitions between the molecular states. The transition rate from $m \rightarrow n$ due to de-exciting, nonradiating collisions (superelastic collisions, of 2nd type) may be called C_{mn} and that for transitions from $n \rightarrow m$ (inelastic collisions, of 1st type) as exciting collisions C_{nm}, respectively (see Figure 4).

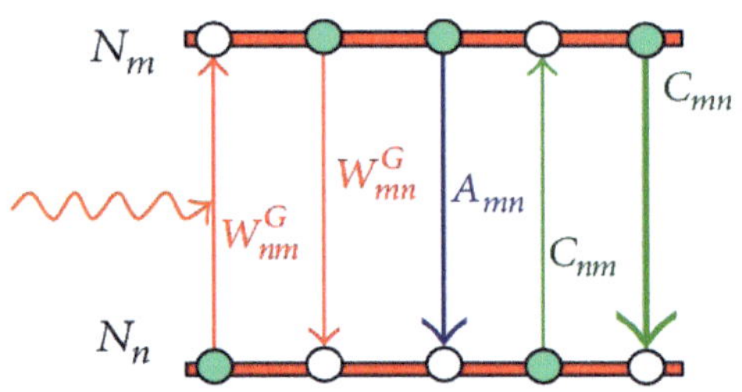

Figure 4: Two-level system with transition rates due to stimulated, spontaneous, and collisional processes.

Rate Equations

Then, with (25) and the abbreviations W^G_{mn} and W^G_{nm} as radiation induced transition rates or transition probabilities (units: s^{-1})

$$W^G_{mn} = B_{mn} u_{\nu_{mn}} = \frac{c}{n} \frac{1}{h\nu_{mn}} \sigma^0_{mn} u_{\nu_{mn}},$$

$$W^G_{nm} = B_{nm} u_{\nu_{mn}} = \frac{c}{n} \frac{1}{h\nu_{mn}} \sigma^0_{nm} u_{\nu_{mn}} = \frac{g_m}{g_n} W^G_{mn} \tag{34}$$

the rate equations as generalization of (12) or (24) and additionally supplemented by the balance of the electromagnetic energy density or photon density (see (22)–(24)) assume the form:

$$\frac{dN_m}{dt} = +\left(W^G_{nm} + C_{nm}\right) N_n - \left(W^G_{mn} + A_{mn} + C_{mn}\right) N_m,$$

$$\frac{dN_n}{dt} = -\left(W^G_{nm} + C_{nm}\right) N_n + \left(W^G_{mn} + A_{mn} + C_{mn}\right) N_m,$$

$$\frac{d\rho_G}{dt} = -W^G_{nm} N_n + W^G_{mn} N_m + A_{mn} N_m. \tag{35}$$

At thermodynamic equilibrium, the left sides of (35) are getting zero. Then, also and even particularly in the presence of collisions the populations of states m and n will be determined by statistical thermodynamics. So, adding the first and third equation of (35), together with (4), it is found some quite universal relationship for the collision rates

$$C_{nm} = \frac{g_m}{g_n} e^{-h\nu_{mn}/kT_G} C_{mn}, \tag{36}$$

showing that transitions due to inelastic collisions are directly proportional to those of superelastic collisions with a proportionality factor given by Boltzmann's distribution. From (35), it also results that states, which are not connected by an allowed optical transition, nevertheless will assume the same populations as those states with an allowed transition.

Radiation Induced Transition Rates

When replacing $u_{\nu_{mn}}$ in (34) by (3), the radiation induced transition rates can be expressed as

$$W^G_{mn} = \frac{A_{mn}}{e^{h\nu_{mn}/kT_G} - 1}, \qquad W^G_{nm} = \frac{g_m}{g_n} \frac{A_{mn}}{e^{h\nu_{mn}/kT_G} - 1}. \tag{37}$$

Inserting some typical numbers into (37), for example, a transition wavelength of 15 μm for the prominent CO^{2-} absorption band and a temperature of $TG = 288$K, we calculate a ratio $W^G{}_{mn}/A_{mn} = 0.037$. Assuming that $gm = gn$, almost the same is found for the population ratio (see (4)) with $N_m/N_n = 0.036$. At spontaneous transition rates of the order of $A_{mn} = 1$ s -1 for the stronger lines in this CO^{2-} band then we get a radiation induced transition rate of only $W^G{}_{mn} = W^G{}_{nm} = 0.03\text{-}0.04$ s $^{-1}$.

Under conditions as found in the troposphere with collision rates between molecules of several 10^9 s^{-1}, any induced transition rate due to the thermal background radiation is orders of magnitude smaller, and even up to the stratosphere and mesosphere, most of the transitions are caused by collisions, so that above all they determine the population of the states and in any case ensure a fast adjustment of a local thermodynamic equilibrium in the gas.

Nevertheless, the absolute numbers of induced absorption and emission processes per volume, scaling with the population density of the involved states (see (35)), can be quite significant.

So, at a CO_2 concentration of 400 ppm, the population in the lower state n is estimated to be about $Nn \sim 7 \times 10^{20}$ m^{-3} (dependent on its energy above the ground level).

Then more than 10^{19} absorption processes per m3 are expected, and since such excitations can take place simultaneously on many independent transitions, this results in a strong overall interaction of the molecules with the thermal bath, which according to Einstein's considerations even in the absence of collisions leads to the thermodynamic equilibrium.

Linewidth and Lineshape of a Transition

The linewidth of an optical or infrared transition is determined by different effects.

Natural Linewidth and Lorentzian Lineshape

For molecules in rest and without any collisions and also neglecting power broadening due to induced transitions, the spectral width of a line only depends on the natural linewidth

$$\Delta\nu_{mn} = \frac{1}{2\pi\tau_m} + \frac{1}{2\pi\tau_n} \approx \frac{A_{mn}}{2\pi} \tag{38}$$

which for a two-level system and $\tau n \gg \tau_m$ is essentially determined by the spontaneous transition rate $A_{mn} \approx 1/\tau_m$. τ_n and τ_m are the lifetimes of the lower and upper state.

The lineshape is given by a Lorentzian $g_L(\nu)$, which in the normalized form can be written as

$$g_L(\nu) = \frac{\Delta\nu_{mn}/2\pi}{(\nu - \nu_{mn})^2 + (\Delta\nu_{mn}/2)^2}, \qquad \int_0^\infty g_L(\nu)\, d\nu = 1. \tag{39}$$

Collision Broadening

With collisions in a gas, the linewidth will considerably be broadened due to state and phase changing collisions. Then the width can be approximated by

$$\Delta\nu_{mn} \approx \frac{1}{2\pi}\left(A_{mn} + C_{mn} + C_{nm} + C_{\text{phase}}\right), \tag{40}$$

where C_{phase} is an additional rate for phase changing collisions. The lineshape is further represented by a Lorentzian, only with the new homogenous width $\Delta\nu_{mn}$ (FWHM) according to (40).

Doppler Broadening

Since the molecules are at an average temperature T_G, they also possess an average kinetic energy

$$\overline{E}_{\text{kin}} = \frac{3}{2}kT_G = \frac{m_G}{2}\overline{v^2}, \tag{41}$$

with m_G as the mass and $\overline{v^2}$ as the velocity of the molecules squared and averaged. Due to a Doppler shift of the moving particles, the molecular transition frequency is additionally broadened and the number of molecules interacting within their homogeneous linewidth with the radiation at frequency ν is limited. This inhomogeneous broadening is determined by Maxwell's velocity distribution and known as Doppler broadening. The normalized Doppler lineshape is given by a Gaussian function of the form:

$$g_D(\nu) = \frac{2\sqrt{\ln 2}}{\sqrt{\pi}\Delta\nu_D}\exp\left(-\frac{(\nu-\nu_{mn})^2}{(\Delta\nu_D/2)^2}\ln 2\right) \tag{42}$$

with the Doppler linewidth

$$\Delta\nu_D = 2\nu_{mn}\left(\frac{2kT_G}{m_G c^2}\ln 2\right)^{1/2} = 7.16\cdot 10^{-7}\nu_{mn}\sqrt{\frac{T_G}{M}}, \tag{43}$$

where M is the molecular weight in atomic units and specified in T_G.

Voigt Profile

For the general case of collision and Doppler broadening, a convolution of $g_D(\nu')g_L(\nu-\nu')$ and $g_V(\nu)$ gives the universal lineshape representing a Voigt profile of the form:

$$g_V(\nu) = \int_0^\infty g_D(\nu') g_L(\nu - \nu') d\nu'$$
$$= \frac{\sqrt{\ln 2}}{\pi\sqrt{\pi}} \frac{\Delta\nu_{mn}}{\Delta\nu_D}$$
$$\times \int_0^\infty \exp\left(-\frac{(\nu' - \nu_{mn})^2}{(\Delta\nu_D/2)^2} \ln 2 \right)$$
$$\cdot \frac{d\nu'}{(\nu - \nu')^2 + (\Delta\nu_{mn}/2)^2}. \tag{44}$$

Spontaneous Emission as Thermal Background Radiation

From the rate equations (35), it is clear that also in the presence of collisions, the absolute number of spontaneously emitted photons per time should be the same as that without collisions.

This is also a consequence of (36) indicating that with an increasing rate C_{mn} for de-exciting transitions (without radiation), also the rate C_{nm} of exiting collisions is growing and just compensating for any losses, even when the branching ratio $R=A_{mn}/C_{mn}$ of radiating to nonradiating transitions decreases.

In other words, when the molecule is in state m, the probability for an individual spontaneous emission act is reduced by the ratio R, but at the same time, the number of occurring decays per time increases with C_{mn}.

Then the spectral power density in the gas due to spontaneous emissions is (see (19))

$$\frac{du_\nu}{dt} = h\nu_{mn} A_{mn} N_m g(\nu), \tag{45}$$

representing a spectral generation rate of photons of energy $h\nu_{mn}$ per volume. Photons emerging from a volume element generally spread out into neighbouring areas, but in the same way, there is a backflow from the neighbourhood, which in a homogeneous medium just compensates these losses. Nevertheless, photons have an average lifetime, before they are annihilated due to an absorption in the gas. With an average photon lifetime

$$\tau_{\rm ph} = l_{\rm ph}\frac{n}{c} = \frac{n}{\bar{\alpha}_{nm} c}, \tag{46}$$

where $l_{ph} = 1/\overline{\alpha}_{nm}$ is the mean free path of a photon in the gas before it is absorbed, we can write for the spectral energy density:

$$u_\nu = h\nu_{mn}A_{mn}\tau_{ph}N_m g(\nu) = \frac{n}{c}h\nu_{mn}\frac{A_{mn}}{\overline{\alpha}_{nm}}N_m g(\nu). \quad (47)$$

The same result is derived when transforming (20) into the time domain and assuming a local thermodynamic equilibrium with $du_\nu/dt = 0$:

$$\begin{aligned}(\sigma_{nm}(\nu)N_n - \sigma_{mn}(\nu)N_m)u_\nu &= \overline{\alpha}_{nm}u_\nu \\ &= \frac{n}{c}h\nu_{mn}A_{mn}N_m g(\nu).\end{aligned}$$

With $\overline{\alpha}_{nm}(\nu)$ from (33) and Boltzmann's relation (4), then the spectral energy density at ν_{mn} is found to be

$$\begin{aligned}u_{\nu_{mn}} &= \frac{8\pi n^3 h\nu_{mn}^3}{c^3}\frac{1}{(g_m N_n/g_n N_m - 1)} \\ &= \frac{8\pi n^3 h\nu_{mn}^3}{c^3}\frac{1}{e^{h\nu_{mn}/kT_G} - 1}.\end{aligned} \quad (49)$$

This is the well-known Planck formula (3) and shows that without an additional external excitation at thermodynamic equilibrium, the spontaneous emission of molecules can be understood as nothing else as the thermal radiation of a gas on the transition frequency.

This derivation differs insofar from Einstein's consideration leading to (16), as he concluded that a radiation field, interacting with the molecules at thermal and radiation equilibrium, just had to be of the type of a Planckian radiator, while here we consider the origin of the thermal radiation in a gas sample, which exclusively is determined and rightfully defined by the spontaneous emission of the molecules themselves. This is also valid in the presence of molecular collisions. Because of this origin, the thermal background radiation only exists on discrete frequencies, determined by the transition frequencies and the linewidths of the molecules, as long as no external radiation is present. But on these frequencies, the radiation strength is the same as that of a blackbody radiator.

Since this spontaneous radiation is isotropically emitting photons into the full solid angle 4π, in average half of the radiation is directed upward and half downward.

INTERACTION OF MOLECULES WITH THERMAL RADIATION FROM AN EXTERNAL SOURCE

In this section, we consider the interaction of molecules with an additional blackbody radiation emitted by an external source like the earth's surface or adjacent atmospheric layers.

We also investigate the transfer of absorbed radiation to heat in the presence of molecular collisions, causing a rise of the atmospheric temperature. And vice versa, we study the transfer of a heat flux to radiation resulting in a cooling of the gas. An appropriate means to describe the mutual interaction of these processes is to express this by coupled rate equations which are solved numerically.

Basic Quantities

Spectral Radiance

The power radiated by a surface element dA on the frequency ν in the frequency interval $d\nu$ and into the solid angle element $d\Omega$ is also determined by Planck's radiation law:

$$I_{\nu,\Omega}(T_E)\, dA \cos\beta d\nu d\Omega$$

$$= \frac{2h\nu^3 n^2}{c^2} \frac{1}{e^{h\nu/kT_E} - 1} dA \cos\beta d\nu d\Omega, \tag{54}$$

with $I_{\nu,\Omega}(T_E)$ as the spectral radiance (units: W/m^2 / Hz/sterad) and TE as the temperature of the emitting surface of the source (e.g., earth's surface). The cosine term accounts for the fact that for an emission in a direction given by the azimuthal angle φ and the polar angle β, only the projection of dA perpendicular to this direction is efficient as radiating surface (Lambertian radiator).

Spectral Flux Density—Spectral Intensity

Integration over the solid angle Ω gives the spectral flux density. Then, representing $d\Omega$ in spherical coordinates as $d\Omega = \sin\beta d\beta d\varphi$ with $d\varphi$ as the azimuthal angle interval and $d\beta$ as the polar interval (see Figure 5), this results in:

$$I_\nu = \int_\Omega^{2\pi} I_{\nu,\Omega}(T_E)\cos\beta d\Omega$$

$$= \int_{\varphi=0}^{2\pi}\int_{\beta=0}^{\pi/2} I_{\nu,\Omega}(T_E)\cos\beta\sin\beta d\beta d\varphi$$

$$= \frac{2\pi h\nu^3 n^2}{c^2}\frac{1}{e^{h\nu/kT_E}-1} \quad (51)$$

or in wavelengths units

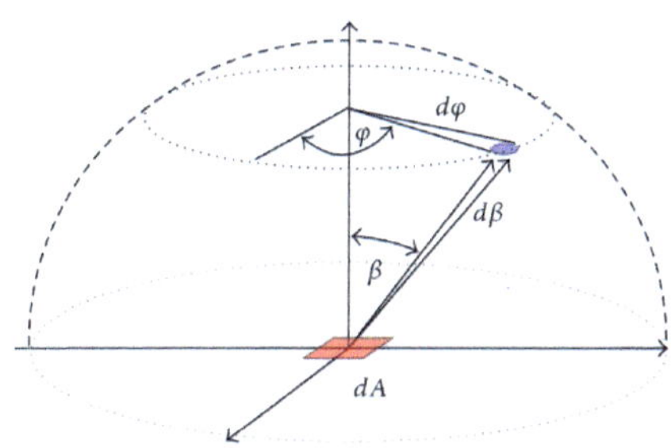

Figure 5: Radiation from a surface element dA into the solid angle $d\Omega = \sin\beta d\beta d\varphi$.

$$I_\lambda = \frac{cn}{\lambda^2} I_\nu = \frac{2\pi hc^2 n^3}{\lambda^5}\frac{1}{e^{hc/kT_E\lambda}-1}. \quad (52)$$

I_λ and I_ν, also known as spectral intensities, are specified in units of $W/m^2/\mu m$ and $W/m^2/Hz$, respectively. They represent the power flux per frequency or wavelength and per surface area into that hemisphere, which can be seen from the radiating surface element. The spectral distribution of a Planck radiator of $T = 299K$ as a function of wavelength is shown in Figure 6.

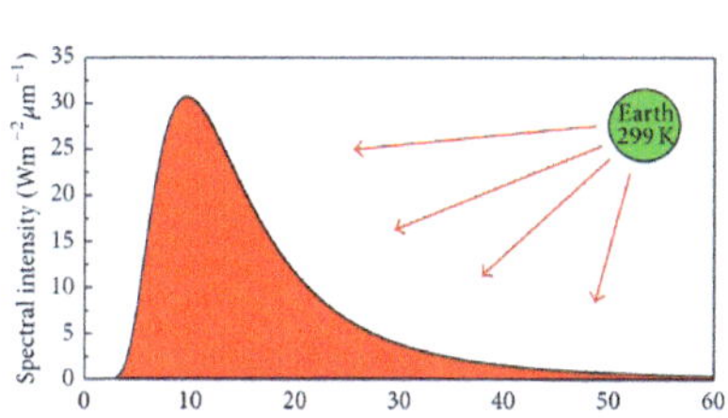

Figure 6: Spectrum of a Planck radiator at 299 K temperature.

Ray Propagation in a Lossy Medium

Spectral Radiance

Radiation passing an absorbing sample generally obeys Lambert-Beer's law, as already applied in (18) for the spectral energy density u_ν. The same holds for the spectral radiance with

$$\frac{dI_{\nu,\Omega}}{dr} = -\bar{\alpha}_{nm}(\nu)\, I_{\nu,\Omega}, \tag{53}$$

where dr is the propagation distance of $I_{\nu,\Omega}$ in the sample. This is valid independent of the chosen coordinate system. The letter r is used when not explicitly a propagation perpendicular to the earth's surface or a layer (z-direction) is meant.

Spectral Intensity

For the spectral intensity, which due to the properties of a Lambertian radiator consists of a bunch of rays with different propagation directions and brightness, some basic deviations have to be recognized. So, because of the individual propagation directions spreading over a solid angle of 2π, only for a homogeneously absorbing sphere and in a spherical coordinate system with the radiation source in the center of this sphere the distances to pass the sample, and thus the individual contributions to the overall absorption, would be the same.

But radiation, emerging from a plane parallel surface and passing an absorbing layer of thickness Δz, is better characterized by its average expansion perpendicular to the layer surface in z-direction. This means, that with respect to this direction an individual ray, propagating under an angle β to the surface normal covers a distance $\Delta l = \Delta z/\cos\beta$ before leaving the layer (see Figure 7). Therefore, such a ray on one side contributes to a larger relative absorption, and on the other side, it donates this to the spectral intensity only with weight $I_{\nu,\Omega}\cos\beta d\Omega$ due to Lambert's law

This means that to first order, each individual beam direction suffers from the same absolute attenuation, and particularly the weaker rays under larger propagation angles β waste relatively more of their previous spectral radiance. Thus, especially at higher absorption strengths and longer propagation lengths, the initial Lambertian distribution will be more and more modified. A criterion for an almost unaltered distribution may be that for angles $\beta \leq \beta_{max}$ the inequality $\alpha_{nm} \cdot \Delta z/\cos\beta \ll 1$ is satisfied. The absorption loss for the spectral intensity as the integral of the spectral radiance $I_{\nu,\Omega}$ (see (51)) then is given by

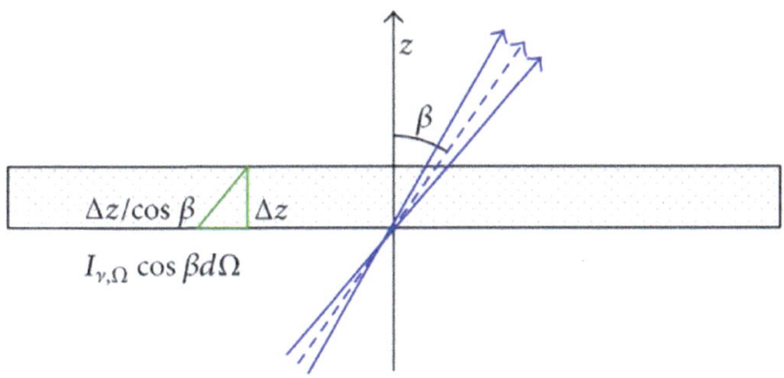

Figure 7: Radiation passing a gas layer.

$$\frac{\int dI_{\nu,\Omega}(z)\cos\beta d\Omega}{dz} = -\overline{\alpha}_{nm}(\nu)\int_0^{2\pi}\int_0^{\pi/2} I_{\nu,\Omega}\frac{\cos\beta}{\cos\beta}\sin\beta d\beta d\varphi. \tag{54}$$

The cosine terms under the integral sign just compensate and (54) can be written as

$$\frac{dI_\nu(z,T_E)}{dz} = -2\overline{\alpha}_{nm}(\nu)I_\nu(z,T_E). \tag{55}$$

This differential equation for the spectral intensity shows that the effective absorption coefficient is twice that of the spectral radiance, or in other words, the average propagation length of the radiation to pass the layer is twice the layer thickness.This last statement means that we also can assume radiation, which is absorbed at the regular absorption coefficient $\overline{\alpha}_{nm}$, but is propagating as a beam under an angle of 60∘ to the surface normal (1/cos 60∘ = 2). In practice, it even might give sense to deviate from an angle of 60∘ , to compensate for deviations of the earth's or oceanic surface from a Lambertian radiator, and to account for contributions due to Mie and Rayleigh scattering.

The absorbed spectral power density $\overline{L}_\nu$ (power per volume and per frequency) with respect to the z-direction then is found to be

$$\overline{L}_\nu(z) = \frac{dI_\nu(z)}{dz} = -2\overline{\alpha}_{nm}(\nu)\, I_\nu(0)\, e^{-2\overline{\alpha}_{nm}(\nu)z}, \tag{56}$$

with $I_\nu(0)$ as the initial spectral intensity at z=0 as given by (51).

An additional decrease of $\overline{L}_\nu$ with z due to a lateral expansion of the radiation over the hemisphere can be neglected, since any propagation of the radiation is assumed to be small compared to the earth's radius (consideration of an extended radiating parallel plane).

Integration of (56) over the lineshape $g(\nu)$ within the spectral interval $\Delta\nu$ gives thc absorbed power density L (units: W/m^3), which similar to (23) may also be expressed as loss or annihilation of photons per volume (ρE in m^{-3}) and per time. Since the average propagation speed of I_ν in z-direction is $c/2n$, also the differentials transform as $dz = c/2n{\cdot}dt$. With I_ν from (51), this results in

$$\begin{aligned}\overline{L}(z) &= \frac{2n}{c}\int_{\Delta\nu}\frac{dI_\nu}{dt}d\nu = \frac{2n}{c}\frac{dI_{\nu_{mn}}}{dt}\Delta\nu = 2\frac{d\rho_E}{dt}h\nu_{mn} \\ &= -\frac{A_{mn}}{2}N_n\left(\frac{g_m}{g_n} - \frac{N_m}{N_n}\right)\frac{h\nu_{mn}}{e^{h\nu_{mn}/kT_E}-1}e^{-2\overline{\overline{\alpha}}_{nm}z},\end{aligned} \tag{57}$$

or for the photon density ρE in terms of the induced transition rates W^E_{nm} and W^E_{mn} caused by the external field:

$$\begin{aligned}\frac{d\rho_E}{dt} &= -W^E_{nm}N_n + W^E_{mn}N_m \\ &= -\frac{A_{mn}}{4}\left(\frac{g_m}{g_n}N_n - N_m\right)\frac{1}{e^{h\nu_{mn}/kT_E}-1}e^{-2\overline{\overline{\alpha}}_{nm}z},\end{aligned} \tag{58}$$

with $\overline{\overline{\alpha}}_{nm} = \overline{\alpha}^0_{nm}/\Delta\nu = S_{nm}N/\Delta\nu$ as an averaged absorption coefficient over the linewidth (see also (32)) and with

$$\begin{aligned}W^E_{mn}(z) &= \frac{A_{mn}\lambda^2_{mn}}{8\pi n^2 h\nu_{mn}}I_{\nu_{mn}}(z) = \frac{\sigma^0_{mn}}{h\nu_{mn}}I_{\nu_{mn}}(z) \\ &= \frac{A_{mn}}{4}\frac{e^{-2S_{nm}Nz/\Delta\nu}}{e^{h\nu_{mn}/kT_E}-1}, \\ W^E_{nm} &= \frac{g_m}{g_n}W^E_{mn}.\end{aligned} \tag{59}$$

Similar to (37), these rates are again proportional to the spontaneous transition probability (rate) A_{mn}, but now they depend on the temperature T_E of the external source and the propagation depth z.

Due to the fact that radiation from a surface with a Lambertian distribution and only from one hemisphere is acting on the molecules, different to (37), a factor of appears in this equation.

At a temperature K and wavelength μm, for example, the ratio of induced to spontaneous transitions due to the external field at is

$$\frac{W_{mn}^{E}}{A_{mn}} = 0.0093 \approx 1\%. \qquad (60)$$

For the case of a two-level system, A_{mn} can also be expressed by the natural linewidth of the transition with (see (38))

$$\Delta\nu_{mn}^{N} = \frac{A_{mn}}{2\pi}. \qquad (61)$$

Then (59) becomes

$$W_{mn}^{E}(z) = \frac{\pi}{2}\Delta\nu_{mn}^{N}\frac{e^{-2S_{nm}Nz/\Delta\nu}}{e^{h\nu_{mn}/kT_E} - 1}. \qquad (62)$$

With a typical natural width of the order of only 0.1 Hz in the 15 μm band of CO^2, then the induced transition rate will be less than 0.01 s^{-1}. Even on the strongest transitions around 4.2 μm, the natural linewidths are only ~100 Hz, and therefore the induced transition rates are about 10 s^{-1}.

Despite of these small rates, the overall absorption of an incident beam can be quite significant. In the atmosphere, the greenhouse gases are absorbing on hundred thousands of transitions over long propagation lengths and at molecular number densities of 10^{19}–10^{23} per m^3 . On the strongest lines in the 15 μm band of CO^2, the absorption coefficient at the center of a line even goes up to 1 m−1. Then already within a distance of a few m, the total power will be absorbed on these frequencies. So, altogether about 85% of the total IR radiation emerging from the earth's surface will be absorbed by these gases.

Alternative Calculation for the Spectral Intensity

For some applications, it might be more advantageous, first to solve the differential equation (53) for the spectral radiance as a function of z and also β, before integrating over Ω. Then an integration in z-direction over the length $z/\cos\beta$ with a z-dependent absorption coefficient gives

$$I_{\nu,\Omega}(z) = I_{\nu,\Omega}(0)\, e^{-(1/\cos\beta)\int_0^z \bar{\alpha}_{nm}(z')dz'}, \tag{63}$$

and a further integration over Ω results in

$$\begin{aligned} I_\nu(z) &= \int_0^{2\pi}\int_0^{\pi/2} I_{\nu,\Omega}(z)\cos\beta\sin\beta d\beta d\varphi \\ &= I_{\nu,\Omega}(0)\int_0^{2\pi}\int_0^{\pi/2} e^{-(1/\cos\beta)\int_0^z \bar{\alpha}_{nm}(z')dz'} \\ &\qquad\times\cos\beta\sin\beta d\beta d\varphi. \end{aligned} \tag{64}$$

First integrating over φ and introducing the optical depth $\int_0^z \bar{\alpha}_{nm}(z')dz' = \tau$ as well as the substitution $u = \tau/\cos\beta$, we can write

$$\begin{aligned} I_\nu(\tau) &= 2\pi I_{\nu,\Omega}(0)\,\tau^2\int_\tau^\infty \frac{e^{-u}}{u^3}du \\ &= I_\nu(0)\,\tau^2\left(E_1(\tau) + e^{-\tau}\left(\frac{1}{\tau^2} - \frac{1}{\tau}\right)\right) \end{aligned} \tag{65}$$

with $I_\nu(0) = \pi I_{\nu,\Omega}(0)$ (see (51)) and the exponential integral

$$E_1(\tau) = \int_\tau^\infty \frac{e^{-u}}{u}du = -0.5772 - \ln\tau + \sum_{k=1}^{\infty}\frac{(-1)^{k+1}\tau^k}{k\cdot k!}. \tag{66}$$

Since the further considerations in this paper are concentrating on the radiation transfer in the atmosphere under the influence of thermal background radiation, it is more appropriate to describe the interaction of radiation with the molecules by a stepwise propagation through thin layers of depth Δz as given by (55).

Rate Equations under Atmospheric Conditions

As a further generalization of the rate equations (35), in this subsection, we additionally consider the influence of the external radiation, which together with the thermal background radiation is acting on the molecules in the presence of collisions (see Figure 8).

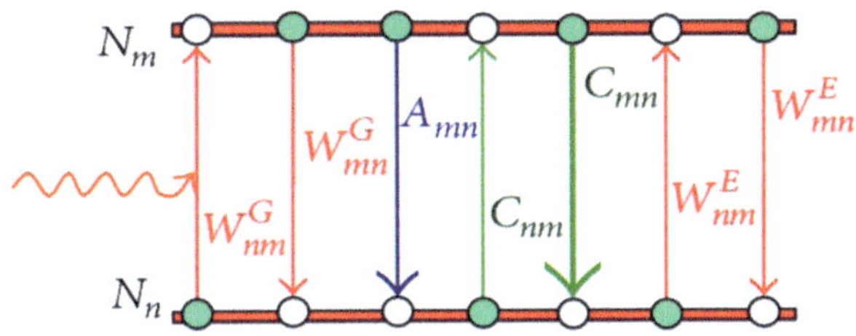

Figure 8: Two-level system with transition rates including an external excitation.

And different to Section 2, the infrared active molecules are considered as a trace gas in an open system, the atmosphere, which has to come into balance with its environment. Then molecules radiating due to their temperature and thus loosing part of their energy have to get this energy back from the surrounding by IR radiation, by sensitive or latent heat or also by absorption of sunlight. This is a consequence of energy conservation.

The radiation loss is assumed to be proportional to the actual photon density ρ_G and scaling with a flux rate ϕ. This loss can be compensated by the absorbed power of the incident radiation, for example, the terrestrial radiation, which is further expressed in terms of the photon density ρ_E (see (57)), and it can also be replaced by thermal energy. Therefore, the initial rate equations have to be supplemented by additional relations for these two processes.

It is evident that both, the incident radiation and the heat flux, will be limited by some genuine interactions. So, the radiation can only contribute to a further excitation, as long as it is not completely absorbed. At higher molecular densities, however, the penetration depth of the radiation in the gas is decreasing and therewith also the effective excitation over some longer volume element. Integrating (55) over the linewidth while using the definitions of (31) and (32) and then integrating over z results in

$$I_{\nu_{mn}}(z) = I_{\nu_{mn}}(0)\, e^{-2S_{nm}Nz/\Delta\nu} \qquad (67)$$

which due to Lambert-Beer's law describes the averaged spectral intensity over the linewidth as a function of the propagation in z-direction. From this, it is quite obvious to define the penetration depth as the length LP, over which the initial intensity reduces to $1/e$ and thus the exponent in (67) gets unity with

$$L_P = \frac{\Delta \nu}{2S_{nm}N_S}. \tag{68}$$

From (68), also the molecular number density *NS* is found, at which the initial intensity just drops to $1/e$ after an interaction length L_P. Since N_S characterizes the density at which an excitation of the gas gradually comes to an end and in this sense saturation takes place, it is known as the saturation density, where N_S and also S_{nm} refer to the total number density *N* of the gas.

Since at constant pressure, the number density in the gas is changing with the actual temperature due to GayLussac's law and the molecules are distributed over hundreds of states and substates, the molecular density, contributing to an interaction with the radiation on the transition $n \rightarrow m$, is given by [15]

$$N_n = N_0 \frac{T_0}{T_G} \frac{g_n e^{-E_n/kT_G}}{P(T_G)}, \tag{69}$$

where N_0 is the molecular number density at an initial temperature T_0, E_n is the energy of the lower level n above the ground state, and $P(T_G)$is the total internal partition sum, defined as [15, 16]

$$P(T_G) = \sum_i g_i e^{-E_i/kT_G}. \tag{70}$$

In the atmosphere typical propagation, lengths are of the order of several km. So, for a stronger CO^2 transition in the 15 μm band with a spontaneous rate A=1 s^{-1}, $Nn/N = Nn/N_0 = 0.07$, a spectral line intensity S_{nm} or integrated cross section of

$$S_{nm} = \overline{\sigma}^0_{nm} = \frac{A_{mn}\lambda^2}{8\pi n^2}\frac{N_n}{N}\left(1 - \frac{g_n}{g_m}\frac{N_m}{N_n}\right)$$
$$= 6 \times 10^{-13}\ \text{m}^2\ \text{Hz} = 2 \times 10^{-19}\ \text{cm}^{-1}/\left(\text{molecules}\cdot\text{cm}^{-2}\right) \tag{71}$$

and a spectral width at ground pressure of $\Delta\nu_{:}$ = 5 GHz, the saturation density over a typical length of L_P = 1 km assumes a value of $N_S \approx 4 \times 10^{18}$ molecules/m^3 . Since the number density of the air at 10^{13} hPa and 288 K is $N_{air} = 2.55 \times 10^{25}$ m^{-3}, saturation on the considered transition already occurs at a concentration of less than one ppm. It should also be noticed that at higher altitudes and thus lower densities, also the linewidth $\Delta\nu_{:}$ due to pressure broadening reduces. This means that to first order, also the saturation density decreases with $\Delta\nu$;, while the gas concentration in the atmosphere, at which saturation appears, almost remains constant. The same applies for the penetration depth.

For the further considerations, it is adequate to introduce an average spectral intensity $\bar{I}_{\nu_{mn}}$ as

$$\bar{I}_{\nu_{mn}} = \frac{1}{L_P}\int_0^{L_P} I_{\nu_{mn}}(z)\,dz$$

$$= \frac{\Delta\nu}{2S_{nm}NL_P} I_{\nu_{mn}}(0)\cdot\left(1 - e^{-2S_{nm}NL_P/\Delta\nu}\right) \tag{72}$$

or equivalently an average photon density

$$\bar{\rho}_E = \frac{n}{c}\frac{\Delta\nu}{h\nu_{mn}}\bar{I}_{\nu_{mn}} = \frac{\Delta\nu}{2S_{nm}NL_P}\rho_E(0)\cdot\left(1 - e^{-2S_{nm}NL_P/\Delta\nu}\right) \tag{73}$$

which characterizes the incident radiation with respect to its average excitation over L_p in the rate equations. In general, the incident flux $\rho(0)$ onto an atmospheric layer consists of two terms, the up- and the downwelling radiation.

Any heat flux, supplied to the gas volume, contributes to a volume expansion, and via collisions and excitation of molecules, this can also be transferred to radiation energy. Simultaneously absorbed radiation can be released as heat in the gas. Therefore, the rate equations are additionally supplemented by a balance for the heat energy density of the air Q [J/m^3], which under isobaric conditions and making use of the ideal gas equation in the form $p_{air} = N_{air} \cdot k \cdot TG$ can be written as

$$\Delta Q = c_p \rho_{air}\Delta T_G = \left(\frac{f}{2} + 1\right)\frac{R_G}{M_{air}}\frac{M_{air}}{N_A}\frac{p_{air}}{kT_G}\Delta T_G$$

$$= \frac{7}{2}\frac{p_{air}}{T_G}\Delta T_G, \tag{74}$$

or after integration

$$Q(T_G) = \frac{7}{2}p_{air}\int_{T_0}^{T_G}\frac{1}{T'_G}dT'_G = \frac{7}{2}p_{air}\ln\left(\frac{T_G}{T_0}\right). \tag{75}$$

Here cp represents the specific heat capacity of the air at constant pressure with $cp = (f/2 + 1) \cdot RG/M_{\text{air}}$, where f are the degrees of freedom of a molecule (for N_2 and O_2: f=5) and $RG = k \cdot N_A$ is the universal gas constant (at room temperature: cp = 1.01 kJ/kg/K). ρair is the specific weight of the air (at room temperature and ground pressure: ρ_{air} = 1.29 kg/m³), which can be expressed as $\rho_{\text{air}} = m_{\text{air}} \cdot N_{\text{air}} = M_{\text{air}}/NA \cdot N_{\text{air}} = M_{\text{air}}/NA \cdot p_{\text{air}}/(kTG)$ with $mair$ as the mass of an air molecule, M_{air} the mol weight, N_A Avogadro's number, N_{air} the number density, and $pair$ the pressure of the air.

The thermal energy can be supplied to a volume element by different means. So, thermal convection and conductivity in the gas contribute to a heat flux j_H [J/(s·m^2)], causing a temporal change of Q proportional to div j_H, or for one direction proportional to $\partial_{jHz}/\partial_z$. j_H is a vector and points from hot to cold. Since this flux is strongly dominated by convection, it can be well approximated by $j_{Hz} = h_C \cdot (T_E - T_G)$ with h_C as the heat transfer coefficient, typically of the order of $hC \approx 10^{-15}$ W/m^2 /K. At some more or less uniform distribution of the incident heat flux over the troposphere (altitude $HT \approx$ 12 km), the temporal change of Q due to convection may be expressed as $\partial_{jHz}/\partial_z = hC/H_T \cdot (T_E - T_G)$ with $hC/H_T \approx 1$ mW/m^3 /K.

$$\begin{aligned}
\frac{dN_m}{dt} &= +\left(\frac{h\nu_{mn}}{\Delta\nu}B_{nm}\left(\rho_G+\bar{\rho}_E\right)+C_{nm}\right)N_n \\
&\quad -\left(\frac{h\nu_{mn}}{\Delta\nu}B_{mn}\left(\rho_G+\bar{\rho}_E\right)+C_{mn}+A_{mn}\right)N_m, \\
\frac{dN_n}{dt} &= -\left(\frac{h\nu_{mn}}{\Delta\nu}B_{nm}\left(\rho_G+\bar{\rho}_E\right)+C_{nm}\right)N_n \\
&\quad +\left(\frac{h\nu_{mn}}{\Delta\nu}B_{mn}\left(\rho_G+\bar{\rho}_E\right)+C_{mn}+A_{mn}\right)N_m, \\
\frac{d\rho_G}{dt} &= -\frac{h\nu_{mn}}{\Delta\nu}\left(B_{nm}N_n-B_{mn}N_m\right)\rho_G+A_{mn}N_m-\phi\rho_G, \\
\frac{d\bar{\rho}_E}{dt} &= +\frac{c/n}{2L_P}\left(1-e^{-(2h\nu_{mn}/(\Delta\nu c/n))(B_{nm}N_n-B_{mn}N_m)L_P}\right)\rho_E(0) \\
&\quad -\frac{h\nu_{mn}}{\Delta\nu}\left(B_{nm}N_n-B_{mn}N_m\right)\bar{\rho}_E, \\
\frac{dQ}{dt} &= +\frac{h_C}{H_T}\left(T_E-T_G\right)+q_{SL}-h\nu_{mn}\left(C_{nm}N_n-C_{mn}N_m\right).
\end{aligned} \tag{76}$$

Another contribution to the thermal energy originates from the absorbed sunlight, which in the presence of collisions is released as kinetic or rotational energy of the molecules. Similarly, latent heat can be set free in the air. In the rate equations, both contributions are represented by a source term q_{SL} [J/(s·m^3)]. Finally, the heat balance is determined by exciting and de-exciting collisions, changing the population of the states n and , and reducing or increasing the energy by $h \cdot \nu_{mn}$. Altogether, this results in a set of coupled differential equations, which describe the simultaneous interaction of molecules with their self-generated thermal background radiation as well as with radiation from the earth's surface and/or a neighbouring layer, this all in the presence of collisions and under the influence of heat transfer processes:

To symbolize the mutual coupling of these equations, here we use a notation where the radiation induced transition rates are represented by the Einstein coefficients and the respective photon densities. A change to the other representation is easily performed applying the identities:

$$W_{mn}^{G} = \frac{h\nu_{mn}}{\Delta\nu} B_{mn}\rho_G = B_{mn}u_{\nu_{mn}},$$

$$W_{mn}^{E} = \frac{h\nu_{mn}}{\Delta\nu} B_{mn}\rho_E = B_{mn}\frac{n}{c}I_{\nu_{mn}},$$

$$W_{nm}^{G,E} = \frac{g_m}{g_n} W_{mn}^{G,E}. \tag{77}$$

It should be noticed that the rate equation for $\overline{\rho}_E$ could also be replaced by the incident radiation, as given by (73). However, since $\overline{\rho}_E$ approaches its equilibrium value within a time constant $\tau = \Delta\nu/(h\nu_{mn}(B_{nm}N_n - B_{mn}N_m))$ short compared with other processes, for a uniform representation the differential form was preferred.

For the special case of stationary equilibrium with $dN_n/dt = dN_m/dt = 0$ from the first rate equation, we get for the population ratio:

$$\frac{N_m}{N_n} = \frac{h\nu_{mn}B_{nm}\left(\rho_G + \overline{\rho}_E\right)/\Delta\nu + C_{nm}}{h\nu_{mn}B_{mn}\left(\rho_G + \overline{\rho}_E\right)/\Delta\nu + C_{mn} + A_{mn}}$$

$$= \frac{W_{nm}^{G} + W_{nm}^{E} + C_{nm}}{W_{mn}^{G} + W_{mn}^{E} + A_{mn} + C_{mn}} \tag{78}$$

which at $T_G \sim T_E = 288$K and $\lambda = 15\ \mu$m due to $W^G_{mn} = 4W^E_{mn} = 0.037A_{mn}$ (see (37) and (60)) in the limit of pure spontaneous decay processes reaches its maximum value of 4.4%, while in the presence of collisions with collision rates in the troposphere of several 10^9 s $^{-1}$, it rapidly converges to 3.5%, given by the Boltzmann relation at temperature T_G and corresponding to a local thermodynamic equilibrium.

For the general case, the rate equations have to be solved numerically, for example, by applying the finite element method. While (76) in the presented form is only valid for a two-level system, a simulation under realistic conditions comparable to the atmosphere requires some extension, particularly concerning the energy transfer from the earth's surface to the atmosphere. Since the main trace gases CO_2, water vapour, methane, and ozone are absorbing the incident infrared radiation simultaneously on thousands of transitions, as an acceptable approximation for this kind of calculation, we consider these transitions to be similar and independent from each other, each of them contributing to the same amount to the energy balance. In the rate equations, this can easily be included by multiplying the last term in the equation for the energy density Q with an effective number M_{tr} of transitions.

In this approximation, the molecules of an infrared active gas and even a mixture of gases are represented by a "standard transition" which reflects the dynamics and time evolution of the molecular populations under the influence of the incident radiation, the background radiation, and the thermal heat transfer. M_{tr} is derived as the ratio of the total absorbed infrared intensity over the considered propagation length to the contribution of a single standard transition. An example for a numerical simulation in the troposphere, more precisely for a layer from ground level up to 100 m, is represented in Figure 9. The graphs show the evolution of the photon densities ρ_G and ρ_E, the population densities N_n and N_m of the states n and m, the accumulated heat density Q in the air, and the temperature TG of the gas (identical with the atmospheric temperature) as a function of time and as an average over an altitude of 100 m.

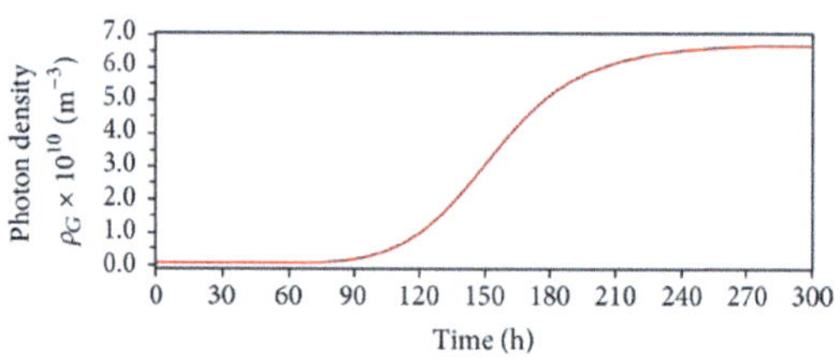

(a)

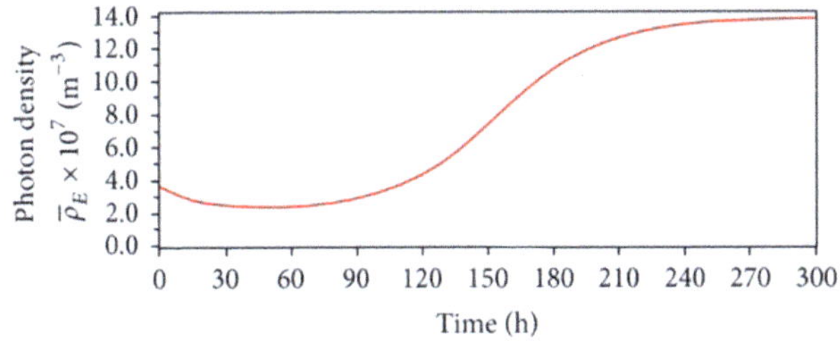
Photon density
$\overline{\rho}_E \times 10^7$ (m^{-3})
14.0
12.0
10.0
8.0
6.0
4.0
2.0
0.0
0 30 60 90 120 150 180 210 240 270 300
Time (h)

(b)

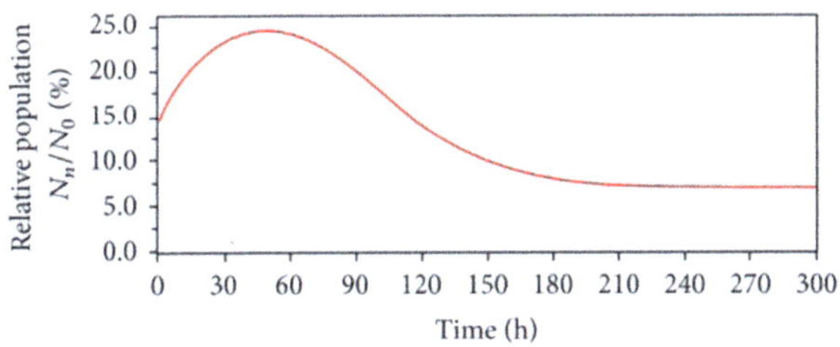
Relative population
N_n/N_0 (%)
25.0
20.0
15.0
10.0
5.0
0.0
0 30 60 90 120 150 180 210 240 270 300
Time (h)

(c)

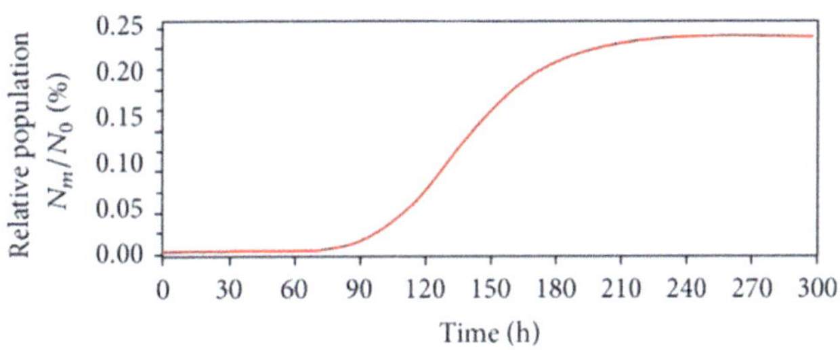
Relative population
N_m/N_0 (%)
0.25
0.20
0.15
0.10
0.05
0.00
0 30 60 90 120 150 180 210 240 270 300
Time (h)

(d)

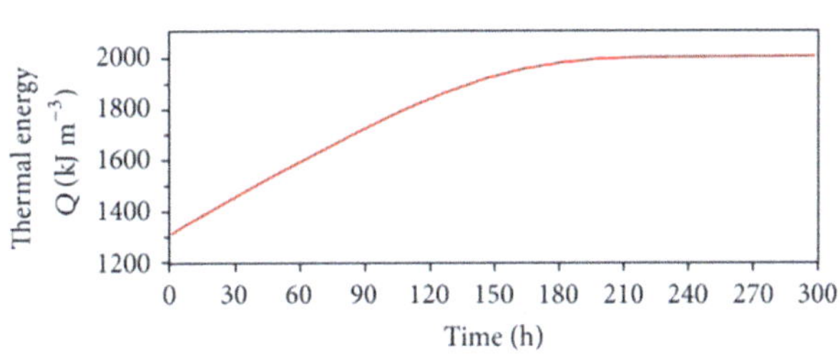
Thermal energy
Q (kJ m^{-3})
2000
1800
1600
1400
1200
0 30 60 90 120 150 180 210 240 270 300
Time (h)

(e)

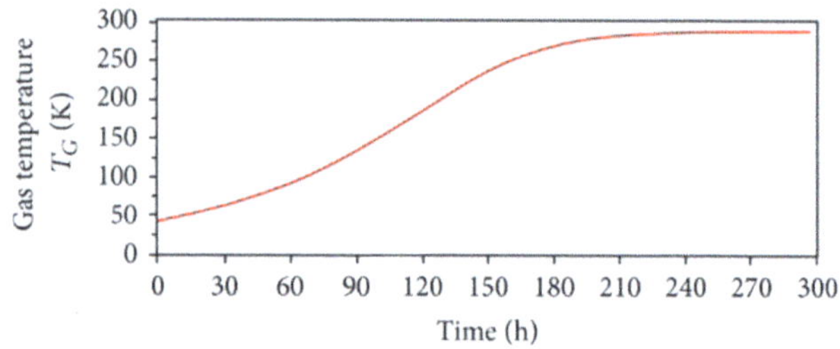

(f)

Figure 9: Solution of the rate equations for the troposphere (close to the earth's surface) representing the evolution of (a) the photon density $\overline{\rho}_E$ (b) the photon density (c) the lower and (d) the upper population density, (e) the thermal energy density, and (f) the gas- (air)-temperature as a function of time.

As initial conditions we assumed a gas and air temperature of $T_G = 40$K, a temperature of the earth's surface of $T_E = 288$K, an initial heat flux due to convection of $jHz = 3$ kW/m^2, a latent heat source of qSL = 3 mW/m^3 , and an air pressure of $pair$ = 10^{13} hPa. The interaction of the terrestrial radiation with the infrared active gases is demonstrated for CO^2 with a concentration in air of 380 ppm corresponding to a molecular number density of $N0$ = 9.7×10^{21} m^{-3} at 288 K. The simulation was performed for a "standard transition" as discussed before with a transition wavelength λ = 15 μm, a spontaneous transition rate Amn = 1 s −1, statistical weights of $g_m = 35$ and $g_n = 33$, a spectral width of Δ] = 5 GHz, a collisional transition rate C_{mn} = 10^8 s^{-1} (we do not distinguish between rotational and vibrational rates [10]), a penetration depth $L_p = 100$ m, and assuming a photon loss rate of $\phi = 7.5 \times 10^5$ s^{-1}. For this calculation, a density and temperature dependence of $\Delta\nu$ and C_{mn} was neglected. Since under these conditions a single transition contributes 0.069 W/m^2 to the total IR-absorption of 212 W/m^2 over 100 m, Mtr = 3095 transitions represent the radiative interaction with all active gases in the atmosphere.

The evolution for another gas and transition can easily displayed by replacing the respective parameters for the other transition and recalibrating the effective transition number.

To ensure reproducible results and to avoid any instabilities, the time interval for the stepwise integration of the coupled differential equation system has to be chosen small enough to avoid changes comparable to the size of a computed quantity itself, or the calculated changes have to be restrict by some upper boundaries, taking into account some smaller repercussions on the absolute time scale. The latter procedure was applied which accelerates the calculations by several orders of magnitude.

A simulation is done in such a way, that starting from the initial conditions for the populations, the photon densities, and the temperature, in a first step, modified population and photon densities as well as a change of the heat density over the time interval Δt are determined. The heat change causes a temperature change ΔT_G (see (74)) and gives a new temperature, which is used to calculate a new collision transfer rate C_{nm} by (36), a corrected population N_n by (69) and a new population N_m for the upper state by (4). The total internal partition sum for CO_2 is deduced from data of the HITRAN-database [13] and approximated by a polynomial of the form:

$$P(T_G) = 0.894 \cdot T_G + 3 \times 10^{-9} T_G^4. \tag{79}$$

The new populations are used as new initial conditions for the next time step, over which again the interaction of the molecules with the radiation and any heat transfer is computed. In this way, the time evolution of all coupled quantities shown in Figure 9 is derived. For this simulation, it was assumed that all atmospheric components still exist in gas form down to 40 K.

The relatively slow evolution of the curves is due to the fact, that the absorbed IR radiation as well as the thermal flux is mostly used to heat up the air volume which at constant air pressure is further expanding and the density decreasing.

Population changes of the CO_2 states, induced by the external or the thermal background radiation, are coupled to the heat reservoir via the collisional transition rates C_{mn} and C_{nm} (see last rate equation), which are linked to each other by (36). As long as the calculated populations differ from a Boltzmann distribution at the gas-temperature T_G, the two terms in the parenthesis of this rate equation do not compensate, and their difference contributes to an additional heating, amplified by the effective number of transitions. Reasons for smaller deviations from a local thermal equilibrium can be the induced population changes due to the external radiation, as well as temperature and density variations with altitude and daytime, or some local effects in the atmosphere. Also vice versa, when the gas is supplied with thermal energy, the heat can be transferred to electromagnetic energy and reradiated by the molecules (radiative cooling).

The photon density ρ_G (Figure 9(a)) shows the spontaneous emission of the molecules and determines the thermal background radiation of the gas.

It is mostly governed by the population of the upper state and in equilibrium is selfadjusting on a level where the spontaneous generation rate $A_{mn} \cdot N_m$ is just balanced by the induced transition rates (optical and collision induced) and the loss rate $\phi\rho_G$, the latter dependent on the density and temperature variations in the atmosphere.

Figure 9(b) represents the effective photon density $\bar{\rho}_E$ available to excite the gas over the layer depth L_p by terrestrial radiation and back radiation from the overlaying atmosphere. The downwelling part was assumed to originate from a layer of slightly lower temperature than the actual gas temperature T_G with a lapse rate also varying with T_G. The curve first declines, since with slightly increasing temperature, the lower state, which is not the ground state of CO_2, is stronger populated (see also Figure 9(c)), and therefore the absorption on this transition increases, while according to saturation effects, the effective excitation over L_p decreases. With further growing temperature $\bar{\rho}_E$ again increases, caused by the decreasing gas density and also reduced lower population, accompanied by a smaller absorption.

Figures 9(c) and 9(d) show the partial increase as well as the depletion of the lower state and the simultaneous refilling of the upper state due to the radiation interaction and thermal heating. The populations are displayed in relative units, normalized to the CO_2 number density N_0 at 288 K.

The heat density of the air and the gas (air) temperature, are plotted in Figures 9(e) and 9(f). Up to about 80 K (first 70 h), heating is determined by convection, while at higher temperatures radiative heating more and more dominates. An initial heat flux of 3 kW/m^2 seems relatively large but quickly reduces with increasing temperature and at a difference of 3 K between the surface and lower atmosphere, similar to latent heat, does not contribute to more than 40 W/m^2, representing only 10% of the terrestrial radiation.

In principle, it is expected that a gas heated by an external source cannot get warmer than the source itself. This is valid for a heat flux caused by convection or conduction in agreement with the second law of thermodynamics, but it leads to some contradiction with respect to a radiation source.

In a closed system as discussed in Section 2, thermal radiation interacting with the molecules is in unison with the population of the molecules, and both the radiation and the gas can be characterized by a unique temperature T_G. In an open system with additional radiation from an external source of temperature T_E, the prevailing radiation, interacting with the molecules, consists of two contributions, which in general are characterized by different temperatures and different loss processes. So, T_E determines the spectral intensity and distribution of the incident radiation and in this sense defines the power flux which can be transferred to the molecules, when this radiation is absorbed. But it does not describe any direct heat transfer, which always requires a medium for the transport and by no means defines the temperature of the absorbing molecules. At thermal equilibrium and in the presence of collisions, the molecules are better described by a temperature, deduced from the population ratio N_m/N_n given by the Boltzmann relation (4). This temperature also determines the spontaneous emission and together with the losses defines the thermal background radiation.

Since an absorption on a transition takes place as long as the population difference $N_n - N_m$ is positive (equal populations signify infinite temperature), molecules can be well excited by a thermal radiation source, which has a temperature lower than that of the absorbing gas. Independent of this statement, it is clear that at equilibrium the gas can only reradiate that amount of energy that was previously absorbed from the incident radiation and/or sucked up as thermal energy.

In any way, for the atmospheric calculations, it seems reasonable, to restrict the maximum gas temperature to values comparable to the earth's surface temperature. An adequate condition for this is a radiation loss rate

$$\phi \geq \frac{c/n}{4L_P}\left(1 - e^{-(2h\nu_{mn}/(\Delta\nu c/n))(B_{nm}N_n - B_{mn}N_m)L_P}\right), \tag{80}$$

as derived from the rate equations at stationary equilibrium and with up- and downwelling radiation of the same magnitude.

A calculation for conditions as found in the upper stratosphere at about 60 km altitude is shown in Figure 10. The air pressure is 0.73 hPa, the regular temperature at this altitude is 242.7 K, and the number density of air is $N_{air} = 2.2\times10^{22}$ m^{-3}.

The density of CO_2 reduces to $N = 8.3 \times 10^{18}$ m^{-3} at a concentration of 380 ppm. At this lower pressure, the collision induced linewidth is three orders of magnitude smaller and declines to about Δ] = 5 MHz. Under these conditions, the remaining width is governed by Doppler broadening with 34 MHz. At this altitude, any terrestrial radiation for an excitation on the transition is no longer available. The only radiation interacting with the molecules originates from neighbouring molecules and the own background radiation.

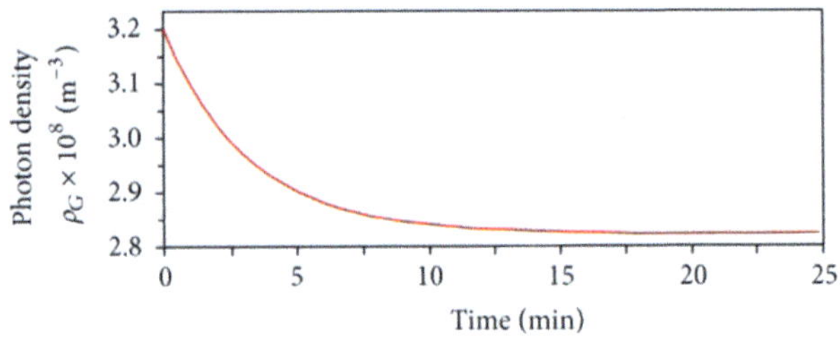
Photon density
$\rho_G \times 10^8$ (m^{-3})
3.2
3.1
3.0
2.9
2.8
0
5
10
15
20
25
Time (min)

(a)

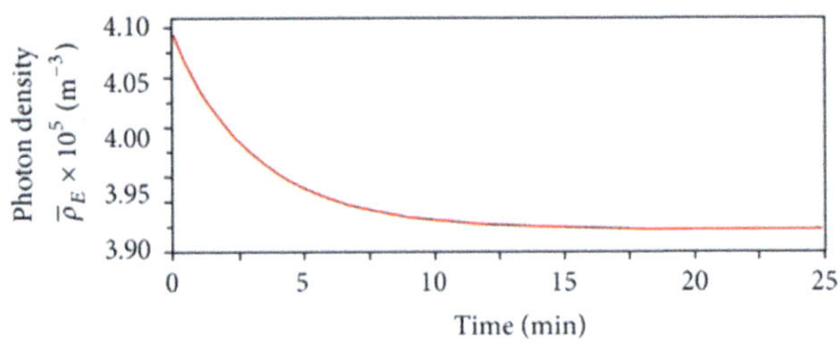
Photon density
$\overline{\rho}_E \times 10^5$ (m^{-3})
4.10
4.05
4.00
3.95
3.90
0
5
10
15
20
25
Time (min)

(b)

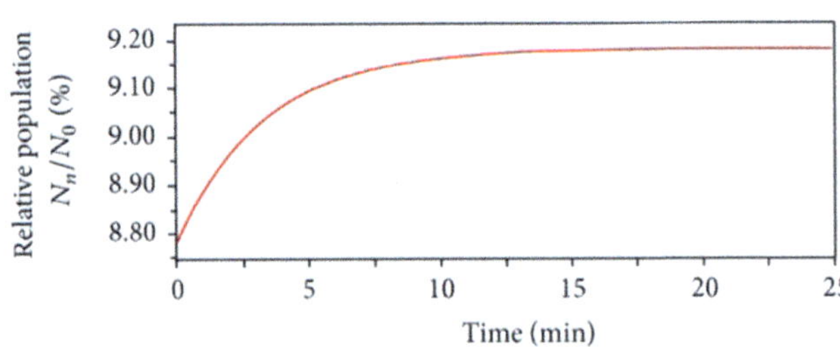
Relative population
N_n/N_0 (%)
9.20
9.10
9.00
8.90
8.80
0
5
10
15
20
25
Time (min)

(c)

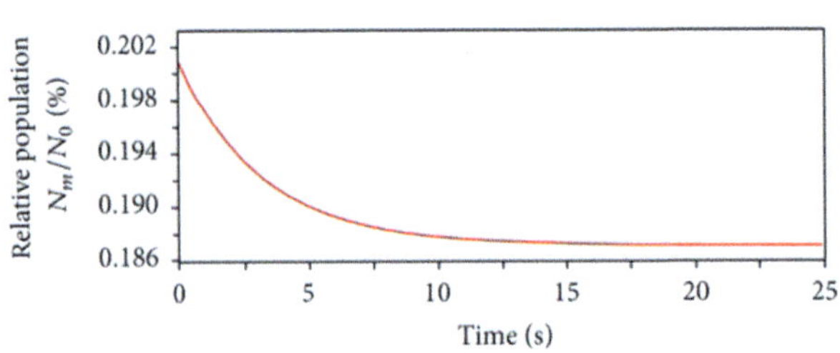
Relative population
N_m/N_0 (%)
0.202
0.198
0.194
0.190
0.186
0
5
10
15
20
25
Time (s)

(d)

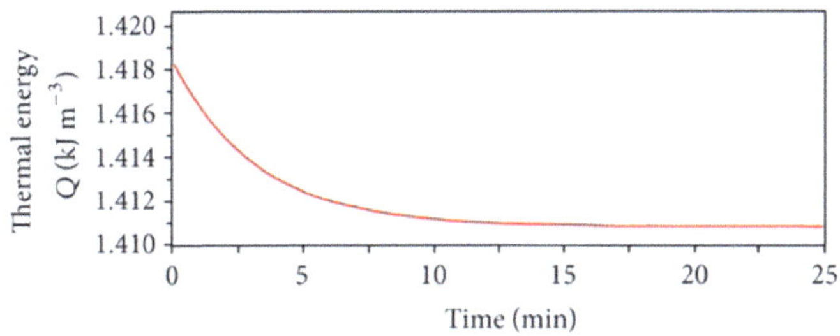

(e)

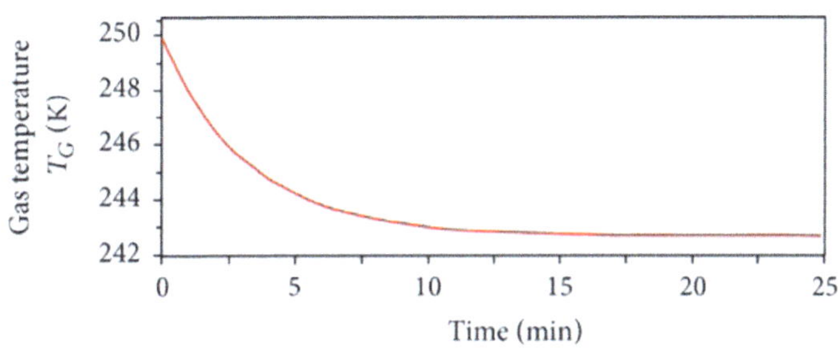

(f)

Figure 10: Solution of the rate equations for the stratosphere (60 km altitude) representing the evolution of (a) the photon density $\overline{\rho}_E$, (b) the photon density , (c) the lower and (d) the upper population density, (e) the thermal energy density, and (f) the gas (atmospheric) temperature as a function of time.

As initial conditions we used a local gas and air temperature $T_G = 250$K, slightly higher than the environment with $T_E = 242.7$ K, the latter defining the radiation (up and down) from neighbouring molecules. The heat flux j_{Hz} from or to adjacent layers was assumed to be negligible, while a heat well due to sun light absorption by ozone with ~15 W/m^2 over ~50 km in this case is the only heat source with $q_{\mathrm{SL}} = 0.3$ mW/m^3 . The calculation was performed for a collisional transition rate $C_{mn} = 10^5$ s^{-1}, a photon loss rate of $\phi = 7.5 \times 10^4$ s^{-1}, a layer depth of $L_p = 1$ km, and an effective number of transitions $M_{\mathrm{tr}} = 1430$ (absorption on standard transition: 2.8×10^{-4} W/m^2 ; total IR-absorption over 1 km: 0.4 W/m^2).

Similar to Figure 9, the graphs show the evolution of the photon densities ρ_G and $\overline{\rho}_E$, the population densities N_n and N_m of the states n and m, the heat density Q in air, and the temperature T_G of the gas as a function of time. But different to the conditions in the troposphere, where heating of the atmosphere due to terrestrial heat and radiation transfer was the dominating effect, the simulations in Figure 10 demonstrate the effect of radiative cooling in the stratosphere, where incident radiation from adjacent layers and locally released heat is transferred to radiation and radiated to space. An increased reradiation will be observed, as represented by the upper graph till the higher temperature diminishes and the radiated energy is in equilibrium with the supplied energy flux.

Due to the lower density and heat capacity of the air at this reduced pressure, the system comes much faster to a stationary equilibrium.

Thermodynamic and Radiation Equilibrium

Thermodynamic equilibrium in a gas sample means that the population of the molecular states is given by the Boltzmann distribution (4) and the gas is characterized by an average temperature T_G. On the other hand, spectral radiation equilibrium requires that the number of absorptions in a considered spectral interval is equal to the number of emissions. While without an external radiation generally, any thermodynamic equilibrium in a sample will be identical with the spectral radiation equilibrium, in the presence of an additional field, these cases have to be distinguished, since the incident radiation also induces transitions and modifies the populations of the molecular states, deviating from the Boltzmann distribution.

Therefore, it is appropriate to investigate the population ratio more closely and to discover how it looks like under some special conditions. This is an adequate means to determine what kind of equilibrium is found in the gas sample.

For this, we consider the total balance of absorption and emission processes as given by the rate equations (76). Under stationary equilibrium conditions (see (78)), it is

$$\left(W_{nm}^{G} + W_{nm}^{E} + C_{nm}\right) N_n = \left(W_{mn}^{G} + W_{mn}^{E} + A_{mn} + C_{mn}\right) N_m. \tag{81}$$

With (17), (34) and (59) we can write

$$\left(B_{nm}\left(u_\nu + \frac{n}{c} \cdot I_\nu\right) + C_{nm}\right) N_n$$

$$= \left(B_{mn}\left(u_\nu + \frac{n}{c} \cdot I_\nu\right) + A_{mn} + C_{mn}\right) N_m. \tag{82}$$

At local thermodynamic equilibrium,$_{mn}$ and C_{nm} are linked to each other by (36). But even more generally, we can conclude that (36) is valid, as long as the molecules can be described by a Maxwellian velocity distribution and can be characterized by an average temperature (see also [12]).

Thus, applying (36) and introducing some abbreviations:

$$u_\nu + \frac{n}{c} \cdot I_\nu = U_\nu, \qquad \frac{C_{nm}}{B_{nm}} = \varepsilon,$$

$$\frac{A_{mn}}{B_{nm}} = \frac{g_n}{g_m} \frac{8\pi n^2}{\lambda_{mn}^2} \frac{h\nu_{mn}}{c/n} = \frac{g_n}{g_m}\kappa \tag{83}$$

after elementary rearrangement, (82) becomes:

$$\frac{N_m}{N_n} = \frac{g_m}{g_n} \frac{U_\nu + \varepsilon}{U_\nu + \kappa + \varepsilon \cdot e^{h\nu_{mn}/(kT_G)}}. \tag{84}$$

It is easy to be seen that for large ε and thus high collision rates, (84) approaches to a Boltzmann distribution. Also for ε=0 (no collisions) and I_ν = 0 (no external radiation), this gives the Boltzmann relation, while for ε=0 and $I_\nu \gg (c/n) \cdot u_\nu$, (84) changes to

$$\frac{N_m}{N_n} \approx \frac{g_m}{g_n} \frac{I_\nu}{I_\nu + (c/n)\kappa} = \frac{g_m}{g_n} \frac{1}{1 + 4\left(e^{h\nu_{mn}/(kT_E)} - 1\right) e^{2S_{nm}Nz/\Delta\nu}}$$
$$\approx \frac{1}{4}\frac{g_m}{g_n} e^{-h\nu_{mn}/kT_E} e^{-2S_{nm}Nz/\Delta\nu}. \tag{85}$$

This equation shows that under conditions as found in the upper atmosphere (small collision rates), the populations are approximated by a modified Boltzmann distribution (right hand side: for $h\nu_{mn} > kT_E$), which with increasing spectral intensity is more and more determined by the temperature T_E of the external radiation source. Under such conditions, the normal thermodynamic equilibrium according to (4) is violated and has to be replaced by the spectral radiation equilibrium.

RADIATION BALANCE AND RADIATION TRANSFER IN THE ATMOSPHERE

A more extensive analysis of the energy and radiation balance of an absorbing sample not only accounts for the net absorbed power from the incident radiation, as considered in (53)–(66), but it also includes any radiation originating from the sample itself as well as any re-radiation due to the external excitation. This was already investigated in some detail in Section 3.3, however, with the view to a local balance and its evolution over time. In this section, we are particularly concentrating on the propagation of thermal radiation in an infrared active gas and how this radiation is modified due to the absorption and re-emission of the gas.

Two perspectives are possible, an energy balance for the sample or a radiation balance for the in- and outgoing radiation. Here we discuss the latter case. Scattering processes are not included.

We consider the incident radiation from an external source with the spectral radiance $I_{\nu,\Omega}(T_E)$ as defined in (50). Propagation losses in a thin gas layer obey Lambert-Beer's law. Simultaneously, this radiation is superimposed by the thermal radiation emitted by the gas sample itself, and as already discussed in Section 2.5, this thermal contribution has its origin in the spontaneous emission of the molecules. For small propagation distances dr in the gas, both contributions can be summed up yielding

$$\frac{dI_{\nu,\Omega}(r)}{dr} = -\bar{\alpha}_{nm}(\nu)\, I_{\nu,\Omega}(r, T_E) + \frac{h\nu_{mn}}{4\pi} A_{mn} N_m g(\nu)\,. \tag{86}$$

The last term is a consequence of (19) or (45) when transforming the spectral energy density u_ν of the gas to a spectral intensity $(c/n)\cdot u_\nu$ and considering only that portion $(c/4\pi n)\cdot u_\nu$ emitted into the solid angle interval $d\Omega$. Equation (86) represents a quite general form of the radiation balance in a thin layer, and it is appropriate to derive the basic relations for the radiation transfer in the atmosphere under different conditions. The differential dr indicates that the absolute propagation length in the gas is meant.

Schwarzschild Equation in Dense Gases

In the troposphere, the population of molecular states is almost exclusively determined by collisions between the molecules even in the presence of stronger radiation. Then, due to the previous discussion, a well-established local thermodynamic equilibrium can be expected, and from (47) or (48), we find

$$\frac{h\nu_{mn}}{4\pi} A_{mn} N_m g(\nu) = \frac{c}{4\pi n}\bar{\alpha}_{nm} u_\nu = \bar{\alpha}_{nm} B_{\nu,\Omega}(T_G) \tag{87}$$

with $B_{\nu,\Omega}(T_G) = (c/4\pi n) \cdot u_\nu$ as the spectral radiance of the gas, also known as Kirchhoff-Planck-function and given by (50), but here at temperature T_G. To distinguish between the external radiation and the radiating gas, we use the letter *B* (from background radiation), which may not be mixed with the Einstein coefficients B_{nm} or B_{mn}. Then we get as the final result:

$$\frac{dI_{\nu,\Omega}(r)}{dr} = -\overline{\alpha}_{nm}(\nu)\, I_{\nu,\Omega}(r, T_E) + \overline{\alpha}_{nm}(\nu)\, B_{\nu,\Omega}(T_G(r)). \tag{88}$$

This equation is known as the Schwarzschild equation [5–7], which describes the propagation of radiation in an absorbing gas and which additionally takes into account the thermal background radiation of the gas. While deriving this equation, no special restrictions or conditions concerning the density of the gas or collisions between the molecules have to be made, only that (36), is valid and the thermodynamic equilibrium is given. In general, this is the case in the troposphere up to the stratosphere.

Another Derivation. Generally, the Schwarzschild equation is derived from pure thermodynamic considerations. The gas in a small volume is called to be in local thermodynamic equilibrium at temperature T_G, when the emitted radiation from this volume is the same as that of a blackbody radiator at this temperature. Then Kirchhoff's law is strictly valid, which means that the emission is identical to the absorption of a sample.

The total emission in the frequency interval $(\nu, \nu + d\nu)$ during the time dt may be $4\pi\varepsilon(\nu)$.The emissivity $\varepsilon(\nu)$ may not be mixed with $\varepsilon = C_{nm}/B_{nm}$ in (83). Then with the spectral radiance $B_{\nu,\Omega}(T_G)$ as given by (50) and with the absorption coefficient $\overline{\alpha}_{nm}(\nu)$, it follows from Kirchhoff 's law (see, e.g., [8–11])

$$\varepsilon(\nu) = \overline{\alpha}_{nm}(\nu) \cdot B_{\nu,\Omega}(T_G). \tag{89}$$

For an incident beam with the spectral radiance $I_{\nu,\Omega}$ and propagating in r-direction we then get:

$$\begin{aligned} \frac{dI_{\nu,\Omega}}{dr} &= -\overline{\alpha}_{nm}(\nu)\, I_{\nu,\Omega}(r, T_E) + \varepsilon(\nu) \\ &= -\overline{\alpha}_{nm}(\nu)\, I_{\nu,\Omega}(r, T_E) + \overline{\alpha}_{nm}(\nu)\, B_{\nu,\Omega}(T_G), \end{aligned} \tag{90}$$

which is identical with (88).

Radiation Transfer Equation for the Spectral Intensity

Normally the radiation and by this the energy, emitted into the full hemisphere, is of interest, so that (88) has to be integrated over the solid angle $\Omega = 2\pi$. As already discussed in Section 3.2, for this integration, we have to have in mind that a beam, propagating under an angle β to the layer normal (see Figure 11), only contributes an amount $I_{\nu,\Omega} \cos \beta d\Omega$ to the spectral intensity (spectral flux density) due to Lambert's law. The same is assumed to be true for the thermal radiation emitted by a gas layer under this angle.

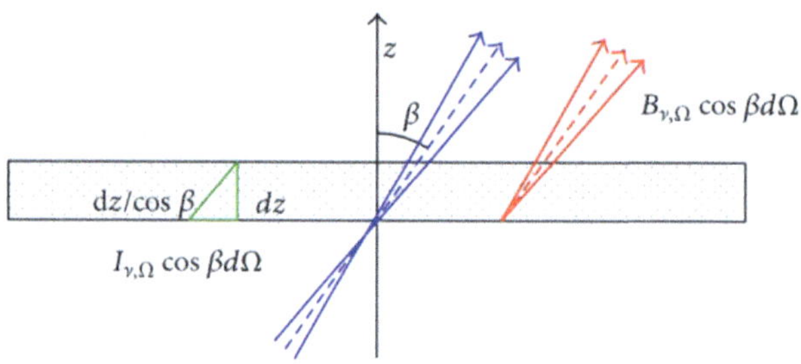

Figure 11: Radiation passing a gas layer.

On the other hand, the propagation length through a thin layer of depth dz is increasing with $dl = dz/\cos \beta$, so that the β-dependence for both terms $I_{\nu,\Omega}$ and $B_{\nu,\Omega}$ disappears.

Therefore, analogous to (54) and (55) integration of (88) over Ω gives for the spectral intensity, propagating in z-direction:

$$\frac{\int dI_{\nu,\Omega}(z) \cos \beta d\Omega}{dz} = \bar{\alpha}_{nm}(\nu) \int_0^{2\pi} \int_0^{\pi/2} \left(-I_{\nu,\Omega}(z, T_E) + B_{\nu,\Omega}(T_G(z))\right) \times \sin \beta d\beta d\varphi, \tag{91}$$

or with the definition for $I_\nu = \pi \cdot I_{\nu,\Omega}$ and analogous $B_\nu = \pi \cdot B_{\nu,\Omega}$:

$$\frac{dI_\nu(z)}{dz} = 2\bar{\alpha}_{nm}(\nu)\left(-I_\nu(z, T_E) + B_\nu(T_G(z))\right). \tag{92}$$

Equation (92) is a 1st-order differential equation with the general solution:

$$I_\nu(z) = e^{-2\int_0^z \overline{\alpha}_{nm}(\nu,z'')dz''} \times \left[2\int_0^z \overline{\alpha}_{nm}(\nu,z') B_\nu(T_G(z')) \times e^{2\int_0^{z'} \overline{\alpha}_{nm}(\nu,z'')dz''} dz' + I_\nu(0) \right] \tag{93}$$

or after some transformation:

$$I_\nu(z) = I_\nu(0)\, e^{-2\int_0^z \overline{\alpha}_{nm}(\nu,z'')dz''} + 2\int_0^z \overline{\alpha}_{nm}(\nu,z') B_\nu(T_G(z'))\, e^{-2\int_{z'}^z \overline{\alpha}_{nm}(\nu,z'')dz''} dz'. \tag{94}$$

Usually the density of the gas, the total pressure and the temperature are changing over the propagation path of the radiation. Therefore, (94) has to be solved stepwise for thin layers of thickness Δz, over which, $\overline{\alpha}_{nm}$ and B_ν can be assumed to be constant (see Figure 12). With the running index i for different layers, then (94) can be simplified as follows:

$$I_\nu^i(\Delta z) = I_\nu^{i-1} e^{-2\overline{\alpha}_{nm}^i(\nu)\Delta z} + B_\nu^i(T_G^i) \cdot \left(1 - e^{-2\overline{\alpha}_{nm}^i(\nu)\Delta z}\right). \tag{95}$$

The intensity in the ith layer is calculated from the previous intensity I_ν^{i-1} of the $(i-1)$th layer with the values $\overline{\alpha}_{nm}^i(\nu)$ and $B_\nu^i(T^i)$ of the ith layer. In this way, the propagation over the full atmosphere is stepwise calculated. The first term in (95) describes the transmission of the incident spectral intensity over the layer thickness, while the second term represents the self-absorption of the thermal background radiation into forward direction and is identical with the spontaneous emission of the layer into one hemisphere.

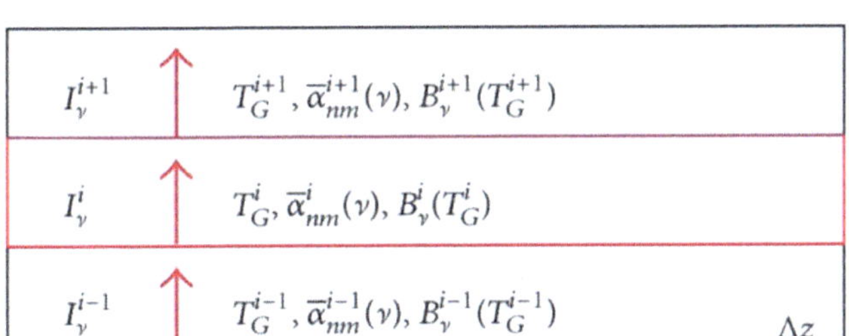

Figure 12: Stepwise calculation of the spectral intensity by the radiation transfer equation.

Schwarzschild Equation in a Thin Gas

Equation (88) was derived under the assumption of a local thermodynamic equilibrium in the gas, which is generally the case at higher gas pressures and therefore higher collision rates of the molecules. At low pressures, however, the population of the molecular states can also significantly be determined by induced transitions caused by the incident radiation (see Section 3.4), and then it gives more sense to consider the monochromatic or spectral radiation equilibrium, which means that the gas re-emits all monochromatic radiation it has absorbed.

Since this absorption is caused by the total spectral radiation incident within the solid angle $\Omega = 2\pi$, $I_{\nu,\Omega}$ has to be integrated over Ω. Then the absorbed spectral intensity over the interval dr is found from (51) together with (53).This must be equal to the spectral power density emitted by the gas as spontaneous emission into the full solid angle 4π. Thus, we obtain

$$\bar{\alpha}_{nm}(\nu)\int_{\Omega} I_{\nu,\Omega}(r,T_E)\cos\beta d\Omega$$
$$= \bar{\alpha}_{nm}(\nu)\, I_{\nu}(r,T_E) \equiv h\nu_{mn}A_{mn}N_m g(\nu)\,. \tag{96}$$

Only $1/4\pi$ of (96) contributes to the spectral radiance, and (86) changes to

$$\frac{dI_{\nu,\Omega}}{dr} = -\bar{\alpha}_{nm}(\nu)\, I_{\nu,\Omega}(r,T_E) + \frac{1}{4\pi}\bar{\alpha}_{nm}(\nu)\, I_{\nu}(r,T_E)\,. \tag{97}$$

With $I_{\nu} = \pi I_{\nu,\Omega}$, (97) becomes

$$\frac{dI_{\nu,\Omega}}{dr} = -\frac{3}{4}\bar{\alpha}_{nm}(\nu)\, I_{\nu,\Omega}(r,T_E)\,. \tag{98}$$

This differential equation for the spectral radiance in thin gases was derived, only assuming the absorption and subsequent emission due to the external radiation and in this respect orientated at the standard considerations as found in the literature (see, e.g., [12]). A more rigorous derivation of this equation, however, also has to take into account the emission originating from the thermal background radiation of the molecules, as long as their temperature and thus their kinetic energy are not zero. This is considered in the next subsection.

Extended Schwarzschild Equation in a Thin Gas. We start from (86), which already describes quite generally the propagation of radiation under conditions when a sample not only absorbs but also emits on the same frequencies as the incident radiation. Again requesting radiation equilibrium, the total balance of absorption and emission processes is given by the rate equations (76), which under stationary conditions results in (see (81))

$$\left(W_{nm}^{G} + W_{nm}^{E} + C_{nm}\right) N_n = \left(W_{mn}^{G} + W_{mn}^{E} + A_{mn} + C_{mn}\right) N_m. \quad (99)$$

With (34) and (59) we find:

$$\left(\frac{g_m}{g_n}\frac{c/n}{h\nu_{mn}}\frac{A_{mn}\lambda_{mn}^2}{8\pi n^2}\left(u_\nu + \frac{n}{c}I_\nu\right) + C_{nm}\right) N_n = \left(\frac{c/n}{h\nu_{mn}}\frac{A_{mn}\lambda_{mn}^2}{8\pi n^2}\left(u_\nu + \frac{n}{c}I_\nu\right) + A_{mn} + C_{mn}\right) N_m. \quad (100)$$

Some rearrangement of (100) and multiplication with $h\nu_{mn}$ gives

$$\frac{A_{mn}\lambda_{mn}^2}{8\pi n^2}\left(\frac{g_m}{g_n}N_n - N_m\right)\left(\frac{c}{n}u_\nu + I_\nu\right) + h\nu_{mn}\left(C_{nm}N_n - C_{mn}N_m\right) = h\nu_{mn}A_{mn}N_m. \quad (101)$$

At local thermodynamic equilibrium, we know that due to (36), the term $C_{nm}N_n - C_{mn}N_m$ will be zero, and trivially it is also zero in the absence of any collisions ($C_{mn} = C_{nm} = 0$). Thus, we can expect that in a thin gas with only few collisions even under radiation equilibrium, this term is quite small and therefore can be neglected. The general case, including this term, will be discussed in the next paragraph. So, with (33) and dividing (101) by 4π as well as using the relations $(c/4\pi n)\cdot u_\nu = B_{\nu,\Omega}$ and $I_\nu = \pi I_{\nu,\Omega}$, we get

$$\frac{1}{4\pi}\bar{\alpha}_{nm}\left(\frac{c}{n}u_\nu + I_\nu\right) = \bar{\alpha}_{nm}\left(B_{\nu,\Omega} + \frac{1}{4}I_{\nu,\Omega}\right) = \frac{h\nu_{mn}}{4\pi}A_{mn}N_m g(\nu). \quad (102)$$

Inserting (102) in (86) then gives the extended Schwarzschild equation in thin gases:

$$\frac{dI_{\nu,\Omega}(r)}{dr} = -\frac{3}{4}\bar{\alpha}_{nm}(\nu)\, I_{\nu,\Omega}(r, T_E) + \bar{\alpha}_{nm}(\nu)\, B_{\nu,\Omega}(T_G(r)). \quad (103)$$

Generalized Schwarzschild Equation

The objective is to find an expression which describes the radiation transfer for both limiting cases, at high collision rates in a dense gas, as given by (88), as well as at negligible collisions in a thin gas, represented by (103). For similar considerations, however, different results; see, for example, [12].

We start from (81) or with the abbreviations of (83) from (84). These equations again contain the balance of all up and down transitions. Elementary transformation yields

$$\left(U_\nu + \kappa + \varepsilon \cdot e^{h\nu_{mn}/(kT_G)}\right) N_m = \frac{g_m}{g_n}\left(U_\nu + \varepsilon\right) N_n, \tag{104}$$

and some further rearrangements give (see on the left side the expansion by +1-1)

$$\left(\kappa + \varepsilon\left(e^{h\nu_{mn}/(kT_G)} + 1 - 1\right)\right) N_m$$

$$= \left(\frac{g_m}{g_n} N_n - N_m\right) U_\nu + \frac{g_m}{g_n} \varepsilon N_n,$$

$$\left(\kappa + \varepsilon\left(e^{h\nu_{mn}/(kT_G)} - 1\right)\right) N_m = \left(\frac{g_m}{g_n} N_n - N_m\right)\left(U_\nu + \varepsilon\right),$$

$$N_m = \frac{N_n\left(g_m/g_n - N_m/N_n\right)\left(U_\nu + \varepsilon\right)}{\kappa + \varepsilon\left(e^{h\nu_{mn}/(kT_G)} - 1\right)}. \tag{105}$$

With the identity (see (33) and (83))

$$\overline{\alpha}_{nm} = \frac{h\nu_{mn}}{c/n}\frac{A_{mn}}{\kappa} N_n\left(\frac{g_m}{g_n} - \frac{N_m}{N_n}\right) g(\nu) \tag{106}$$

and multiplying (105) with $(h\nu_{mn}/4\pi) \cdot A_{mn} g(\nu)$, we obtain

$$\frac{h\nu_{mn}}{4\pi} A_{mn} N_m g(\nu) = \frac{c/n}{4\pi} \frac{\overline{\alpha}_{nm}\left(U_\nu + \varepsilon\right)}{1 + (\varepsilon/\kappa)\left(e^{h\nu_{mn}/(kT_G)} - 1\right)}$$

$$= \overline{\alpha}_{nm} \frac{B_{\nu,\Omega} + (1/4)\, I_{\nu,\Omega} + ((c/n)/4\pi)\,\varepsilon}{1 + (\varepsilon/\kappa)\left(e^{h\nu_{mn}/(kT_G)} - 1\right)}. \tag{107}$$

Defining a relative material parameter

$$\eta = \frac{c/n}{4\pi}\frac{\varepsilon}{B_{\nu,\Omega}} = \frac{c/n}{4\pi}\frac{C_{nm}}{B_{nm}}\frac{c^2}{2h\nu_{mn}^3 n^2}\left(e^{h\nu_{mn}/(kT_G)} - 1\right)$$

$$= \frac{\varepsilon}{\kappa}\left(e^{h\nu_{mn}/(kT_G)} - 1\right) \tag{108}$$

and inserting (107) and (108) into (86), we get

$$\frac{dI_{\nu,\Omega}(r)}{dr} = -\bar{\alpha}_{nm}(\nu)\, I_{\nu,\Omega}(r, T_E) + \bar{\alpha}_{nm} \frac{B_{\nu,\Omega} + (1/4)\, I_{\nu,\Omega} + \eta B_{\nu,\Omega}}{1+\eta},$$

$$\frac{dI_{\nu,\Omega}(r)}{dr} = -\bar{\alpha}_{nm}(\nu)\left(1 - \frac{1/4}{1+\eta}\right) I_{\nu,\Omega}(r, T_E) + \bar{\alpha}_{nm}(\nu)\, B_{\nu,\Omega}(T_G(r)). \tag{109}$$

With the abbreviation

$$\chi = 1 - \frac{1/4}{1+\eta} = \frac{3+4\eta}{4+4\eta}, \tag{110}$$

where χ assumes values from $3/4 \leq \chi \leq 1$ for $0 \leq \eta \leq \infty$, (109) can be represented in the more compact form

$$\frac{dI_{\nu,\Omega}(r)}{dr} = \bar{\alpha}_{nm}(\nu, r)\left(-\chi \cdot I_{\nu,\Omega}(r, T_E) + B_{\nu,\Omega}(T_G(r))\right). \tag{111}$$

This is the generalized Schwarzschild equation, which is valid at low as well as at high gas densities and where the transition from one to the other limiting case is controlled by the density dependent parameter η. According to (108) and (83) together with (36), η may be expressed as

$$\eta = \frac{\varepsilon}{\kappa}\left(e^{h\nu_{mn}/(kT_G)} - 1\right) = \frac{g_n}{g_m}\frac{C_{nm}}{A_{mn}}\left(e^{h\nu_{mn}/(kT_G)} - 1\right) = \frac{C_{mn}}{A_{mn}}\frac{\left(e^{h\nu_{mn}/(kT_G)} - 1\right)}{e^{h\nu_{mn}/(kT_G)}} \approx \frac{C_{mn}}{A_{mn}}. \tag{112}$$

Therefore, any variations of η and thereby also of χ originate from the pressure dependent collision transfer rate C_{mn}, which is changing with the altitude above the earth's surface, so that all these parameters become a function of z. Without any collisions, η gets zero, and thus $\chi = 3/4$.

Then (111) is identical with (103). On the other hand, at high gas densities and high collision rates with $\eta \gg 1$ and $\chi \rightarrow 1$, (111) converges towards (88).

Generalized Radiation Transfer Equation for the Spectral Intensity

In order to derive the basic radiation transfer equation (RTE) for the spectral intensity expanding in z- direction, analogous to (91), we have to integrate (111) over the solid angle $\Omega = 2\pi$:

$$\frac{\int dI_{\nu,\Omega}(z)\cos\beta d\Omega}{dz} = \bar{\alpha}_{nm}(\nu, z)\int_0^{2\pi}\int_0^{\pi/2}\left(-\chi(z) I_{\nu,\Omega}(z, T_E) + B_{\nu,\Omega}(T_G(z))\right) \times \sin\beta d\beta d\varphi \tag{113}$$

or with the definitions of I_ν and B_ν:

$$\frac{dI_\nu(z)}{dz} = 2\bar{\alpha}_{nm}(\nu, z)\left(-\chi(z) I_\nu(z, T_E) + B_\nu(T_G(z))\right). \tag{114}$$

This is the generalized RTE for the spectral intensity originating from a Planckian radiator of temperature T_E and propagating through an absorbing gas layer of temperature T_G.

Similar to (92)–(94), the solution of (114) is found to be

$$I_\nu(z) = I_\nu(0)\, e^{-2\int_0^z \chi(z'')\bar{\alpha}_{nm}(\nu,z'')dz''} + 2\int_0^z \bar{\alpha}_{nm}(\nu, z') B_\nu(T_G(z'))\, e^{-2\int_{z'}^z \chi(z'')\bar{\alpha}_{nm}(\nu,z'')dz''} dz', \tag{115}$$

and a stepwise numerical integration can be obtained by

$$I_\nu^i(\Delta z) = I_\nu^{i-1} e^{-2\chi^i \bar{\alpha}_{nm}^i(\nu)\Delta z} + \frac{1}{\chi^i} B_\nu^i(T_G^i)\cdot\left(1 - e^{-2\chi^i\bar{\alpha}_{nm}^i(\nu)\Delta z}\right). \tag{116}$$

It should be recalled that under conditions of stronger absorption, and longer path lengths radiation emerging from the individual layers may deviate from the assumed Lambertian radiation law (see Section 3.2). This can result in a slightly too strong absorption caused by those rays propagating under larger angles to the layer normal. However, since these rays only provide relatively small contributions to the total absorption, this only causes relatively small deviations.

Another effect may result from deviations of the rays from geometric optics, caused by inhomogeneities and scattering processes in the atmosphere.

Therefore, altogether it seems reasonable to apply a slightly smaller effective absorption coefficient $2\chi^i\overline{\alpha}^i_{nm}$ in (116). The factor of two had its origin in an effective average path length through a layer of $2\Delta z$, corresponding to an average propagation direction $\beta = 60°$. Therefore, assuming a slightly smaller average angle of $\beta \approx 50°$ might approximate more realistically the conditions in the atmosphere.

SIMULATION OF RADIATION TRANSFER IN THE ATMOSPHERE

Infrared active gases in the atmosphere have a significant influence on the energy and radiation balance of the earth-atmosphere system, as this was already discussed in some detail in Section 3.3. They absorb short wavelength radiation from the sun and long wavelength radiation from the earth. This absorbed energy as well as thermal energy (after conversion to internal energy) is again reradiated uniformly into all directions. This emission is the origin of the up- and downwelling radiation of the atmosphere itself, which together with any terrestrial or cloud radiation determines the total radiation observed in up- or downward direction.

The most important infrared active gases are water vapour, CO_2, CH_4, and O_3, which have stronger absorption bands in the infrared, distributed over a wide spectral range from about 3 to 200 μm or 50 to 3330 cm^{-1}.

Particularly, CO_2 is believed to be the main culprit of anthropogenic global warming and is made responsible for the increasing temperature over the last 150 years. Therefore, it is worthwhile to look more carefully how this gas influences the absorption and radiation balance in the atmosphere. At the same time, this is an instructive example for the useful applicability of the theoretical background presented in this paper.

Absorption and Emission Spectrum of CO_2

The strongest absorption band of this molecule is located around 4.2 μm or 2350 cm^{-1}. But for the absorption-emission balance in the atmosphere, even more important is the ten times weaker absorption band at 15 μm (670 cm^{-1}), which almost coincides with the spectral maximum of the terrestrial radiation (see Figure 6), and therefore primarily determines the interaction of CO_2 molecules with the radiation.

Figure 13 shows the respective absorption spectrum for 380 ppm of CO_2 in dry air for different propagation lengths at a pressure of 1013 hPa and a temperature of 288 K. The red lines are markers indicating the center and strength of an individual transition, while the blue curve is the superposition of all line profiles and represents the degree of transmission—or inversely the degree of absorption—of incident radiation at these wavenumbers. On the Q-branch (band center), radiation is almost completely absorbed after a length of only 1 m, and with increasing propagation distance, even the weaker lines of the P- and R-branches reflect totally saturated absorption. At 1 km or more, only the far wings of the band structure still show some rest transmission.

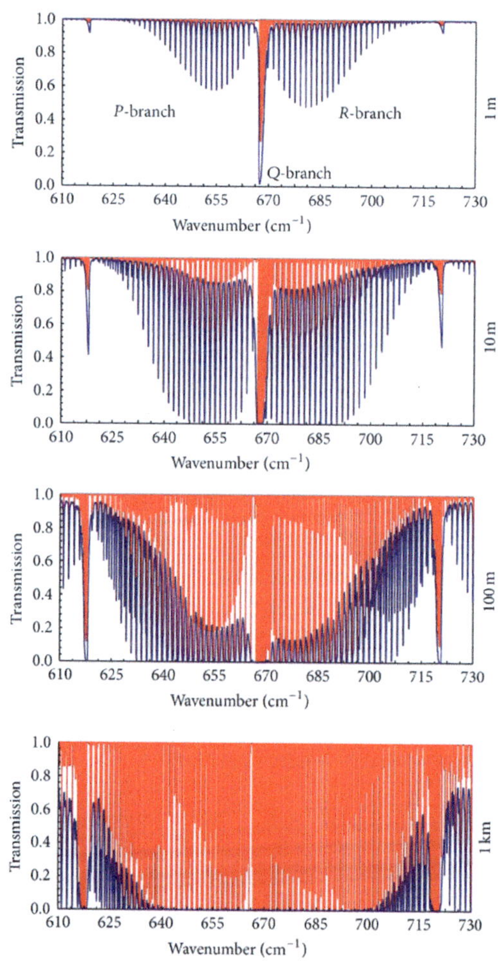

Figure 13: Absorption spectra for 380 ppm of CO2 in dry air at different propagation lengths (air pressure: 1013 hPa, temperature: 288 K).

On the other hand, from Sections 3 and 4, we know that an absorbing gas with a temperature comparable to that of an external radiation source—here the earth's surface—will act as own radiator and modify the incident radiation noticeably. Figure 14 displays a simulation for the upwelling radiation emitted by only 3.8 ppm CO_2 in dry air (1/100th of the regular concentration) at different altitudes above ground.

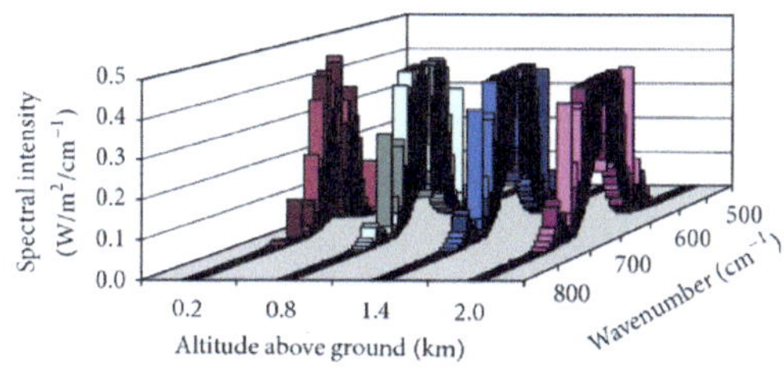

Figure 14: Spectral intensity radiated by 3.80 ppm of CO2 in dry air at different altitudes (ground pressure: 1013 hPa, ground temperature of the gas: 288 K).

This calculation is based on (116) and uses for the vertical air pressure and temperature profile the *US Standard Atmosphere* [17]. On the band center, the radiation already rapidly attains a maximum spectral intensity of 0.47 W/($m^2 \cdot cm^{-1}$), and with increasing altitude a broader plateau is gradually developing, which at higher altitudes again slightly decreases. This can be observed more pronounced in Figure 15 for a series of calculations up to 10 km altitude with a concentration of 380 ppm of CO_2 in dry air.

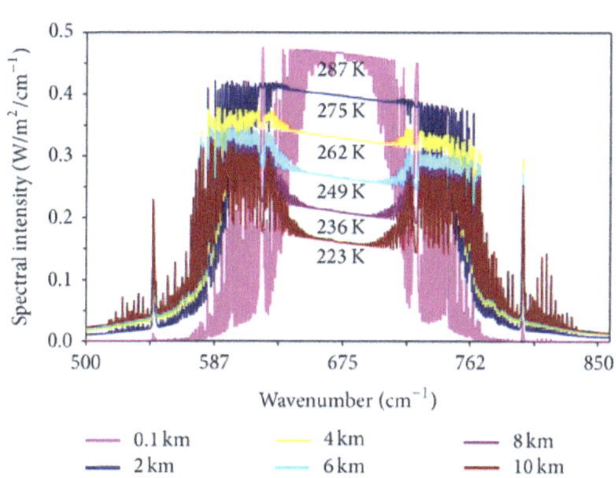

Figure 15: Spectral intensity emitted by 380 ppm of CO2 in dry air at different altitudes (ground pressure: 1013 hPa, ground temperature of the gas: 288 K).

The broad plateaus essentially reflect the actual temperature of the gas at the specific altitudes. The higher shoulders of the curves from 4–10 km are an indication that with reducing pressure and thus decreasing linewidths the wings of the absorption band become more transparent, and also radiation from lower, therefore warmer, and more intensively radiating layers can contribute to the observed spectra.

All radiative transfer calculations were performed on the program-platform *MolExplorer* [18] as line-by-line calculations and rely on the actual *HITRAN08-database* [13]. For the numerical integration of (116), a layer thickness of 100 m was used, over which the pressure, density, and temperature were considered to be constant. According to (112), η is determined by the ratio C_{mn}/A_{mn}. Since up to the mesopause, the collision rate C_{mn} is still larger than 1 s−1 and the spontaneous emission rate on the CO2 transitions $A_{mn} \leq 1$ s −1, η is larger 1 and thus $\chi \approx 1$. Therefore, for this absorption band the original Schwarzschild equation (88) is still a good approximation up to an altitude of about 90 km.

Influence of Water Vapour

The strongest impact on the radiation in the atmosphere originates from water in its different aggregate states. So, thicker cumulus or altostratus cloud layers can totally absorb and reradiate infrared radiation, or ice crystals in the tropopause may cause scattering of the radiation. But also atmospheric water vapour has a significant influence, which with respect to the scope of this paper will be considered in more detail. This strong impact results from the fact that H_2O molecules possess strong absorption bands and single absorption lines which are almost continuously spread over broader regions of the infrared spectrum.

Figure 16 shows the spectral transmission through wet air on a logarithmic scale over an altitude of 10 km. The water vapour concentration at the ground at 288 K is assumed to be 1.46% as an average over the three climate zones tropics, mid-latitudes, and the polar regions. It is deduced from GPS measurements [19] and decreases exponentially with altitude because of the temperature dependence of the vapour pressure (for details, see [20]).

From 10–360 cm^{-1} and 1400–1800 cm^{-1}, the atmosphere is completely opaque. Only between 800 and 1100 cm^{-1} and above 2200 cm^{-1}, the so-called atmospheric windows, it becomes quite transparent for radiation from the earth's surface.

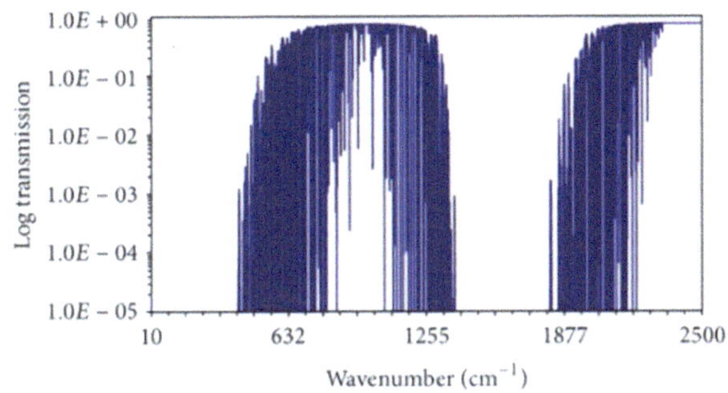

Figure 16: Transmission through wet air over 10 km altitude. Water vapour concentration at ground is 1.46% at 288 K.

But at the same time, the water molecules are strong emitters, radiating at their respective temperature, as this is represented in Figure 17. As direct comparison are also shown Planck distributions of different temperatures. Although water molecules are ideal blackbody radiators on their stronger transitions, their spectral intensity is significantly lower than the surface emission of the earth. Again this results from the fact that the spectral components observed at the tropopause in 12.5 km height originate from layers of different altitude and therefore different temperature. These layers may be characterized by an optical depth $\tau_{nm}(\nu) = \overline{\alpha}_{nm}(\nu)\Delta z \approx 1$ at frequency ν.. Otherwise a new absorption-emission cycle would occur within the layer thickness or over the further path up to the observation altitude.

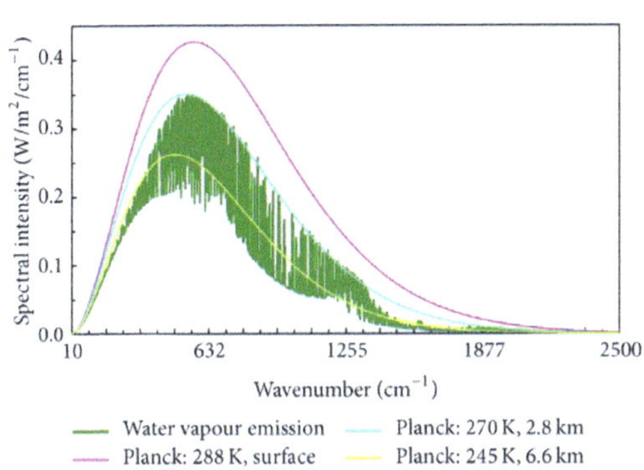

Figure 17: Water vapour emission at 12.5 km altitude. Concentration at ground is 14,615 ppm. As comparison are shown Planck curves at 288 K, 270 K, and 245 K.

So, within spectral regions of medium or lower absorption the envelope of the emission curve can be represented quite well by a Planck distribution of 270 K, corresponding to an altitude of 2.8 km, while in the spectral intervals of very high opacity (10–400 cm^{-1} and 1350–1800 cm^{-1}), a distribution with 245 K and an altitude of 6.6 km fits much better.

Radiation Transfer under Atmospheric Conditions

While it was the objective of the previous two subsections, to present the individual spectral properties of the two most important infrared active gases in the atmosphere and to identify their interaction with the atmospheric radiation, in this paragraph we investigate their common influence on the up- and downwelling radiation. Figure 18 shows the impact of both gases on the emitted radiation at 12.5 km altitude.

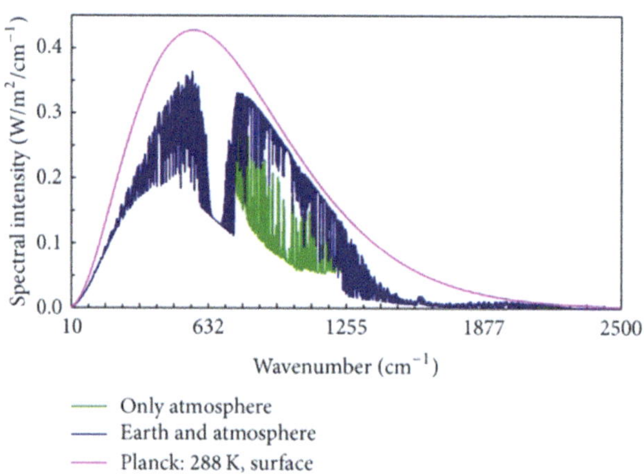

Figure 18: Total upwelling radiation in 12.5 km altitude (blue) and emission of only the atmosphere by water vapour and CO2 (green). Surface radiation is shown as pink line.

The blue curve represents the total radiation, emitted from the earth's surface and the atmosphere, while the contribution only from the atmosphere is depicted as green graph. Below 750 cm^{-1} and above 1300 cm^{-1} both curves are completely identical and appear as a unique line. Only over the central atmospheric window, the influence of the underlying terrestrial radiation, shown as pink graph, can be identified as increased intensity, while in the outer regions, this radiation is completely absorbed by water vapour and CO_2. Similar to Figure 17, the dominant influence of the water lines over the whole spectral range can be seen, but now superimposed by the strong absorption band of CO_2 around 670 cm^{-1}. The water vapour concentration at ground again was assumed to be 14,615 ppm.

To assess the more or less strong influence of increasing CO_2 concentrations in the atmosphere, as heavily discussed with respect to anthropogenic global warming, Figure 19 compares two simulations, which were performed with 380 ppm (blue line) and 760 ppm (red curve) of CO_2 at average global conditions. Due to the strongly saturated absorption and emission on the CO_2 band around 670 cm^{-1}, the broad funnel changes neither its depth nor its width. Only the wings and the weak absorption bands at 970 and 1070 cm^{-1} which are all strongly superimposed by intense water lines contribute to a slightly higher absorption.

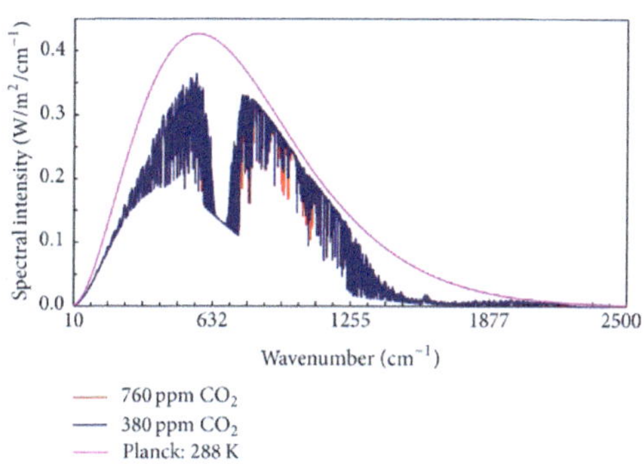

Figure 19: Upwelling radiation at 12.5 km altitude with 380 ppm (blue line) and 760 ppm CO2 (red line). Water vapour concentration is 14,615 ppm and temperature at ground 288 K.

Subtraction of the integrated spectra yields an intensity difference of 5.8 W/m^2; so, this amount is less reradiated by the troposphere at doubled CO_2 concentration. While these calculations were done under clear sky conditions, at an average cloudiness of 50% and a cloud altitude of 5 km, this reduces to about 4.3 W/m^2. Under new quasiequilibrium conditions, it is found that 61% of this amount amplifies the back radiation, the other 39% go to space (see also [20]). An enhanced downward emission from the stratosphere at higher CO_2 concentration [21, 22] is partly shielded by the cloudiness and in the radiation balance further compensated by increased losses of the sun flux, which is absorbed by the earth's surface (shortwave radiative forcing). So, altogether about 2.6 W/m^2 would contribute to the radiative forcing and thus to global heating at doubled CO_2 concentration.

An example for the downwelling radiation is shown in Figure 20. The calculations start at the intersection of the mesosphere to the mesopause in 86 km altitude (TOA) and track the emission as well as the absorption down to the earth's surface. The simulations again include water vapour in the troposphere with the ground concentration of 14,615 ppm, CO_2, and CH_4 as well mixed gases over the whole altitude with concentrations for CO_2 of 380 ppm (blue line) and 760 ppm (red line), for CH_4 with 1.8 ppm and O_3 spread over the stratosphere and tropopause with a maximum of 7 ppm around 38 km altitude. Except the smaller contribution from O_3 around 1020 cm^{-1}, all the radiation originates from the lower troposphere and is over wide regions identical with a blackbody source of 288 K. The integral radiation in downward direction is 312 W/m^2, representing 80% of a blackbody source at that temperature and corresponding to 61% of the total intensity, radiated only by the atmosphere. The difference between ordinary and doubled CO_2 concentrations at clear sky is 4.4 W/m^2, while at average cloudiness that amount, which contributes to an additional heating of the earth, is the same as discussed above.

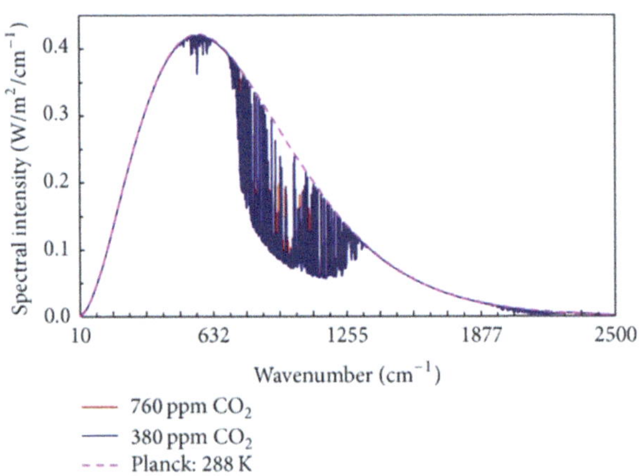

Figure 20: Downwelling radiation from TOA to ground with 380 ppm (blue line) and 760 ppm CO2 (red line). Water vapour concentration is 14,615 ppm and temperature at ground 288 K.

The calculated radiative forcing of 2.6 W/m^2 at doubled CO_2 concentration is 1.1 W/m^2 smaller than that of Myhre et al. [22]. Up to some degree, this discrepancy is expected to result from different atmospheres used for the calculations. So, most of the spectroscopic computations in the literature are based on the *US Standard 1976 Atmosphere*. However, the corresponding water vapour concentration of this data basis is only half of that found in the average global atmosphere (AGA) and also only half of that derived from actual GPS measurements [19], as used in our calculations. Further deviations might result from different concepts, for example, not assuming thermal equilibrium or calculating with different effective path lengths of the radiation through the atmosphere. To demonstrate the reliability and validity of the presented simulations, we compare our calculations with two measurements, which were recorded under quite different conditions. Figure 21(a) shows a measurement of the outgoing infrared emission spectrum of the cloud-free atmosphere as detected by a satellite in 20 km altitude looking downward over the polar ice sheet [23].

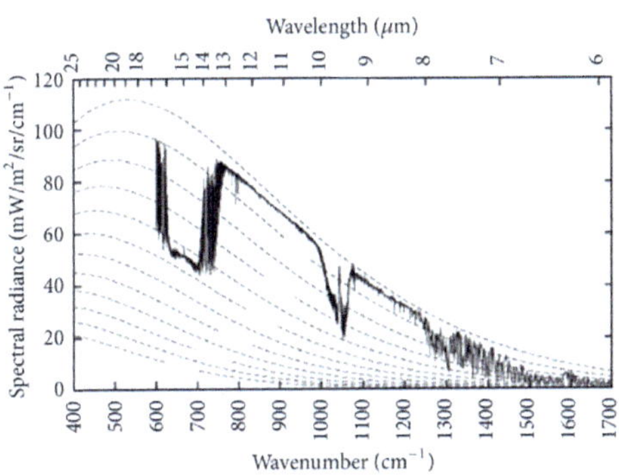

(a)

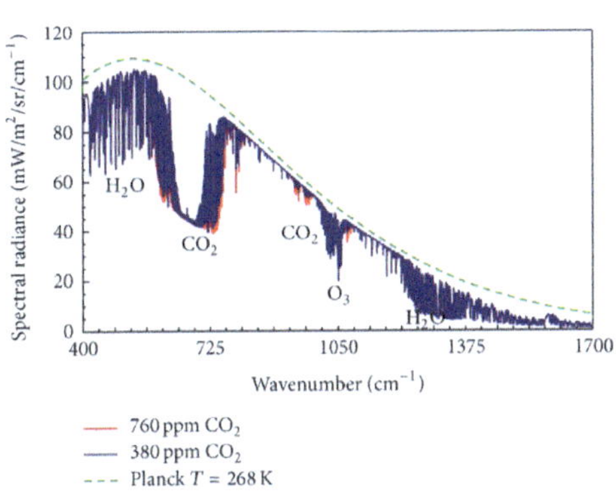

(b)

Figure 21: Upwelling radiance (a) measured from a satellite over the polar ice sheet [23], (b) calculated with 380 ppm (blue line) and 760 ppm CO2 (red line). Water vapour concentration is 2100 ppm and ground temperature 268 K.

The simulation shown in Figure 21(b) was performed for a ground temperature of 266 K, a water vapour content at the surface of 2100 ppm, a CO_2 concentration of 380 ppm (blue graph) and 760 ppm (red line), a CH_4 concentration of 1.8 ppm, and a slightly varying O_3 concentration of 1–3 ppm over the tropopause.

On the other hand, Figure 22(a) gives an example for the downwelling radiance recorded by a ground station in Reno over one day with strongly varying cloud coverage [24]. Figure 22(b) shows the simulations based on (111) but superimposed by a Planckian distribution, which accounts for the emitted radiation of the clouds. This Planck radiation is defined by the cloud temperature and thus the cloud altitude, and it contributes to the spectrum with a weighting factor given by the cloud coverage.

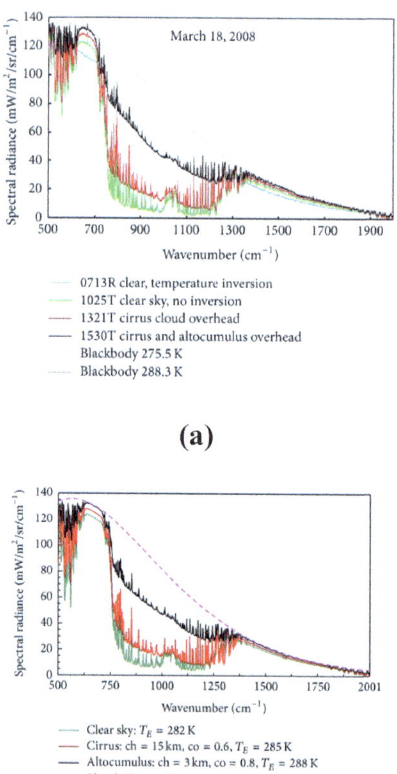

(a)

(b)

Figure 22: Downwelling radiance (a) measured at ground with varying cloud coverage [24], (b) calculated with water vapour concentration of 3000 ppm and different temperatures.

The ground temperatures varied from 282 to 288 K over the day, and the water vapour concentration was assumed to be 3000 ppm, while for the other gases, the standard conditions were applied.

All characteristic features of the measured spectra are well reproduced. The smaller deviations in the shape of the CO_2 funnel as well as in the form of the downward radiation at 670 cm^{-1} and also discrepancies in the O_3 band structure around 1050 cm^{-1} mainly originate from a lack of knowledge of some experimental parameters. So, the resolution of the spectrometers and also the acceptance angle of the detection optics have significant influence on the shape, widths, and intensity of the measured spectra.

But also inaccurate or missing atmospheric parameters like the pressure and temperature profile over the path length during the measurements, or the vapour and gas concentrations, will limit the exact reproducibility of the measured spectra.

It should be emphasized that the measurements in Figures 21 and 22 show the up- and downwelling spectral radiance, that is, radiation, which is collected within a restricted solid angle. This angle is normally determined by the receiver optic, which causes a spatial filtering and essentially detects rays, transmitting the atmosphere on a shorter way (mostly perpendicular to the earth's surface).

In contrast to this, Figures 19 and 20 were calculated for the spectral intensity, which spreads out over the whole hemisphere in upward or downward direction and corresponds to an average path length about twice that of the radiance. Therefore, due to the shorter effective interaction length particularly the wings of the spectral bands and lines in Figures 21 and 22 appear less broadened and saturated. Additionally, at the lower temperatures and dryer air, the water lines are less dominant, so that under clear sky conditions the atmosphere gets much more transparent, particularly within the spectral window from 800 to 1200 cm^{-1}.

These examples clearly demonstrate that the basic interaction processes of infrared molecules with thermal radiation are well understood and that the presented radiation transfer calculations are in excellent agreement with the observations, even under the quite complex conditions in the atmosphere. From this, we can also conclude that simulations of the type as shown in Figures 19–21 with different concentrations of CO_2 in the atmosphere are quite reliable and give an important assessment for the impact of greenhouse gases on our climate.

Absorption of Terrestrial Radiation

The dominant influence of water vapour on the radiation and energy budget in the atmosphere can also be deduced from an analysis of pure absorption spectra. So, under conditions as discussed before with an average global water vapour content of 1.46% at the surface and a temperature of 288 K, the overall absorption of the terrestrial radiation, only caused by water vapour, is already 77.7%. Together with methane and ozone in the atmosphere, the absorption increases to 80.7%. On the other hand, CO_2 alone at a concentration of 380 ppm would contribute to 24.1%, but in the presence of other gases, its influence is repelled to 4.6%, which is due to the strong spectral overlap with the other components, particularly with water vapour, and the total absorption only climbs up to 85.3%.

As already outlined in Sections 3.3 and 5.1, the centre of the CO_2 absorption band around 670 cm^{-1} rapidly goes into saturation with increasing propagation length and also with increasing concentration. Only the wings and the weaker absorption bands at 970 and 1070 cm^{-1} still contribute to a slightly rising absorption with increasing CO_2 concentration. This is shown in Figure 23, demonstrating that due to the spectral overlap and the strong saturation behaviour, a doubling of CO_2 under clear sky conditions only donates additional 1.2% to the total absorption.

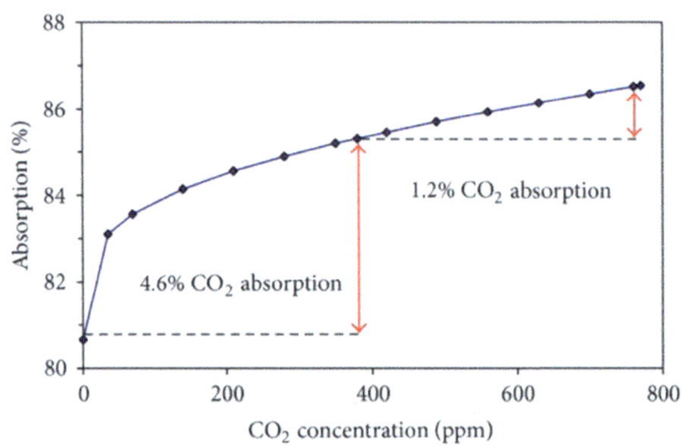

Figure 23: Absorption of terrestrial radiation as a function of the CO2 concentration.

With cloudiness this contribution is even further pushed down to less than 1%, and from this 39% will be reradiated to space. So, altogether only about 0.6%, corresponding to 2.4 W/m^2 of the total terrestrial radiation

(391 W/m^2 at 288.15 K) or 0.8% with respect to the total back radiation of 312 W/m^2 (see Section 5.3), can contribute to an additional global heating at doubled CO_2 concentration. This result is in quite good agreement with the previous calculation of the radiative forcing and demonstrates why a further increase in the CO_2 concentration only gives marginal corrections in the radiation budget.

Based on the simple assumption that due to the back radiation, the natural greenhouse effect causes a global temperature increase of approximately 33 K and that the surface temperature linearly responds to any changes in the energy or radiation budget, an increase in this budget of less than 1% should only contribute to a temperature increase of about 0.3 K. This is in clear contradiction to the IPCC, which issues an official climate sensitivity (temperature increase at doubled CO_2) of $C_S = 3.2$K [25].. But even a more sophisticated consideration, based on an energy balance model for the surface-atmosphere system and distinguishing between three climate zones, including the shortwave radiation budget as well as water vapour and lapse rate feedback, only gives a climate sensitivity of 0.6 K [20], which is still 5x smaller than the IPCC value.

CONCLUSION

The paper reviews the basic relations for the interaction of radiation with infrared active molecules and presents a new theoretical concept for the description of the radiation transfer in the atmosphere. Starting from Einstein's fundamental considerations on the quantum nature of radiation, it is shown that the thermal radiation of a gas can be well understood only in terms of the spontaneous emission of the molecules. The interaction with radiation from an external blackbody source in the presence of the background radiation and under the influence of molecular collisions is elaborately investigated. The energy and radiation balance in the atmosphere, including heat transfer processes, are described by rate equations which are solved numerically for typical conditions as found in the troposphere and stratosphere. These simulations demonstrate the continuous and dominant conversion process of heat to radiation and vice versa.

Consideration of the radiation interaction on a molecular basis allows to derive the Schwarzschild equation, describing the radiation transfer in the atmosphere. This equation is investigated under conditions of only few intermolecular collisions as found in the upper mesosphere as well as at high collision rates as observed in the troposphere. A generalized form of the radiation transfer equation is presented, which covers both limiting cases of thin and dense atmospheres and allows a continuous transition from low to high densities, controlled by a density dependent parameter. This equation is derived for the spectral radiance as well as for the spectral flux density (spectral intensity).

The radiation transfer equation is applied to simulate the up- and downwelling radiation and its interaction with the most prominent greenhouse gases water vapour, carbon dioxide, methane, and ozone in the atmosphere. The radiation in upward direction is the only possibility to render any absorbed energy to space and by this to keep the atmospheric temperature in balance with all the supplied direct and indirect energy from the sun. On the other hand, the downwelling part determines the back radiation from the atmosphere to the earth's surface, which is the source term of the so heavily discussed atmospheric greenhouse or atmospheric heating effect. While the upwelling radiation only contributes to cooling, as long as the radiation can escape to space, the downwelling part only causes heating, when it reaches the surface.

Propagation of radiation in the atmosphere is not the result of a single event. It underlies continuous absorption and reemission processes in the gas, which are determined by the mean free absorption length and strongly depend on the frequency of the radiation as well as on the gas density. Radiation also converts into molecular kinetic energy and via exciting collisions changes back to radiation. This kinetic energy of the molecules results from absorbed long- and shortwave radiation as well as from sensible and latent heat due to convection and evapotranspiration from the surface to the atmosphere.

So, any propagation of the radiation in the atmosphere can be described similar to a diffusion process of particles (photons), which are partly annihilated and again created while propagating through the atmosphere. This propagation in a gas is just described by the radiation transfer equation (111) for the spectral radiance and by (116) for the spectral intensity (spectral flux density), and these equations are valid independent, if an external radiation or the eigen radiation of the atmosphere in up or down direction is calculated.

An asymmetry in the up- and downwelling eigen radiation (only from atmosphere) will be found with a stronger contribution in downward direction, which is caused by the lapse rate as well as the density profile over the atmosphere with higher temperature and pressure at lower atmospheric layers and therefore a higher net downward radiation. Calculations for the radiative forcing at doubled CO_2 concentration and at an average cloudiness give a 30% smaller forcing than applied by the IPCC.

Thermal radiation is electromagnetic radiation and no heat. Therefore, in the same way, as radio waves can propagate from a colder antenna to a warmer receiver, microwaves can be absorbed by a hot chicken, or CO_2-laser radiation ($10.6 \mu m$) can be used for welding and melting of metals up to several thousand °C, so any back radiation from colder and higher atmospheric layers can be absorbed by the lower and warmer layers, and this back radiation can also be absorbed by a warmer surface of the earth without violating the 2nd law of thermodynamics. As long as the surface is assumed to be a black or gray absorber, it does not filter any frequencies of the incoming radiation, in the same way as it does not reject any frequencies of the broad Planck spectrum of a thermal radiator, independent, if it has a higher or lower temperature than the earth. Radiation converts to heat after an absorption, followed by an emission in accordance with a newly adjusting thermodynamic equilibrium, which only requires that the net energy transfer is in balance.

REFERENCES

1. A. Einstein, "Quantentheorie der Strahlung," *Zeitschrift für Physik*, vol. 18, p. 121, 1917.
2. M. Planck, "Über eine Verbesserung der Wienschen Spektralgleichung," *Verhandlungen der Deutschen Physikalischen Gesellschaft*, vol. 2, no. 5/6, pp. 519–525, 1900.
3. L. Boltzmann, "allgemeine Sätze über Wärmegleichgewicht," *Wiener Berichte*, vol. 63, pp. 679–711, 1871.
4. L. Boltzmann, "Weitere Studien über das Wärmegleichgewicht unter Gasmolekülen," *Wiener Berichte*, vol. 66, pp. 275–370, 1872.
5. A. Schuster, "Radiation through a Foggy Atmosphere," *The Astrophysical Journal*, vol. 21, pp. 1–22, 1905.
6. K. Schwarzschild, "Über das Gleichgewicht in der solaren Atmosphäre—on the equilibrium of the sun's atmosphere," *Nachrichten der Koniglichen Gesellschaft der Wissenschaften zu Gottingen*, vol. 195, p. 41, 1906.
7. K. Schwarzschild, "Diffusion and absorption in the sun's atmosphere," *Sitz. K. Preuss. Akad. Wiss*, p. 1183, 1914.
8. G. C. Wick, "über ebene Diffusionsprobleme," *Zeitschrift für Physik*, vol. 121, no. 11-12, pp. 702–718, 1943.
9. S. Chandrasekhar, "On the radiative equilibrium of a stellar atmosphere. II," *Astrophysical Journal*, vol. 100, article 76, 1944.
10. R. M. Goody and Y. L. Yung, *Radiation: Theoretical Basis*, Oxford University Press, Oxford, UK, 1989.
11. M. L. Salby, *Physics of the Atmosphere and Climate*, Cambridge University Press, Cambridge, UK, 2012.
12. E. A. Milne, "Effect of collisions on monochromatic radiative equilibrium," *The Royal Astronomical Society*, vol. 88, article 493, 1928.
13. L. S. Rothman, I. E. Gordon, A. Barbe et al., "The HITRAN 2008 molecular spectroscopic database," *Journal of Quantitative Spectroscopy and Radiative Transfer*, vol. 110, no. 9-10, pp. 533–572, 2009.
14. N. Jacquinet-Husson, N. A. Scott, A. Chédin et al., "The 2003 edition of the GEISA/IASI spectroscopic database," *Journal of Quantitative Spectroscopy and Radiative Transfer*, vol. 95, no. 4, pp. 429–467, 2005.

15. S. S. Penner, *Molecular Spectroscopy and Gas Emissivities*, Addison-Wesley, Reading, Mass, USA, 1959.
16. C. H. Townes and A. L. Schawlow, *Spectroscopy*, Dover, New York, NY, USA, 1975.
17. US Standard Atmosphere, 1976.
18. H. Harde and J. Pfuhl, *MolExplorer—A Program-Platform for the Calculation of Molecular Spectra and Radiation Transfer in the Atmosphere*, Helmut-Schmidt-University Hamburg, Hamburg, Germany, 2013.
19. S. Vey, *Bestimmung und Analyse des atmosphärischen Wasserdampfgehaltes aus globalen GPS-Beobachtungen einer Dekade mit besonderem Blick auf die Antarktis [Doctoral theses]*, Technische Universität Dresden, Dresden, Germany, 2007.
20. H. Harde, *Was trägt CO_2 wirklich zur globalen Erwärmung bei? Spektroskopische Untersuchungen und Modellrechnungen zum Einfluss von H_2O, CO_2, CH_4 und O_3 auf unser Klima*, Books on Demand, Norderstedt, Germany, 2011.
21. V. Ramanathan, L. Callis, R. Cess et al., "Chemical interactions and effects of changing atmospheric trace gases," *Reviews of Geophysics*, vol. 25, pp. 1441–1482, 1987.
22. G. Myhre, E. J. Highwood, K. P. Shine, and F. Stordal, "New estimates of radiative forcing due to well mixed greenhouse gases," *Geophysical Research Letters*, vol. 25, no. 14, pp. 2715–2718, 1998.
23. *Data Courtesy of D. Tobin*, Space Science and Engineering Center, University of Wisconsin-Madison, Madison, Wis, USA, 1986.
24. *Graphic from Pat Arnott*, University of Nevada, Reno, Nev, USA, 2008, ATMS 749.
25. D. A. Randall, R. A. Wood, S. Bony et al., "Climate models and their evaluation, in Climate Change 2007: The Physical Science Basis," in *Contribution of Working Group I to the Fourth Assessment Report of the Intergovernmental Panel on Climate Change*, S. Solomon, D. Qin, M. Manning et al., Eds., Cambridge University Press, Cambridge, UK, 2007.

CHAPTER 20

Atmospheric Ionization and Cloud Radiative Forcing

Henrik Svensmark,[1] Jacob Svensmark,[2] Martin Bødker Enghoff,[1] and Nir J. Shaviv[3]

[1]National Space Institute, Technical University of Denmark, Elektrovej 327, 2800 Kongens Lyngby, Denmark

[2]Niels Bohr Institute, University of Copenhagen, Niels Bohr Building, Jagtvej 128, 2. floor, 2200 Copenhagen N., Denmark

[3]Racah Institute of Physics, Hebrew University of Jerusalem, Giv'at Ram, Jerusalem, 91904 Israel

Citation: Svensmark, H., Svensmark, J., Enghoff, M.B. et al. "Atmospheric ionization and cloud radiative forcing". Sci Rep 11, 19668 (2021). https://doi.org/10.1038/s41598-021-99033-1.

ABSTRACT

Atmospheric ionization produced by cosmic rays has been suspected to influence aerosols and clouds, but its actual importance has been questioned. If changes in atmospheric ionization have a substantial impact on clouds, one would expect to observe significant responses in Earth's energy budget. Here it is shown that the average of the five strongest week-long decreases in atmospheric ionization coincides with changes in the average net radiative balance of 1.7 W/m2 (median value: 1.2 W/m2) using CERES satellite observations. Simultaneous satellite observations of clouds show that these variations are mainly caused by changes in the short-wave radiation of low liquid clouds along with small changes in the long-wave radiation, and are almost exclusively located over the pristine areas of the oceans. These observed radiation and cloud changes are consistent with a link in which atmospheric ionization modulates aerosol's formation and growth, which survive to cloud condensation nuclei and ultimately affect cloud formation and thereby temporarily the radiative balance of Earth.

Subject terms: Climate sciences, Astronomy and planetary science

INTRODUCTION

A fundamental question during the last two decades has been whether changes in atmospheric ionization can perturb aerosols and thereby cloud properties. Both aerosols and clouds are an essential part of the terrestrial atmosphere, influencing weather and climate[1–3]. Ionization in the atmosphere is mainly caused by cosmic ray particles, which have their origin outside our solar system. Solar activity modulates the flux of cosmic ray particles on time scales from days to millennia, whereas on geological time scales it is the position of the solar system in our Galaxy, which is important for ionization[4,5]. A link involving ionization, aerosols, and clouds would be an exciting interconnection between Earth and the Galaxy. Indeed, laboratory experiments demonstrated that ions assist the nucleation of new (~ 1–2 nm) aerosol particles[6,7], and evidence from airborne observations also document that ion nucleation is an important source of aerosols in the free troposphere[8]. However, these results do not by themselves assure that aerosols survive to cloud condensation nuclei (CCN) such that variations in ionization can change CCN concentrations and subsequently clouds. In fact, studies testing the role of ion-nucleation using global numerical aerosol models indicate that the response of CCN to changes in ionization is too small to be of any significance[9–14]. The reason being that any changes in ion-nucleated aerosol number density get attenuated by absorption on existing aerosols before the particles can grow to CCN. However, there is evidence contrasting the numerical results. Recent theoretical and experimental results detail how ions can accelerate the growth of small aerosols by increasing the mass-flux from the gas phase to aerosols[15],[16], and this ion-condensation mechanism is not included in the numerical modeling. A mechanism accelerating aerosol growth in the real atmosphere would lead to a higher survival rate of aerosols growing to CCN sizes.

Observational support for the link connecting atmospheric ionization with cloud changes has been pursued using naturally occurring week-long suppression of atmospheric ionization of the order 10–20%. Such events are called Forbush Decreases (FDs) [17] and are caused by a magnetized plasma cloud from the Sun hitting Earth, thereby shielding part of the cosmic ray flux. Initially, FD studies gave conflicting results[18]–[23] (see section 7.4 in[23]) but by sorting the FDs according to their strength a significant response was found in both aerosols and clouds in the case of the strong FDs[23].

However, for the ion-aerosol-cloud link to be important, the changes in atmospheric ionization should significantly influence Earth's energy balance. Here we address this question by using satellite data from Clouds and the Earth's Radiant System (CERES) [24] of top-of-atmosphere (TOA CERES daily SYN1deg) radiative forcing during strong FDs together with simultaneous cloud observation from the Moderate Resolution Imaging Spectroradiometer (MODIS) [25] (See Methods). The thirteen strongest FDs after 2000, where the CERES and MODIS instruments have been active, have been ranked according to their ionization strength in the atmosphere (see Table 1). It is shown that a ~10% decrease in cosmic ray ionization of the five strongest FDs results in a global top-of-atmosphere radiative forcing of 1–2 W/m2 with a delay of ~ 5–7 days caused by mainly low liquid clouds over the oceans. Deep convective clouds also react to the FD, but here the net effect on the radiative budget is muted due to opposing effects of short-wave and long-wave forcing.

Table 1: The 13 most influential FD events since 2000 during which the CERES and MODIS instruments have been active, sorted by strength.

Order	Date	Decrease (%)
1	31/10/2003	119
2	19/1/2005	83
3	13/9/2005	75

4	16/7/2000	70
5	12/4/2001	64
6	10/11/2004	53
7	26/9/2001	50
8	17/7/2005	47
9	27/7/2004	45
10	31/5/2003	44
11	25/11/2001	39
12	15/5/2005	38
13	28/8/2001	37

The first and second columns are, respectively, the order and date of minimum cosmic ray flux. The third column shows the percentage decrease in the ion production *relative* to the reduction in cosmic rays from solar minimum to solar maximum over a solar cycle, i.e., FD 1 changed the GCR flux more than the average change over the solar cycle[23].Note that while there have been several FDs since the latest on the list (Sept. 13th 2005) none of them have been strong enough to enter the top 5 of the list.

RESULTS

One of the main problems in demonstrating significant responses in aerosols and clouds to sudden changes in ionization during a FD event is the dominant meteorological noise that tends to mask any signal. Another problem is large differences in the resulting decrease in ionization between individual FD events. A possible solution to this problem is to rank the FDs according to their strength and use the strongest FD events to search for significant signals. This strategy has been used previously, where in addition to using the strength, the response was integrated over 10 days following the FD minimum in cosmic rays[23]. This procedure takes possible auto-correlation into account, i.e., the situation where the signal is distributed over a few days. The significance of the obtained signal can then be evaluated relative to a Monte-Carlo simulation using random FD dates (details in Methods). This approach will now be used on the CERES data.

Globally Averaged Responses

For a given FD event we consider a period beginning 15 days before the FD minimum in the galactic cosmic ray (GCR) flux to 20 days after. After global averaging, all 36-day time series have a linear trend removed (see Methods) to account for seasonal effects. The left column of Fig. 1 shows the responses in (a) NET radiation, (b) SW radiation, and (c) LW radiation in the CERES instrument at TOA superposed for the five strongest FD events. Only the five strongest events are used as adding more events decreases the signal-to-noise ratio due to the weaker strength of the rest of the events (see Methods). The right column shows the distribution function and the achieved significance level of the integrated signals (gray area from day 0 to 10 in the left panels of Fig. 1) based on 104 Monte Carlo bootstrap samples of the integrated/summed response between day 0 to 10 using random FD dates in the full CERES time series 2000–2006[23] (see Methods). Figure 1a shows the NET radiation for three relevant situations: Areas with liquid cloud fraction > 0.25 (red curve), areas with high cloud fraction > 0.25 (cyan curve), and globally (purple curve). The areas of high or liquid clouds are defined as an average cloud fraction of more than 0.25 in the corresponding MODIS parameter averaged over the period 2000-2006. The fractional areas of Earth with a given cloud type above the threshold level are: Area (Liquid clouds) = 0.52, area (High clouds) = 0.44, and area (Ice clouds) = 0.40. Note in Fig. 1a that approximately 5 days after the GCR minimum (blue curve) an increased flux of ~1–2 W/m2 is entering the Earth system.

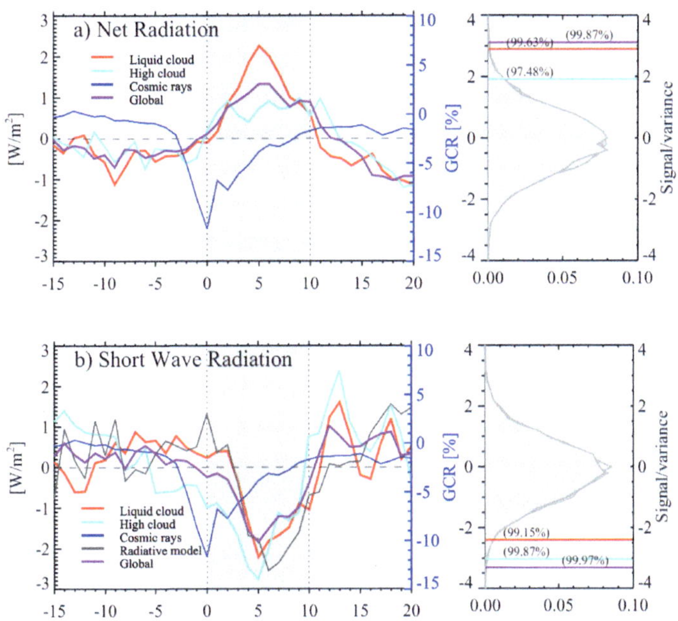

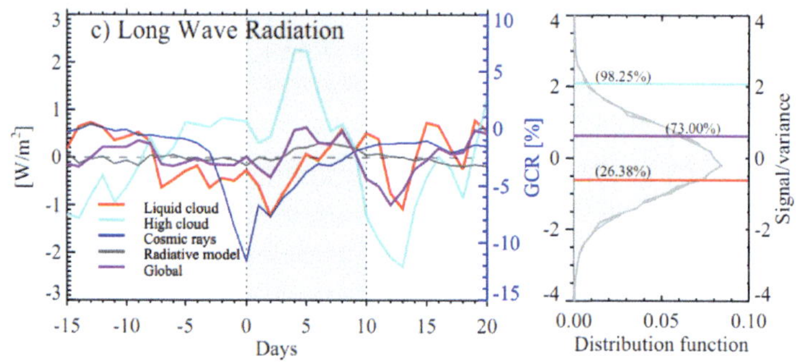

Figure 1: The top of atmosphere radiative balance (relative to the mean of the time series, left column) from CERES data superposed for the five strongest FDs since 2000. The panel **(a)** is the change in net radiative forcing for three situations: geographical areas with predominantly, (1) liquid clouds (red curve), (2) areas with high clouds (cyan curve), (3) and global averages (purple curve). The gray area signifies the period used to integrate the response (day 0–10). The blue curve is the variation in cosmic rays as measured by the Oulu neutron monitor. Right-hand columns are bootstrap distribution functions of the integrated signal. The red, cyan, and purple lines denote the significance of the FD signal obtained from the three distribution functions. Panel **(b)** left-hand side shows changes in TOA short wave radiation for areas with liquid clouds (red curve), high clouds (cyan curve), and globally (purple curve). Results of the Fu-Liou radiative transfer model of changes in TOA SW radiative forcing obtained using the observed changes in MODIS cloud parameters are described by the gray curve. The right-hand panel **(b)** shows the achieved significance of the FD signals. Finally, panel **(c)** is the change in TOA LW radiative forcing similarly as for SW in panel **(b)**.

Figure 1b displays the SW radiation at the TOA, again separated into areas with predominantly liquid clouds, areas with mostly high clouds, and globally. Here a minimum about five days after the GCR minimum is seen where ~2–3 W/m2 less SW radiation is reflected from Earth. The Fu-Liou radiative transfer model[26],[27] is used to compare the change in SW and LW at TOA using built-in response functions for the CERES SW and LW instruments. The inputs to the model are simultaneous MODIS observations of liquid cloud changes of cloud fraction, optical thickness, and effective droplet radius, superposed over the same 5 FDs (grey curve), using a low cloud height between 900 and 700 hPa. The right-hand panel of Fig. 1b depicts the achieved significance levels. Finally, Fig. 1c is similar to Fig. 1b but for the change in LW. The gray curve is the LW result of the Fu-

Liou model. See Methods for how TOA Net, SW, and LW are measured. Figure 2 illustrates the relevance of using the FD strength by superposing data for the five strongest FD as is done in the above Fig. 1. In Fig. 2 the largest deviation in global SW radiation between day 0 and day 10 is plotted as a function of the strength of the individual 13 FDs from Table 1, with a linear fit to the data. Using all available CERES data with random FD dates it is possible to estimate the mean of the largest deviation over a 10 day period and compare this value with the actual FD dates. The mean of the largest deviations is shown as the horizontal dash-dotted line in Fig. 2. It is seen that black points with the lowest strength (<60%) are mainly below this value and therefore in the noise. The group of black points with strength larger than 60% are mainly above this level. The right-hand panels of Fig. 2 are the distribution functions of the slope and intercept of linear fits using all 13 random FD dates based on 104 Monte Carlo simulation. Note that the slope and intercept of the real FD data have a significance corresponding to three σ, demonstrating a large statistical signicance between FD strength and the response as defined above.

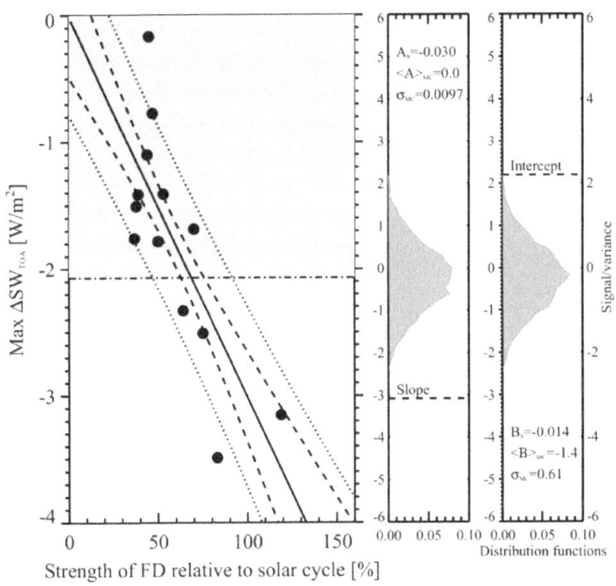

Figure 2: The response of CERES short-wave radiation as a function of FD strength. The left panel shows the most substantial deviation between day 0 and day 10 in short wave radiation globaly averaged for each Forbush decrease as a function of the FD strength (dots). The black line is a linear fit to the 13 data points and the dashed lines and dotted lines represent two and four standard

deviations of the fit, respectively. The dash-dotted line is the mean of the most substantial deviation between day 0 and day 10 using 104 Monte Carlo simulations with random FD dates. The black circles are the 13 FDs. The two panels to the right display the distribution functions of the slope and intercept of the linear fit, based on 104 Monte Carlo simulations using CERES SW data with random dates for the FD. Note that the significance of the above linear slope is ~3σ, and that the Monte Carlo simulations using random dates for the FD events result in an average zero slope as expected.

Although the linear trend has a high significance, it does not exclude the possibility that a single strong event dominates the overall signal. As can be seen in Fig. 2 FD 2, which occurred January 19, 2003, has the largest response of all the FDs. Consequently, statistical tests were performed to see if FD 2 is an outlier. Although the definition of outliers is a quarrelsome issue, there are statistical methods to test if a data point classifies as an outlier[28]. First, the deviation, Δy, of the FD response from the linear regression line seen in Fig. 2, was calculated. The 13 Δy data points underwent several tests: Z-score, Pierce test, Grubbs's test, and Dixon's Q-test. They all predicted that FD2 is not an outlier and it is therefore kept in the sample. As can be seen in Fig. 2 the weak FDs have comparatively weak responses. It is therefore important to optimize the signal to noise ratio. Due to the varying FD strength (see Methods), the signal-to noise-ratio has a local maximum when superposing the five strongest FDs. The maximum in signal-to-noise ratio is the argument for using the superposition of the five strongest FDs to demonstrate significance of FD signals in CERES and MODIS.

Zonal Responses

Figure 3 illustrates the zonal (latitudes −90∘ −90∘) and temporal changes (36-day time series) in clouds together with simultaneous changes in the global TOA radiative balance. The zonal maps in the left column are again the average of the five strongest FDs. The column on the right-hand side shows the relative change in GCR (black curve), and the 1-σ scatter (gray band) using a box-car average of 3 days, measured in (%) relative to the base level between day −15 and −5. Figure 3a is the change in MODIS liquid cloud fraction in (%). Figure 3b are ice-cloud-fraction observations from MODIS in (%). Figure 3c shows the TOA CERES NET radiation in (W/m^2). Note the response approximately 5 days after the FD minimum in the latitude range −50∘ to 40∘. Figure 3d shows TOA CERES SW radiation in (W/m2), and finally Fig. 3e is the change in TOA LW in (W/m2).

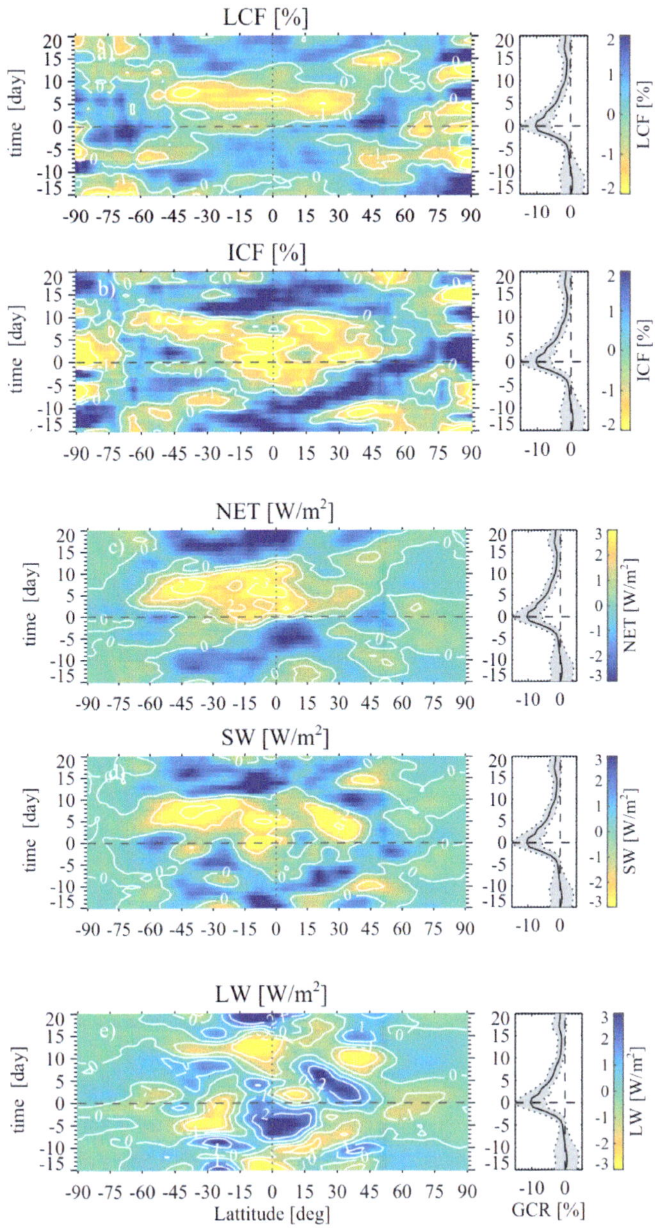

Figure 3: Maps depicting changes in clouds and TOA radiative balance as a function of latitude and time (centered on the FD minimum). The maps are superposed data from the five strongest FDs after 2000. **(a)** The left panel is based on liquid cloud fraction observations from MODIS in (%). The right

panel in all rows displays the change in cosmic rays during the same period (black curve). **(b)** Left panel is ice-cloud-fraction observations from MODIS in (%). **(c)** Left panel is TOA CERES NET radiation in (W/m2). **(d)** Left panel is TOA CERES SW radiation in (W/m2). **(e)** Shows the LW similar to the SW in **(d)**.

Global Maps

It is possible to obtain a global map of changes in the radiative balance averaged over the five strongest FDs. Figure 4's top panels show on the left-hand side a global map of the NET radiation before the FD-minimum, averaged over the period day −15 to day −6 before the minimum in FD ionization. It represents variations in the average NET radiative state before the FD (for details see "Methods"). The global map on the right-hand side is the NET radiation change after the FD-minimum averaged over day 3 to day 12 to accentuate the response. Similarly, the middle panel graphs the SW response of Fig. 4, while the bottom panel shows the LW response of Fig. 4. Although FD2 is the strongest, by excluding it from the map in Fig. 4 we show that the remaining four strongest FD still produce quite a similar pattern of responses, as can be seen from Fig. 5. Note that the Monte Carlo statistics used in the globally averaged signals, as in Fig. 1, are one way of testing the significance of the maps in Fig. 4, i.e. corresponding to the summation of the above maps.

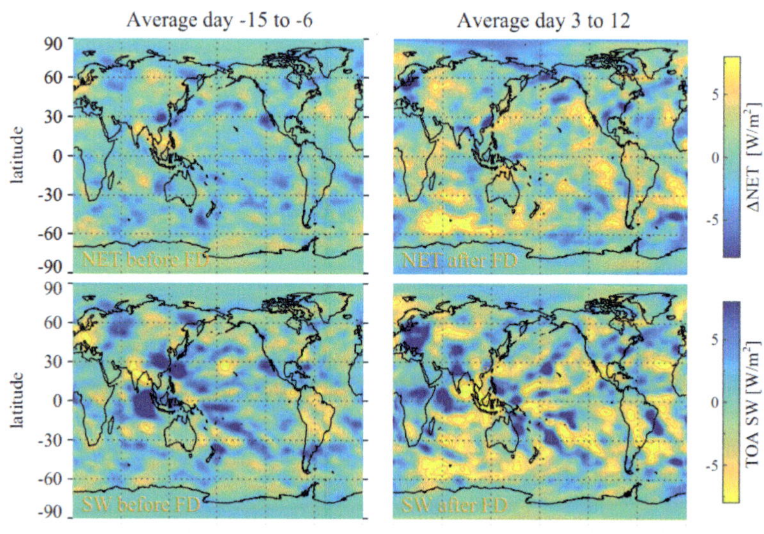

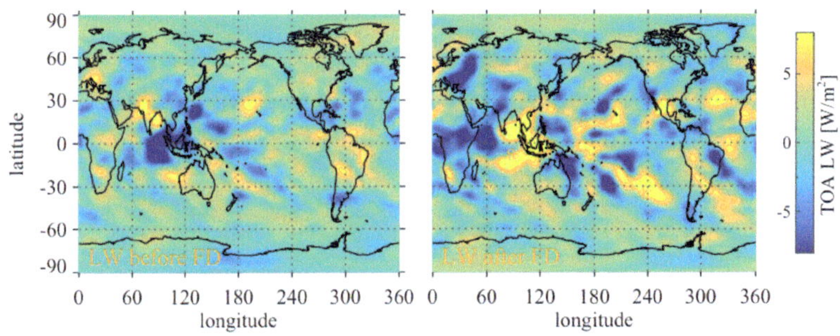

Figure 4: The left-hand column represents variations in the average radiative state before the FD (for details see Methods) averaged over the first nine days before the FD minimum for the NET, SW, and LW, respectively, and the right-hand column is the variations in the average radiative state after the FD averaged over nine days after the FD starting on day 3. Top panels are NET radiation. Middle panels are for SW radiation, and finally, the bottom panel is LW radiation.

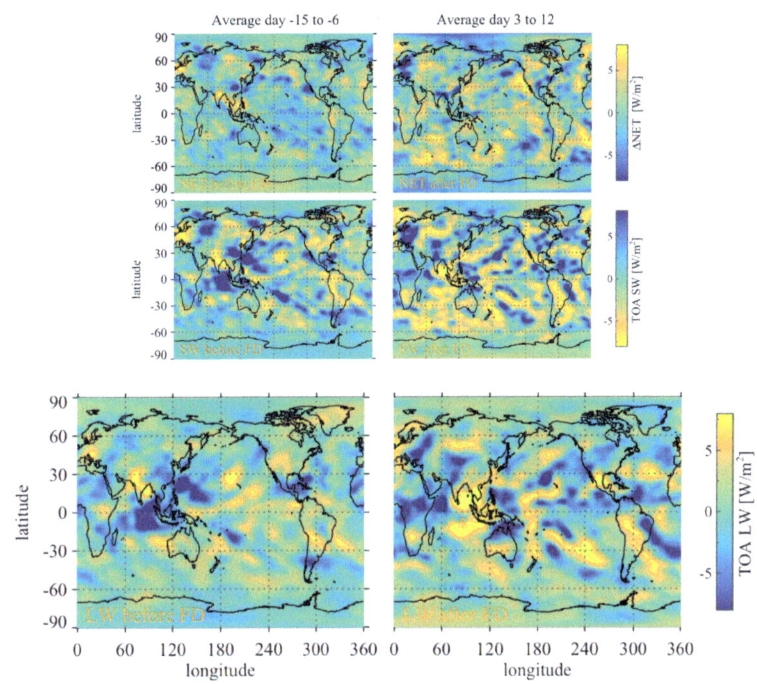

Figure 5: Same figure as Fig. 4 but without the strong FD2 event. Note that the overall patterns are similar to Fig. 4 but as expected a little weaker.

Figure 6 displays three panels of Earth with liquid, high, and ice cloud frequency. Superposed on all three sub-figures are the regions with the largest responses in NET radiation (see Fig. 4 top right panel). The top panel demonstrates that the strongest signal in NET radiation is almost exclusively regions with low clouds.

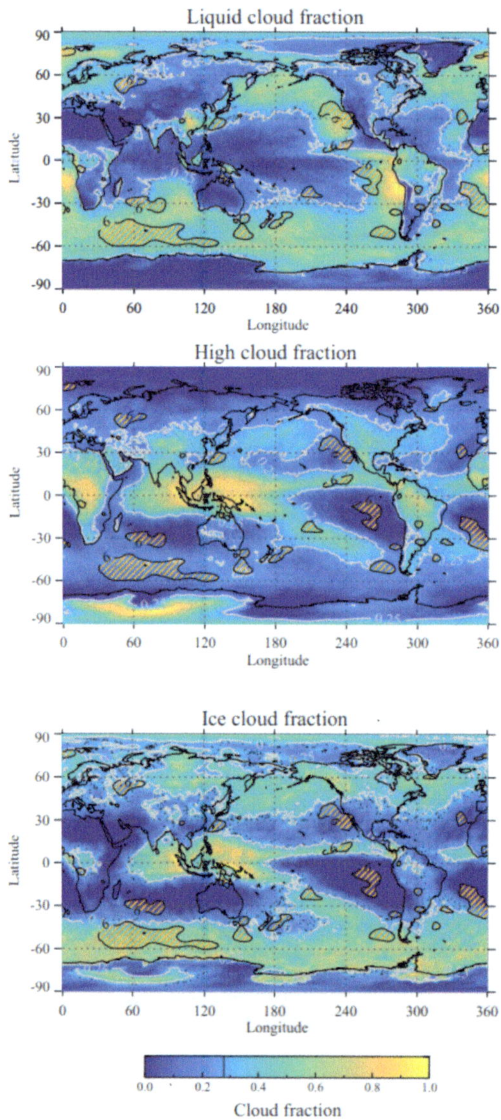

Figure 6: Global maps of three cloud types based on MODIS observations. The top panel is regions with liquid clouds (mainly the green regions). The

middle panel is high clouds. The bottom panel is the ice-cloud-fraction. The gray contour lines on all three maps show threshold cloud fraction 0.25 used for the cloud masks. The fractional area of Earth with cloud types exceeding the threshold are: a (Liquid clouds) = 0.52, a (High clouds) = 0.44 and, a (Ice clouds) = 0.40. Superposed on all three maps are the (orange hatched) regions with large responses found in CERES TOA net radiation following the minimum in cosmic rays (black contour lines are 6 W/m2). Note that these regions are almost exclusively confined to areas dominated by liquid clouds.

DISCUSSION

The present results combine statistical correlations with expectations from atmospheric physics. If a link between ionization, aerosols, and clouds exists, it is reasonable to expect that the strength of an FD should influence the size of an atmospheric response in clouds and radiative forcing. Figure 2 gives a demonstration of a linear relationship between the strength of 13 FD and the global SW responses. Testing the significance by a Monte Carlo simulation using random dates showed that the slope of the linear relationship was significant at the three σ level. As can be observed from Fig. 2 FD2 has the strongest response, and it could be that the event is so dominant that it defines the response. Statistical tests did not classify the FD2 SW response as an outlier relative to the other 12 FDs and was therefore kept in the sample.

Figure 1 offers insight into the response of the terrestrial cloud system during strong FDs. The global signals (purple curves) of the TOA NET, and SW radiation in Fig. 1 show a significant response, reaching a 2–3 W/m2 perturbation around 5 days after the FD minimum. An increase is seen in the NET radiation, while the SW exhibits a complementary decrease. Notably, the LW component in the bottom panel shows no response to the FDs. By applying cloud masks selecting for either liquid cloud or for high clouds, it is possible to understand the response.

Considering first the liquid clouds (red curves), the signals in the NET and SW radiation in Fig. 1 persist and even strengthen, while the LW component remains almost unaffected. The explanation is that liquid clouds are low clouds, and therefore the difference between cloud top temperature and Earth's surface temperature is relatively small. Thus, a decrease in liquid cloud fraction would manifest as a nearly unchanged LW signal. In contrast, liquid clouds affect SW radiation, which increases NET-absorbed radiation. This scenario is corroborated by using the Fu-Liou radiative transfer model[26],[27] using input of simultaneous measured MODIS liquid cloud parameters (cloud fraction, cloud optical thickness, and effective cloud droplet radius) as displayed in Fig. 1b for SW changes (gray curve) and Fig. 1c for LW (gray curve).

Applying a cloud mask to extract high cloud regions leads to a different result (Fig. 1, cyan curves). Here, the response in NET radiation is relatively smaller and less significant. However, the effects in the SW and LW radiation are both significant and closely mirror each other, which is consistent with expectations from a decrease in the high cloud fraction. Fewer high clouds will lead to less SW radiation reflected out to space (negative SW at TOA). However, since a large fraction of the high clouds are deep convective clouds, there will be a large LW contrast between the cold cloud top and the warm terrestrial surface. A decrease in high clouds will, therefore, increase the LW radiation escaping to space. As a consequence[29], the change in SW and LW will approximately cancel each other resulting in the smaller NET radiation signal ~1 W/m2, as seen in Fig. 1.

The connection between CERES TOA radiation balance and clouds can be further consolidated when studying the zonal temporal responses, as shown in Fig. 3. Figure 3a shows the response in MODIS liquid cloud fraction with minimum 5–7 days after day 0 with a longer response time seemingly in the southern hemisphere. The zonal distribution of the response is almost symmetric around the equator ranging from −50o to 40o, but suggestive of a stronger response in the southern hemisphere. The average solar zenith angle for the 5 FD events was −0.6o, which does not in itself indicate a solar angle preference for either hemisphere. Figure 3b expresses the response in ice clouds. The main difference between ice clouds and the liquid clouds is that the temporal response close to the equator starts earlier in the ice clouds. This difference may be due to the much larger vertical velocities in the deep convective clouds, which activate aerosols at even smaller sizes than in the low liquid clouds due to a higher supersaturation of water vapor. Alternatively, the time difference could be due to nucleation happening high in the troposphere followed by aerosol growth as they descend[30] making them available as CCN later at the heights were liquid cloud formation takes place.

NET and SW response in Fig. 3c,d, have similarities with maximum changes of 3 and −4 W/m2, respectively, and both have a well defined maximum/minimum around day 5. A response extremum at day 5 in the clouds and NET radiation as seen in Fig. 3a,c indicate that the observed effect can not be a result of meridional transport since typical meridional wind speeds are of the order of 1 m/s. Therefore an ion-effect on the time scales of days has to be relatively confined to a particular latitude.

The effect of FDs on the overall radiative balance is visible from the global maps shown in Fig. 4. Figure 4 illustrates the radiative responses in NET, SW, and LW (visible in the right-hand panels). Comparing these responses with the cloud distributions of liquid clouds, high clouds, and ice clouds of Fig. 6 provide valuable information on the workings of the ion-cloud link. From Fig. 4 (top right panel) one see the response in NET radiation and comparing the maximum responses with the distribution of liquid clouds Fig. 6 (top panel) the NET responses are mainly over the clean southern oceans and where low clouds dominate. Furthermore, comparing Fig. 6 (middle right panel) with Fig. 4 LW (bottom right panels) it is seen that large responses in LW are mainly confined to regions with a high frequency of deep convective clouds. The same LW regions can be seen in the SW map in Fig. 4 (middle panels) with a near cancellation of LW and SW signals, in agreement with the well-known effect of deep convective clouds having significant but compensating effects in the LW and SW that largely cancel[29]. These observations therefore imply that the main NET effect of FDs on the radiative budget is from low liquid clouds caused by the change in the cloud fraction with SW albedo changes dominating. Although FD 2 is the strongest, by excluding it from the map in Fig. 4 we demonstrate that the remaining four strongest FD still produced quite a similar pattern of responses, as can be seen from Fig. 5.

A further complication to FD 2 is that one of the strongest ground level events occurred within the FD decrease, which has the opposite effect of the FD by increasing the ionization. The acceleration of particles in a ground-level event is due to solar activity. The particle energy is rarely more than 10 GeV (much lower than Galactic cosmic rays) and with a peak lasting less than an hour. The solar accelerated particles are therefore usually too weak to penetrate Earth's magnetic field at low latitudes. Neutron monitor counts of daily-mean show that the ground level event does not dominate the FD2 event for latitudes below approximately 55 deg., corresponding to more than 80% of Earth. However, the ground-level event dominates in the polar regions, with a significant excess of atmospheric ionization. From the cosmic ray, aerosol, and cloud link, one would expect a decrease in cosmic rays, aerosol, and cloud over most of Earth, except in polar regions, where the increased ionization should produce an excess of aerosols. Such an increase has been seen[31]. Consequently, this unique feature of FD2 provides an additional independent consistency test of the cosmic-ray-aerosol link.

In summary, the results presented show that

- A statistically significant linear relationship exists between the strength of the 13 FDs and the globally averaged SW responses. The presented results are mainly based on an average of the five strongest Forbush events, which ensured that the signal-to-noise ratio was a maximum. By only having five strong Forbush decreases, there is a limit to the robustness of the presented results. Nonetheless, removing Forbush event 2 from the calculation of the global net radiation change still produces a significant result of 1.2 W/m2 compared to 1.7 W/m2 with FD2 included.
- Following the minimum in cosmic rays, one observes a delay of 5-7 days in clouds and radiative response as expected from the time it takes aerosols to grow from nucleated sizes (1–2 nm) to cloud condensation nuclei. Clouds in the tropics which contain ice are mainly deep convective clouds. Here the updraft velocity is much larger and, therefore, also the supersaturation[32]. Higher supersaturation activates much smaller sizes of aerosols as CCN. As a consequence, a shorter delay is expected in the response, as seen.
- The second strongest FD 2 is exceptional by the occurrence of a ground-level event within the period of the FD 2. The consequence is a decrease in ionization in more than 80% of the area of Earth and an increase in polar regions. Observation in polar regions and the rest of the Earth demonstrates a decrease of aerosols[21],[23] and an increase in polar regions[31]. These and the present observations are consistent with a cosmic ray-aerosol-cloud link.
- The zonal responses in liquid clouds show a delay in minimum SW between 50° to 45° with a slight tendency towards a longer delay in the southern hemisphere (see Fig. Fig.3a,d).3a,d). A longer delay in the southern hemisphere is consistent with a lower trace-gas concentration (from which particles grow) due to a more clean atmosphere.
- From the spatial maps, it is clear that the primary responses are over the oceans and that low liquid clouds are mainly responsible for the change in net radiative forcing. The global response in net radiation to the average of the five strongest FD is approximately 2 W/m2.

A consistent picture is emerging, suggesting that variations in ionization are connected to aerosols and clouds[21,23], and now also the energy budget. In fact, the aerosol–cloud observations get a simple explanation by the ion–aerosol–cloud hypothesis. A decrease in atmospheric ionization decreases the number of ion-nucleated aerosols (1–1.5 nm). This generation of fewer small aerosols then has to grow to CCN sizes (~50 nm) where they can influence clouds. In the free troposphere it typically takes ~5 days (depending on the supersaturation) for aerosols to grow to CCN[33], which is consistent with the observed delay seen in Figs. 1 and 3. One implication of a cosmic ray-aerosol-cloud link is that numerical models are missing relevant physics related to ion and aerosol interactions. Specifically, it was shown that the electrostatic interactions between ions and aerosols continuously add mass to the aerosols at a rate depending on the ion-concentration[16]. Therefore, by changing the ion concentration the aerosol growth rate will also change, which is important for the survival of aerosols to CCN. This ion-aerosol growth acceleration mechanism was found to be important under the following conditions: (1) aerosols are smaller than ~25 nm, (2) the total number of aerosol particles is small, and (3) the condensing gas concentration (e.g. H2SO4) is low[16,34]. The first condition is important for aerosol survival and the last two conditions are typically fulfilled over the remote oceans[8]. For example, low CCN concentrations result in low droplet concentrations in marine stratus clouds ($N < 100$ cm−3) as can be found over large parts of the remote oceans[35–37]. Ion–aerosol interactions and aerosol survival to CCN is therefore suggested to be sensitive to the prevailing ion densities and are capable of modifying clouds and the radiative energy budget in the secluded atmosphere over the oceans.

METHODS

Data

The data employed in this work originate from three sources: (1) Daily GCR counts are obtained from the Oulu neutron monitor (http://cosmicrays.oulu.fi/). (2) The ordered Forbush decrease (FD) list is obtained from Ref.[23], and listed in Table 1 for reference. (3) Data on Earth's radiative balance is obtained from the NASA satellite program, "Clouds and the Earth's Radiant Energy System" (CERES)[24]. The data are "Observed Radiative Fluxes" (SYN1deg-Day) product of daily averages (https://eosweb.larc.nasa.gov/project/ceres/syn1deg-day_ed3a_table). The specific parameters used are, top-of-atmosphere (TOA) outgoing (reflected) short wave radiation between 0.3μm and 5.0μm (SW), this observed broadband shortwave reflected (upwelling) flux at the top of the atmosphere (TOA—around 20 km altitude) is defined as positive. TOA outgoing longwave between 5.0μm and 125μm (LW) is the (upwelling) thermal outgoing LW flux at the top of the atmosphere and is defined as positive. The Net TOA Flux, is the broadband incoming solar (downwelling) minus the reflected shortwave (SW) (upwelling) and longwave (LW) emitted (up-welling) flux at the top of the atmosphere. If the Net Flux is positive the Earth is warming, while when negative the Earth is cooling. (4) Cloud data are obtained from the Moderate Resolution Imaging Spectroradiometer (MODIS)[25]. Daily averages of the following key cloud parameters are extracted: (1) Cloud effective emissivity, (2) Cloud optical thickness, (3) Liquid water path (LWP), (4) Liquid water cloud fraction, (5) Liquid cloud effective radius (Reff), (6) Column density of CCNs, (7) Ice-cloud-fraction, (8) High cloud inferred fraction. All data from the "MOD08_D3" product.

Statistics

A Monte Carlo bootstrap method determines the statistical significance of the globally averaged signals. CERES data of SW, LW and NET radiation are in the form of global maps, U(i,j,tn) where index i∈[1;180], j∈[1;360] (1×1∘ resolution) and a temporal resolution of 1 day, covering the period 1st March 2000 to 31st December 2005. The same holds for the MODIS cloud parameters. Global spatial averages over 36 days consisting of 15 days before the FD minimum (on day 0) and 20 days after are given by

$$\langle U(t_n)\rangle = \sum_{j=1}^{360}\sum_{i=-90}^{90} U(i,j,t_n)\cos[\phi(i)] \Big/ \sum_{i=-90}^{90}\cos[\phi(i)],\quad t_n \in [-15,20], \tag{1}$$

where cos[ϕ(i)] is the area correction as a function of latitude ϕ(i). The zonal averages are functions of both time and latitude:

$$\langle U(i,t_n)\rangle = \sum_{j=1}^{360} U(i,j,t_n)/360. \tag{2}$$

A linear trend is removed from the temporal variability in all data. For global averages, each 36-day time series is de-trended:

$$\langle U(t_n)\rangle_0 = \langle U(t_n)\rangle - at_n - b, \tag{3}$$

where a, b are best linear fit coefficients. For a zonal averages, we de-trend temporally on a latitude basis, such that

$$\langle U(i, t_n)\rangle_0 = \langle U(i, t_n)\rangle - a_i t_n - b_i, \tag{4}$$

where ai,bi are the best linear fit coefficients for the time-series at latitude i. Finally for global maps we de-trend on a pixel-by-pixel level, such that

$$\langle U(i, j, t_n)\rangle_0 = \langle U(i, j, t_n)\rangle - a_{i,j} t_n - b_{i,j}, \tag{5}$$

where ai,j,bi,j are the best linear fit coefficients for the time-series at latitude i and longitude j.

The above 36 days time series for global averages, zonal means and maps, abbreviated as Uk for the k'th strongest FD, may then be averaged over the five strongest events:

$$\langle U\rangle_{FD} = \frac{1}{5}\sum_{k=1}^{5}\langle U_k\rangle_0. \tag{6}$$

Here k refers to each of the individual five strongest FD in Table 1.

Next, a statistic for the global response signal in an analysed satellite parameters can be defined. The time series in Eq. [6] for the global mean $\langle U(t_n)\rangle_{FD}$ can be integrated over the dates $tn \in [t_1, t_2]$ where a signal is physically expected:

$$FS = \sum_{n=t_1}^{t_2}\langle U(t_n)\rangle_{FD}. \tag{7}$$

For the present analysis we take $t_1=0$ and $t_2=10$. Then a Monte Carlo bootstrap statistic from similar constructs however with the GCR mininum date of the FD replaced with a random date within the full time series can be made. NB sets of 5 de-trended time series ⟨Bi,j(tn)⟩0 can be constructed in the same way as the FD time series as

$$\langle B_j(t_n)\rangle_0 = \frac{1}{5}\sum_{i=1}^{5}\langle B_{i,j}(t_n)\rangle_0, \tag{8}$$

and their signal can be integrated (summed)

$$BS_j = \sum_{t=t_1}^{t_2} \langle B_j(t_n) \rangle_0. \tag{9}$$

Here j is the bootstrap sample number. We want to know if

$$DFS = FS - \frac{1}{N_B} \sum_{j=1}^{N_B} BS_j \tag{10}$$

is drawn from the same distribution as the bootstrap samples

$$DBS_j = BS_j - \frac{1}{N_B} \sum_{j=1}^{N_B} BS_j. \tag{11}$$

Then, we define the statistic of the achieved significance level (26) as

$$ASL = \left[\text{Number of bootstrap samples where } \mathrm{DBS_j} \geq \mathrm{DFS}\right]/N_B. \tag{12}$$

Statistical effect of having varying signal strength

If the n FD signals all have the same strength S1 then the signal to noise ratio would grow as

$$\frac{S_n}{\sigma} = \sqrt{n}\ \frac{S_1}{\sigma_1} \tag{13}$$

where σ is the variance of the added noise, and σ1 the noise of a single time series. This is the standard result if the signals all have the same strength. However, if the signals are of decreasing relative strength fn, ($f_1 = 1$), then

$$\frac{S_n}{\sigma} = \sum_{i=1}^{n} \sqrt{n}\ \frac{f_i S_1}{\sigma_1} \tag{14}$$

This function will in general not grow as $\sqrt{n}$ as in the general case, but can decrease and even have a local minimum. In the case of TOA SW data the signal to noise ratio has a local maximum at n=5 and decreases when more FD are added. This help the justification of using the five strongest FD in the present work.

Estimates of the strength of the individual FDs is detailed in[23]. The result was a list of 26 FD in the period 1983–2007, all ranked according to strength. There the change in cosmic rays primary spectrum at 1 AU was calculated using available neutron monitors and muon detectors from Nagoya. From this variation, one can estimate the change in atmospheric ionization using CORSIKA—COsmic Ray SImulations for KAscade, a Monte Carlo code to simulate extensive air showers. Table 1 show the ranking and timing of 13 FD events after 2000. Interestingly, there has been no strong FD since 2006. The first column is the date of the minimum in neutron counts during the FD, and the second column is the strength relative to the change in cosmic rays over a solar cycle. The FD event in 2003 had a larger drop in cosmic rays than the reference solar cycle modulation and is why the strength is set to 119%[23].

ACKNOWLEDGEMENTS

JS is funded by the Carlsberg Foundation and the Danish council for independent research under the project "Fundamentals of Dark Matter Structures", DFF-6108-00470, and thanks them both for their support. NJS wishes to thank the generous support of the Israel Ministry of Energy.

Author Contributions

H.S. conceived the project, H.S., J.S., M.B.E. and N.J.S. analysed the results. All authors reviewed the manuscript.

Competing Interests

The authors declare no competing interests.

Footnotes

Publisher's note

Springer Nature remains neutral with regard to jurisdictional claims in published maps and institutional affiliations.

These authors contributed equally: Jacob Svensmark, Martin Bødker Enghoff and Nir J. Shaviv.

REFERENCES

1. Twomey S. The influence of pollution on the shortwave Albedo of clouds. *J. Atmos. Sci.* 1977;**34**:1149–1152. doi: 10.1175/1520-0469(1977)034<1149:TIOPOT>2.0.CO;2.
2. Albrecth, B. A. Aerosols, cloud microphysics, and fractional cloudiness. *Science***245**:1227–1230. 10.1126/science.245.4923.1227 (1989). https://science.sciencemag.org/content/245/4923/1227.full.pdf.
3. Hartmann, D. L. Radiative effects of clouds on Earth's climate. in *Aerosol-Cloud-Climate Interactions* (Hobbs, P. V. ed.). 151 (1993).
4. Shaviv NJ. Cosmic ray diffusion from the galactic spiral arms, iron meteorites, and a possible climatic connection. *Phys. Rev. Lett.* 2002;**89**:051102. doi: 10.1103/PhysRevLett.89.051102.
5. Svensmark H. Evidence of nearby supernovae affecting life on Earth. *MNRAS.* 2012;**423**:1234–1253. doi: 10.1111/j.1365-2966.2012.20953.x.
6. Svensmark, H., Pedersen, J. O. P., Marsh, N. D., Enghoff, M. B. & Uggerhøj, U. I. Experimental evidence for the role of ions in particle nucleation under atmospheric conditions. *Proc. R. Soc. London, Ser. A***463**, 385–396. 10.1098/rspa.2006.1773 (2007).
7. Kirkby J, et al. Role of sulphuric acid, ammonia and galactic cosmic rays in atmospheric aerosol nucleation. *Nature.* 2011;**476**:429–433. doi: 10.1038/nature10343.
8. Lee S-H, et al. Particle formation by ion nucleation in the upper troposphere and lower stratosphere. *Science.* 2003;**301**:1886–1889. doi: 10.1126/science.1087236.
9. Pierce JR, Adams PJ. Can cosmic rays affect cloud condensation nuclei by altering new particle formation rates? *Geophys. Res. Lett.* 2009;**36**:9820. doi: 10.1029/2009GL037946.
10. Snow-Kropla EJ, Pierce JR, Westervelt DM, Trivitayanurak W. Cosmic rays, aerosol formation and cloud-condensation nuclei: Sensitivities to model uncertainties. *Atmos. Chem. Phys.* 2011;**11**:4001–4013. doi: 10.5194/acp-11-4001-2011.
11. Kazil J, et al. The present-day decadal solar cycle modulation of Earth's radiative forcing via charged H2SO4/H2O aerosol nucleation. *Geophys. Res. Lett.* 2012;**39**:L02805. doi: 10.1029/2011GL050058.

12. Yu F, Luo G. Effect of solar variations on particle formation and cloud condensation nuclei. *Environ. Res. Lett.* 2014;**9**:045004. doi: 10.1088/1748-9326/9/4/045004.
13. Dunne, E. M. *et al.* Global atmospheric particle formation from cern cloud measurements. *Science***354**, 1119–1124. 10.1126/science.aaf2649 (2016). http://science.sciencemag.org/content/354/6316/1119.full.pdf.
14. Gordon, H. *et al.* Causes and importance of new particle formation in the present-day and preindustrial atmospheres. *J. Geophys. Res. Atmos.***122**, 8739–8760. 10.1002/2017JD026844 (2017). https://agupubs.onlinelibrary.wiley.com/doi/pdf/10.1002/2017JD026844.
15. Svensmark H, Enghoff MB, Pedersen JOP. Response of cloud condensation nuclei (> 50 nm) to changes in ion-nucleation. *Phys. Lett. A.* 2013;**377**:2343–2347. doi: 10.1016/j.physleta.2013.07.004.
16. Svensmark H, Enghoff MB, Shaviv NJ, Svensmark J. Increased ionization supports growth of aerosols into cloud condensation nuclei. *Nat. Commun.* 2017;**8**:2199. doi: 10.1038/s41467-017-02082-2.
17. Forbush S. A model for particle formation and growth in the atmosphere with molecular resolution in size. *Phys. Rev.* 1937;**51**:1108. doi: 10.1103/PhysRev.51.1108.3.
18. Kristjánsson JE, et al. Cosmic rays, cloud condensation nuclei and clouds—A reassessment using MODIS data. *Atmos. Chem. Phys.* 2008;**8**:7373–7387. doi: 10.5194/acp-8-7373-2008.
19. Sloan, T. & Wolfendale, A. W. Testing the proposed causal link between cosmic rays and cloud cover. *Environ. Res. Lett.***3**, 044001–+. 10.1088/1748-9326/3/2/024001 (2008).
20. Laken, B., Wolfendale, A. & Kniveton, D. Cosmic ray decreases and changes in the liquid water cloud fraction over the oceans. *Geophys. Res. Lett.***36**. 10.1029/2009GL040961 (2009).
21. Svensmark, H., Bondo, T. & Svensmark, J. Cosmic ray decreases affect atmospheric aerosols and clouds. *Geo. Phys. Res. Lett.***36**, 5101–+. 10.1029/2009GL038429. (2009).
22. Calogovic, J. *et al.* Sudden cosmic ray decreases: No change of global cloud cover. *Geophys. Res. Lett.***37**, 3802–+. 10.1029/2009GL041327 (2010).

23. Svensmark, J., Enghoff, M. B., Shaviv, N. J. & Svensmark, H. The response of clouds and aerosols to cosmic ray decreases. *J. Geophys. Res. Space Phys.* 8152–8181. 10.1002/2016JA022689(2016).
24. Wielicki, B. A. *et al.* Clouds and the earth's radiant energy system (ceres): An earth observing system experiment. *Bull. Am. Meteorol. Soc.*77, 853–868. 10.1175/1520-0477(1996)077%3c0853:CATERE%3e2.0.CO;2 (1996).
25. Salomonson VV, Barnes WL, Maymon PW, Montgomery HE, Ostrow H. MODIS—Advanced facility instrument for studies of the earth as a system. *IEEE Trans. Geosci. Remote Sens.* 1989;**27**:145–153. doi: 10.1109/36.20292.
26. Fu, Q. & Liou, K. N. Parameterization of the radiative properties of cirrus clouds.*J. Atmos. Sci.***50**,2008–2025.10.1175/1520-0469(1993)050%3c2008:POTRPO%3e2.0.CO;2 (1993).
27. Kato S, Rose FG, Charlock TP. Computation of domain-averaged irradiance using satellite-derived cloud properties. *J. Atmos. Ocean. Technol.* 2005;**22**:146–164. doi: 10.1175/JTECH-1694.1.
28. Hawkins, D. M. *Identification of Outliers* (1980).
29. Kiehl, J. On the observed near cancellation between longwave and shortwave cloud forcing in tropical regions. *J. Clim.* 7, 559–565. 10.1175/1520-0442(1994)007%3c0559:OTONCB%3e2.0.CO;2 (1994).
30. Williamson CJ, et al. A large source of cloud condensation nuclei from new particle formation in the tropics. *Nature.* 2019;**574**:399–403. doi: 10.1038/s41586-019-1638-9.
31. Mironova IA, Usoskin IG, Kovaltsov GA, Petelina SV. Possible effect of extreme solar energetic particle event of 20 January 2005 on polar stratospheric aerosols: Direct observational evidence. *Atmos. Chem. Phys.* 2012;**12**:769–778. doi: 10.5194/acp-12-769-2012.
32. Fan, J. *et al.* Substantial convection and precipitation enhancements by ultrafine aerosol particles. *Science***359**, 411–418. 10.1126/science.aan8461(2018). https://science.sciencemag.org/content/359/6374/411.full.pdf.
33. Clarke AD, et al. Free troposphere as a major source of ccn for the equatorial pacific boundary layer: long-range transport and teleconnections. *Atmos. Chem. Phys.* 2013;**13**:7511–7529. doi: 10.5194/acp-13-7511-2013.

34. Svensmark, J., Shaviv, N. J., Enghoff, M. B. & Svensmark, H. The ion and charged aerosol growth enhancement (ion-cage) code: A numerical model for the growth of charged and neutral aerosols. arXiv:1909.12784 (2019).
35. Bennartz, R. Global assessment of marine boundary layer cloud droplet number concentration from satellite. *J. Geophys. Res. Atmos.* **112**. 10.1029/2006JD007547(2007). https://agupubs.onlinelibrary.wiley.com/doi/pdf/10.1029/2006JD007547.
36. Rausch, J., Meyer, K., Bennartz, R. & Platnick, S. Differences in liquid cloud droplet effective radius and number concentration estimates between modis collections 5.1 and 6 over global oceans. *Atmos. Meas. Tech.* **10**, 2105–2116. 10.5194/amt-10-2105-2017(2017).
37. Bennartz R, Rausch J. Global and regional estimates of warm cloud droplet number concentration based on 13 years of aqua-Modis observations. *Atmos. Chem. Phys.* 2017;**17**:9815–9836. doi: 10.5194/acp-17-9815-2017.

INDEX

H

I

J

K

L

M

N

T

U

V

W

Z